This volume provides a detailed account of bosonization. This important technique represents one of the most powerful nonperturbative approaches to many-body systems currently available.

The first part of the book examines the technical aspects of bosonization. Topics include one-dimensional fermions, the Gaussian model, the structure of Hilbert space in conformal theories, Bose–Einstein condensation in two dimensions, non-Abelian bosonization, and the Ising and WZNW models. The second part presents applications of the bosonization technique to realistic models including the Tomonaga–Luttinger liquid, spin liquids in one dimension and the spin-1/2 Heisenberg chain with alternative exchange. The third part addresses the problems of quantum impurities. Chapters cover potential scattering, the X-ray edge problem, impurities in Tomonaga–Luttinger liquids and the multi-channel Kondo problem. This book will be an excellent reference for researchers and graduate students working in theoretical physics, condensed matter physics and field theory.

Alexander Gogolin was born in Tbilisi, Georgia, in 1965. He received his Ph.D. in 1991 from the Lebedev Physical Institute. Since then he has been a permanent scientific member of staff at the Landau Institute for Theoretical Physics in Moscow. In 1995 he was appointed as EPSRC Advanced Fellow in the Mathematics Department, Imperial College London. His main research interests are in the area of condensed matter physics and field theory.

Alexander Nersesyan was born in Tbilisi, Georgia in 1943. He was awarded his Ph.D. from Tbilisi State University in 1971 and has been a Senior Research Fellow at the Institute of Physics, Georgia, since 1976. His main research interests also cover condensed matter physics and field theory.

Alexei Tsvelik was born in Samara, Russia in 1954. In 1980 he was awarded his Ph.D. from the Kurchatov Institute for Atomic Energy. He took up his current position as lecturer in Physics at the University of Oxford in 1992. Dr Tsvelik is also the author of *Quantum Field Theory for Condensed Matter Physicists* (Cambridge University Press 1995, paperback 1996).

Bosonization and Strongly Correlated Systems

Alexander O. Gogolin

Imperial College of Science and Technology, London

Alexander A. Nersesyan

Institute of Physics, Tbilisi, Georgia

Alexei M. Tsvelik

Brasenose College, University of Oxford

CAMBRIDGE
UNIVERSITY PRESS

PUBLISHED BY THE PRESS SYNDICATE OF THE UNIVERSITY OF CAMBRIDGE
The Pitt Building, Trumpington Street, Cambridge CB2 1RP, United Kingdom

CAMBRIDGE UNIVERSITY PRESS
The Edinburgh Building, Cambridge CB2 2RU, United Kingdom
40 West 20th Street, New York, NY 10011-4211, USA
10 Stamford Road, Oakleigh, Melbourne 3166, Australia

First published 1998

Printed in the United Kingdom at the University Press, Cambridge

Typeset in 11pt Times

A catalogue record for this book is available from the British Library

Library of Congress Cataloguing in Publication data

Gogolin, A. O. (Alexander Olegovich), 1965–
Bosonization approach to strongly correlated systems /
A. O. Gogolin, A. A. Nerseyan, A. M. Tsvelik.
p.   cm.
Includes bibliographical references.
ISBN 0 521 59031 0 (hb)
1. Bosons. 2. Many-body problem. 3. Condensed matter. 4. Mathematical physics.
I. Nersesyan, A. A. (Alexander Artyomovich) 1943– .
II. Tsvelik, Alexei M. III. Title.
QC793.5.B62G64   1998
530.4′16–dc21   97-35519   CIP

ISBN 0 521 59031 0 hardback

Contents

Preface

We used to think that if we know one, we knew two, because one and one are two. We are finding that we must learn a great deal more about 'and'.
Sir Arthur Eddington, from *The Harvest of a Quiet Eye*, by A. Mackay

The behaviour of large and complex aggregations of elementary particles, it turns out, is not to be understood in terms of a simple extrapolation of the properties of a few particles. Instead, at each level of complexity entirely new properties appear, and the understanding of the new behaviours requires research which I think is as fundamental in its nature as any other.
P. W. Anderson, from *More is Different* (1972)

High energy physics continues to fascinate people inside and outside of science, being perceived as the 'most fundamental' area of research. It is believed somehow that the deeper inside the matter we go the closer we get to the truth. So it is believed that 'the truth is out there' – at high energies, small distances, short times. Therefore the ultimate theory, Theory of Everything, must be a theory operating at the smallest distances and times possible where there is no difference between gravitational and all other forces (the Planck scale). All this looks extremely revolutionary and complicated, but once a condensed matter physicist has found time and courage to acquaint himself with these ideas and theories, these would not appear to him utterly unfamiliar. Moreover, despite the fact that the two branches of physics study objects of vastly different sizes, the deeper into details you go, the more parallels you will find between the concepts used. In many cases the only difference is that models are called by different names, but this has more to do with funding than with the essence. Sometimes differences are more serious, but similarities still remain, for example, the Anderson–Higgs phenomenon in particle theory is very similar to the Meissner effect in superconductivity; the concept of 'inflation' in cosmology is taken from the physics of first order phase transitions; the hypothetical 'cosmic strings' are similar to magnetic field vortex lines in type II superconductors; the Ginzburg–Landau theory of superfluid He^3 has many features in common with the theory of hadron–meson interaction etc. When you realize the existence of this astonishing parallelism, it is very difficult not to think that there is something very deep about it, that here you come across a general principle of Nature according to which same ideas are realized on different space-time scales, on different hierarchical 'layers', as a Platonist would put it. This view puts things in a new perspective where truth is no longer 'out there', but may be seen equally well in a 'grain of sand' as in an elementary particle.

In this book we are going to deal with the area of theoretical physics where the parallels between high energy and condensed matter physics are especially strong. This area is the theory of strongly correlated low-dimensional systems. Below we will briefly go through these parallelisms and discuss the history of this discipline, its main concepts, ideas and also the features which excite interest in different communities of physicists.

The problems of strongly correlated systems are among the most difficult problems of physics we are now aware of. By definition, strongly correlated systems are those ones which cannot be described as a sum of weakly interacting parts. So here we encounter a situation when the whole is greater than its parts, which is always difficult to analyse. The well-known example of such a problem in particle physics is the problem of strong interactions – that is a problem of formation and structure of heavy particles – hadrons (with proton and neutron being the examples) and mesons. In popular literature, which greatly influences minds outside physics, one may often read that particles constituting atomic nuclei consist in their turn of 'smaller', or 'more elementary', particles called quarks, coupled together with gluon fields. However, invoking images and using language quite inadequate for the essence of the phenomenon in question this description more confuses than explains. The confusion begins with the word 'consist' which here does not have the same meaning as when we say that a hydrogen atom consists of a proton and an electron. This is because a hydrogen atom is formed by electromagnetic forces and the binding energy of the electron and proton is small compared to their masses: $E \sim -\alpha^2 m_e c^2$, where $\alpha = e^2/hc \approx 1/137$ is the fine structure constant and m_e is the electron mass. The smallness of the dimensionless coupling constant α obscures the quantum character of electromagnetic forces, yielding a very small cross section for processes of transformation of photons into electron–positron pairs. Thus α serves as a small parameter in a perturbation scheme where in the first approximation the hydrogen atom is represented as a system of just two particles. Without small α quantum mechanics would be a purely academic discipline.* One cannot describe a hadron as a quantum mechanical bound state of quarks, however, because the corresponding fine structure constant of the strong interactions is not small: $\alpha_G \sim 1$. Therefore gluon forces are of essentially quantum nature, in the sense that virtual gluons constantly emerge from vacuum and disappear, so that the problem involves an infinite number of particles and therefore is absolutely non-quantum-mechanical. It turns out, however, that the proton and neutron have the same quantum numbers as a quantum mechanical bound state of three particles of a certain kind. Only in this sense can one say that a 'proton consists of three quarks'. The reader would probably agree that this is a very nontraditional use of this word. So it is not actually a statement about the material content of a proton (as a wave on a surface of the sea, it does not have any *permanent* material content), but about its symmetry properties, that is to what representation of the corresponding symmetry group it belongs.

It turns out that reduction of dimensionality may be of a great help in solving models of strongly correlated systems. Most nonperturbative solutions presently known (and only nonperturbative ones are needed in physics of strongly correlated systems) are related to $(1 + 1)$-dimensional quantum or two-dimensional classical models. There are two ways to relate such solutions to reality. One way is that you imagine that reality on some level is also two dimensional. If you believe in this you are a string theorist. Another way is to study systems where the dimensionality is artificially reduced. Such systems are known in condensed matter physics; these are mostly materials consisting of well separated chains, but there are other examples of effectively one-dimensional problems such as problems of solitary magnetic impurities (Kondo effect) or of edge states in the quantum Hall effect. So if you are a theorist who is interested in seeing your predictions fulfilled during your life time, condensed matter physics gives you a chance.

At present, there are two approaches to strongly correlated systems. One approach, which will be only very briefly discussed in this book, operates with exact solutions of many-body theories. Needless to say not every model can be solved exactly, but fortunately many interesting ones can. So this method can provide a treasury of valuable information.

The other approach is to try to reformulate complicated interacting models in such a way that they become weakly interacting. This is the idea of bosonization which was pioneered by Jordan and Wigner in 1928 when they established equivalence between the spin $S = 1/2$ anisotropic Heisenberg chain and the model of interacting fermions (we shall discuss

---

* With only bodiless spirits to discuss it, for sure, because there would not be stable complex atoms to form bodies.

this solution in detail in the text). Thus in just two years after introduction of the exclusion principle by Pauli it was established that in many-body systems the wall separating bosons from fermions might become penetrable. The example of the spin-1/2 Heisenberg chain has also made it clear that a way to describe a many-body system is not unique, but is a matter of convenience.

If the anisotropy is such that the coupling between the z-components of spins vanishes, the fermionic model becomes noninteracting. Thus, at least at this point, the excitation spectrum (and hence thermodynamics) can be easely described. However, since spins are expressed in terms of the fermionic operators in a nonlinear and nonlocal fashion, the problem of correlation functions remains nontrivial to the extent that it took another 50 years to solve it.

The transformation from spins to fermions completes the solution only for the special value of anisotropy; at all other values fermions interact. Interacting fermions in $(1 + 1)$-dimensions behave very differently from noninteracting ones. It turns out however, that in many cases interactions can be effectively removed by the second transformation – in the given case from the fermions to a scalar massless bosonic field. This transformation is called bosonization and holds in the continuous limit, that is for energies much smaller than the bandwidth. So at such energies the spin $S = 1/2$ Heisenberg chain can be reduced to a bunch of oscillators.

The spin $S = 1/2$ Heisenberg chain has provided the first example of 'particles transmutation'. We use these words to describe a situation when low-energy excitations of a many-body system differ drastically from the constituent particles. Of course, there are elementary cases when constituent particles are not observable at low energies, for example, in crystalline bodies atoms do not propagate and at low energies we observe propagating sound waves – phonons; in the same way in magnetically ordered materials instead of individual spins we see magnons etc. These examples, however, are related to the situation where the symmetry is spontaneously broken, and the spectrum of the constituent particles is separated from the ground state by a gap. Despite the fact that continuous symmetry cannot be broken spontaneously in $(1 + 1)$-dimensions and therefore there is no finite order parameter even at $T = 0$, spectral gaps may form. This nontrivial fact, known as dynamical mass generation, was discovered by Vaks and Larkin in 1961.

However, one does not need spectral gaps to remove single electron excitations since they can be suppressed by overdamping occurring when $T = 0$ is a critical point. In this case propagation of a single particle causes a massive emission of soft critical fluctuations. Both scenarios will be discussed in detail in the text.

The fact that soft critical fluctuations may play an important role in

Fig. 1.   A. I. Larkin and I. E. Dzyaloshinskii.

(1 + 1)-dimensions became clear as soon as theorists started to work with such systems. It also became clear that the conventional methods would not work. Bychkov, Gor'kov and Dzyaloshinskii (1966) were the first who pointed out that instabilities of one-dimensional metals cannot be treated in a mean-field-like approximation. They applied to such metals an improved perturbation series summation scheme called 'parquet' approximation (see also Dzyaloshinskii and Larkin (1972)). Originally this method was developed for meson scattering by Diatlov, Sudakov and Ter-Martirosian (1957) and Sudakov (1957).

It was found that such instabilities are governed by quantum interference of two competing channels of interaction – the Cooper and the Peierls ones. Summing up all leading logarithmic singularities in both channels (the *parquet* approximation) Dzyaloshinskii and Larkin obtained differential equations for the coupling constants which later have been identified as Renormalization Group equations (Solyom (1979)). From the flow of the coupling constants one can single out the leading instabilities of the system and thus conclude about the symmetry of the ground state. It turned out that even in the absence of a spectral gap a coherent propagation of single electrons is blocked. The charge–spin separation – one of the most striking features of a one-dimensional liquid of interacting electrons – had already been captured by this approach.

The problem the diagrammatic perturbation theory could not tackle

Fig. 2. A. A. Abrikosov.

was that of the strong coupling limit. Since phase transition is not an option in $(1 + 1)$-dimensions, it was unclear what happens when the renormalized interaction becomes strong (the same problem arises for the models of quantum impurities as the Kondo problem where similar singularities had also been discovered by Abrikosov (1965)). The failure of the conventional perturbation theory was sealed by P. W. Anderson (1967) who demonstrated that it originates from what he called 'orthogonality catastrophy': the fact that the ground state wave function of an electron gas perturbed by a local potential becomes orthogonal to the unperturbed ground state when the number of particles goes to infinity.[†] That was an indication that the problems in question cannot be solved by a partial summation of perturbation series. This does not prevent one from trying to sum the entire series which was brilliantly achieved by Dzyaloshinskii and Larkin (1974) for the Tomonaga–Luttinger model using the Ward identities. In fact, the subsequent development followed the spirit of this work, but the change in formalism was almost as dramatic as between the systems of Ptolemy and Copernicus.

As it almost always happens, the breakthrough came from a change of the point of view. When Copernicus put the Sun in the centre of the coordinate frame, the immensely complicated host of epicycles was transformed into an easily intelligeble system of concentric orbits. In a similar way a transformation from fermions to bosons (hence the term *bosonization*) has provided a new convenient basis and leads to a radical

---

[†] Particle transmutation includes orthogonality catastrophy as a particular case.

Fig. 3.   Alan Luther.

simplification of the theory of strong interactions in $(1 + 1)$-dimensions. The bosonization method was conceived in 1975 independently by two particle and two condensed matter physicists – Sidney Coleman and Sidney Mandelstam, and Daniel Mattis and Alan Luther respectively.[‡] The focal point of their approach was the property of Dirac fermions in $(1 + 1)$-dimensions. They established that correlation functions of such fermions can be expressed in terms of correlation functions of a free bosonic field. In the bosonic representation the fermion forward scattering became trivial which made a solution of the Tomonaga–Luttinger model a simple exercise.

The new approach had been immediately applied to previously untreatable problems. The results by Dzyaloshinskii and Larkin were rederived for short range interactions and generalized to include effects of spin. It was then understood that the low-energy sector in one-dimensional metallic systems might be described by a universal effective theory later christened 'Luttinger-' or 'Tomonaga–Luttinger liquid'. The microscopic description of such a state was obtained by Haldane (1981), the original idea, however, was suggested by Efetov and Larkin (1975). Many interesting applications of bosonization to realistic quasi-one-dimensional metals had been considered in the 1970s by many researches.

Another quite fascinating discovery was also made in the 1970s and

---

[‡] The first example of bosonization was considered earlier by Schotte and Schotte (1969).

Fig. 4. Victor Emery.

concerns particles with fractional quantum numbers. Such particles appear as elementary excitations in a number of one-dimensional systems, with a typical example being spinons in the antiferromagnetic Heisenberg chain with half-integer spin. A detailed description of such systems will be given in the main text; here we just present an impressionistic picture.

Imagine that you have a magnet and wish to study its excitation spectrum. You do it by flipping individual spins and looking at propagating waves. Naturally, since the minimal change of the total spin projection is $|\Delta S^z| = 1$ you expect that a single flip generates a particle of spin-1. In measurements of dynamical spin susceptibility $\chi''(\omega, q)$ an emission of this particle is seen as a sharp peak. This is exactly what we see in conventional magnets with spin-1 particles being magnons.

However, in many one-dimensional systems instead of a sharp peak in $\chi''(\omega, q)$, we see a continuum. This means that by flipping one spin we create at least two particles with spin-1/2. Hence fractional quantum numbers. However, excitations with fractional spin are subject to topological restriction – in the given example this restriction tells us that the particles can be produced only in pairs. Therefore one can say that the elementary excitations with fractional spin (spin-1/2 in the given example) experience

Fig. 5.   P. W. Anderson.

'topological confinement'. Topological confinement puts restriction only on the overall number of particles leaving their spectrum unchanged. Therefore it should be distinguished from dynamical confinement which occurs, for instance, in a system of two coupled spin-1/2 Heisenberg chains (see Chapter 21). There the interchain exchange confines the spinons back to form $S = 1$ magnons giving rise to a sharp single-magnon peak in the neutron cross section which spreads into the incoherent two-spinon tail at high energies.

An important discovery of non-Abelian bosonization was made in 1983–4 by Polyakov and Wiegmann (1983), Witten (1984), Wiegmann (1984) and Knizhnik and Zamolodchikov (1984). The non-Abelian approach is very convenient when there are spin degrees of freedom in the problem. Its application to the Kondo model done by Affleck and Ludwig in their series of papers (see references in Part III) has drastically simplified our understanding of the strong coupling fixed point.

The year 1984 witnessed another revolution in low-dimensional physics. In this year Belavin, Polyakov and Zamolodchikov published their fundamental paper on conformal field theory (CFT). CFT provides a unified approach to all models with gapless linear spectrum in $(1 + 1)$-dimensions.

Fig. 6. Andreas Ludwig.

It was established that if the action of a $(1 + 1)$-dimensional theory is quantizable, that is its action does not contain higher time derivatives, the linearity of the spectrum guarantees that the system has an infinite dimensional symmetry (conformal symmetry). The intimate relation between CFT and the conventional bosonization had became manifest when Dotsenko and Fateev represented the CFT correlation functions in terms of correlators of bosonic exponents (1984). In the same year Cardy (1984) and later Blöte, Cardy and Nightingale (1986) found the important connection between finite size scaling effects and conformal invariance.

Both non-Abelian bosonization and CFT are steps from the initial simplicity of the bosonization approach towards complexity of the theory of integrable systems. Despite the fact that correlation functions can in principle be represented in terms of correlators of bosonic exponents, the Hilbert space of such theories is not equivalent to the Hilbert space of free bosons. In order to make use of the bosonic representation one must exclude certain states from the bosonic Hilbert space. It is not always convenient to do this directly; instead one can calculate the correlation functions using the Ward identities. It is the most important result of CFT that correlation functions of critical systems obey an infinite number of

Fig. 7. Alexander Belavin.

the Ward identities which have the form of differential equations. Solving these equations one can uniquely determine all multi-point correlation functions. This approach is a substitute for the Hamiltonian formalism, since the Hamiltonian is effectively replaced by Ward identities for correlation functions. Conformally invariant systems being systems with infinite number of conservation laws constitute a subclass of exactly solvable (integrable) models.

After many years of intensive development the theory of strongly correlated systems became a vast and complicated area with many distingushed researchers working in it. Different people have different styles and different interests – some are concerned with applications and some with technical developments. There is certainly a gap between those who develop new methods and those who apply them. As an example we can mention the Ising model which has been very extensively studied, but scarcely used in applications. Meanwhile, as will be demonstrated later in the text, this is the simplest theory among those which remain solvable outside of criticality.

This book is an attempt to breach the gap between mathematics of strongly correlated systems and its applications. In our work we have been inspired by the idea that the theory in $(1 + 1)$-dimensions, though being but a small subsector of a global theory of strongly correlated systems, may give an insight for more important and general problems and give the reader a better vision of 'the Universe as a great idea'. The reader will judge whether our attempt is successful.

In conclusion we say several words about the structure and style of this book. The reader should keep in mind that we shall frequently and without much discussion switch between Hamiltonian and Lagrangian formalisms. As a consequence the same notations will stand for operators in the first case and for number (or Grassmann number) fields in the second case. Please beware of this and watch what formalism is used to avoid confusion. We shall also frequently use the field theory jargon: for example, electronic densities are often called currents. Bear in mind that the essence of things does not depend on how they are called, and be indulgent.

The book contains three parts – in the first one we discuss the method, in the second part describe its applications to some interesting $(1 + 1)$-dimensional systems, and in the third part discuss nonlinear quantum impurities. There are important topics which we do not cover; some being even very important – such as applications of bosonization in more than one spatial dimension and the boundary conformal theory. The only reason for omitting these topics is our ignorance.

References

A. A. Abrikosov, *Physica* **2**, 5 (1965).

P. W. Anderson, *Phys. Rev. Lett.* **18**, 1049 (1967).

A. A. Belavin, A. M. Polyakov and A. B. Zamolodchikov, *Nucl. Phys.* **B241**, 333 (1984).

H. W. J. Blöte, J. L. Cardy and M. P. Nightingale, *Phys. Rev. Lett.* **56**, 742 (1986).

Yu. A. Bychkov, L. P. Gor'kov and I. E. Dzyaloshinskii, *Sov. Phys. JETP,* **23**, 489 (1966).

J. L. Cardy, *J. Phys.* **A17**, L385; L961 (1984).

S. Coleman, *Phys. Rev.* **D11**, 2088 (1975).

I. T. Diatlov, V. V. Sudakov and K. A. Ter-Martirosian, *Sov. Phys. JETP* **5**, 631 (1957).

Vl. S. Dotsenko and V. A. Fateev, *Nucl. Phys.* **B240**, 312 (1984).

I. E. Dzyaloshinskii and A. I. Larkin, *Sov. Phys. JETP* **34**, 422 (1972).

I. E. Dzyaloshinskii and A. I. Larkin, *Sov. Phys. JETP* **38**, 202 (1974).

K. B. Efetov and A. I. Larkin, *Sov. Phys. JETP* **42**, 390 (1975).

F. D. M. Haldane, *J. Phys.* C **14**, 2585 (1981).

P. Jordan and E. Wigner, *Z. Phys.* **47**, 631 (1928).

V. G. Knizhnik and A. B. Zamolodchikov, *Nucl. Phys.* **B247**, 83 (1984).

A. Luther and I. Peschel, *Phys. Rev.* **B9**, 2911 (1974).

S. Mandelstam, *Phys. Rev.* **D11**, 3026 (1975).

D. Mattis, *J. Math. Phys.* **15**, 609 (1974).

A. M. Polyakov and P. B. Wiegmann, *Phys. Lett.* **B131**, 121 (1983).

K. D. Schotte and U. Schotte, *Phys. Rev.* **182**, 479 (1969).

I. Solyom, *Adv. Phys.* **28**, 201 (1979).

V. V. Sudakov, *Sov. Phys. Doklady* **1**, 662 (1957).

V. Vaks and A. I. Larkin, *Sov. Phys. JETP* **40**, 282 (1961).

P. B. Wiegmann, *Phys. Lett.* **B141**, 217; *Ibid.*, **142**, 173 (1984).

E. Witten, *Comm. Math. Phys.* **92**, 455 (1984).

Acknowledgements

We are grateful to all those who shared with us their experience and views on the ideas and methodology of bosonization, and who provided support and advice in the process of writing of this book. Our special thanks go to Natan Andrei, Fabian Essler, Andrew Green, Michele Fabrizio, David Khmelnitskii, Ian Kogan, Ilya Krive, Alan Luther, Konstantin Matveev, Philippe Nozières, Nabuhiko Taniguchi, Yu Lu, Feodor Smirnov, David Shelton and Nikolay Prokof'ev.

During his work on this book, A. O. G. was supported by the EPSRC of the United Kingdom under the Advanced Fellowships' programme. In living memory of Arkadii Aronov, A. O. G. wishes to acknowledge their numerous inspiring discussions which motivated much of his activity in the field.

A. A. N. is grateful to the Department of Theoretical Physics of the University of Oxford for kind hospitality and acknowledges the support of ESRSC grant No. GR/K4 1229. He would also like to express his sincere gratitude to Yu Lu for his hospitality and friendly assistance during A. A. N.'s visit to the International Centre for Theoretical Physics, Trieste.

A. M. T. acknowledges precious conversations with Gilbert Lonzarich and Piers Coleman which have always been an invaluable source of inspiration. A. M. T. is also much obliged to Dr Richard Cooper, Brasenose College cellararius, without whose services he would hardly have had energy to work on this book.

Part I

Technical aspects of bosonization

1

A simple case of Bose–Fermi equivalence: Jordan–Wigner transformation

The simplest example of Fermi–Bose equivalence is provided by the spin $S = 1/2$ Heisenberg model. This model in one dimension describes a chain of spins interacting via exchange forces. For many practical cases it is sufficient to consider only nearest neighbour exchange. Then the Heisenberg Hamiltonian extended to include exchange anisotropy Δ, the so-called Heisenberg–Ising or XXZ model, has the following form

$$\hat{H} = J \sum_j \left[\frac{1}{2}(S_j^+ S_{j+1}^- + S_j^- S_{j+1}^+) + \Delta S_j^z S_{j+1}^z \right] \tag{1.1}$$

where J fixes the energy scale of the problem, and S_j^+, S_j^-, S_j^z are operators of spin $S = 1/2$ defined on the site j. These operators have the following remarkable property: taken on the same site they anticommute

$$\{S_j^+, S_j^-\} = 1$$

which suggests an analogy with Fermi creation and annihilation operators:

$$S_j^+ = \psi_j^+, \quad S_j = \psi_j, \quad S_j^z = \psi_j^+ \psi_j - 1/2 \tag{1.2}$$

This analogy fails for spins on a lattice: spin operators on different sites do not anticommute, but **commute**. It turns out, however, that it is possible to modify the definitions and reproduce the correct spin algebra in terms of fermions. Such a transformation was discovered by Jordan and Wigner back in 1928:

$$S_j^+ = \psi_j^+ \exp\left(i\pi \sum_{k<j} \psi_k^+ \psi_k \right)$$

$$S_j^- = \exp\left(-i\pi \sum_{k<j} \psi_k^+ \psi_k \right) \psi_j \tag{1.3}$$

$$S_j^z = \psi_j^+ \psi_j - 1/2$$

The lattice Fermi operators satisfy the standard anticommutation relations

$$\{\psi_j, \psi_{j'}^\dagger\} = \delta_{jj'}, \qquad \{\psi_j, \psi_{j'}\} = 0$$

Substituting expressions (1.3) into Eq. (1.1), we get an equivalent form of the spin-chain Hamiltonian (1.1) in terms of interacting spinless fermions:

$$\hat{H} = J \sum_j [\frac{1}{2}(\psi_j^+ \psi_{j+1} + \psi_{j+1}^+ \psi_j) + \Delta(\psi_j^+ \psi_j - 1/2)(\psi_{j+1}^+ \psi_{j+1} - 1/2)] \quad (1.4)$$

This equivalence will play an important role in the further discussion of the Heisenberg chain. Later on we shall consider this model in great detail, but now we just want to illustrate a simple physical point. As we know, magnetic ordering does not occur in one-dimensional systems with a continuous symmetry. The XXZ spin-chain (1.1) is invariant under arbitrary global rotations of the spins in the basal (xy)-plane (the full SU(2) symmetry is recovered at $\Delta = \pm 1$). Any magnetic ordering in the basal plane would break the continuous U(1) symmetry of the Hamiltonian (1.1), which is forbidden. Does the fermionized version of the Heisenberg model make it easier to see this basic fact? Yes, it does, at least for a special value of the anisotropy parameter $\Delta = 0$ when the model (1.4) significantly simplifies, reducing to noninteracting fermions. In the absence of external magnetic field the tight-binding spectrum is given at $\Delta = 0$ by

$$\epsilon(k) = J \cos k \quad (1.5)$$

and the minimum of the ground state energy corresponds to the half-filled band with a *fixed* number of particles equal to $N/2$. Therefore $\langle S_j^z \rangle = \langle \psi_j^+ \psi_j \rangle - 1/2 = 0$, as well as $\langle S_j^x \rangle = \langle S_j^y \rangle = 0$.

We can easily calculate the longitudinal spin–spin correlation function $\langle S_j^z S_{j'}^z \rangle$, because at $\Delta = 0$ it coincides with the density–density correlation function for noninteracting fermions. It decays slowly with distance:

$$\langle S_j^z S_{j'}^z \rangle = -\frac{1}{2\pi^2} \frac{1 - (-1)^{j-j'}}{(j - j')^2}, \quad (1.6)$$

which means that $T = 0$ is a point of second order phase transition, i.e. a critical point (this is not forbidden in one dimension, and later we explore this possibility in more detail). We shall also demonstrate that the system remains critical throughout the region $|\Delta| < 1$ and all its correlation functions decay at $T = 0$ as power laws. At $T \neq 0$ a finite correlation length appears: $\xi \propto 1/T$. However, having only conventional calculational tools at our disposal we cannot go further than Eq. (1.6). In order to calculate other correlation functions and also to discover many other interesting things, we have to learn more about one-dimensional fermions.

2
One-dimensional fermions.
States near the Fermi points

In the previous chapter we have discovered that the spin $S = 1/2$ Heisenberg magnet can be described by the theory of spinless fermions. This is by no means a single fermionic model used in physics of one-dimensional systems. Therefore we are compelled to study fermions in one dimension. As we shall see, this is not a trivial subject and even noninteracting fermions may reveal some surprising features.

The first surprising feature is the appearence of Lorentz invariance. It turns out that this important symmetry exists not only in high energy physics, but may appear for *low energy* states in a one-dimensional metal! Of course, the light speed c in this case is replaced by the Fermi velocity v. There are also other subtleties, such as the presence of spin which will be discussed later. Now we shall consider one-dimensional spinless fermions and show that their low-energy sector (that is states close to the Fermi points) are described by the Dirac Lagrangian. Let us consider the Euclidean action of nonrelativistic fermions interacting with an Abelian gauge field. The fermionic part of the action is (we put $c = 1$)[*]

$$S = \int d\tau dx [\psi_\sigma^*(\partial_\tau + ie\phi - \epsilon_F)\psi_\sigma + \psi_\sigma^* \hat\epsilon(-i\partial_x - eA)\psi_\sigma] \qquad (2.1)$$

The dispersion law $E(p) = \epsilon(p) - \epsilon_F$ has the qualitative form depicted in Fig. 2.1.

Let us assume now that external fields ϕ, A have small amplitudes. Then all relevant processes take place near the Fermi points. There are two consequences of this, the first is that the only relevant Fourier components of the external fields are those with small wave vectors, and wave vectors of the order of $\pm 2p_F$. The former leave electrons near their Fermi points and the latter transfer them from one Fermi point to another. In real systems such fields are generated, for example, by lattice displacements

---

[*] Here we use the Lagrangian formalism; however, see the warning at the end of Preface.

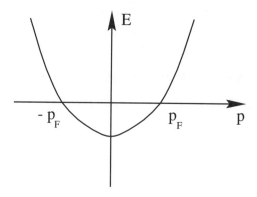

Fig. 2.1.   The qualitative form of electronic dispersion in a one-dimensional metal.

(phonons). Because of the gauge symmetry we can move all fast terms into the scalar potential and write

$$\phi(\tau,x) = A_0(\tau,x) + \exp(2ip_Fx)\Delta(\tau,x) + \exp(-2ip_Fx)\Delta^*(\tau,x)$$
$$A(\tau,x) = A_1(\tau,x) \tag{2.2}$$

where the fields $A_0, A_1, \Delta, \Delta^*$ are slow in comparison with $\exp[\pm 2ip_Fx]$. The second fact is that since the relevant fermions are situated near the Fermi points we can linearize their spectrum:

$$E(p + p_F) \approx vp, \quad E(p - p_F) \approx -vp, \quad |p| << p_F \tag{2.3}$$

where $v = \partial E/\partial p|_{p=p_F}$ is the Fermi velocity. In real space notations this looks as follows:

$$\exp(\pm ip_Fx)\hat{E}(-i\partial_x - eA)\left[\exp(\mp ip_Fx)f(x)\right] \approx \pm v(-i\partial_x - eA)f(x) \tag{2.4}$$

Here we assume that the Fourier expansion of $f(x)$ contains only harmonics with small wave vectors $|p| << 2p_F$.

In the vicinity of the Fermi level the Fermi fields exist as wave packets with average wave vectors $\pm p_F$. This can be expressed as follows (we omit the spin index):

$$\psi(\tau,x) \approx R(\tau,x)\exp(ip_Fx) + L(\tau,x)\exp(-ip_Fx) \tag{2.5}$$

where $R(x), L(x)$ are fields slow in comparison with $\exp(\pm ip_Fx)$. These fields represent right- and left-moving fermions near the right and left Fermi points, respectively. Substituting this expression together with Eq. (2.2) into the original action (2.1), using Eq. (2.4) and taking into account only non-oscillatory terms, we arrive at the Dirac action

$$S = \int d\tau dx[\bar{\psi}\gamma^\mu(\partial_\mu + ieA_\mu)\psi + M_1\bar{\psi}\psi + iM_2\bar{\psi}\gamma_5\psi] \tag{2.6}$$

with $M_1 = \Re e\Delta$, $M_2 = \Im m\Delta$ (see, for example, Chapter A5 of the book by Zinn-Justin). The gamma matrices are chosen as

$$\gamma^0 = \sigma^x, \; \gamma^1 = \sigma^y, \; \gamma_S = \sigma^z$$

As usual $\bar\psi = \psi^+\gamma^0$, and ψ is a two component fermion spinor:

$$\psi = \begin{pmatrix} L \\ R \end{pmatrix}, \; \bar\psi = (R^+, L^+)$$

It is suitable here to make several comments about the Dirac theory. Despite their vector-like appearance, the fermion fields ψ are not vectors. The reason for this is that the name 'vector' is reserved for entities transforming as space-time coordinates under Lorentz transformations. Spinors transform differently, according to their own representation of the Lorentz group, which is called 'spinor' representation. γ^μ-matrices transform as vectors.

An important difference between relativistic and nonrelativistic fermions comes from the fact that the role of the conjugate field in the path integral is played not by ψ^+, but by $\bar\psi$. The reason for this is that it is

$$\bar\psi\psi$$

and not

$$\psi^+\psi$$

which is a Lorentz invariant object. Since the measure of integration must be Lorentz invariant, it cannot be

$$D\psi^+ D\psi$$

as it is for nonrelativistic fermions, but

$$D\bar\psi D\psi$$

In $(1+1)$-dimensions it is $d\bar\psi d\psi = dR^+dLdL^+dR$. Correspondingly, the irreducible Green's function for relativistic electrons in Euclidean space-time at $A_\mu = 0$ is defined as

$$G_{ab}(1,2) \equiv \langle\langle\psi_a(1)\bar\psi_b(2)\rangle\rangle = [(\gamma^\mu\partial_\mu + M_1 + iM_2\gamma_S)^{-1}]_{ab} \quad (2.7)$$

(here and below we reserve the double angle brackets for *irreducible* correlation functions and single angle brackets for simple time-ordered averages). The Dirac action in two-dimenional Euclidean space-time is worth writing in the explicit form:

$$S = \int d\tau dx (\bar L, \bar R) \begin{pmatrix} \Delta & 2\partial + i\bar A \\ 2\bar\partial + iA & \Delta^* \end{pmatrix} \begin{pmatrix} L \\ R \end{pmatrix} \quad (2.8)$$

where

$$z = \tau + ix/v, \partial = \frac{1}{2}(\partial_\tau - iv\partial_x)$$

$$\bar{z} = \tau - ix/v, \bar{\partial} = \frac{1}{2}(\partial_\tau + iv\partial_x)$$

and $A = A_0 - iA_1, \bar{A} = A_0 + iA_1$.

In order to find the Green's function in zero field and constant Δ, we make a Fourier transformation of the fermionic fields. Then from Eq. (2.7) we find

$$G(\omega_n, p) = \frac{1}{\omega_n^2 + v^2p^2 + |\Delta|^2} \begin{pmatrix} \Delta^* & i\omega_n - vp \\ (i\omega_n + vp) & \Delta \end{pmatrix} \qquad (2.9)$$

For future purposes we need to have this expression in real space-time. At $T = 0$ putting $v = 1$ we have

$$\int \frac{d^2p}{(2\pi)^2} \frac{e^{ipx}}{p^2 + |\Delta|^2} = \frac{1}{2\pi} K_0(|\Delta||\mathbf{x}|) \qquad (2.10)$$

where $K_0(x)$ is the zeroth Bessel function of imaginary argument. Therefore we can write

$$G(1,2) = \frac{1}{2\pi} \begin{pmatrix} \Delta^* & -2\partial \\ -2\bar{\partial} & \Delta \end{pmatrix} K_0(|\Delta||z_{12}|) \qquad (2.11)$$

It is especially instructive to calculate the zero mass limit of this expression. At small arguments we have

$$K_0(|\Delta||z_{12}|) \approx -\ln(|\Delta||z_{12}|) \qquad (2.12)$$

and therefore

$$G_{\Delta=0}(1,2) = \frac{1}{2\pi} \begin{pmatrix} 0 & 1/z_{12} \\ 1/\bar{z}_{12} & 0 \end{pmatrix} \qquad (2.13)$$

I   Chiral anomaly

It turns out that the case of massless electrons is especially interesting. This should be expected because massless electrons are conducting and therefore have a nontrivial effect on the electromagnetic field. This field excites electron–hole pairs through the Fermi points and becomes screened. Mathematically the screening effect of the relativistic massless fermions is associated with the so-called *chiral anomaly* – a subject of much fascination and surprise for many years. The anomaly appears as a paradox. Let us consider massless fermions in Minkovsky space-time in an external time-independent potential which we parametrize as follows:[†]

$$A_0 + A_1 = \partial_x \Phi(x), \ A_0 - A_1 = \partial_x \Theta(x)$$

where Φ, Θ are some functions of x. Classically, the above potential appears to have no effect on the fermions at all because it can be removed from the action by the canonical transformation

$$\hat{R}(x) \rightarrow e^{-i\Phi(x)} \hat{R}(x),$$
$$\hat{L}(x) \rightarrow e^{i\Theta(x)} \hat{L}(x) \qquad (2.14)$$

It is clear that this result is nonsense. Indeed, the potential is not a pure gauge since the corresponding electric field does not vanish:

$$F_{01} = \partial_x A_0 - \partial_t A_1 = \frac{1}{2} \partial_x^2 (\Phi - \Theta) \neq 0$$

It is unthinkable that charged particles do not respond to application of an electric field! The explanation of the paradox is that at the quantum level the chiral symmetry, i.e. the symmetry with respect to the transformations (2.14), is broken. The transformation (2.14) does not leave the partition function invariant because it affects the combination

$$\bar{\psi}\psi \rightarrow e^{i[\Phi(x)+\Theta(x)]} R^+ L + e^{i[-\Phi(x)-\Theta(x)]} L^+ R$$

and therefore modifies the measure in the path integral:

$$D\bar{\psi}D\psi$$

Technically the latter fact means that single-fermion wave functions

$$\begin{pmatrix} f_{R,E}(x) \\ f_{L,E}(x) \end{pmatrix}, \ \begin{pmatrix} f_{R,E'}(x) \\ f_{L,E'}(x) \end{pmatrix}$$

with energies E, E', being orthogonal before the transformation are not orthogonal after it. Later we shall demonstrate that the paradox is absent in the diagram expansion, which automatically takes all these subtle properties of the measure into account. Hence the diagram expansion is more reliable than operator transformations of Hamiltonians.

In general, *anomalies* appear in theories which allow transformations affecting only the measure of integration and not the action itself. In the particular case of $(1 + 1)$-dimensional massless Quantum Electrodynamics, the anomaly leads to the illusion that the fermions decouple from the vector potential. They do not, as we have just found out. As we shall demonstrate later, the anomaly allows us to integrate over the fermions and obtain an explicit expression for the effective action of the gauge field.

[†] Such parametrization exists for any two functions A_μ ($\mu = 0, 1$).

One-dimensional fermions

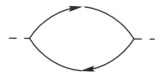

Fig. 2.2.   Current–current correlation function.

II   Anomalous commutators

The chiral anomaly can be rephrased in terms of commutators. Let us
introduce the density operator for right-moving particles:

$$\bar{J}(q) = \sum_p R^+(p+q)R(p)$$

Adopting the terminology of field theory we shall call this operator
the right *current*. Let us consider a commutator of two currents. The
straighforward calculation yields

$$[\bar{J}(q), \bar{J}(p)] = 0 \tag{2.15}$$

with the same result for the left-moving currents $\bar{J}(p)$.

The above result is also absurd because it suggests that the pair correlation function of currents vanishes. On the other hand we can calculate
this correlation function using Feynman diagrams. It is more convenient
to do this in real space-time. The current–current correlation function is
given by the diagram in Fig. 2.2.

According to Eq. (2.13), the single electron Green's functions are given
by

$$\langle\langle R(1)R^+(2)\rangle\rangle = G_R(\bar{z}_{12}) = \frac{1}{2\pi\bar{z}_{12}}, \tag{2.16}$$

$$\langle\langle L(1)L^+(2)\rangle\rangle = G_L(z_{12}) = \frac{1}{2\pi z_{12}} \tag{2.17}$$

Thus Green's function of the left-moving particles depends only on $z = \tau + ix$ and the one for right-moving particles contains only $\bar{z} = \tau - ix$. For
this reason the corresponding components of currents can also be treated
as analytical or antianalytical functions. Substituting these expressions
into the diagram in Fig. 2.2, we get

$$\langle\langle J(z_1)J(z_2)\rangle\rangle = \frac{1}{4\pi^2 z_{12}^2} \tag{2.18}$$

We can use this expression to calculate the commutator. In fact, the
procedure described below is quite general and will be used throughout
the book. Therefore we discuss it in detail. Let A and B be two Bose

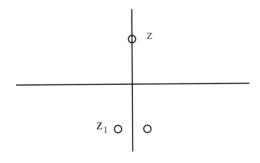

Fig. 2.3.

operators whose correlation function is known. Then one can calculate their commutator at equal times according to the following rule:

$$\langle\langle[A(x), B(y)]\rangle\rangle = \lim_{\tau\to+0} [\langle\langle A(\tau, x)B(0, y)\rangle\rangle - \langle\langle A(-\tau, x)B(0, y)\rangle\rangle] \quad (2.19)$$

Let us show that it follows from the definition of the time-ordered correlation function. Let us imagine that the operators $\hat{A}(z)$ and $\hat{B}(z_1)$ stand inside of some correlation function, perhaps surrounded by other operators which we denote as $X(\{z_i\})$:

$$G(z, z_1, \{z_i\}) = \langle\langle A(z)B(z_1)X(\{z_i\})\rangle\rangle \quad (2.20)$$

Recall, that in our notations $z = \tau + ix$. Let us consider the situation when $\Re ez = \Re ez_1 \pm \delta$ where δ is a positive infinitesimal (see Fig. 2.3). Without loss of generality we can put $z = ix \pm 0$, $z_1 = iy$, where x, y are real. Then due to the time-ordering present in the correlation function the following is valid:

$$G(\tau-\tau' = +0; x, y) - G(\tau-\tau' = -0; x, y) = \langle\langle[A(x), B(y)]X(\{z_i\})\rangle\rangle \quad (2.21)$$

which leads to Eq. (2.19).

Substituting (2.18) into Eq. (2.19), we get

$$\langle\langle[J(x), J(y)]\rangle\rangle = \frac{1}{4\pi^2} \left\{ \frac{1}{[0 + i(x - y)]^2} - \frac{1}{[-0 + i(x - y)]^2} \right\}$$

$$= \frac{1}{4\pi^2}\partial_x \left[\frac{1}{(-i0 + x - y)} - \frac{1}{(i0 + x - y)} \right] = \frac{i}{2\pi}\partial_x\delta(x - y) \quad (2.22)$$

A similar procedure for the left currents yields

$$\langle\langle[\bar{J}(x), \bar{J}(y)]\rangle\rangle = -\frac{i}{2\pi}\partial_x\delta(x - y) \quad (2.23)$$

One can say that there is a difference in definitions. The original commutator does not include any averaging and the one which we have just calculated does. Here we have an example of a commutator which is

Fig. 2.4.   Eduardo Fradkin.

nonzero only for a system of infinite number of particles and vanishes for any finite system. Such commutators are called *anomalous*. There is a nice and clear discussion of the anomalous commutator of currents in Sub-section 4.3.1 of Fradkin's book. An **important lesson** which follows from the present discussion is the following. In order not to miss anomalous commutators, it is always better not to calculate them straightforwardly, but rather use a diagram expansion and the identity (2.19).

Sometimes it is more convenient to use instead of commutators the *Wilson operator product expansion* (OPE):

$$J(z)J(z') = \frac{1}{4\pi^2(z-z')^2} + \dots \tag{2.24}$$

where dots denote nonsingular terms. This expansion means that one can substitute the product of currents *inside of a correlation function* by the right-hand side of Eq. (2.24).

Now consider a massless Bose field Φ governed by the action

$$S_B = \frac{1}{2} \int d^2x (\nabla\Phi)^2 \tag{2.25}$$

and a model of free massless Dirac fermions

$$S_F = \int d^2x \; \bar{\psi}\gamma_\mu \partial_\mu \psi \tag{2.26}$$

with a Noether vector current $J_\mu = \bar{\psi}\gamma_\mu\psi$. Conservation of this current $\partial_\mu J_\mu = 0$ will follow automatically, if we express it in terms of derivatives of the field Φ:

$$J_\mu(x) \equiv \frac{1}{\sqrt{\pi}} \, \epsilon_{\mu\nu} \partial_\nu \Phi \tag{2.27}$$

According to (2.27), the chiral components of the fermionic current, J and $\bar{J}$, are then represented as

$$\bar{J} = \frac{1}{\sqrt{4\pi}} (\partial_x \Phi - \Pi), \quad J = \frac{1}{\sqrt{4\pi}} (\partial_x \Phi + \Pi) \tag{2.28}$$

where Π is the momentum conjugate to the field Φ. It is readily seen that, at quantum level, the anomalous current algebra (2.22) and (2.23) is correctly reproduced, if Φ and Π satisfy the canonical commutation relation

$$[\Phi(x), \Pi(x')] = i\delta(x - x') \tag{2.29}$$

Thus, the identification (2.27) gives us the first example of *bosonization*.

A pedantic reader may ask a question: Why, after all, the anomalous current algebra is an exclusive feature of one space dimension? Why it does not appear in higher dimensions? Here we present simple arguments to answer these questions.

Consider noninteracting (for simplicity, spinless) fermions with isotropic spectrum $\epsilon(\mathbf{k})$ in arbitrary space dimensions d. The Hamiltonian is given by

$$H = \sum_{\mathbf{k}} [\epsilon(\mathbf{k}) - \mu] \, c_{\mathbf{k}}^\dagger c_{\mathbf{k}}$$

where μ is the chemical potential. Introduce the density-fluctuation operator

$$\delta\rho(\mathbf{x}) = \rho(\mathbf{x}) - \bar{\rho} = \frac{1}{L^d} \sum_{\mathbf{q} \neq 0} e^{i\mathbf{q}\mathbf{x}} \rho(\mathbf{q}), \quad \rho(\mathbf{q}) = \sum_{\mathbf{k}} c_{\mathbf{k}}^\dagger c_{\mathbf{k}+\mathbf{q}},$$

$\bar{\rho}$ being the average density of the particles. Consider the retarded density–density Green's function

$$D^{(R)}(\mathbf{x}, t) = -i\theta(t) \langle [\delta\rho(\mathbf{x}, t), \delta\rho(\mathbf{0}, 0)] \rangle \tag{2.30}$$

Its frequency–momentum Fourier transform is given by the well-known Lindhardt formula:

$$D^{(R)}(\mathbf{q}, \omega) = \int \frac{d^d\mathbf{k}}{(2\pi)^d} \frac{n_{\mathbf{k}} - n_{\mathbf{k}+\mathbf{q}}}{\omega - \epsilon(\mathbf{k} + \mathbf{q}) + \epsilon(\mathbf{k}) + i\delta} \tag{2.31}$$

where $n_{\mathbf{k}}$ is the Fermi distribution function. At $|\mathbf{q}| \ll k_F$, $|\omega| \ll \epsilon_F$, the integral in (2.31) is contributed by the momenta $\mathbf{k}$ close to the Fermi surface. Therefore one can first do the radial part of the integral and then express the result as angular integral over the Fermi surface:

$$
\begin{aligned}
D^{(R)}(\mathbf{q}, \omega) &\simeq L^d \int \frac{d^d k}{(2\pi)^d} \frac{(\mathbf{q} \cdot \mathbf{e_F})\delta(|k| - k_F)}{\omega - (\mathbf{q} \cdot \mathbf{e_F})v_F + i\delta} \\
&= \frac{L^d k_F^{d-1}}{(2\pi)^d} \oint d\mathbf{e_F} \frac{(\mathbf{q} \cdot \mathbf{e_F})}{\omega - (\mathbf{q} \cdot \mathbf{e_F})v_F + i\delta}
\end{aligned} \tag{2.32}
$$

where $\mathbf{e_F} = \mathbf{k_F}/k_F$ is a unit vector normal to the Fermi surface.

In space dimensions $d > 1$, the angular integration in (2.32) results in branch cut singularities of $D^{(R)}(\mathbf{q}, \omega)$ at $\omega = \pm q v_F$. For example, in the three-dimensional case

$$
D^{(R)}(\mathbf{q}, \omega) \sim 1 + \frac{\omega}{2 q v_F} \ln \left| \frac{\omega - q v_F}{\omega + q v_F} \right|,
$$

showing no traces of well defined propagating bosonic modes. Thus, for noninteracting fermions in $d > 1$ dimensions, the low-energy spectrum of the density excitations is exhausted by incoherent background of electron–hole pairs.

The situation dramatically changes in one dimension. Since in this case the Fermi surface degenerates into *two points*, $k = k_F$ ($\mathbf{e_F} = \hat{z}$) and $k = -k_F$ ($\mathbf{e_F} = -\hat{z}$), $D(q, \omega)$ turns out to be a sum of two singular terms

$$
D^{(R)}(q, \omega) = \frac{Lq}{2\pi} \left(\frac{1}{\omega - q v_F + i\delta} - \frac{1}{\omega + q v_F + i\delta} \right) \tag{2.33}
$$

Each term has a pole structure and represents the Green's function of one-dimensional massless bosons with a linear spectrum $\omega = \pm q v_F$, propagating along the chain either in the positive or negative direction with the Fermi velocity v_F. The poles clearly indicate that we are dealing with coherent bosonic modes, the right-moving and left-moving Tomonaga bosons.

The important fact that in one-dimensional Fermi systems particle–hole excitations with small energy and momentum constitute well defined bosonic modes is a consequence of tremendous reduction of the available phase space. For a long-wavelength density excitation, both the particle and the hole that constitute the excitation must be in the vicinity of the *same* Fermi point. Since close to $\pm k_F$ the single-fermion spectrum is almost linear, it turns out that, for such particle–hole pairs, conservation of the momentum automatically implies conservation of the energy, and vice versa. (When considering one-dimensional systems of *interacting* fermions in subsequent sections, we shall see that it is this property which gives rise to the so-called infrared catastrophe, leading to disappearance of single-fermion excitations in the sea of gapless bosonic modes.) As a result, the particle–hole excitation energy, $\omega(k, q) = \epsilon(k + q) - \epsilon(k)$, at $|q| \ll k_F$ and

$k \sim \pm k_F$ only depends on q:

$$\omega(k,q) = \epsilon_{R(L)}(k+q) - \epsilon_{R(L)}(k) = \pm v_F q \qquad (2.34)$$

The additive structure of $D^{(R)}(q,\omega)$ in (2.33) immediately suggests decomposition of the long-wavelength component of the density operator into the right and left components (currents):

$$\delta\rho(x) = J(x) + \bar{J}(x) \qquad (2.35)$$

where

$$\bar{J}(x) = \,: R^\dagger(x)R(x) :, \quad J(x) = \,: L^\dagger(x)L(x) : \qquad (2.36)$$

The Fourier transforms of these currents satisfy equations of motion

$$(i\partial_t - v_F q)J(q,t) = 0, \quad (i\partial_t + v_F q)\bar{J}(q,t) = 0 \qquad (2.37)$$

The retarded Green functions of the right and left currents are defined as

$$D^{(R)}(q,\omega)(q,t;q',0) = -i\theta(t)\langle\, [\, J(q,t), J(-q',0)\,]\,\rangle,$$
$$D = D + \bar{D}$$

Using Eqs. (2.37), we get

$$(i\partial_t - qv_F)\bar{D}^{(R)}(q,t;q',0) = \langle\, [\, \bar{J}(q,0), \bar{J}(-q',0)\,]\,\rangle\delta(t)$$

$$(i\partial_t + qv_F)D^{(R)}(q,t;q',0) = \langle\, [\, J(q,0), J(-q',0)\,]\,\rangle\delta(t)$$

or equivalently

$$D^{(R)}(q,q';\omega) = \frac{[\, J(q), J(-q')\,]}{\omega + qv_F + i\delta}\delta_{qq'}, \quad \bar{D}^{(R)}(q,q';\omega) = \frac{[\, \bar{J}(q), \bar{J}(-q')\,]}{\omega - qv_F + i\delta}\delta_{qq'} \qquad (2.38)$$

Comparing (2.38) with (2.33), we arrive at the anomalous commutation relations for the currents:

$$[J(q), J(-q')] = -[\bar{J}(q), \bar{J}(-q')] = \frac{Lq}{2\pi}\delta_{qq'} \qquad (2.39)$$

which, being rewritten in real space, reduce to Eqs. (2.22, 2.23).

3

Gaussian model.
Lagrangian formulation

As we have learnt in the previous chapter, commutation relations of the fermionic currents can be expressed as commutators of the bosonic fields. Since this fact is crucial for the transmutation of statistics, it is logical to discuss the corresponding bosonic model. So in this chapter we discuss the apparently trivial theory of the free bosonic massless scalar field in a two-dimensional Euclidean space. This model is usually called the Gaussian model and its action is defined as

$$S = \frac{1}{2} \int_{\mathscr{A}} d\tau dx [v^{-1}(\partial_\tau \Phi)^2 + v(\partial_x \Phi)^2] \tag{3.1}$$

where $\mathscr{A}$ is some area of the infinite (τ, x)-plane. The most important physical case is, of course, when $\mathscr{A}$ is a rectangle:

$$\mathscr{A} = (0 < \tau < \beta, \, 0 < x < L) \tag{3.2}$$

Here, as usual, β is the inverse temperature. It is important, however, to keep an option of arbitrary $\mathscr{A}$ because the model (3.1) possesses a hidden symmetry which makes itself manifest in transformation properties of its correlation functions with respect to the area change. As usual for Lorentz invariant theories one needs to restrict its energy spectrum to make correlation functions nonsingular (regularization). To regularize the theory we introduce the small distance cut-off a assuming this to be the smallest possible *interval* between two points in (τ, x)-space.

The Gaussian model (3.1) forms a backbone of the bosonization approach. Its simplicity makes the latter approach a great success story. We shall see, however, that this apparent simplicity hides many highly nontrivial features. The first such feature is that the model (3.1) has quite remarkable correlation functions.

The correlation functions of such a simple theory can be most conveniently calculated using the path integral approach. Introduce the generating functional of fields Φ:

$$Z[\eta] = \int D\Phi(x)e^{-S - \int d\tau dx \eta(x)\Phi(x)} \tag{3.3}$$

where S is the action (3.1), and $x = (\tau, x)$. The generating functional can be calculated explicitly for any η which will be very convenient. We can do it for any area $\mathscr{A}$, but to keep the calculations simple, we shall consider a rectangular area (3.2) first. To calculate the path integral one has to write down all functions in terms of their Fourier components. Then we get

$$Z[\eta] = \int \prod_p d\Phi(p) \exp\left[-\frac{1}{2}\sum_p \Phi(-p)(vq^2 + v^{-1}\omega^2)\Phi(p) + \sum_p \eta(-p)\Phi(p)\right] \tag{3.4}$$

where

$$\mathbf{p} = (\omega, q), \quad \omega = 2\pi n_0/\beta, \quad q = 2\pi n_1/L$$

and

$$\Phi(\mathbf{p}) = \int d\tau dx e^{i\omega_n \tau} e^{iqx} \Phi(\tau, x)$$

The integral (3.4) is just a product of simple Gaussian integrals; it can be estimated by the shift of variables, which removes the term linear in Φ from the exponent:

$$\tilde{\Phi}(\mathbf{p}) = \Phi(\mathbf{p}) + G(\mathbf{p})\eta(\mathbf{p}) \tag{3.5}$$

$$\ln Z[\eta] = \ln Z[0] + \frac{1}{2\beta V}\sum_p' \eta(-\mathbf{p})G(\mathbf{p})\eta(\mathbf{p}) \tag{3.6}$$

The prime in the summation means that the term with $\mathbf{p} = 0$ is excluded and $G(\mathbf{p}) = (vq^2 + v^{-1}\omega^2)^{-1}$ is the Fourier transform of the Green's function of the Laplace operator satisfying the equation

$$-(v\partial_x^2 + v^{-1}\partial_\tau^2)G(x, \tau; x', \tau') = \delta(x - x')\delta(\tau - \tau') \tag{3.7}$$

One can easely generalize this formula for an arbitrary area $\mathscr{A}$ writing it in the following form:

$$Z[\eta(x)]/Z[0] = \exp\left[\frac{1}{2}\eta(\xi)G(\xi, \xi')\eta(\xi')\right] \tag{3.8}$$

where $G(\xi, \xi')$ is the Green's function of the Laplace operator on the area $\mathscr{A}$. This general expression will turn out very useful for us later.

It will be more convenient for later purposes to use the complex coordinates $z = \tau + ix/v, \bar{z} = \tau - ix/v$ where

$$\Delta = 4v^{-1}\partial_z\partial_{\bar{z}}, \quad \partial_z = \frac{1}{2}(\partial_\tau - vi\partial_x), \quad \partial_{\bar{z}} = \frac{1}{2}(\partial_\tau + vi\partial_x)$$

Let us consider the case of a very large rectangle or a disk of very large radius R. The words 'very large' in this context mean that we shall be interested only in correlations very far from the boundaries. The Green's function has the following well known form:

$$G(z, \bar{z}) = \frac{1}{4\pi} \ln\left(\frac{R^2}{z\bar{z} + a^2}\right) \tag{3.9}$$

(recall that a is the minimal interval in the theory).

Now let us consider a particular choice of η, namely

$$\eta(\xi) = \eta_0(\xi) \equiv i \sum_{n=1}^{N} \beta_n \delta(\xi - \xi_n) \tag{3.10}$$

where β_n are some numbers.

The generating functional (3.8) for this particular choice of η coincides with the correlation function of bosonic exponents:

$$Z[\eta_0]/Z[0] \equiv \langle \exp[i\beta_1 \Phi(\xi_1)] \dots \exp[i\beta_N \Phi(\xi_N)] \rangle \equiv \mathscr{F}(1, 2, \dots N) \tag{3.11}$$

Substituting Eq. (3.10) into (3.8) we get the expresssion which is, probably, the **most important** formula in the book:

$$\mathscr{F}(1, 2, \dots N) = \exp\left[-\sum_{i>j} \beta_i \beta_j G(\xi_i; \xi_j)\right] \exp\left[-\frac{1}{2} \sum_i \beta_i^2 G(\xi_i; \xi_i)\right] \tag{3.12}$$

The terms with the Green's functions of coinciding arguments are singular in the continuous limit. However, since we have regularized the theory, they are finite. Substituting the expression for G from Eq. (3.9) into Eq. (3.12) we get the following result for the correlation function of bosonic exponents:

$$\mathscr{F}(1, 2, \dots N) = \prod_{i>j} \left(\frac{z_{ij}\bar{z}_{ij}}{a^2}\right)^{(\beta_i\beta_j/4\pi)} \left(\frac{R}{a}\right)^{-\left(\sum_n \beta_n\right)^2/4\pi} \tag{3.13}$$

Two simple points need to be mentioned. Usually we are interested in very large systems $R/a \to \infty$. In this case the expression (3.13) is only nonzero provided the *'electroneutrality'* condition is fulfilled:

$$\sum_n \beta_n = 0 \tag{3.14}$$

For a finite system, however, even a single exponent has a nonzero average:

$$\langle \exp[i\beta\Phi(\xi)] \rangle = (R/a)^{-\beta^2/4\pi} \tag{3.15}$$

Thus we have a general expression for correlation functions of bosonic exponents. In fact, as we have already mentioned, Eq. (3.12) holds not only for a plane, but for any area $\mathscr{A}$, provided G is chosen properly. This expression gives us potentially *complete* knowledge of correlation functions of the Gaussian model (3.1). Indeed, since the bosonic exponents form a basis for local functionals of Φ, one can calculate correlation functions of any local functional $F(\Phi)$ by expanding it as the Fourier integral

$$F(\Phi) = \int d\beta \tilde{F}(\beta) e^{i\beta\Phi}$$

and using Eq. (3.13). In fact, even more can achieved. Let us come back to Eq. (3.13) and rewrite its right-hand side as a product of the analytic and the anti-analytic functions:

$$\langle \exp[i\beta_1\Phi(\xi_1)]... \exp[i\beta_N\Phi(\xi_N)]\rangle$$
$$= G(z_1, ..., z_N)G(\bar{z}_1, ..., \bar{z}_N)\delta_{\Sigma\beta_n,0} \tag{3.16}$$

where

$$G(\{z\}) = \prod_{i>j} \left(\frac{z_{ij}}{a}\right)^{(\beta_i\beta_j/4\pi)}$$

This factorization guarantees that analytic and anti-analytic parts of the correlation functions can be studied **independently**. Since factorization of the correlation functions is a general fact, it can be formally written as factorization of the corresponding fields: *inside the* $\langle ... \rangle$-*sign* one can rewrite $\Phi(z,\bar{z})$ as a sum of independent *analytic* and *anti-analytic* fields (we shall call them *chiral* components of the field Φ):

$$\Phi(z,\bar{z}) = \phi(z) + \bar{\phi}(\bar{z}), \tag{3.17}$$
$$\exp[i\beta\Phi(z,\bar{z})] = \exp[i\beta\phi(z)] \exp[i\beta\bar{\phi}(\bar{z})] \tag{3.18}$$

We emphasize that this decomposition should be understood only as a property of correlation functions and *not* as a restriction on the variables in the path integral (3.4). The attentive reader understands that the expression (3.16) is obtained by integration over *all* fields Φ. The meaning of the chiral fields $\phi, \bar{\phi}$ becomes quite transparent in the Hamiltonian formulations of the Gaussian model which is discussed in the next chapter.

For many purposes it is convenient to use the 'dual' field $\Theta(z,\bar{z})$ defined as

$$\Theta(z,\bar{z}) = \phi(z) - \bar{\phi}(\bar{z}) \tag{3.19}$$

The dual field satisfies the following equations:

$$\partial_\mu\Phi = -i\epsilon_{\mu\nu}\partial_\nu\Theta \tag{3.20}$$

or, in components

$$\partial_z\Phi = \partial_z\Theta$$
$$\partial_{\bar{z}}\Phi = -\partial_{\bar{z}}\Theta \tag{3.21}$$

In order to study correlation functions of the analytic and the anti-analytic fields, we define the fields

$$A(\beta, z) \equiv \exp\left\{\frac{i}{2}\beta[\Phi(z, \bar{z}) + \Theta(z, \bar{z})]\right\}$$

$$\bar{A}(\bar{\beta}, \bar{z}) \equiv \exp\left\{\frac{i}{2}\bar{\beta}[\Phi(z, \bar{z}) - \Theta(z, \bar{z})]\right\} \tag{3.22}$$

with, generally speaking, **different** $\beta, \bar{\beta}$. With the operators $A(\beta, z), \bar{A}(\bar{\beta}, \bar{z})$ one can expand local functionals of mutually nonlocal fields Φ and Θ. Let us construct a complete basis of bosonic exponents for a space of local periodic functionals. Suppose that

$$F(\Phi, \Theta)$$

is a local functional periodic both in Φ and Θ with the periods T_1 and T_2, respectively. This functional can be expanded in terms of the bosonic exponents:

$$F(\Phi, \Theta) = \sum_{n,m} \tilde{F}_{n,m} \exp[(2i\pi n/T_1)\Phi + (2i\pi m/T_2)\Theta]$$

$$\times \sum_{n,m} \tilde{F}_{n,m} A(\beta_{nm}, z)\bar{A}(\bar{\beta}_{nm}, \bar{z}) \tag{3.23}$$

where

$$\beta_{nm} = 2\pi \left(\frac{n}{T_1} + \frac{m}{T_2}\right)$$

$$\bar{\beta}_{nm} = 2\pi \left(\frac{n}{T_1} - \frac{m}{T_2}\right) \tag{3.24}$$

It turns out that the periods T_1, T_2 are not arbitrary, but related to each other. The reason for this lies in the fact that the correlation functions must be uniquely defined on the complex plane. We can see how this argument works using the pair correlation function as an example:

$$\langle A(\beta_{nm}, z_1)\bar{A}(\bar{\beta}_{nm}, \bar{z}_1)A(-\beta_{nm}, z_2)\bar{A}(-\bar{\beta}_{nm}, \bar{z}_2)\rangle$$

$$= (z_{12})^{-\beta_{nm}^2/4\pi}(\bar{z}_{12})^{-\bar{\beta}_{nm}^2/4\pi} = \frac{1}{|z_{12}|^{2d}}\left(\frac{z_{12}}{\bar{z}_{12}}\right)^S \tag{3.25}$$

where we introduce the quantities

$$d = \Delta + \bar{\Delta} = \frac{1}{8\pi}(\beta^2 + \bar{\beta}^2)$$

and

$$S = \Delta - \bar{\Delta} = \frac{1}{8\pi}(\beta^2 - \bar{\beta}^2)$$

which are called the 'scaling dimension' and the 'conformal spin', respectively.

The two branch cut singularities in Eq. (3.25) cancel each other and give a uniquely defined function only if

$$2S = \beta_{nm}^2/4\pi - \bar{\beta}_{nm}^2/4\pi = (\text{integer}) \qquad (3.26)$$

i.e., physical fields with uniquely defined correlation functions must have integer or half-integer conformal spins. This equation suggests the relation

$$T_2 = \frac{4\pi}{T_1} \equiv \sqrt{4\pi K} \qquad (3.27)$$

as the minimal solution. The relationship (3.27) specifies the exponents of the multi-point correlation functions. Here we introduce the new notation K for future convenience. The normalization is such that at $K = 1$ the periods for the field Φ and its dual are equal. The quantities $\Delta, \bar{\Delta}$ are called 'conformal dimensions' or 'conformal weights'. In the case of the Gaussian model (3.1) the conformal dimensions of the basic operators are given by:

$$\Delta_{nm} \equiv \beta_{nm}^2/8\pi = \frac{1}{8}\left(n\sqrt{K} + \frac{m}{\sqrt{K}}\right)^2$$

$$\bar{\Delta}_{nm} \equiv \bar{\beta}_{nm}^2/8\pi = \frac{1}{8}\left(n\sqrt{K} - \frac{m}{\sqrt{K}}\right)^2 \qquad (3.28)$$

I Bosonization

Correlation functions of operators with integer conformal spins are invariant under permutation of coordinates; therefore such operators are bosonic. On the contrary, when one permutes operators with half-integer conformal spins, the correlation function changes its sign. Therefore a field with a half-integer conformal spin is fermionic. As we have seen, both types of fields can be represented by bosonic exponents. Therefore for a fermionic theory with a linear spectrum, the fermionic operators can be represented in terms of exponents of the massless bosonic field and its dual field. This is an extension of the ideas discussed in the previous chapter where we established the equivalence between fermionic currents and derivatives of the bosonic fields. As we have already mentioned, the original idea of this transformation was suggested by Coleman (1975) and Mandelstam (1975). One can also find a thorough and up-to-date discussion of bosonization in Chapter 4 of Fradkin's book.

Let us consider pair correlation functions of the bosonic exponents at the special point $K = 1$:

$$\beta = \sqrt{4\pi}$$

Table 3.1.

Massless Bosons	Massless Fermions
Action	Action
$\frac{1}{2}\int d^2x(\nabla\Phi)^2$	$2\int d^2x(R^+\partial_{\bar{z}}R + L^+\partial_z L)$
Operators	Operators
$(\sqrt{2\pi a})^{-1}\exp[\pm i\sqrt{4\pi}\phi(z)]$	$R,\ R^+$
$(\sqrt{2\pi a})^{-1}\exp[\mp i\sqrt{4\pi}\bar{\phi}(\bar{z})]$	$L,\ L^+$
$(\pi a)^{-1}\cos[\sqrt{4\pi}\Phi(z,\bar{z})]$	$R^+L + L^+R$
$\frac{i}{\sqrt{\pi}}\partial\Phi$	L^+L
$-\frac{i}{\sqrt{\pi}}\bar{\partial}\Phi$	R^+R

Then we reproduce Eq. (2.17):

$$G(z) \equiv \frac{1}{2\pi a}\langle\exp[i\sqrt{4\pi}\phi(z_1)]\exp[-i\sqrt{4\pi}\phi(z_2)]\rangle = \frac{1}{2\pi z_{12}} \quad (3.29)$$

Let us regularize this function:

$$G_L(z,\bar{z};a) = \frac{1}{2\pi z} = \lim_{a\to 0}\frac{1}{2\pi}\partial_z\ln(a^2+|z|^2)$$

which then satisfies the equation

$$2\partial_{\bar{z}}G_L(z,\bar{z};a) = \delta^{(2)}(x,y) \quad (3.30)$$

for the Green's function of the massless fermionic field with right chirality! (Recall the discussion in the previous Chapter.) It is obvious that the corresponding anti-analytic function coincides with the Green's function of relativistic massless fermions with left chirality (2.17). So bringing together these results and the formulas for currents obtained in the end of Chapter 2, we can establish the following correspondence between operators of the two theories wich have the same correlation functions:

The rest of the book (with some marked exceptions!) will be mostly devoted to applications of this bosonization table to various interesting physical models. Just to give a simple example of how bosonization works let us consider the following action:

$$S = \int d^2x\{\frac{1}{2}(\nabla\Phi)^2 + \frac{M}{\pi a}\cos(\sqrt{4\pi}\Phi)\} \equiv \int d^2x(\bar{\psi}\gamma^\mu\partial_\mu\psi + M\bar{\psi}\psi) \quad (3.31)$$

The equivalence follows from the fact that in the two theories the perturbation expansions in M coincide. Thus a model which is nonlinear in the bosonic formulation is a disguised theory of free massive fermions!

II   Interaction with an electromagnetic field; gauge invariance

In order to get a better understanding of the physical picture behind bosonization, let us consider how the bosonization approach describes the interaction of spinless fermions with an electromagnetic field. The interaction term is

$$eJ_\mu A_\mu$$

and according to the bosonization table the bosonized action is

$$S = \int d\tau dx \left[\frac{1}{2}(\partial_\mu \Phi)^2 + \frac{ie}{\sqrt{\pi}} A_\mu \epsilon_{\mu\nu} \partial_\nu \Phi \right] \qquad (3.32)$$

This action must be gauge invariant, that is not to change under the transformation

$$A_\mu \to A_\mu + \partial_\mu \alpha \qquad (3.33)$$

It is easily seen that this is indeed the case, which means that one does not need to do anything about the Φ-field to achieve the gauge invariance. This is an important result meaning that Φ is a *neutral* field, not changing under gauge transformations.

Let us now switch to the dual description. The naive application of Eq. (3.20) leads us into trouble since the resulting action is not gauge invariant. The reason for this is that the Θ-field is charged and does change under gauge transformations, which has to be taken into account in writing Eq.(3.20) in the presence of an electromagnetic field. To find transformation properties of the Θ-field we write down the combination

$$R^+ L^+ = \frac{1}{2\pi a} \exp\left(i\sqrt{4\pi}\Theta \right) \qquad (3.34)$$

Since under the gauge transformation (3.33) the fermionic fields change

$$R(x), L(x) \to e^{-ie\alpha(x)} \{R(x), L(x)\}$$

so the Θ-field must change:

$$\Theta(x) \to \Theta(x) - \frac{e}{\pi}\alpha(x) \qquad (3.35)$$

which suggests that the gauge invariant form of the duality relation (3.20) must be

$$\partial_\mu \Phi = -i\epsilon_{\mu\nu} \left(\partial_\nu \Theta + \frac{e}{\pi} A_\nu \right) \qquad (3.36)$$

We may also conjecture the following gauge invariant action:

$$S_{dual} = \frac{1}{2} \int d^2 x \left(\partial_\nu \Theta + \frac{e}{\pi} A_\nu \right)^2 \tag{3.37}$$

This looks like a Ginzburg–Landau action for a superconductor with $\sqrt{4\pi}\Theta$ being a phase of the order parameter.

Let us make sure that actions (3.32) and (3.37) give the same effective action for the electromagnetic field. To make the integration over the bosonic fields easier, we decompose A_μ into the gradient and curl:

$$A_\mu = \partial_\mu \alpha + \epsilon_{\mu\nu} \partial_\nu \chi \tag{3.38}$$

$$\chi = \partial^{-2} \epsilon_{\mu\nu} \partial_\nu A_\mu \equiv \partial^{-2} F_{01} \tag{3.39}$$

Substituting this expression into Eq. (3.32) and integrating the partition function over Φ we get the effective action for the electromagnetic field:

$$S_{el-mag} = \frac{e^2}{\pi} \int d^2 x (\partial_\mu \chi)^2 = -\frac{e^2}{\pi} \int d^2 x F_{01} \partial^{-2} F_{01} \tag{3.40}$$

Now let us substitute Eq. (3.38) into Eq. (3.37). Here we have no need to integrate – the transformation (3.35) leaves us with the action (3.40).

In conclusion we would like to warn the reader that the duality equations (3.20) can be used only for asymptotics of correlation functions. It is easy to see that, for example,

$$\langle\langle \partial_x \Phi(\tau, x) \partial_x \Phi(0,0) \rangle\rangle + \langle\langle \partial_\tau \Theta(\tau, x) \partial_\tau \Theta(0,0) \rangle\rangle = (2\pi)^2 \delta(\tau)\delta(x)$$

Since the delta function refers to small distances where the continuous approach does not work, we have no control over it.

Exercise.   Imagine that the gauge field A_μ is dynamical, being governed by the standard bare Euclidean action

$$S_0 = \frac{1}{8\pi} \int d^{D+1} x F_{\mu\nu}^2 \tag{3.41}$$

Using the effective action (3.40) calculate correlation functions of $F_{\mu\nu}$ when

(a) $D = 1$, that is both electrons and the field live in one spacial dimension,

(b) $D = 2, 3$, that is the field lives in a space of greater dimensionality. A physical example of such situation would be a problem of screening of three-dimensional Coulomb interaction in a one-dimensional wire.

4

Conformal symmetry
and finite size effects

As we have already demonstrated, the apparently trivial Gaussian model has non-trivial correlation functions. We have also seen that these functions can be used in the bosonization procedure. In fact, in order to use this tool effectively, we need to learn more about properties of the Gaussian model. One of the most important properties is that the Gaussian model possesses a special hidden symmetry – the conformal symmetry. This symmetry plays a great role in the modern field theory. We choose the Gaussian model as the simplest example to illustrate the general concept.

Conformal symmetry reveals itself in special properties of the correlation functions under transformations of the area A on which the field Φ is defined. Let $\mathscr{A}$ be an arbitrary area of the complex plane. As we have said before, the expression (3.16) for multi-point correlation functions of the bosonic exponents is valid for any $\mathscr{A}$:

$$\langle \exp[i\beta_1\Phi(\xi_1)]...\exp[i\beta_N\Phi(\xi_N)]\rangle$$

$$= \exp\left[-\sum_{i>j}\beta_i\beta_j G(\xi_i;\xi_j)\right]\exp\left[-\frac{1}{2}\sum_i \beta_i^2 G(\xi_i;\xi_i)\right] \quad (4.1)$$

where G is the Green's function of the Laplace operator on $\mathscr{A}$. As shown in the theory of the Laplace equation, the Green's function $G(\xi_i;\xi_j)$ can be written explicitly if one knows a transformation $z(\xi)$ which maps $\mathscr{A}$ onto the infinite plane. Then the Green's function of the Laplace operator on $\mathscr{A}$ is given by:

$$G(\xi_1,\xi_2) = -\frac{1}{2\pi}\ln|z(\xi_1)-z(\xi_2)| - \frac{1}{4\pi}\ln[|\partial_{\xi_1}z(\xi_1)\partial_{\xi_2}z(\xi_2)|] \quad (4.2)$$

Let us consider the case of two exponents: $N = 2$. Recall that the correlation functions (4.1) are products of analytic and anti-analytic parts. Substituting expression (4.2) into Eq. (4.1) with $N = 2$ and $\beta_1 = \beta$, $\beta_2 = -\beta$, we get the following expression for the analytic part of the pair

25

correlation function:

$$D(\xi_1;\xi_2) = \frac{1}{[z(\xi_1) - z(\xi_2)]^{2\Delta}} [\partial_{\xi_1} z(\xi_1) \partial_{\xi_2} z(\xi_2)]^{\Delta} \qquad (4.3)$$

where $\Delta = \beta^2/8\pi$.

As we see, the correlation functions transform locally under analytic coordinate transformations. Therefore one can assign these transformations to the corresponding operators – the bosonic exponents $A_{\Delta}(z) \equiv \exp[i\beta\phi(z)]$. As is clear from Eq. (4.3), these operators transform as tensors of rank $(\Delta, 0)$:

$$A_{\Delta}(\xi) = A_{\Delta}[z(\xi)](dz/d\xi)^{\Delta} \qquad (4.4)$$

Correspondingly, the antianalytical exponent $\bar{A}_{\bar{\Delta}}(\bar{z})$ transforms as a tensor of rank $(0, \bar{\Delta})$.

There is one transformation which is particularly important for applications:

$$z(\xi) = \exp(2\pi\xi/L) \qquad (4.5)$$

This transforms a strip of width L in the x-direction into an infinite complex plane. For $(1 + 1)$-dimensional systems this transformation (i) relates correlation functions of finite quantum chains to those of infinite systems, and also (ii) relates correlation functions at $T = 0$ to correlation functions at finite temperatures (in the latter case $T = i/L$). Substituting $z(\xi)$ into Eq. (4.3) we get

$$D(\xi_1;\xi_2) = \langle A_{\Delta}(z_1)A_{\Delta}^+(z_2)\rangle = \left\{\frac{\pi/L}{\sinh[\pi(\xi - \xi')/L]}\right\}^{2\Delta}$$

$$\bar{D}(\xi_1;\xi_2) = \langle \bar{A}_{\bar{\Delta}}(\bar{z}_1)\bar{A}_{\bar{\Delta}}^+(\bar{z}_2)\rangle = \left\{\frac{\pi/L}{\sinh[\pi(\bar{\xi} - \bar{\xi}')/L]}\right\}^{2\bar{\Delta}} \qquad (4.6)$$

Below we continue to treat our system as a quantum $(1 + 1)$-dimensional one at $T = 0$. Let $T = 0$ and the system be a circle of a finite length L. In this case we have

$$\xi = \tau + ix, \quad -\infty < \tau < \infty, \quad 0 < x < L$$

Let us expand the expressions for $D(1,2), \bar{D}(1,2)$ at large τ_{12}. The result is

$$D(\xi_1;\xi_2)\bar{D}(\xi_1;\xi_2)$$

$$= \sum_{n,m=0}^{\infty} C_{nm}(\pi/L)^d \exp\left[-\frac{2\pi}{L}(d+n)\tau_{12}\right] \exp\left[-ix_{12}\frac{2\pi}{L}(S+m)\right] \qquad (4.7)$$

where C_{nm} are universal numerical coefficients (note that $n, m \geq 0$, it is important!).

Exercise. Calculate C_{nm}.

On the other hand this correlation function can be expanded in the Lehmann series:

$$D(\tau_{12}, x_{12})\bar{D}(\tau_{12}, x_{12})$$
$$= \sum_q |\langle 0| \exp[i\beta\phi(0) + i\bar{\beta}\bar{\phi}(0)]|q\rangle|^2 \exp[-E_q\tau_{12} - iP_q x_{12}] \qquad (4.8)$$

where q labels eigenstates of the Hamiltonian and E_q, P_q are eigenvalues of energy and momenta of the state q. Comparing these two expansions we find:

$$E_q = \frac{2\pi}{L}(d + n) \qquad (4.9)$$

$$P_q = \frac{2\pi}{L}(S + m) \qquad (4.10)$$

$$\left(\frac{\pi}{L}\right)^{(d+n)} C_{nm} = \sum_q |\langle 0| \exp[i\beta\phi(0) + i\bar{\beta}\bar{\phi}(0)]|q\rangle|^2$$
$$\times \delta\left[E_q - \frac{2\pi}{L}(d + n)\right] \delta\left[P_q - \frac{2\pi}{L}(S + m)\right] \qquad (4.11)$$

All three expressions are of profound significance. The first two relate conformal dimensions of the correlation functions to the eigenvalues of energy and momentum operators. It is usually much easier to calculate energies than correlation functions. In the latter case the amount of computational work greatly increases since one must calculate also the matrix elements. Therefore it is a tremendous relief that for a quantum system with a gapless spectrum the relationships (4.9) and (4.10) allow us to avoid direct calculations of the correlation functions. Instead one can solve the problem of low-lying energy levels in a finite size system (which can be done numerically or even exactly) and then, using the relationships (4.8), (4.9) and (4.10), restore the correlation functions. From Eqs. (4.9) and (4.10) we see that the problem of conformal dimensions can be formulated as an eigenvalue problem for the following operators:

$$\Delta = \frac{L}{4\pi}(\hat{H} + \hat{P}) \equiv \hat{T}_0 \qquad (4.12)$$

$$\bar{\Delta} = \frac{L}{4\pi}(\hat{H} - \hat{P}) \equiv \hat{\bar{T}}_0 \qquad (4.13)$$

It can be shown that the operators $\hat{T}_0, \hat{\bar{T}}_0$ are related to analytic and anti-analytic components of the stress energy tensor defined as

$$T_{ab}(x) = \frac{\delta S}{\delta g^{ab}(x)} \qquad (4.14)$$

where g^{ab} is the inverse metric tensor (so it is assumed that the system is in a curved space characterized by the metric tensor g_{ab}). Namely,

$$\hat{T}_0 = \int_0^L dx T_{zz}; \quad \bar{\hat{T}}_0 = \int_0^L dx T_{\bar{z}\bar{z}}$$

It is not unnatural that the stress-energy tensor appears in the present context. Indeed, the scaling dimensions are related to coordinate transformations and such transformations change the metric on the surface:

$$dz d\bar{z} \rightarrow (dz/d\xi)(d\bar{z}/d\bar{\xi}) d\xi d\bar{\xi}$$

The relations between scaling dimensions and the stress-energy tensor have a general character and hold for all conformal theories. To make the new concept of the stress-energy tensor more familiar, let us calculate T_{ab} for the Gaussian model explicitly. The action on a curved background is given by

$$S = \frac{1}{2} \int d^2 x \sqrt{g} g^{ab} \partial_a \Phi \partial_b \Phi \qquad (4.15)$$

where $g = \det \hat{g}$ and $g^{ab} = (g^{-1})_{ab}$. In the vicinity of a flat space where the metric tensor is equal to $g_{ab} = \delta_{ab}$ we get:

$$T_{ab} = \frac{1}{2} : \left[\partial_a \Phi \partial_b \Phi - \frac{1}{2} \delta_{ab} (\partial_c \Phi)^2 \right] : \qquad (4.16)$$

(the dots denote the normal ordering, i.e., it is assumed that the vacuum average of the operator is subtracted – $\langle T_{ab} \rangle = 0$).

I   Gaussian model in the Hamiltonian formulation

Since we need the eigenvalues of $\hat{T}_0$ and $\bar{\hat{T}}_0$, it is convenient to express the Hamiltonian and the momentum operator $\hat{P}$ in terms of the creation and the annihilation operators. Thus from this moment we shall work in the Hamiltonian formalism. The Gaussian model (3.1) has the following Hamiltonian:

$$\hat{H}_0 = \frac{v}{2} \int dx [\hat{\Pi}^2 + (\partial_x \Phi)^2]$$

$$[\Pi(x), \Phi(y)] = -i\delta(x - y) \qquad (4.17)$$

We shall assume that the velocity $v = 1$ for simplicity.

We remark that since $\partial_x \Theta = -\Pi$, the latter commutation relations can be written as

$$[\Theta(x), \Phi(y)] = \theta(x - y) \qquad (4.18)$$

The Hamiltonian (4.17) describes a set of coupled oscillators (a string). In order to introduce creation and annihilation operators, we expand the field Φ into the normal modes:

$$\Phi(\tau, x) = \Phi_0 + i\sqrt{\pi}J\tau/L + \sqrt{\pi}Q(x/L)$$
$$+ \sum_{q \neq 0} \frac{1}{\sqrt{2|q|L}}(\hat{a}_q^+ e^{|q|\tau - iqx} + \hat{a}_q e^{-|q|\tau + iqx}) \qquad (4.19)$$

where $q = 2\pi k/L$ (k is an integer). Then the momentum operator is

$$\hat{\Pi}(\tau, x) = i\partial_\tau \Phi_0 = -\sqrt{\pi}J/L$$
$$+ i\sum_{q \neq 0} \sqrt{|q|/2L}(\hat{a}_q^+ e^{|q|\tau - iqx} - \hat{a}_q e^{-|q|\tau + iqx}) \qquad (4.20)$$

The quantities Q and J are the total charge of the system and the total current throught it. This identification follows from the expressions for currents given in the bosonization table.

The operator representation is particularly useful since it makes it obvious that the field $\Phi(z, \bar{z})$ is a sum of the analytic and the anti-analytic components. Indeed, it is straightforward to extract these components from (4.19):

$$\Phi(\tau, x) = \phi(\tau + ix) + \bar{\phi}(\tau - ix)$$
$$\Theta(\tau, x) = \phi(\tau + ix) - \bar{\phi}(\tau - ix) \qquad (4.21)$$

$$\phi(z) = \frac{i\sqrt{\pi}}{2}(J - Q)z/L + \sum_{q>0} \frac{1}{\sqrt{2qL}}(e^{-qz}\hat{a}_{-q} + e^{qz}\hat{a}_{-q}^+)$$

$$\bar{\phi}(\bar{z}) = \frac{i\sqrt{\pi}}{2}(J + Q)\bar{z}/L + \sum_{q>0} \frac{1}{\sqrt{2qL}}(e^{q\bar{z}}\hat{a}_q^+ + e^{-q\bar{z}}\hat{a}_q) \qquad (4.22)$$

It turns out that the total current J and charge Q are related to the scaling dimensions. This follows from the fact that we consider only the operators periodic in Φ and Θ with periods $T_1 = \sqrt{4\pi/K}$ and $T_2 = \sqrt{4\pi K}$, respectively. The fields Φ, Θ are defined on the circle with circumference L, but they are not periodic functions of x: as follows from Eqs. (4.21) and (4.22) $\phi(z + iL) = \phi(z) - \sqrt{\pi}(J - Q)/2$, $\bar{\phi}(\bar{z} - iL) = \bar{\phi}(\bar{z}) + \sqrt{\pi}(J + Q)/2$. Therefore in order to maintain the periodicity of the operators $\exp[2i\pi(\phi + \bar{\phi})n/T_1 + i(\phi - \bar{\phi})T_1 m/2]$ one needs to satisfy the following conditions:

$$Q = T_1 m/2\pi^{1/2} = \frac{m}{\sqrt{K}}, \quad J = 2n\pi^{1/2}/T_1 = n\sqrt{K} \qquad (4.23)$$

Substituting expressions (4.21) and (4.22) for the Φ into the Hamiltonian

Fig. 4.1.   Konstantin Efetov.

(4.17) we get the **Tomonaga–Luttinger liquid** Hamiltonian:

$$\hat{H} = \frac{\pi v}{2L}\left(Kn^2 + \frac{m^2}{K}\right) + \sum_q v|q|(a_q^+ a_q + 1/2) \qquad (4.24)$$

(we have restored the velocity in this expression). The last term has eigenvalues $2\pi v/L\times$(integer), so that the entire expression reproduces correctly those eigenvalues which follow from the correlation functions. The first two terms of the Hamiltonian (4.24) describe the centre of mass motion of the string. Due to the periodicity requirements discussed above, this motion is quantized. The second term in (4.24) can be written in terms of the integer charge m:

$$\frac{m^2}{2\chi L}$$

where χ is the charge susceptibility. Using Eq. (4.23) we obtain the follow-

Fig. 4.2. Duncan Haldane.

ing useful identity (Efetov and Larkin (1975), Haldane (1981)):

$$K = \pi v \chi \qquad (4.25)$$

This identity establishes a direct connection between correlation functions of the spinless Tomonaga–Luttinger liquid and its thermodynamics. Needless to say, thermodynamics quantities are much easier to calculate than correlation functions. Therefore identity (4.25) has proved itself very valuable in many practical problems where the theory of free bosons emerges via a sequence of equivalences (as in the case of the Hubbard model or the Heisenberg chain).

It is instructive also to have expressions for T_0 and $\bar{T}_0$. Since the stress-energy tensor is defined with the normal ordering, we must substract the infinite nonuniversal part of the ground state energy keeping, however, its finite L-dependent universal part:

$$T_0 = -\frac{1}{24L} + \frac{1}{8}\left[n\sqrt{K} + \frac{m}{\sqrt{K}}\right]^2 + \frac{L}{2\pi}\sum_{q>0}qa_q^+a_q$$

$$\bar{T}_0 = -\frac{1}{24L} + \frac{1}{8}\left[n\sqrt{K} - \frac{m}{\sqrt{K}}\right]^2 - \frac{L}{2\pi}\sum_{q<0}qa_q^+a_q \qquad (4.26)$$

The detailed derivation of the $1/L$ correction to T_0 will be given in the next chapter (see the discussion leading to Eq. (5.16)). Here we just remark that its appearance is a particular case of the Kazimir effect – a change of the vacuum energy with a change of geometry.

Naturally, the motion of the centre of mass does not contribute to the bulk free energy in the limit $L \to \infty$. This free energy is determined exclusively by the third term of the Hamiltonian (4.24). It is remarkable, however, that the terms of the Hamiltonian which have a negligible effect on the thermodynamics determine the asymptotic behaviour of the correlation functions! The bulk free energy, being the free energy of free bosons, is equal to

$$\frac{F_b}{L} = \text{const} + T\int_{-\infty}^{\infty}\frac{dp}{2\pi}\ln\left(1 - e^{-v|k|/T}\right) \qquad (4.27)$$

It is easy to check, using explicit expressions for the corresponding table integrals, that

$$\frac{F_b}{L} = \text{const} - T\int_{-\infty}^{\infty}\frac{dp}{\pi}\ln\left(1 + e^{-v|k|/T}\right) = \text{const} - \frac{\pi}{6v}T^2 \qquad (4.28)$$

i.e., F_b coincides with the free energy of free spinless fermions (bosonization!).* This coincidence is just one extra demonstration of the fact that there is a one-to-one correspondence between the eigenstates of both theories. It follows from Eq. (4.28) that the product of the linear coefficient in the specific heat C_V by the velocity is a number:

$$C = \frac{3C_Vv}{\pi T} = 1 \qquad (4.29)$$

This number C is called the central charge and plays an enormous role in conformal field theory. In the above example $C = 1$, in general $C = N$ for a theory with N species of free bosonic fields, but it is also possible to have non-integer central charge. The relationship between thermodynamics and conformal properties was discovered by Blöte, Cardy and Nightingale (1986) and Affleck (1986). We shall return to the concept of central charge in the next chapter.

---

* Here it is convenient to keep the velocity in explicit form, i.e., not putting $v = 1$.

Eqs. (4.9, 4.10, 4.22, 4.29) form a link between exact solutions and numerical methods from one side and conformal field theory on the other. The former methods are usually very good in calculations of thermodynamics quantities and energy levels. The described procedures allow one to extract from these calculations information about asymptotics of correlation functions.

References

I. Affleck, *Phys. Rev. Lett.* **56**, 746 (1986).

H. W. J. Blöte, J. L. Cardy and M. P. Nightingale, *Phys. Rev. Lett.* **56**, 742 (1986).

K. B. Efetov and A. I. Larkin, *Sov. Phys. JETP* **42**, 390 (1975).

F. D. M. Haldane, *Phys. Rev. Lett.* **47**, 1840 (1981); *J. Phys.* C**14**, 2585 (1981).

5

Virasoro algebra

Does the Hamiltonian have a form?
- The question A. I. Larkin used to ask the speakers on the
Landau Institute seminars.

In this chapter we continue to study the group of conformal transformations of the complex plane. Since this is a long and sometimes difficult
enterprise, it is better to explain why we should undertake it. Physics is
full of beautiful mathematics, but a part of its beauty is its efficiency in
solving physical problems. So, why do we need all these extra complications – stress-energy tensors, Virasoro algebras etc., why not to leave
this stuff to mathematicians? Of course, if all interesting physics could
be reduced to the Gaussian model, it would be difficult to save Virasoro
algebras from the Occam's razor.* However rich a hidden symmetry of
this model, one does not need to be very sophisticated to calculate its
correlation functions. However, there are many other important models
possessing conformal symmetry beyond the Gaussian model, which are
not as simple. Examples of such models will be given later, and some
of them (the Wess–Zumino–Novikov–Witten model, for instance) will be
discussed in great detail. It is for these nontrivial theories, not for the
Gaussian model, that we need to learn more about conformal groups. As
we have mentioned elsewhere, the Gaussian model is just a convenient
example to illustrate the concepts whose power extends much futher.

In two dimensions the group of conformal transformations is isomorphic
to the group of analytic transformations of the complex plane. Since the
number of analytic functions is infinite, this group is infinite-dimensional,
i.e. has infinite number of generators. Infinite number of generators generates infinite number of Ward identities for correlation functions. These

---

* 'Do not introduce new entities without necessity'.

34

identities serve as differential equations on correlation functions. No matter how many operators you have in your correlation function, you always have enough Ward identities to specify it. Thus in conformally invariant theories (for brevity we shall call them 'conformal' theories) one can (at least in principle) calculate all multi-point correlation functions. We have already calculated multi-point correlation functions of the bosonic exponents for the Gaussian model (see Chapter 3), below we shall see less trivial examples.

Let us take for good faith that such models exist, that is we postulate that there are models besides the Gaussian one whose spectrum is linear, whose correlation functions factorize into products of analytic and anti-analytic parts and which have operators transforming under analytic (anti-analytic) transformations as (4.4). Such operators will be called *primary* fields. This is almost a definition of what conformal theory is. To make this definition rigorous we have to add one more property, namely, to postulate that the three-point correlation functions of the primary fields have the following form:

$$\langle A_{\Delta_1}(z_1)A_{\Delta_2}(z_2)A_{\Delta_3}(z_3)\rangle = \frac{C_{123}}{(z_{12})^{\Delta_1+\Delta_2-\Delta_3}(z_{13})^{\Delta_1+\Delta_3-\Delta_2}(z_{23})^{\Delta_2+\Delta_3-\Delta_1}} \quad (5.1)$$

where C_{123} are some constants.

Since the eigenvalues of the Hamiltonian and momentum are related to the zeroeth components of the stress-energy tensor, it is logical to suggest that the generators of transformations (4.4) are Fourier components of the stress-energy tensor. In order to find commutation relations between these components and the primary fields we have to consider the infinitely small version of transformation (4.4):

$$z' = z + \epsilon(z) \quad (5.2)$$

According to (4.4) this transformation changes the primary field $A_\Delta(z)$:

$$\begin{aligned} \delta A_\Delta(z) &= A_\Delta(z+\epsilon)(1+\partial\epsilon)^\Delta - A_\Delta(z) \\ &= \epsilon(z)\partial A_\Delta(z) + \partial\epsilon(z)\Delta A_\Delta(z) + O(\epsilon^2) \end{aligned} \quad (5.3)$$

Let us restrict our correlation functions to the line $\tau = 0$ where $z = ix$. This eliminates the time dependence and therefore we can use the operator language of quantum mechanics. As we know from quantum mechanics, if an operator changes under some transformation, this change can be expressed as an action of some other operator – a generator of this transformation. For our case it means that

$$\delta \hat{A}_\Delta(x) = [\hat{Q}_\epsilon, \hat{A}_\Delta(x)] \quad (5.4)$$

where $\hat{Q}_\epsilon$ is an element of the group of conformal transformations corresponding to the infinitesimal transformation (5.2). For a general Lie group

G with generators $\hat{t}_i$ an infinitesimal transformation is given by $\epsilon_i \hat{t}_i$, where ϵ_i are infinitely small parameters – coordinates of the transformation. Since conformal transformations in two dimensions are characterized not by a finite set of parameters, but by an entire function $\epsilon(z)$, the conformal algebra is infinite-dimensional. Therefore a general infinitesimal transformation can be written as an integral:[†]

$$\hat{Q}_\epsilon = \frac{1}{2\pi} \int dy \epsilon(y) \hat{T}(y)$$

Therefore we have

$$\frac{i}{2\pi} \int dy \epsilon(y) [\hat{T}(y), \hat{A}_\Delta(x)] = \epsilon(x) \partial_x A_\Delta(x) + \partial_x \epsilon(x) \Delta A_\Delta(x) \qquad (5.5)$$

which corresponds to the following local commutation relations:

$$\frac{i}{2\pi} [\hat{T}(y), \hat{A}_\Delta(x)] = \delta(x-y) \partial_x A_\Delta(x) - \partial_y \delta(y-x) \Delta A_\Delta(x) \qquad (5.6)$$

These commutation relations must be satisfied in any conformal theory.

Following the general rules explained in Chapter 2, we can rewrite these commutation relations as OPE:

$$T(z) A_\Delta(z_1) = \frac{\Delta}{(z-z_1)^2} A_\Delta(z_1) + \frac{1}{(z-z_1)} \partial_{z_1} A_\Delta(z_1) + \dots \qquad (5.7)$$

where dots stand for terms nonsingular at $(z - z_1) \to 0$.

Let us now study the algebraic properties of the stress-energy tensor $T(z)$. For this purpose we shall study its correlation functions using the Gaussian model stress energy tensor (4.16) as an example. In order to simplify the results, we change notations and introduce the new components[‡]

$$T_{zz} = -\pi(T_{11} - T_{22} - 2iT_{12}) \equiv T = -2\pi :(\partial\phi)^2 :$$
$$T_{\bar{z}\bar{z}} = -\pi(T_{11} - T_{22} + 2iT_{12}) \equiv \bar{T} = -2\pi :(\bar{\partial}\bar{\phi})^2 :$$
$$\mathrm{Tr}\, T = T_{11} + T_{22} = 0 \qquad (5.8)$$

Now it is obvious that correlation functions of operators T and $\bar{T}$ depend on z or $\bar{z}$ only. Since the two-point correlation function is the simplest one, let us calculate it first. A straightforward calculation yields:

$$\langle T(1)T(2) \rangle = \frac{C}{2(z_1 - z_2)^4}$$

$$\langle \bar{T}(1)\bar{T}(2) \rangle = \frac{C}{2(\bar{z}_1 - \bar{z}_2)^4} \qquad (5.9)$$

[†] The factor 2π is introduced to conform to the accepted notations.
[‡] The factor 4π is introduced in order to make the present notations conform to the standard ones.

where $C = 1$ in the given case. We keep C in the relations (5.9) because in this form they hold for all conformal theories. As far as the correlation functions of $\mathrm{Tr}T$ are concerned, it seems that Eq. (5.8) dictates them to be identically zero. The miracle is that it is *wrong*! Here we encounter another example of anomaly which is called *conformal* anomaly.

Let us consider the matter more carefully. Since the stress-energy tensor is a conserved quantity, i.e.,

$$\partial_a T_{ab} = 0$$

which reflects the conservation of energy and momentum, the two-point correlation function must satisfy the identity

$$q_a \langle T_{ab}(-q) T_{cd}(q) \rangle = 0$$

This suggests that the Fourier transformation of Eqs. (5.8) must be modified:

$$\langle T_{ab}(-q) T_{cd}(q) \rangle = \frac{Cq^2}{48\pi} \left(\delta_{ab} - \frac{q_a q_b}{q^2} \right) \left(\delta_{cd} - \frac{q_c q_d}{q^2} \right) \tag{5.10}$$

which gives the nonvanishing correlation function of the traces:

$$\langle \mathrm{Tr}T(-q) \mathrm{Tr}T(q) \rangle = \frac{Cq^2}{48\pi} \tag{5.11}$$

In real space the pair correlation function of the traces is ultralocal:

$$\langle \mathrm{Tr}T(x) \mathrm{Tr}T(y) \rangle = \frac{C}{48\pi} \nabla^2 \delta^{(2)}(\mathbf{x} - \mathbf{y}) \tag{5.12}$$

Since this correlation function is short-range, it is generated at high energies. It is no wonder, therefore, that we have missed it in the straightforward calculations.

As we have seen in the previous chapter, the central charge appears in the expression for the specific heat. Later we shall learn that it also determines the conformal dimensions of the correlation functions. The fact that the specific heat of a conformal theory is proportional to its central charge means that the latter is related to the number of states. Since the conformal dimensions are eigenvalues of the stress-energy tensor components, the relation between C and the correlation functions suggests that C also influences the scaling dimensions. For what follows it is necessary to know the fusion rules for the stress-energy tensor itself. We can find these commutation relations using general properties of the stress-energy tensor. At first, the stress energy tensor is really a tensor, and as such it has definite transformation properties. From this fact it follows that its conformal dimensions are $(2, 0)$, which is consistent with Eq. (5.9). Therefore the fusion rules for $T(z)T(z_1)$ must include the right-hand side of (5.7) with $\Delta = 2$. This is not all, however; the expansion must contain a term with the identity operator, since otherwise the pair correlation function $\langle T(z)T(z_1)\rangle$ would vanish. This term can be deduced from Eq. (5.9); collecting all these terms together we get the following OPE (the Virasoro fusion rules):

$$T(z)T(z_1) = \frac{C}{2(z-z_1)^4} + \frac{2}{(z-z_1)^2}T(z_1) + \frac{1}{(z-z_1)}\partial_{z_1}T(z_1) + \dots \quad (5.13)$$

A comparison of Eq. (5.13) with Eq. (5.7) shows that the stress-energy tensor *is not a primary field*, that is, despite its name, the stress-energy tensor does not transform as a *tensor* under conformal transformations. It can be shown that instead of the tensorial law (4.4) the stress-energy tensor transforms under finite conformal transformations as follows:

$$T(\xi) = T[z(\xi)](dz/d\xi)^2 + \frac{C}{12}\{z,\xi\}$$

$$\{z,\xi\} = \frac{z'''}{z'} - \frac{3}{2}\left(\frac{z''}{z'}\right)^2 \quad (5.14)$$

In particular, it follows from this expression that the stress energy tensor on a strip of width L in the x-direction ($z = \exp(2\pi\xi/L)$) is equal to

$$T_{\text{strip}}(\xi) = (2\pi/L)^2\left[z^2 T_{\text{plane}}(z) - C/24\right] \quad (5.15)$$

Since $\langle T_{\text{plane}}(z)\rangle = 0$, we find

$$\langle T_{\text{strip}}(z)\rangle = -\frac{C}{24}(2\pi/L)^2 \quad (5.16)$$

Now recalling the relationship between stress-energy tensor and Hamiltonian (4.13) we reproduce from the latter equation the $1/L$ correction to the stress-energy tensor eigenvalues (4.26).

It is also convenient to have the Virasoro fusion rules (5.13) in operator form with commutators:

$$\frac{1}{2i\pi}[\hat{T}(x),\hat{T}(y)] = \delta(x-y)\partial_x T(x) - 2\partial_x\delta(x-y)T(x) + \frac{C}{6}\partial_x^3\delta(x-y) \quad (5.17)$$

Since the commutation relations (5.17) are local, they do not depend on the global geometry of the problem.

It is customary to write the Virasoro fusion rules using the Laurent components of the stress-energy tensor and the operators defined on the infinite complex plane:

$$\hat{T}(z) = \sum_{-\infty}^{\infty}\frac{L_n}{z^{n+2}}, \quad \hat{A}_{\Delta}(z) = \sum_{-\infty}^{\infty}\frac{A_{\Delta,n}}{z^{n+2}}$$

The transformations laws (5.6) and (5.14) establish a connection between this expansion and the Fourier expansion which we use in a strip geometry:

$$A_\Delta^{\text{strip}}(x) = \left(\frac{2\pi}{L}\right)^\Delta \sum_n A_{\Delta,n} e^{-2\pi i x(n-\Delta)/L},$$

$$T^{\text{strip}}(x) = \left(\frac{2\pi}{L}\right)^2 \left(\sum_n L_n e^{-2\pi i n x/L} - C/24\right) \tag{5.18}$$

Substituting these expansions into the fusion rules (5.13) and (5.5) we get the following commutation relations for the Laurent components:

$$[L_n, L_m] = (n-m)L_{n+m} + \frac{C}{12}n(n^2-1)\delta_{n+m,0} \tag{5.19}$$

and

$$[L_n, A_{\Delta,m}] = [\Delta(1-n) + m + n]A_{\Delta,m+n} \tag{5.20}$$

From Eq. (5.19) we see that the set of operators L_n, I (I is the identity operator) is closed with respect to the operation of commutation. Therefore components of the stress-energy tensor in conformal theories together with the identity operator form an algebra (Virasoro algebra). As we have said, this fact opens up the possibility of formal studies of conformal theories as representations of the Virasoro algebra.

I   Ward identities

The identity (5.5) can be used to derive Ward identities for correlation functions of stress-energy tensor with primary fields. Namely, let us insert this identity into a correlation function:

$$\frac{1}{2\pi i} \int_C dz \epsilon(z) \langle T(z) A_{\Delta_1}(1)...A_{\Delta_N}(N)\rangle$$

$$= \sum_{i=1}^N [\epsilon(z_i)\partial_{z_i}\langle A_{\Delta_1}(1)...A_{\Delta_N}(N)\rangle + \partial_{z_i}\epsilon(z_i)\Delta_i\langle A_{\Delta_1}(1)...A_{\Delta_N}(N)\rangle] \tag{5.21}$$

and assume that $\epsilon(z)$ is analytic in the domain enclosing the points z_i and the contour C encircles these points. Then Eq. (5.21) suggests that as a function of z the integrand of the left-hand side has simple and double poles at $z = z_i$:

$$\langle T(z) A_{\Delta_1}(1)...A_{\Delta_N}(N)\rangle$$

$$= \sum_{i=1}^N \left[\frac{\Delta_i}{(z-z_i)^2} + \frac{1}{z-z_i}\partial_{z_i}\right] \langle A_{\Delta_1}(1)...A_{\Delta_N}(N)\rangle \tag{5.22}$$

This is the Ward identity for $T(z)$ we had in mind.

II   Subalgebra sl(2)

It follows from Eq. (5.19) that three operators L_0, $L_{\pm 1}$ compose a subalgebra of the Virasoro algebra isomorphic to the algebra sl(2):

$$[L_0, L_{\pm 1}] = \mp 2L_{\pm 1}, \quad [L_1, L_{-1}] = 2L_0 \tag{5.23}$$

The corresponding group is the group of all rational transformations of the complex plane

$$w(z) = \frac{az + b}{cz + d}, \quad ad - bc = 1 \tag{5.24}$$

Let us show that all correlation functions in critical two-dimensional theories are invariant with respect to the SL(2) transformations (5.24). This fact follows from the Ward identity (5.22). Indeed, since $\langle T(z)\rangle = 0$ on an infinite plane, the correlation function in the left hand side of Eq. (5.22) must decay at infinity as

$$\langle T(z)A_{\Delta_1}(1)...A_{\Delta_N}(N)\rangle \sim z^{-4}D(z_1,...z_N) \tag{5.25}$$

Expanding the right-hand side of Eq. (5.21) in powers of z^{-1} we find that the terms containing z^{-1}, z^{-2} and z^{-3} must vanish, which gives rise to the following identities:

$$\sum_i \partial_{z_i}\langle A_{\Delta_1}(1)...A_{\Delta_N}(N)\rangle = 0 \tag{5.26}$$

$$\sum_i (z_i\partial_{z_i} + \Delta_i)\langle A_{\Delta_1}(1)...A_{\Delta_N}(N)\rangle = 0 \tag{5.27}$$

$$\sum_i (z_i^2\partial_{z_i} + 2\Delta_i z_i)\langle A_{\Delta_1}(1)...A_{\Delta_N}(N)\rangle = 0 \tag{5.28}$$

It is easy to see that the operators

$$L_{-1} = \partial_z, \quad L_0 = (z\partial_z + \Delta), \quad L_1 = (z^2\partial_z + 2\Delta z) \tag{5.29}$$

satisfy the commutation relations (5.23) thus composing a representation of the sl(2) algebra.

Exercise. Show that the SL(2) invariance dictates that a four-point correlation function of four primary fields with the same conformal dimension has the following general form:

$$\langle \Phi(1)...\Phi(4)\rangle = \left(\left|\frac{z_{13}z_{24}}{z_{12}z_{14}z_{23}z_{34}}\right|\right)^{4\Delta} G(x, \bar{x}) \tag{5.30}$$

where

$$x = \frac{z_{12}z_{34}}{z_{13}z_{24}}$$

6

Structure of Hilbert space in conformal theories

In this chapter we shall generate eigenvectors of the Hilbert space of conformal theories using the Virasoro operators. One may wonder, why we use Virasoro operators whose commutation relations are so awkward? The answer is that we want to make sure that the conformal symmetry is preserved, and the best way to guarantee it is to use as creation and annihilation operators the generators of the conformal group.

In fact, the procedure we are going to follow is a standard one in group theory, where it is used for constructing representations. In order to feel ourselves on a familiar ground, let us recall how one constructs representations of spin operators in quantum mechanics. Consider a three-dimensional rotator whose Hamiltonian is the sum of squares of generators of the SU(2) group – components of the spin:

$$\hat{H} = \sum_{i=x,y,z} \hat{S}_i^2 = \hat{S}_z^2 + \frac{1}{2}(\hat{S}_+\hat{S}_- + \hat{S}_-\hat{S}_+)$$

Since $[\hat{H}, \hat{S}^z] = 0$, $\hat{S}^z$ is diagonal in the basis of eigenfunctions of $\hat{H}$. The key point is to write down the operator algebra in the proper form:

$$[\hat{S}_\pm, \hat{S}_z] = \mp \hat{S}_\pm, \quad [\hat{S}_+, \hat{S}_-] = 2\hat{S}_z$$

In doing so we separate the diagonal generator $\hat{S}_z$ from the raising and lowering operators. As follows from the commutation relations, the operators $\hat{S}_+$ ($\hat{S}_-$) increase (decrease) an eigenvalue of $\hat{S}_z$ by one. Let us suppose that there exists an eigenvector of $\hat{S}^z$ which is annihilated by $\hat{S}_-$:

$$\hat{S}_-|-S\rangle = 0; \quad \hat{S}_z|-S\rangle = -S|-S\rangle$$

Then $(\hat{S}_+)^j|-S\rangle$ is also an eigenvector of $\hat{S}_z$ with the eigenvalue $j-S$. As we know, it can also be shown that if $2S$ is a positive integer, the state

$$|\chi\rangle = (\hat{S}_+)^{(2S+1)}|-S\rangle \tag{6.1}$$

is a null vector, that is representations of the SU(2) group with integer $(2S + 1)$ are finite dimensional.

Let us show that in the case of the Virasoro algebra the state similar to $|-S\rangle$ is

$$|\Delta, \bar{\Delta}\rangle = A_{\Delta,0} A_{\bar{\Delta},0}|0\rangle$$

According to Eq. (4.7), $\langle n|A_\Delta(z)|0\rangle = 0$ for all negative n, which according to the definition of $A_{\Delta,n}$ is equivalent to

$$A_{\Delta,n}|0\rangle = 0, \quad n > 0 \tag{6.2}$$

Similar considerations carried out for the two-point correlation function of the stress-energy tensor lead to

$$L_n|0\rangle = 0 \quad (n > 0) \tag{6.3}$$

Now we can prove that operators L_n $(n > 0)$ annihilate the state $|\Delta, \bar{\Delta}\rangle$. Indeed, according to Eq. (5.20)

$$L_n A_{\Delta,0}|0\rangle = [\Delta(1 - n) + n]A_{\Delta,n}|0\rangle + A_{\Delta,0}L_n|0\rangle = 0 \tag{6.4}$$

Therefore we can use L_n with positive n as lowering operators and L_n with negative n as raising operators. Thus we look for the eigenvectors in the following form:

$$|\bar{n}_1, \bar{n}_2, ..., \bar{n}_M ; n_1, n_2, ..., n_M ; \Delta, \bar{\Delta}\rangle$$
$$= \bar{L}_{-\bar{n}_1} \bar{L}_{-\bar{n}_2} ... \bar{L}_{-\bar{n}_M} L_{-n_1} L_{-n_2} ... L_{-n_M}|\Delta, \bar{\Delta}\rangle \tag{6.5}$$

We see that a tower of states is built on each primary field, including the trivial identity operator. It follows from the Virasoro commutation relations (5.19), that these vectors are eigenstates of the operators L_0 and $\bar{L}_0$ with eigenvalues

$$\Delta_n = \Delta + \sum_j n_j,$$

$$\bar{\Delta}_n = \bar{\Delta} + \sum_j \bar{n}_j \tag{6.6}$$

Operators obtained from primary fields by action of the Virasoro generators L_{-n} are called *descendants* of the primary fields. It can be shown that for the Gaussian model all primary fields and their descendants are linearly independent and their Hilbert space is isomorphic to the Hilbert space built up by the bosonic creation operators. This, however, is only one particular representation of the Virasoro algebra. There are many others and different methods have been suggested for finding nontrivial representations of the Virasoro algebra. Some representations can be constructed by truncation of the Gaussian theory Hilbert space, i.e. by postulating that there are linearly dependent states among the states (6.5).

In this chapter we consider only the simplest case when the truncation occurs on the second level. Suppose that the states

$$L_{-2}|\Delta\rangle, \quad L^2_{-1}|\Delta\rangle$$

are linearly dependent, i.e., there is a number α such that

$$|\chi\rangle = (L_{-2} + \alpha L^2_{-1})|\Delta\rangle = 0 \tag{6.7}$$

that is $|\chi\rangle$ is a null vector similar to (6.1).

Since all other Virasoro generators L_{-n} with $n > 2$ are generated by these two via the commutation relations (5.20), the fact that L_{-2} is proportional to L^2_{-1} means that the only linearly independent eigenvectors from the set (6.5) are

$$|n, \Delta\rangle = (L_{-1})^n|\Delta\rangle$$

Let us consider conditions for existence of the null state $|\chi\rangle$ (6.7). Due to their mutual independence the left and right degrees of freedom can be considered separately. Since we want to preserve conformal invariance of the theory, condition (6.7) must survive conformal transformations, i.e.,

$$L_n|\chi\rangle = 0 \ (n > 0) \tag{6.8}$$

Using the Virasoro commutation relations (5.20) and the properties

$$L_0|\Delta\rangle = \Delta|\Delta\rangle, \quad L_n|\Delta\rangle = 0 \ (n > 0)$$

we get from Eq. (6.8) the following two equations:

$$3 + 2\alpha + 4\alpha\Delta = 0 \tag{6.9}$$
$$4\Delta(2 + 3\alpha) + C = 0 \tag{6.10}$$

whose solution is

$$\alpha = -\frac{3}{2(1 + 2\Delta)}, \quad \Delta = \frac{1}{16}\left(5 - C \pm \sqrt{(5 - C)^2 - 16C}\right) \tag{6.11}$$

I   Differential equations for correlation functions

We can use Ward identities for the stress-energy tensor to derive differential equations for the correlation functions. As the first step let us write down the complete OPE for the stress-energy tensor with a primary field $A_\Delta(z_1)$ which is a generalization of Eq. (5.7):

$$T(z)A_\Delta(z_1) = \sum_{n=0}^{\infty}(z - z_1)^{n-2}L_{-n}A_\Delta(z_1) \tag{6.12}$$

The absence in this expansion of the Virasoro generators with positive index follows from the Ward identity (5.22). Comparing this OPE with

Eq. (5.7) or Eq. (5.22) we obtain the following identities:

$$L_0 A_\Delta(z_1) = \Delta A_\Delta(z_1), \quad L_{-1} A_\Delta(z_1) = \partial_{z_1} A_\Delta(z_1) \tag{6.13}$$

The next step is to use these identities to extract the action of L_{-2}. This we do by subtracting the singular part of the stress-energy tensor at $z \to z_1$ and using (5.22):

$$\langle L_{-2} A_{\Delta_1}(1)...A_{\Delta_N}(N) \rangle = \langle \left[T(z) - \frac{L_0}{(z-z_1)^2} - \frac{L_{-1}}{z-z_1} \right] A_{\Delta_1}(1)...A_{\Delta_N}(N) \rangle |_{z \to z_1}$$

$$= \sum_{j \neq 1} \left[\frac{\Delta_j}{(z_1 - z_j)^2} + \frac{1}{(z_1 - z_j)} \partial_j \right] \langle A_{\Delta_1}(1)...A_{\Delta_N}(N) \rangle \tag{6.14}$$

According to Eq. (6.7) L_{-1}^2 is identified with $-\alpha \partial^2$, and we obtain from (6.14) the linear differential equation for a multi-point correlation function of primary fields $A_j(z_j, \bar{z}_j)$ with analytic conformal dimensions Δ_j:

$$\left\{ \frac{3}{2(1 + 2\Delta_1)} \partial_1^2 - \sum_{j \neq 1} \left[\frac{\Delta_j}{(z_1 - z_j)^2} + \frac{1}{(z_1 - z_j)} \partial_j \right] \right\}$$

$$\times \langle A_1(z_1)...A_N(z_N) \rangle = 0 \tag{6.15}$$

We emphasize that this equation holds only for the simplest representation of the Virasoro algebra, namely the one defined by Eq. (6.7).

For the four-point correlation function the presence of the SL(2) symmetry helps to reduce Eq. (6.15) to an ordinary differential equation. In the case when all conformal dimensions are equal the general expression for a four-point correlation function (5.30) acquires the following form:

$$\langle \Phi(1)...\Phi(4) \rangle = \left(|\frac{z_{13} z_{24}}{z_{12} z_{14} z_{23} z_{34}}| \right)^{4\Delta} G(x, \bar{x}) \tag{6.16}$$

where

$$x = \frac{z_{12} z_{34}}{z_{13} z_{24}}, \quad 1 - x = \frac{z_{14} z_{23}}{z_{13} z_{24}} \tag{6.17}$$

are the so-called anharmonic ratios.

Substituting Eq. (6.16) into Eq. (6.15) we obtain the following conventional differential equation for $G(x, \bar{x})$:

$$x(1-x)G'' + \frac{2}{3}(1 - 4\Delta)(1 - 2x)G' + \frac{4\Delta}{3}(1 - 4\Delta)G = 0 \tag{6.18}$$

(and the same equation for $\bar{x}$).

Equations of this type appear frequently in conformal field theory and therefore we shall discuss the solution in detail. Eq. (6.18) is a particular case of the hypergeometric equation. Its special property is invariance with respect to the transformation $x \to 1 - x$ which reflects the invariance of the correlation function with respect to permutation of the coordinates $(2 \rightarrow 4$, for instance). Thus if $\mathcal{F}(x)$ is a solution, $\mathcal{F}(1 - x)$ is also a solution. If $\frac{2}{3}(1 - 4\Delta)$ is not an integer number, two linearly independent solutions of Eq. (6.18) nonsingular in the vicinity of $x = 0$ are given by the hypergeometric functions:

$$\mathcal{F}^{(0)}(x) = F(a, b, c; x)$$

$$\mathcal{F}^{(1)}(x) = x^{1-c} F(a - c + 1, b - c + 1, 2 - c; x)$$

$$a, b = \frac{2}{3}(1 - 4\Delta) - 1/2 \pm \sqrt{2/9 + (1/2 - 4\Delta/3)^2}; \quad c = \frac{2}{3}(1 - 4\Delta) \quad (6.19)$$

When c is integer, the second solution is not a hypergeometric function. We shall discuss this case later.

A general solution for $G(x, \bar{x})$ is

$$G(x, \bar{x}) = W_{ab} \mathcal{F}^{(a)}(x) \mathcal{F}^{(b)}(\bar{x}) \quad (6.20)$$

This solution must satisfy two requirements. The first is that it must be invariant with respect to interchange of $x \rightarrow 1 - x$ (*crossing* symmetry) which reflects the fact that the entire correlation function does not change when any two coordinates are interchanged. The second requirement is that the correlation function must be single-valued on the $(x, \bar{x})$ plane. These requirements fix the matrix W (up to a factor).

If c is not an integer, we use the known identities for hypergeometric functions and the fact that $a + b = 2c - 1$, to find

$$\mathcal{F}^{(a)}(1 - x) = A_{ab} \mathcal{F}^{(b)}(x)$$

$$A_{00} = \frac{\Gamma(c)\Gamma(1 - c)}{\Gamma(c - a)\Gamma(1 + a - c)}, \quad A_{01} = \frac{\Gamma(c)\Gamma(c - 1)}{\Gamma(a)\Gamma(b)}$$

$$A_{10} = (1 - A_{00}^2)/A_{01}, \quad A_{11} = -A_{00} \quad (6.21)$$

The crossing symmetry condition gives

$$W_{ab} A_{ac} A_{bd} = W_{cd} \quad (6.22)$$

The monodromy matrix $\hat{A}$ has eigenvalues $\lambda = \pm 1$. Therefore a general solution of Eq. (6.22) is

$$W_{ab} = C_+ e_a^{(+)} e_b^{(+)} + C_- e_a^{(-)} e_b^{(-)} \quad (6.23)$$

where $e^{(\pm)}$ are the eigenvectors of $\hat{A}^t$. Substituting the explicit expressions for these eigenvectors into the expression for $G(x, \bar{x})$, we find

$$G(x, \bar{x}) =$$
$$C_+ [A_{01} \mathcal{F}^{(1)}(x) + (1 + A_{00}) \mathcal{F}^{(0)}(x)][A_{01} \mathcal{F}^{(1)}(\bar{x}) + (1 + A_{00}) \mathcal{F}^{(0)}(\bar{x})]$$
$$+ C_- [A_{01} \mathcal{F}^{(1)}(x) + (A_{00} - 1) \mathcal{F}^{(0)}(x)][A_{01} \mathcal{F}^{(1)}(\bar{x}) + (A_{00} - 1) \mathcal{F}^{(1)}(\bar{x})]$$
$$(6.24)$$

To determine $C_\pm$ we recall that the correlation function must be a single-valued function. From Eq. (6.19) we see that $\mathscr{F}^{(0)}(x)$ is analytic in the vicinity of $x = 0$, but $\mathscr{F}^{(1)}(x)$ has a branch cut. Therefore there should be no terms containing cross products of these two solutions. Such cross products vanish if

$$C_+(1 + A_{00}) = C_-(1 - A_{00}) \tag{6.25}$$

Finally we get

$$G(x, \bar{x}) \sim A_{01}^2 \mathscr{F}^{(1)}(x)\mathscr{F}^{(1)}(\bar{x}) + (1 - A_{00}^2)\mathscr{F}^{(0)}(x)\mathscr{F}^{(0)}(\bar{x}) \tag{6.26}$$

For the future purposes it is convenient to express the solution in terms of $\mathscr{F}^{(0)}$. Using Eq. (6.21), we get

$$G(x, \bar{x}) \sim [\mathscr{F}^{(0)}(x)\mathscr{F}^{(0)}(1 - \bar{x}) + \mathscr{F}^{(0)}(\bar{x})\mathscr{F}^{(0)}(1 - x)]$$
$$- A_{00}^{-1}[\mathscr{F}^{(0)}(1 - x)\mathscr{F}^{(0)}(1 - \bar{x}) + \mathscr{F}^{(0)}(x)\mathscr{F}^{(0)}(\bar{x})] \tag{6.27}$$

(here we have changed the normalization factor). The purpose of the latter exercise is to obtain a limit of integer c. In this case, according to Eq. (6.21), $A_{00}^{-1} \to 0$ and the cross-symmetric and single-valued solution is

$$G(x, \bar{x}) = \mathscr{F}(x)\mathscr{F}(1 - \bar{x}) + \mathscr{F}(\bar{x})\mathscr{F}(1 - x) \tag{6.28}$$

Now we shall consider several particular cases. As we know from Eq. (6.11), there are two possible conformal dimensions for each choice of C. For $C = 1/2$ we get

- Ising model

$$\Delta_- = 1/16, \ \Delta_+ = 1/2 \tag{6.29}$$

which correspond to conformal dimensions of the Ising model primary fields (see Chapter 12 for details). Fields with the conformal dimensions $(1/16, 1/16)$ are called order (disorder) parameter fields and are denoted as $\sigma(z, \bar{z})$ ($\mu(z, \bar{z})$). These two fields are mutually nonlocal and one can choose either of them for the basis of primary fields.

Exercise. Check that the following correlation function satisfies Eq. (6.18) with $\Delta = 1/16$ and $C = 1/2$:

$$G(x, \bar{x}) = \left\{[1 + (1 - x)^{1/2}][1 + (1 - \bar{x})^{1/2}]\right\}^{1/2}$$
$$+ \left\{[1 - (1 - x)^{1/2}][1 - (1 - \bar{x})^{1/2}]\right\}^{1/2} \tag{6.30}$$

There are two other interesting cases: $C = 0$ and $C = -2$. In the first case we have $\Delta_+ = 5/8$ and $\Delta_- = 0$. In the second case we have $\Delta_- = -1/8$ and $\Delta_+ = 1$. In both cases conformal dimensions of some primary fields are integers, thus being degenerate with dimensions of the descendants of the unity operator. This leads to interesting modifications in the OPE (see Caux *et al.* (1996), (1997) and references therein). It goes without saying that both models are nonunitary. The model $C = 0$ describes classical percolation and the model with $C = -2$ describes dense polymers (Rozansky and Saleur (1992)). In the $\Delta = 5/8$ case we have

$$\mathscr{F}(x) = x^2 F(-1/2, 3/2, 3; \ x)$$

and for $\Delta = -1/8$ we have

$$\mathscr{F}(x) = F(1/2, 1/2, 1; \ x)$$

II Dotsenko–Fateev bosonization scheme for the minimal models

Let us consider how we can modify the Gaussian model without breaking the conformal symmetry. As we have once mentioned, the Gaussian model can be viewed as a model of a two-dimensional classical Coulomb gas where the bosonic exponents generate charges. One may expect that adding an extra charge on infinity (or spreading it over the background) will not break the conformal symmetry. For the sake of brevity we shall denote

$$V_\alpha = \exp[i\sqrt{4\pi}\alpha\Phi]$$

With the extra charge $-Q$ added to the system we shall adopt a new definition for correlation functions:

$$\langle V_{\alpha_1}(1)...V_{\alpha_N}(N)\rangle_Q = \lim_{R\to\infty} R^{2Q^2} \langle V_{\alpha_1}(1)...V_{\alpha_N}(N)V_{-Q}(R)\rangle \qquad (6.31)$$

In particular, the only nonvanishing two-point correlation function is now given by

$$\langle V_{-\alpha}(1)V_{(\alpha-Q)}(2)\rangle_Q$$

$$\equiv \lim_{R\to\infty} R^{2Q^2} \langle \exp[i\sqrt{4\pi}\alpha\Phi(1)] \exp[i\sqrt{4\pi}(Q-\alpha)\Phi(2)] \exp[-i\sqrt{4\pi}Q\Phi(R)]\rangle$$

$$= (|z_{12}|)^{-4\Delta} \qquad (6.32)$$

$$\Delta = \frac{1}{2}\alpha(\alpha - Q) \qquad (6.33)$$

So, as expected, the correlation function still decays as a power-law, but the exponents are modified thus indicating that the stress-energy tensor is changed.

Notice that the bosonic exponents having the same conformal dimension are now V_α and $V_{Q-\alpha}$. If our deformation generates a realistic critical theory, this theory must have no vanishing four-point correlation functions

of these exponents. It is obvious, however, that one cannot combine four
exponents with charges α and $Q-\alpha$ to satisfy the electroneutrality condition

$$\sum_i \alpha_i = Q \qquad (6.34)$$

Therefore such simple modification of the Gaussian model is not sufficient
to obtain a nontrivial theory. The way out of this difficulty was found by
Dotsenko and Fateev (1984) (DF), who added marginal operators to the
action of the Gaussian model. Following their method, let us consider the
modified action

$$S = \int d^2x \left[\frac{1}{2}(\partial_\mu \Phi)^2 + i\sqrt{4\pi}QR^{(2)}\Phi + \sum_{\sigma=\pm} \mu_\sigma \exp(i\sqrt{4\pi}\alpha_\sigma\Phi) \right] \qquad (6.35)$$

where the 'charges' $\alpha_\pm$ are chosen in such a way that the dimensions of
the corresponding exponents (*screening* operators) given by Eq. (6.33) are
equal to one:

$$\alpha_\pm(\alpha_\pm - Q) = 2$$

or

$$\alpha_\pm = Q/2 \pm \sqrt{Q^2/4 + 2} \qquad (6.36)$$

The quantity $R^{(2)}$ represents the Riemann curvature of the surface. In this
representation the extra charge Q is spread over the entire surface. Such
representation makes it easier to calculate the stress-energy tensor, but is
not very convenient for calculations of the correlation functions.

The action (6.35) may appear odd, because the presence of $i = \sqrt{-1}$
in the action raises doubts whether the theory is unitary. We shall prove,
however, that for certain values of Q the unitarity is preserved. It may also
appear that by adding nonlinear terms to the action we make the theory
intractable. It is not the case, however, because multiparticle correlation
functions of $V_{\alpha_\pm}$-operators vanish and thus the perturbation expansion in
$V_{\alpha_\pm}$ contains only a finite number of terms.

Later in this section we shall consider how to apply this procedure for
calculations of correlation functions in more detail. Now let us discuss
some more elementary problems. We know that the model (6.35) supports
only those correlation functions where the 'charges' satisfy the 'electroneutrality' condition (6.34). As we have already seen, this condition is easily
satisfied for two operators and we have concluded that exponents with the
charges $-\alpha$ and $\alpha-Q$ have the same conformal dimensions. Now we want
to have nontrivial multi-point correlation functions of such operators.
Without screening charges such functions will always vanish. However, if

α belongs to the set

$$\alpha_{n,m} = -\frac{1}{2}(n-1)\alpha_- - \frac{1}{2}(m-1)\alpha_+ \tag{6.37}$$

the four-point correlation function is not zero:

$$\langle V_{\alpha_{n,m}}(1)V_{\alpha_{n,m}}(2)V_{\alpha_{n,m}}(3)V_{Q-\alpha_{n,m}}(4)\rangle_Q$$

$$= \lim_{R\to\infty} R^{2Q^2}\mu^{(n+m-2)}\int d^2\xi_1...d^2\xi_{n+m-2}\langle V_{\alpha_{n,m}}(1)V_{\alpha_{n,m}}(2)V_{\alpha_{n,m}}(3)$$
$$V_{Q-\alpha_{n,m}}(4)V_{\alpha_+}(\xi_1)...V_{\alpha_+}(\xi_{n-1})V_{\alpha_-}(\xi_n)...V_{\alpha_-}(\xi_{n+m-2})V_Q(R)\rangle \tag{6.38}$$

Notice, that the perturbation expansion in powers of μ contributes only a single term unless $\alpha_+ p + \alpha_- q = 0$ for some integer p, q. According to Eq. (6.37) the latter would amount to an existence of a primary field with zero conformal dimension (this occurs, for instance, in the theory with $C = 0$ mentioned in the previous Section). If there are no such operators, however, the model (6.35) remains tractable despite being nonlinear.

Quantization condition (6.37) determines the spectrum of conformal dimensions of the theory (6.35). Substituting (6.37) into Eq. (6.33) we get the formula for the permitted conformal dimensions:

$$\Delta_{n,m} = \frac{1}{2}\left[(\alpha_- n - \alpha_+ m)^2 - (\alpha_- + \alpha_+)^2\right] \tag{6.39}$$

Now we have to calculate the central charge. The stress-energy tensor can be calculated straightforwardly differentiating the action (6.35) with respect to the metric. The answer is

$$T(z) = -2\pi : \partial\Phi\partial\Phi : +i\sqrt{4\pi}Q\partial^2\Phi \tag{6.40}$$

Calculating the correlation function $\langle T(z)T(0)\rangle = C/2z^4$ we find the value of the central charge:

$$C = 1 - 3Q^2 \tag{6.41}$$

Exercise. We leave it to the reader to check that thus defined $T(z)$ has the correct OPE (5.13) with the bosonic exponent V_α, which reproduces the conformal dimension (6.33).

The family of models described by the action (6.35) contains a special subset of unitary models with

$$C = 1 - \frac{6}{p(p+1)}; \quad \Delta_{n,m} = \frac{[pn - (p+1)m]^2 - 1}{4p(p+1)} \tag{6.42}$$

where $p = 3, 4....$ These models are called *minimal*. The number of primary fields in these models is finite. The minimal model with $p = 3$ has $C = 1/2$ and is equivalent to the critical Ising model, the model with $p = 4$ has $C = 7/10$ and is equivalent to the Ising model at the tricritical point, and the model with $p = 5$ has $C = 4/5$ and is equivalent to the Z_3 Potts model at criticality.

In order to get a better insight into how the DF scheme works, let us consider a four-point correlation function of $O_{1,2}$ fields. This field can be represented by the exponents $V_{-\alpha_+/2}$ and $V_{Q+\alpha_+/2}$. The nonvanishing four-point correlation function is obtained if one adds one screening operator:

$$\lim_{R \to \infty} R^{2Q^2} \langle V_{-\alpha_+/2}(1) V_{-\alpha_+/2}(2) V_{-\alpha_+/2}(3) V_{Q+\alpha_+/2}(4) V_{\alpha_+}(\xi) V_{-Q}(R) \rangle$$

Again, one can check that there are no contributions from other screening operators. As a result we have (from now on we shall drop the subscript Q in notations of correlation functions)

$$\langle \Phi(1)...\Phi(4) \rangle \equiv \langle O_{1,2}(1) O_{1,2}(2) O_{1,2}(3) O_{1,2}(4) \rangle$$

$$= \mu \frac{|z_{12} z_{13} z_{23}|^{\alpha_+^2/2}}{|z_{14} z_{24} z_{34}|^{\alpha_+(\alpha_++2Q)/2}} \int d^2\xi \frac{|z_4 - \xi|^{\alpha_+(\alpha_++2Q)}}{|(z_1 - \xi)(z_2 - \xi)(z_3 - \xi)|^{\alpha_+^2}} \quad (6.43)$$

Let us make sure that the obtained expression is conformally invariant; that is, it can be written in the canonical form (6.16). To achieve this we perform a transformation of the variable in the integral:

$$\xi = \frac{z_{13} z_4 \eta + z_{34} z_1}{z_{13} \eta + z_{34}} \quad (6.44)$$

This transformation maps the points $\xi = z_1, z_2, z_3, z_4$ onto $0, x, 1$ and ∞ respectively. Substituting this expression into (6.43) and taking into account the fact that $\alpha_+(\alpha_+ - Q) = 2$ we obtain

$$\langle \Phi(1)...\Phi(4) \rangle = \frac{1}{|z_{13} z_{24}|^{4\Delta}} |x(1-x)|^{\alpha_+^2/2} \int d^2\eta \left[|\eta(1-\eta)(x-\eta)| \right]^{-\alpha_+^2/2}$$

$$(6.45)$$

where $\Delta = \alpha_+(\alpha_+ + 2Q)/4$. The latter expression coincides with Eq. (6.16) with

$$G(x, \bar{x}) = |x(1-x)|^{\alpha_+^2/2+4\Delta} \int d^2\eta \left[|\eta(1-\eta)(x-\eta)| \right]^{-\alpha_+^2/2} \quad (6.46)$$

The integral over η may diverge at certain α. In this case it should be treated as analytic continuation from the area of α where it is convergent. The problem of calculation of such integrals was solved by DF who have obtained the following expression for a general four-point correlation function:

Fig. 6.1. Ian Kogan.

$$\langle O_{n_1,m_1}(1)O_{n_2,m_2}(2)O_{n_3,m_3}(3)O_{n_4,m_4}(4)\rangle$$

$$= \frac{|z_{13}|^{2[\Delta(\alpha_1+\alpha_3+\alpha_+)-\Delta_1-\Delta_3+\alpha_+\alpha_2]}|z_{24}|^{2[\Delta(\alpha_2+\alpha_4)-\Delta_2-\Delta_4+\alpha_+\alpha_2]}}{|z_{12}|^{2[\Delta_1+\Delta_2-\Delta(\alpha_1+\alpha_2)]}|z_{23}|^{2[\Delta_2+\Delta_3-\Delta(\alpha_2+\alpha_3)]}}$$

$$\times|z_{34}|^{-2[\Delta(\alpha_3+\alpha_4+\alpha_+)-\Delta_3-\Delta_4]}|z_{14}|^{-2[\Delta(\alpha_1+\alpha_4+\alpha_+)-\Delta_1-\Delta_4]}$$

$$\times\{\sin[\pi\alpha_-(\alpha_1+\alpha_2+\alpha_3)]\sin[\pi\alpha_-\alpha_2]|I_1(x)|^2$$
$$+\sin[\pi\alpha_-\alpha_1]\sin[\pi\alpha_-\alpha_3]|I_2(x)|^2\} \qquad (6.47)$$

where $\Delta(\alpha)$ is determined by the formula (6.33) and

$$I_1(x) = \int_1^\infty \mathrm{d}t\, t^{\alpha_-\alpha_1}(t-1)^{\alpha_-\alpha_2}(t-x)^{\alpha_-\alpha_3}$$

$$= \frac{\Gamma[-1-\alpha_-(\alpha_1+\alpha_2+\alpha_3)]\Gamma(\alpha_-\alpha_2)}{\Gamma[-\alpha_-(\alpha_1+\alpha_3)]}$$

$$\times F[-\alpha_-\alpha_3,-1-\alpha_-(\alpha_1+\alpha_2+\alpha_3),-\alpha_-(\alpha_1+\alpha_3);x]$$

Structure of Hilbert space

$$I_2(x) = \int_0^x dt\, t^{\alpha-\alpha_1}(t-1)^{\alpha-\alpha_2}(t-x)^{\alpha-\alpha_3}$$

$$= x^{1+\alpha-(\alpha_1+\alpha_3)} \frac{\Gamma(1+\alpha-\alpha_1)\Gamma(1+\alpha-\alpha_3)}{\Gamma[2+\alpha-(\alpha_1+\alpha_3)]}$$

$$\times F[-\alpha-\alpha_2, 1+\alpha-\alpha_1, 2+\alpha-(\alpha_1+\alpha_3); x] \qquad (6.48)$$

References

J.-S. Caux, I. I. Kogan and A. M. Tsvelik, *Nucl. Phys.* **B466**, 444 (1996).

J.-S. Caux, A. Lewis, I. I. Kogan and A. M. Tsvelik, *Nucl. Phys.* **B489**, 469 (1997).

V. S. Dotsenko and V. A. Fateev, *Nucl. Phys.* **B240**, 312 (1984).

L. Rozansky and H. Saleur, *Nucl. Phys.* **B376**, 461 (1992).

7
Current (Kac–Moody) algebras; the first assault

In this chapter we continue the discussion of fermionic current operators started in Chapter 2. We shall also give an example of nontrivial conformal theory. As we have demonstrated in Chapter 2, the Hamiltonian of spinless fermions can be written in terms of currents which leads to the Gaussian model and bosonization. A similar procedure exists for fermions with spin and is called *non-Abelian* bosonization. To be adequately equipped for it, we have to discuss algebraic properties of the corresponding currents.

Let $R^+_{\alpha,n}, R_{\alpha,n}$ ($L^+_{\alpha,n}, L_{\alpha,n}$) be the right- (left-) moving components of *free* massless fermions, with α and n ($n = 1, 2, ...k$) being the spin and 'flavour' indices, respectively. It is assumed that the fermionic fields transform according to some Lie groups G and F, operating in the spin and flavour spaces. The reason for introducing the second index n will become clear later. The current operators on group G are defined as follows:

$$\bar{J}^a(z) = \sum_{n=1}^{k} R^+_{\alpha,n}(\bar{z})\tau^a_{\alpha\beta}R_{\beta,n}(\bar{z})$$

$$J^a(z) = \sum_{n=1}^{k} L^+_{\alpha,n}(z)\tau^a_{\alpha\beta}L_{\beta,n}(z) \tag{7.1}$$

where τ^a are matrices – generators of the Lie algebra G. They satisfy the following relations:

$$[\tau^a, \tau^b] = if^{abc}\tau^c$$

$$\mathrm{Tr}\tau^a\tau^b = \frac{1}{2}\delta^{ab} \tag{7.2}$$

For the case of the SU(2) group $\tau^a = S^a$ are the matrices of spin $S = 1/2$ and $f^{abc} = \epsilon^{abc}$. In a similar way one can define the current operators on the group F, using for this purpose its generators T^j.

Since the fermionic operators with different chiralities commute, so do the currents with different chiralities; currents with the same chirality

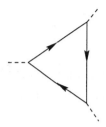

Fig. 7.1. The correlation function of three currents.

satisfy the following Wilson operator expansion:

$$J^a(z)J^b(z') = \frac{k}{8\pi^2(z-z')^2}\delta^{ab} + if^{abc}\frac{J^c(z')}{2\pi(z-z')} + ... \qquad (7.3)$$

To check the validity of this very important identity, one should consider two diagrams: one for the current–current correlation function depicted on Fig. 2.2 and another for the correlation function of three currents (see Fig. 7.1).

It is also very convenient to rewrite the operator expansion (7.3) in terms of commutators:

$$[J^a(x), J^b(y)] = \frac{ik}{4\pi}\delta^{ab}\delta'(x-y) + if^{abc}J^c(y)\delta(x-y) \qquad (7.4)$$

In this form it is called a *Kac–Moody* algebra.

As for the spinless currents, the term with a derivative of the delta function is anomalous; it appears only in infinite systems. The second term in the right-hand side of Eq. (7.4) being absent in the spinless case is an ordinary commutator and can be obtained by a straightforward calculation.

Often the Kac–Moody algebra is written for the Fourier components of current operators. In this case we assume that the system is placed on the strip $0 < x < L$ with periodic boundary conditions and expand the current operators into the series

$$J^a(x) = \frac{1}{L}\sum_n e^{-2i\pi nx/L}J_n^a$$

Substituting this expansion into Eq. (7.4) we get the Kac–Moody algebra for the Fourier components:

$$[J_n^a, J_m^b] = \frac{nk}{2}\delta^{ab}\delta_{n+m,0} + if^{abc}J_{n+m}^c \qquad (7.5)$$

The Kac–Moody algebra includes the original algebra G as its subalgebra;

this is the algebra of the zeroth components of the currents:

$$[J_0^a, J_0^b] = i f^{abc} J_0^c \tag{7.6}$$

Therefore one can think of these zeroth components as about matrices – generators of the group G.

Let us now follow the example of spinless fermions and consider the operators defined as G-invariant quadratic forms of chiral currents.

$$T(z) = \frac{1}{k + c_v} : J^a(z) J^a(z) :$$

$$\bar{T}(\bar{z}) = \frac{1}{k + c_v} : \bar{J}^a(\bar{z}) \bar{J}^a(\bar{z}) : \tag{7.7}$$

The numerical value of the coefficient $1/(k + c_v)$, where c_v is the Kasimir operator in the adjoint representation, i.e., is defined by the identity

$$f_{abc} f_{\bar{a}bc} = c_v \delta_{a\bar{a}}$$

Exercise. We leave it to the reader to show that the operators such defined are indeed chiral components of a stress energy tensor, i.e. they satisfy the Virasoro algebra (5.19) with the central charge

$$C = \frac{\dim G}{k + c_v} \tag{7.8}$$

The easiest way to prove it would be to calculate the two- and three-point correlation functions of T making use of the free fermion representation of current operators (7.1).

I   Sugawara Hamiltonian for Wess–Zumino–Novikov–Witten model

According to the general property of conformal theories a sum of the zeroth components of traceless stress-energy tensors T_0, $\bar{T}_0$ gives a Hamiltonian. From Eq. (7.7) we extract the corresponding Hamiltonian which traditionally carries the name of Sugawara:

$$\hat{H}(k, G) = \frac{2\pi v}{L}(L_0 + \bar{L}_0) = \frac{2\pi v}{(k + c_v)L} \left[J_0^a J_0^a + 2 \sum_{n>0} J_{-n}^a J_n^a + (J \to \bar{J}) \right] \tag{7.9}$$

The Sugawara Hamiltonian defines the Hamiltonian of the Wess–Zumino–Novikov–Witten (WZNW) model whose action will be discussed later in Chapter 14.

Since we have introduced the Hamiltonian (7.9) axiomatically, the reader may remain perplexed about its physical meaning. It is clear that the WZNW model represents some conformal theory, but we have not explained relations between this theory and more conventional models. These

relations are established by the following identity which we give only for
the case G = SU(N):

$$\int dx[-iR^+_{\alpha,n}\partial_x R_{\alpha,n} + iL^+_{\alpha,n}\partial_x L_{\alpha,n}] = H[U(1)] + H[k, SU(N)] + H[N, SU(k)]$$

(7.10)

$$H[U(1)] = 2\pi \int dx[: J(x)J(x) : + : \bar{J}(x)\bar{J}(x) :]$$ (7.11)

$$H[k, SU(N)] = \frac{2\pi}{(N+k)} \sum_{a=1}^{D_N} \int dx[: J^a(x)J^a(x) : + : \bar{J}^a(x)\bar{J}^a(x) :]$$ (7.12)

$$H[N, SU(k)] = \frac{2\pi}{(N+k)} \sum_{\lambda=1}^{D_k} \int dx[: \mathscr{J}^\lambda(x)\mathscr{J}^\lambda(x) : + : \bar{\mathscr{J}}^\lambda(x)\bar{\mathscr{J}}^\lambda(x) :]$$ (7.13)

where $\mathscr{J}^\lambda$ are currents of the SU(k) group. This representation corresponds to the decomposition of the nonsimple symmetry group of the free
fermions into the product of simple groups U(1)× SU(N)× SU(k). It has
been used here that for the SU(N) group $c_v = N$ and dim SU(N) = N^2-1.
The decomposition (7.10) is a generalization of the decomposition considered later in Chapter 15 for a case of fermions with U(1)×SU(2)
symmetry. It would be too complicated to give a complete rigorous proof
of the validity of this identity. A simple argument in favour of this identity
is equality of the central charges. The central charge of the theoretical
free Dirac fermions is equal to the number of fermion species, so it is
Nk. According to (7.8) the central charge of the SU(N)-invariant WZNW
model is given by

$$C = \frac{k(N^2 - 1)}{k + N}$$ (7.14)

where the number of generators is $D_N = N^2 - 1$. Thus we have

$$Nk = 1 + \frac{k(N^2 - 1)}{k + N} + \frac{N(k^2 - 1)}{k + N}$$

which is identically satisfied.

From Eq. (7.10) we see that the WZNW model on group G can be
understood as a G-invariant subsector of the model of free fermions. So,
take free fermions and project some states out. If you are careful and do
this projection intelligently, preserving the conformal symmetry, you will
obtain a model with interesting properties. As we shall see later in Part
II, many interactions do this job perfectly thus leaving the WZNW model
to be an effective theory for low-lying excitations.

Now we have to discuss the WZNW theory in more detail. It is instructive to construct the full basis of its eigenstates. One should do this using
the current operators *only*, because only in this way can one be sure that

all eigenstates are G-invariant. To fulfil this task, we need to know how the current operators commute with the Hamiltonian. The corresponding commutation relations follow from the definition of the stress-energy tensor (7.7):

$$[L_n, J_m^a] = -mJ_{n+m}^a \tag{7.15}$$

Since $H = 2\pi v(L_0 + \bar{L}_0)/L$, we get

$$[H, J_m^a] = -\frac{2\pi v m}{L} J_m^a \tag{7.16}$$

Now we have everything we need to construct the eigenstates. Let us define the vacuum vectors $|h\rangle$ as states annihilated by positive Fourier components of the currents:

$$J_m^a|h\rangle = 0; \quad \bar{J}_m^a|h\rangle = 0 \tag{7.17}$$

The Hamiltonian (7.9) acting on these states is reduced to

$$H_{\text{reduced}} = \frac{2\pi v}{(k + c_v)L}(J_0^a J_0^a + \bar{J}_0^a \bar{J}_0^a) \tag{7.18}$$

Therefore $|h\rangle$ are eigenstates of the Hamiltonian if they realize irreducible representations of the left and right G groups:

$$J_0^a J_0^a |h\rangle = C_{\text{rep}}|h\rangle; \quad \bar{J}_0^a \bar{J}_0^a |h\rangle = \bar{C}_{\text{rep}}|h\rangle \tag{7.19}$$

Now we can use the negative components of the currents as creation operators. According to Eq. (7.16), the following vectors

$$\bar{J}_{-m_1}^{a_1} \dots \bar{J}_{-m_p}^{a_p} J_{-n_1}^{a_1} \dots J_{-n_q}^{a_q} |h\rangle \tag{7.20}$$

are eigenvectors of the Hamiltonian (7.9) with the energies

$$E_{pq}(h) = \frac{2\pi v}{L} \left(\frac{C_{\text{rep}}}{k + c_v} + \sum_{i=1}^q n_i + \frac{\bar{C}_{\text{rep}}}{k + c_v} + \sum_{i=1}^p m_i \right) \tag{7.21}$$

As we have seen in Chapter 4, each eigenstate in conformal field theories is associated with some conformal field, whose conformal dimensions are related to the eigenvalues of energy and momentum via Eqs. (4.9) and (4.10). In the given case the eigenstates of the Hamiltonian are eigenstates of the Kazimir operators of the group G. The conformal dimensions are

$$\Delta = \frac{C_{\text{rep}}}{k + c_v} + \sum_{i=1}^q n_i$$

$$\bar{\Delta} = \frac{\bar{C}_{\text{rep}}}{k + c_v} + \sum_{i=1}^p m_i \tag{7.22}$$

It turns out that the basis of states (7.20) is overcomplete. To make it complete one has to restrict the number of vacuum states $|h\rangle$ by choosing a finite number of irreducible representations of G. For example, in the case G = SU(2) where the irreducible representations are representations of spin operators with $C_{rep} = S(S+1), S = 1/2, 1, ...$, the basis is composed by $S = 1/2, 1, ...k/2$ (Fateev and Zamolodchikov (1986)). Each vacuum vector $|h\rangle$ can be considered as a state created from the lowest vacuum $|0\rangle$ by a corresponding primary field $g_h(\tau, x)$. Indeed, according to Eq. (4.7) we have

$$|h\rangle = \lim_{\tau \to +\infty} e^{\tau(E_h - E_0)}(L/2\pi)^{(2\Delta_h + 2\bar{\Delta}_h)} g_h(\tau, x = 0)|0\rangle \qquad (7.23)$$

Thus the primary fields are matrices g_{ab} – tensors realizing irreducible representations of the group G.

There is one more restriction on the vacuum states, namely a requirement that physical fields must have integer or half-integer conformal spins. If these fields are built exclusively from the WZNW fields, this means that

$$\frac{C_{rep}}{k + c_v} - \frac{\bar{C}_{rep}}{k + c_v} = n/2 \qquad (7.24)$$

where n is an integer. In general, this requirement is difficult to satisfy except for $n = 0$. In this latter case $C_{rep} = \bar{C}_{rep}$ and the primary fields are dim $C_{rep} \times$ dim C_{rep}. This is the case considered in the original paper by Knizhnik and Zamolodchikov (1984). There are other cases, however, where physical fields are products of fields of several WZNW models (the corresponding examples will be given in the chapters where we discuss spin–charge separation). Then the restriction on right and left representations is different.

II   Knizhnik–Zamolodchikov (KZ) equations

Now we shall derive differential equations for multi-point correlation functions of the WZNW primary fields. The derivation is based on the fact that the WZNW stress energy tensor is quadratic in currents (7.7). Writing this expression in components, we get

$$L_n = \frac{1}{c_v + k} \sum_m : J_m^a J_{n-m}^a : \qquad (7.25)$$

where the normal ordering assumes that the operators J_m^a with a positive subscript (annihilation operators) stay on the right. Among the Virasoro generators there is one whose action on operators is particularly simple – it is $L_{-1} \equiv \partial_z$. For $n = -1$ we have

$$\partial_z \equiv L_{-1} = \frac{2}{c_v + k}\left[J_{-1}^a J_0^a + \sum_{m=1}^{\infty} J_{-m-1}^a J_m^a \right] \qquad (7.26)$$

Since primary fields are vacuum states, they are annihilated by positive components of current operators (7.17). Therefore from (7.26) it follows that any primary field satisfies the identity

$$\left(\partial_z - \frac{2}{c_v + k} J^a_{-1} J^a_0\right) \phi_h(z) = 0 \tag{7.27}$$

We already know that J^a_0 acts simply as the generator of the group (see Eq. (7.19)). To proceed further we need to know the action of J_{-1}. This can be extracted from the Ward identity

$$\langle J^a(z)\phi(1)...\phi(N)\rangle = \sum_j \frac{\tau^a_j}{z - z_j} \langle \phi(1)...\phi(N)\rangle \tag{7.28}$$

Thus we have

$$\langle J^a_{-1}\phi(1)...\phi(N)\rangle = \frac{1}{2\pi i} \int_C dz(z_1 - z)^{-1} \langle J^a(z)\phi(z_1)...\phi(N)\rangle$$

$$= \sum_{j \neq 1} \frac{\tau^a_j}{z_1 - z_j} \langle \phi(1)...\phi(N)\rangle \tag{7.29}$$

where the contour C encircles z.

Combining these results we conclude that the N-point functions of primary fields satisfy the following system of equations:

$$\left[\frac{1}{2}(c_v + k)\partial_{z_i} - \sum_{j \neq i} \frac{\tau^a_i \tau^a_j}{z_i - z_j}\right] \langle \phi(1)...\phi(N)\rangle = 0 \tag{7.30}$$

where a matrix τ^a_i acts on the indices of the i-th operator.

Let us consider this equation for the case $N = 2$. Then we have

$$\left[\frac{1}{2}(c_v + k)\partial_{z_1} - \frac{\tau^a_1 \tau^a_2}{z_1 - z_2}\right] \langle \phi(1)\phi(2)\rangle = 0$$

$$\left[\frac{1}{2}(c_v + k)\partial_{z_2} + \frac{\tau^a_1 \tau^a_2}{z_1 - z_2}\right] \langle \phi(1)\phi(2)\rangle = 0 \tag{7.31}$$

From these two equations it follows that $\langle \phi(1)\phi(2)\rangle = G(z_{12})$ and the function $G(z)$ satisfy the following equation:

$$\partial_z G^{\beta,\beta'}_{\alpha,\alpha'}(z) - \frac{2\tau^a_{\alpha,\gamma} \tau^a_{\beta,\delta}}{(c_v + k)z} G^{\delta,\beta'}_{\gamma,\alpha'}(z) = 0 \tag{7.32}$$

It is natural to suggest that a nonzero solution exists only if the second field is the Hermitian conjugate of the first one: $\phi(2) = \phi^+(1)$. It is essential that τ^a_2 acting on this operator gives a minus sign. Then $\tau^a_1 \tau^a_2$ becomes a Kazimir operator:

$$\tau^a_1 \tau^a_2 = -(\tau^a)^2 = -C_{rep}$$

and we get

$$\langle \phi(1)\phi^+(2) \rangle = z_{12}^{-2\Delta_h} \tag{7.33}$$

with the correct conformal dimension (7.22).

Solutions of KZ equations for four-point correlation functions of primary fields in the fundamental representation of the SU(N) group were found by Knizhnik and Zamolodchikov (1984). Solutions for four-point functions of all primary fields of the SU$_k$(2) WZNW model were found by Fateev and Zamolodchikov (1986). There is a regular procedure for finding multi-point correlation functions, called the Wakimoto construction, which is based on the representation of the WZWN operators in terms of free bosonic fields (Wakimoto (1986)). The details of this procedure can be found in Dotsenko (1990).

References

I. Affleck and A. W. W. Ludwig, *Nucl. Phys.* **B352**, 849 (1991); *Phys. Rev. Lett.* **67**, 3160 (1991).

Vl. S. Dotsenko, *Nucl. Phys.* **B338**, 747 (1990); **358**, 547 (1990).

V. A. Fateev and A. B. Zamolodchikov, *Yad. Fiz. (Sov. Nucl. Phys.)* **43**, 657 (1986).

V. G. Knizhnik and A. B. Zamolodchikov, *Nucl. Phys.* **B247**, 83 (1984).

A. M. Tsvelik, *Sov. Phys. JETP* **66**, 221 (1987).

M. Wakimoto, *Commun. Math. Phys.* **104**, 605 (1986).

E. Witten, *Commun. Math. Phys.* **92**, 455 (1984).

8

Relevant and irrelevant fields

In this chapter we continue to study the massless scalar field described by the Gaussian model action (3.1). We have seen that this model has a gapless excitation spectrum and the correlation functions of bosonic exponents follow power laws. This behaviour implies that the correlation length is infinite and the system is in its critical phase. It is certainly very important to know how stable this critical point is with respect to perturbations. Any model is only an idealization; when one derives it certain interactions are neglected. How do we decide which interactions are important and which are not? Obviously, we can neglect interactions whose effect on the correlation functions is small. The trouble is, however, that usually the correlation functions are affected differently on different scales, the long distance asymptotics being affected the most. Therefore it can happen that a certain perturbation causes only tiny changes at short distances, but changes the large distance behaviour profoundly. In the renormalization group picture this is observed as a growth of the coupling constant associated with the perturbing operator. Such growth is a frequent phenomenon in critical theories; a slow decay of their correlation functions gives rise to divergences in their diagram series.

Operators whose influence grows on large scales (small momenta) are called 'relevant'. The problem of relevancy of perturbations can be formulated and solved in a general form. Suppose we have a system at criticality (i.e., all its correlation functions decay as a power law at large distances). Let it be described at the critical point by the action S_0 and let the physical field $A_d(\mathbf{r})$ be a field with scaling dimension d, that is

$$\langle A_d(\mathbf{r})A_d^+(\mathbf{r}')\rangle \sim |\mathbf{r} - \mathbf{r}'|^{-2d}$$

Now let us consider the perturbed action

$$S = S_0 + g \int \mathrm{d}^D r A_d(\mathbf{r})$$

Does it describe the same critical point, i.e., does the perturbation affect the scaling dimensions of the correlation functions?

Theorem. *A perturbation with zero conformal spin and scaling dimension d is relevant if*

$$d < D$$

and irrelevant if

$$d > D$$

The case $d = D$ is the marginal one and the answer depends on the sign of g.

It is very instructive to discuss the consequences of the above theorem in the framework of the Gaussian model. This will provide a nice illustration and at the same time give us an opportunity to appreciate the importance of the previous discussions. Let us consider some perturbation of our bosonic theory. Since we have the operator basis of bosonic exponents we can expand all perturbations local in Φ and Θ. Since only perturbations with small scaling dimensions are really important, it is not necessary to consider fields containing derivatives of the bosonic exponents – all such fields have scaling dimensions larger than 2 and are irrelevant. Therefore it is sufficient to study the problem of relevance for the following simple perturbation

$$g \int d^2x \cos(\beta\Phi)$$

where β belongs to the content of conformal dimensions of our theory and the scaling dimension of the perturbation is equal to

$$d = \beta^2/4\pi$$

The perturbed model has the following action:

$$S = \int d^2x \left[\frac{1}{2}(\nabla\Phi)^2 + g \cos(\beta\Phi)\right] \tag{8.1}$$

and is called the sine-Gordon model. In fact, this model is one of the most important models of quantum $(1 + 1)$-dimensional and classical two-dimensional physics, with many interesting applications. We postpone a detailed description of the solution until Chapter 10 and now restrict ourselves to a simplified approach.

Let us consider the perturbation expansion of the simplest two-point correlation function of bosonic exponents:

$$\langle\exp[i\beta_0\Phi(1)]\exp[-i\beta_0\Phi(2)]\rangle$$

where β_0 is one of the permitted βs. The first nonvanishing correction to this correlation function is equal to

$$g^2 \int d^2x_3 d^2x_4 \langle \exp[i\beta_0\Phi(1)] \exp[-i\beta_0\Phi(2)] \exp[i\beta\Phi(3)]$$

$$\exp[-i\beta\Phi(4)]\rangle$$

$$\sim \int d^2z_3 d^2z_4 |z_1 - z_2|^{-\beta_0^2/2\pi} \left(\frac{|z_1 - z_3||z_2 - z_4|}{|z_1 - z_4||z_2 - z_3|} \right)^{\beta_0\beta/2\pi} |z_3 - z_4|^{-\beta^2/2\pi} \quad (8.2)$$

The double integral in this expression converges at large distances when

$$2d \equiv \beta^2/2\pi > 4$$

i.e., if $d > 2$. This means that at $d > 2$ the perturbation expansion does not contain infrared singularities; so if the bare coupling constant is small, its effect will remain small and will not be amplified in the process of renormalization.

Let us consider what happens if $d < 2$, i.e., for the case when the perturbation is relevant. In general, there are two possibilities. A relevant perturbation can drive the system to another critical point.[*] Another possibility is that the conformal symmetry is totally lost and the system acquires a finite correlation length. For a $(1 + 1)$-dimensional system this means that elementary excitations become massive. The latter occurs in the sine-Gordon model. As we shall see below, the detailed mass spectrum is strongly β-dependent, but the overall scale of mass gaps can be estimated from a simple argument.

Suppose we know that a relevant field $A_\Delta(x)$ with a scaling dimension 2Δ gives rise to a finite correlation length. Naturally, this correlation length is such that the contribution of the field A_Δ to the action is of order of 1. That is we have

$$gv^{-1} \int_0^\xi d^2x |x/a|^{-2\Delta} \sim 1 \quad (8.3)$$

(v being the velocity), which gives

$$\xi \sim v/m \sim a(ga^2/v)^{-1/(2-2\Delta)} \quad (8.4)$$

Now we can estimate the average value of the operator in the ground state:

$$\langle A_\Delta(x) \rangle \sim (\xi/a)^{-2\Delta} \sim (ga^2/v)^{\Delta/(1-\Delta)} \quad (8.5)$$

Let us return to the sine-Gordon model. In fact, we already know the spectrum at one point, namely for $d = 1$ ($\beta = \sqrt{4\pi}$): according to the

---

bosonization scheme the sine-Gordon model with this β is equivalent to the model of free massive fermions (see Eq. (3.31)) with the mass $m \propto g$. It is also easy to solve the model (8.1) at $d \ll 1$, where one can expand the cosine term around $\Phi = \pi/\beta$:

$$\cos(\beta\Phi) \approx -1 + \frac{1}{2}\beta^2(\Phi - \pi/\beta)^2$$

As a result we get the theory of massive bosons:

$$S_{sg} = \frac{1}{2} \int d^2x[(\nabla\Phi')^2 + g\beta^2\Phi'^2], \ \Phi' = \Phi - \pi/\beta \qquad (8.6)$$

with the following excitation spectrum and the free energy:

$$\omega(p) = \sqrt{p^2 + m^2}, \ m^2 = g\beta^2 \qquad (8.7)$$

$$F/L = T \int \frac{dp}{(2\pi)} \ln\{1 - \exp[-\omega(p)/T]\} \qquad (8.8)$$

Thus the spectrum has a gap as in the $d = 1$ case; the elementary excitations, however, are different: they are not fermions, but bosons. It is not difficult to calculate correlation functions of the bosonic exponents in this approximation. They are still given by Eq. (3.12), but with

$$G(\xi_1, \xi_2) = \frac{1}{(2\pi)^2} \int d^2k \frac{e^{ik\xi_{12}}}{k^2 + m^2} = \frac{K_0(m|\xi_{12}|)}{\pi}$$

For the pair correlation function, for instance, we have

$$D_{12} = \langle \exp[i\beta_0\Phi(1)] \exp[-i\beta_0\Phi(2)]\rangle - |\langle \exp[i\beta_0\Phi(1)]\rangle|^2$$

$$\approx (ma)^{8\Delta}\{\exp[-4\Delta K_0(m|z_{12}|)] - 1\}$$

where $\Delta = \beta_0^2/8\pi$. At small distances $|z_{12}| \ll m^{-1}$ where $K_0(x) \approx \frac{1}{2}\ln(1/x)$ one recovers the unperturbed expression Eq. (3.13). At large distances $|z_{12}| \gg m^{-1}$ where $K_0(x) \approx \sqrt{\pi/2x}\exp(-x)$ we have

$$D_{12} = -8\Delta(ma)^{4\Delta}\exp(-m|z_{12}|)\sqrt{\pi/2|z_{12}|} + \dots$$

We conclude this chapter with the discussion of a nontrivial case when the perturbation contains two operators with *integer* conformal spins $\pm S$. Let us consider the following perturbation:

$$V = g \int dz d\bar{z} \cos(\beta\Phi)\cos(\bar{\beta}\Theta)$$

$$= \frac{g}{2} \int dz d\bar{z}\{\cos[(\beta + \bar{\beta})\phi + (\beta - \bar{\beta})\bar{\phi}] + \cos[(\beta - \bar{\beta})\phi + (\beta + \bar{\beta})\bar{\phi}]\}$$

$$(8.9)$$

where $\beta, \bar{\beta}$ belong to the set (3.29). Since $\beta\bar{\beta}/\pi = 2S(\text{even})$, we get for the scaling dimensions of both operators:

$$d = \frac{1}{4\pi}(\beta^2 + \bar{\beta}^2) = \frac{1}{4\pi}(\beta^2 + 4S^2\pi^2/\beta^2) \geq S \qquad (8.10)$$

The criterion $d < 2$ suggests that the operator V is irrelevant (marginal) for $S > 2$ ($S = 2$) and for $S = 1$ restricts the area of its relevance by

$$\frac{\beta^2}{2\pi} < 2 - \sqrt{3} \qquad (8.11)$$

(since the action in this case is invariant under the duality transformation $\Phi \to \Theta$, $\beta \to \bar{\beta}$, we shall consider $\beta^2 < 2\pi$ only). We shall see in a moment that these statements are not correct and the criterion of relevance must be modified.

To establish a new criterion, let us consider the perturbation series for the free energy. The first divergency in the expansion of the partition function appears in the fourth order in g:

$$Z^{(4)} \sim g^4 \int d^2x_1...d^2x_4 \langle \exp[i(\beta + \bar{\beta})\phi(1)] \exp[i(\beta - \bar{\beta})\phi(2)]$$

$$\times \exp[i(-\beta + \bar{\beta})\phi(3)] \exp[i(-\beta - \bar{\beta})\phi(4)]\rangle \langle \exp[i(\beta - \bar{\beta})\bar{\phi}(1)]$$
$$\times \exp[i(\beta + \bar{\beta})\bar{\phi}(2)] \exp[i(-\beta - \bar{\beta})\bar{\phi}(3)] \exp[i(-\beta + \bar{\beta})\bar{\phi}(4)]\rangle \qquad (8.12)$$

$$+(\beta \to \bar{\beta})$$

Suppose that

$$(\bar{\beta}^2 - \beta^2)/4\pi > 1 \qquad (8.13)$$

which corresponds to

$$\frac{\beta^2}{2\pi} < -1 + \sqrt{S^2 + 1} \qquad (8.14)$$

Then the leading contribution to the integrals in x_{12} and x_{34} comes from small distances of order of the ultraviolet cut-off. Calculating these integrals first we can rewrite approximately

$$\int dz_2 \exp[i(\beta + \bar{\beta})\phi(1)] \exp[i(\beta - \bar{\beta})\phi(2)] \sim \exp[2i\beta\phi(1)]a^{1+(\beta^2 - \bar{\beta}^2)/4\pi}$$
$$(8.15)$$

The remaining integral in x_1 and x_3 is

$$Z^{(4)} \sim g_1^2 \int d^2x_1 d^2x_3 \langle \cos[2\beta\Phi(1)] \cos[2\beta\Phi(3)]\rangle;$$

$$g_1 = g^2 a^{2+(\beta^2 - \bar{\beta}^2)/2\pi} \qquad (8.16)$$

In the opposite case $\beta > \bar{\beta}$ one can repeat the above derivation interchanging β and $\bar{\beta}$. The integrals divergent at small distances are now integrals in x_{13} and x_{24}. In this case the procedure generates the perturbation $g_2 \cos[2\bar{\beta}\Theta]$ where $g_2 = g^2 a^{2+(\bar{\beta}^2-\beta^2)/2\pi}$. The criterion (8.13) is replaced by

$$(\beta^2 - \bar{\beta}^2)/4\pi > 1 \qquad (8.17)$$

Thus the long distant behaviour of the correlation functions is the same as if we had the perturbations

$$V_1 = g_1 \cos(2\beta\Phi), \ (\beta < \bar{\beta}) \ ; V_2 = g_2 \cos(2\bar{\beta}\Theta), \ (\beta > \bar{\beta}) \qquad (8.18)$$

Thus we see that the original perturbation with nonzero conformal spin generates the perturbations (8.18) with zero conformal spin. The renormalization group analysis described in Chapter 20 shows that these perturbations are always generated. However, their behaviour is simple only if additional restrictions (8.13) for V_1 or (8.17) for V_2 are imposed. Then the original perturbation (8.9) dies out and does not appear in the effective action. The remaining interactions (8.18) have zero conformal spin and are liable to the conventional criterion of relevance. Otherwise one has to deal with all three interactions (8.9) and (8.18) simultaneously.

The mechanism of generation of relevant perturbations outlined above was first described by Brazovskii and Yakovenko (1985) and later by Kusmartsev, Luther and Nersesyan (1992).

References

S. A. Brazovskii and V. M. Yakovenko, *J. Phys. Lett. (Paris)* **69**, 46 (1985); *Sov. Phys. JETP* **62**, 1340 (1985).

F. V. Kusmartsev, A. Luther and A. A. Nersesyan, *JETP Lett.* **55**, 692 (1992); *Phys. Lett.* A**176**, 363 (1993).

A. B. Zamolodchikov, *JETP Lett.* **43**, 730 (1986). See also in the book *Conformal Invariance and Applications to Statistical Mechanics*, edited by C. Itsykson, H. Saleur and J.-B. Zuber, World Scientific (1988).

9

Bose–Einstein Condensation in two dimensions; Beresinskii–Kosterlitz–Thouless transition

Since the Pauli matrices also describe hard core bosons, the anisotropic spin-1/2 Heisenberg chain in a magnetic field is a good model for a one-dimensional Bose liquid. The magnetic field in this case represents chemical potential. Later we shall show that a one-dimensional system of repulsing bosons does not condense even at $T = 0$, but has a critical point at zero temperature with power-law correlations. However, the situation in two dimensions is different; namely there the Bose condensate exists at $T = 0$, power law correlations persist to finite temperature T^* and then disappear abruptly via a phase transition.

Let us consider the action describing repulsive bosons in two dimensions:

$$S = \int d\tau d^2x [\Psi^+ \partial_\tau \Psi + \frac{1}{2m} \nabla \Psi^+ \nabla \Psi - \mu \Psi^+ \Psi + \frac{1}{2} g (\Psi^+ \Psi)^2] \qquad (9.1)$$

At $\mu > 0$ the bosons condense. For $D > 2$ such condensation corresponds to the appearence of a nonzero average $\langle \Psi \rangle$ corresponding to a spontaneously broken U(1) symmetry. For $D = 2$ a continuous symmetry cannot be broken at finite temperatures, and this average is zero, but the correlation function $\langle \Psi(x) \Psi^+(y) \rangle$ at $T < T^*$ decays as a power-law.

Let us first rewrite the path integral in the radial and angular variables. The correct choice of variables corresponding to the unit Jacobian is

$$\Psi = \sqrt{\rho} e^{i\alpha} \qquad (9.2)$$

Substituting this expression into Eq. (9.1) we get

$$S = \int d\tau d^2x \left[i\rho \partial_\tau \alpha + \frac{1}{8m\rho} (\nabla \rho)^2 + \frac{1}{2m} \rho (\nabla \alpha)^2 + \frac{1}{2} g (\rho - \rho_0)^2 \right] \qquad (9.3)$$

where $\rho_0 = \mu/g$. When μ is large enough, the radial component of Ψ fluctuates weakly around its nonzero average value $\sqrt{\rho_0}$. In this case we can expand the action in powers of $\tilde{\rho} = \rho - \rho_0$.

Exercises.   (a) Keeping only quadratic terms in $\tilde{\rho} = \rho - \rho_0$ in Eq. (9.3), obtain the Gaussian effective action for α and $\tilde{\rho}$. Find the pair correlation functions of $\tilde{\rho}$ and α. The answer is

$$\langle\langle\tilde{\rho}(-\omega_n, -\mathbf{k})\tilde{\rho}(\omega_n, \mathbf{k})\rangle\rangle = \frac{\rho_0\mathbf{k}^2/m}{\omega_n^2 + \frac{1}{4m}\mathbf{k}^2(\mathbf{k}^2/m + 4\rho_0 g)}$$

$$\langle\langle\alpha(-\omega_n, -\mathbf{k})\tilde{\rho}(\omega_n, \mathbf{k})\rangle\rangle = \frac{\omega_n}{\omega_n^2 + \frac{1}{4m}\mathbf{k}^2(\mathbf{k}^2/m + 4\rho_0 g)}$$

$$\langle\langle\alpha(-\omega_n, -\mathbf{k})\alpha(\omega_n, \mathbf{k})\rangle\rangle = \frac{g + \mathbf{k}^2/4m\rho_0}{\omega_n^2 + \frac{\mathbf{k}^2}{4m}(\mathbf{k}^2/m + 4\rho_0 g)} \qquad (9.4)$$

(b) Derive from these expressions the excitation spectrum.

(c) Obtain an explicit expression for the pair correlation function of the phase field $\langle\langle\alpha(\tau, \mathbf{x})\alpha(0, 0)\rangle\rangle$ in real space. What term in the sum over frequencies gives the most singular contribution? What effect does it have on the behaviour of $\langle\Psi(\mathbf{x})\Psi^+(0)\rangle$?

It is instructive to rewrite the correlation function of phases separating the zeroth Matsubara frequency:

$$\langle\langle\alpha(-\omega_n, -\mathbf{k})\alpha(\omega_n, \mathbf{k})\rangle\rangle$$

$$= \delta_{n,0}\frac{m}{\rho_0\mathbf{k}^2} + (1 - \delta_{n,0})\frac{m}{\rho_0[\mathbf{k}^2 + 16\pi^2 n^2 T^2 m^2/(4g\rho_0 m + \mathbf{k}^2)]} \qquad (9.5)$$

Now it is clear that the $n = 0$ component has a special status having a singular correlation function. For

$$|k| < a^{-1} \equiv 2\pi T\sqrt{m/\rho_0 g} \qquad (9.6)$$

the static component of α can be considered in isolation. Substituting time-independent α into Eq. (9.3) we get the Gaussian effective action

$$S = \frac{\rho_0}{2mT}\int d^2x(\partial_\mu\alpha)^2 \qquad (9.7)$$

The action (9.7) apparently coincides with the action of the free scalar field (3.1) after rescaling $\alpha = \sqrt{T/m\rho_0}\Phi$. The fact that operators of the original model (9.7), being local functionals of Ψ, are periodic functionals of Φ with a period

$$T_1 = 2\pi\sqrt{\rho_0/mT} \qquad (9.8)$$

specifies the set of scaling dimensions. They are given by Eq. (3.28) with

$$K = \frac{mT}{\pi\rho_0}$$

The pair correlation function of Ψ-fields decays at large distances $|x_{12}| > a$ as a power-law:

$$\langle \Psi(\xi_1)\Psi^+(\xi_2)\rangle = \rho_0 \left(\frac{a}{|\xi_{12}|}\right)^{mT/2\pi\rho_0}$$

The above simple-minded derivation misses the important 2π-periodicity in phase α present in the original Hamiltonian as follows from the definition of the field Ψ (9.2). In the Gaussian action (9.7), however, this periodicity is lost and the field α is defined on the entire real axis. It turns out that to recover the periodicity in α one should add to the action (9.7) the dual exponents:

$$S = \frac{1}{2}\int d^2x \left[(\nabla\Phi)^2 + \sum_{k=1} A_k \cos\left(2\pi k\Theta\sqrt{\rho_0/mT}\right)\right] \qquad (9.9)$$

where A_k are some coefficients.* These perturbations are relevant at large temperatures $T > \pi\rho_0/2m$ where a finite correlation length $\xi(T)$ arises and the correlation functions decay exponentially at distances larger than ξ. In order to see this, let us recall the path integral procedure. In a path integral one integrates over continuous functions; in the present case over $\Phi(x)$. It seems that all singular functions are automatically excluded since, giving an infinitely large contribution to the exponent, they give a zero contribution to the path integral. This is not correct, however, because we have regularized our theory and all singularities are smoothed out over a distance $\sim a$. Therefore contributions from singular configurations to the action are never zero, they rather contain some negative powers of a. In particular, there are configurations of fields which give logarithmically divergent contributions $S \propto \ln a$. The simplest configuration is

$$\mathbf{n} = (\cos\gamma, \sin\gamma) = \frac{\mathbf{r} - \mathbf{r}_0}{|\mathbf{r} - \mathbf{r}_0|} \qquad (9.10)$$

where γ is the angle between the $\hat{x}$-axis and a radius vector drawn from some point $\mathbf{r}_0$ (see Fig. 9.1). This configuration is singular because the vector field has an indefinite direction at $\mathbf{r} = \mathbf{r}_0$. Graphically the vector field looks like a plane flow current with a source at $\mathbf{r}_0$.

The configuration (9.10) gives the following contribution to the action:

$$S_0 = \frac{\rho_0}{2mT}\int \frac{d^2r}{r^2} \approx \frac{\pi\rho_0}{mT}\ln(R/a)$$

This contribution is infinite for an infinite system. A finite contribution comes from configurations containing a source and a 'drain' (Fig. 9.2). For

---

* Note that these perturbations contain dual exponents with *even* numbers only.

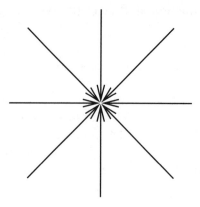

Fig. 9.1.   The vector field looks as a plane flow current with a source at $\mathbf{r}_0$.

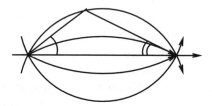

Fig. 9.2.   A field configuration containing a source and a 'drain'.

this configuration we have $\Phi\sqrt{\rho_0/mT} = \alpha_1 - \alpha_2$ (the angles are defined in Fig. 9.2) and the action is estimated as:

$$S_{\text{dip}} \approx \frac{2\pi\rho_0}{mT}\ln(l/a) \qquad (9.11)$$

where l is the distance between the source and the drain.

It is easy to see that all singular configurations of the $\Phi(z,\bar{z})$-field can be represented by branch cut singularities of a logarithmic function. One can write down a general function $\Phi(z,\bar{z})$ as a sum of singular and regular pieces:

$$\Phi = i\sum_i e_i \ln\left(\frac{z - z_i}{\bar{z} - \bar{z}_i}\right) + \Phi_{\text{reg}} \qquad (9.12)$$

where

$$e_i = \pm\frac{1}{2}\sqrt{\rho_0/mT}$$

Exercise.   Check that the configuration in Fig. 9.1 corresponds to

$$\Phi_1 = -ie\ln\left(\frac{z - z_0}{\bar{z} - \bar{z}_0}\right)$$

and the configuration in Fig. 9.2 corresponds to

$$\Phi_2 = ie \ln \left(\frac{z - z_1}{\bar{z} - \bar{z}_1} \right) - ie \ln \left(\frac{z - z_2}{\bar{z} - \bar{z}_2} \right)$$

Substituting Eq. (9.12) into Eq. (3.1) we get

$$\partial\Psi\bar{\partial}\Psi = \left(i \sum_i e_i \frac{1}{z - z_i} + \partial\Phi_{\text{reg}} \right)\left(-i \sum_i e_i \frac{1}{\bar{z} - \bar{z}_i} + \bar{\partial}\Phi_{\text{reg}} \right)$$

To proceed further we first replace the derivative of Φ_{reg} by the derivative of Θ using the definition of the Θ-field (3.20) and then integrate the obtained expressions by parts using the identity

$$\partial_z \frac{1}{\bar{z}} + \partial_{\bar{z}} \frac{1}{z} = 2\pi\delta^{(2)}(x, y) \tag{9.13}$$

The resulting expression is given by

$$S_N \equiv 2 \int dz d\bar{z} \partial\Phi\bar{\partial}\Phi = S_{\text{cl}} + 4\pi i \sum_i e_i \Theta_{\text{reg}}(z_i, \bar{z}_i) + 2 \int dz d\bar{z} \partial\Phi_{\text{reg}} \bar{\partial}\Phi_{\text{reg}}$$

$$S_{\text{cl}} = 2 \int dz d\bar{z} \sum_{i,j} \frac{e_i e_j}{(z - z_i)(\bar{z} - \bar{z}_j)} = 2\pi \sum_{i,j} e_i e_j \ln \left(\frac{R^2}{|z_i - z_j|^2 + a^2} \right) \tag{9.14}$$

Thus the entire partition function has the following form:

$$Z = \sum_{N=0}^{\infty} \frac{1}{N!} \int \prod_{i=0}^{N} dz_i d\bar{z}_i \int D\Phi_{\text{reg}} \exp\left(-S_N\right) = Z_0$$

$$+ Z_0 \sum_{N=1}^{\infty} \frac{1}{N!} \prod_{i=1}^{N} \int dz_i d\bar{z}_i \exp(-S_{\text{cl}})$$

$$\times \langle \exp[4\pi i e_1 \Theta(1)] \dots \exp[4\pi i e_N \Theta(N)] \rangle \tag{9.15}$$

Where Z_0 is the partition function without singularities which includes integration over Φ_{reg}. Notice that this equation represents a perturbation series expansion of the partition function for the modified action (9.9).

Let us first consider the classical term S_{cl}, which appears in the exponent in front of the correlation function. For the infinite system S_{cl} is finite only for neutral configurations containing an equal number of sources and drains, which implies the condition $\sum_i e_i = 0$. According to the analysis given in Chapter 8, the same condition is necessary for the correlation function of the exponents to be nonzero. As is obvious from Eq. (9.14), S_{cl} represents the classical energy of a two-dimensional plasma of electric charges. This energy is finite for electrically neutral configurations and its magnitude is given by:

$$S_{\text{cl}} \approx N \frac{\pi \rho_0}{mT} \ln(l/a) \tag{9.16}$$

where $l \sim R/\sqrt{N}$ is the average distance between the charges. Since the classical energy for each even N (electroneutrality!) is finite the relevance of the singular configurations depends on convergence of the integrals of the correlation functions of the bosonic exponents. It is not difficult to see that correlation functions of the dual and the direct exponents are given by the same expression (3.16):

$$\langle \exp[4\pi i e_1 \Theta(1)]... \exp[4\pi i e_N \Theta(N)]\rangle = \prod_{i>j} (|z_i - z_j|/a)^{8\pi e_i e_j} \quad (9.17)$$

Now by induction we can figure out the leading divergence of the correlation function at large distances. For this purpose we multiply all $z_i, \bar{z}_i$ in the Nth term in the expansion of the partition function by λ. Since the entire expression acquires a factor

$$\lambda^{N(2-4\pi e^2)}$$

we conclude that the integral converges only if the power of λ is negative, i.e.,

$$2 - 4\pi e^2 < 0 \rightarrow \frac{\pi \rho_0}{2mT} > 1 \quad (9.18)$$

Thus at temperatures lower than

$$T^* = \frac{\pi \rho_0}{2m}$$

the singular configurations can be taken into account perturbatively. The phenomenon described here was discovered by Beresinskii (1971) and Kosterlitz and Thouless (1973). On the contrary, at $T > T^*$ when the integrals diverge, the singular configurations of the order parameter make essential contributions, as has just been demonstrated. It can be shown that at $T > T^*$ the correlation functions decay exponentially at distances larger than the correlation length

$$\xi \sim \exp \left[\frac{\text{const}}{\sqrt{(T - T^*)}} \right] \quad (9.19)$$

At $T = T^*$ where $mT^*/2\pi\rho_0 = 1/4$ the correlation functions of the Ψ-field abruptly change their asymptotic behaviour from exponential to power law:

$$\langle \Psi(x)\Psi^+(y)\rangle|_{T=T^*-0} \sim |x - y|^{-1/4} \quad (9.20)$$

References

V. L. Beresinskii, *Sov. Phys. JETP* **32**, 493 (1971).

J. M. Kosterlitz, *J. Phys.* **C7**, 1046 (1974).

J. M. Kosterlitz and D. J. Thouless, *J. Phys.* **C6**, 1181 (1973).

10

The sine-Gordon model

I   The renormalization group analysis

The analysis of relevancy of the cosine perturbation to the free massless scalar field theory carried out in Chapter 8 has demonstrated the existence of two qualitatively different infrared regimes in the sine-Gordon model. At $\beta^2 > 8\pi$, the perturbation is irrelevant, and the system maintains the gapless spectrum. At $\beta^2 < 8\pi$, the perturbation is relevant and drives the system to a strong-coupling, massive phase. In the vicinity of the transition between the two phases, $|\beta^2/8\pi - 1| \sim g^2$, the phase diagram is very nontrivial and is determined by renormalization of *two* parameters, the amplitude g *and* the coupling constant β. The renormalization group (RG) equations for the two coupling constants which we shall now derive are known as Kosterlitz–Thouless equations. These define the so-called Kosterlitz–Thouless universality class for a wide class of $(1 + 1)$-dimensional quantum physical systems and two-dimensional models of classical statistics (Wiegmann (1978, 1980); Amit *et al.* (1980); Jose *et al.* (1977); Kogut (1979)).

We consider a two-dimensional Euclidean action

$$S = S_0 + S_1$$
$$= \frac{1}{2} \int d^2\mathbf{x} \, (\nabla\Phi)^2 + z \int \frac{d^2\mathbf{x}}{a^2} \cos \beta\Phi(\mathbf{x}) \tag{10.1}$$

$z = ga^2$ being a small dimensionless amplitude, and construct the partition function

$$Z = \int D\Phi e^{-S[\Phi]}, \tag{10.2}$$

We put the theory on a lattice assuming that the Fourier transform of the field, $\Phi_{\mathbf{k}}$, is defined in the Brillouin zone. Technically, it is most convenient to choose a circular cut-off in momentum space, $|\mathbf{k}| < \Lambda$, $\Lambda \sim 1/a$. The

73

most important information about the infrared properties of the model is incorporated in the long-wavelength part of the field determined by Fourier components $\Phi_{\mathbf{k}}$ with $|\mathbf{k}| \ll \Lambda$. The idea of renormalization is to start moving towards larger distances (lower energies) by integrating out the fields with shorter and shorter wavelengths. Technically, the RG procedure is based on a decomposition of the original field into short-wavelength and long-wavelength parts, performing partial path integration in (10.2) over the short-wavelength component and representing the result in terms of an effective model for the long-wavelength field. If the model under consideration is renormalizable, which is known to be the case for the sine-Gordon model, the effective action will have the same structure as the original one, but with a new set of coupling constants. Such a procedure is repeated many times, and at each RG step the form of the original model is reproduced (up to irrelevant terms). Relations between the 'bare' and renormalized couplings then lead to RG equations. Their solution reveals the dominant tendencies developing in the system on increasing the length scale (or, equivalently, decreasing the characteristic energy, frequency, temperature, etc.).

Let us split the Brillouin zone $|k| < \Lambda$ into a wide region $0 < |k| < \Lambda'$ and narrow slice $\Lambda' < |k| < \Lambda$, where $\Lambda' = \Lambda - d\Lambda$. Represent the original field as

$$\Phi_\Lambda(\mathbf{x}) = V^{-1/2} \sum_{k<\Lambda'} e^{i\mathbf{k}\mathbf{x}} \Phi_{\mathbf{k}} + V^{-1/2} \sum_{\Lambda'<k<\Lambda} e^{i\mathbf{k}\mathbf{x}} \Phi_{\mathbf{k}}$$

$$\equiv \Phi_{\Lambda'}(\mathbf{x}) + h(\mathbf{x}) \qquad (10.3)$$

where $\Phi_{\Lambda'}(\mathbf{x})$ and $h(\mathbf{x})$ are the slow and fast components of Φ_Λ. The Gaussian part S_0 of the total action (10.1) is additive under decomposition (10.3). Therefore

$$Z_\Lambda = \int D\phi_{\Lambda'} \, Dh \exp\{-S_0[\Phi_{\Lambda'}] - S_0[h] - S_1[\Phi_{\Lambda'} + h]\}$$

$$= Z_h \int D\Phi_{\Lambda'} \, e^{-S_0[\Phi_{\Lambda'}]} \langle e^{-S_1[\Phi_{\Lambda'}+h]} \rangle_h \qquad (10.4)$$

Here

$$Z_h = \int Dh \, \exp\{-S_0[h]\} \qquad (10.5)$$

is a contribution of short-wavelength fluctuations to the partition function, which is apparently nonsingular. In Eq. (10.4) we used notation

$$\langle \mathscr{F}[h] \rangle_h \equiv Z_h^{-1} \int Dh \, e^{-S_0[h]} \, \mathscr{F}[h] \qquad (10.6)$$

Now we can rewrite (10.4) by exponentiating $\langle ... \rangle_h$ in the integrand. In this way we obtain the effective action $S_{eff}[\Phi_{\Lambda'}]$, which involves only the slow component of the field:

$$S_{eff}[\Phi_{\Lambda'}] = S_0[\Phi_{\Lambda'}] - \ln\langle e^{-S_1[\Phi_{\Lambda'}+h]}\rangle_h \qquad (10.7)$$

Clearly, $S_{eff}[\Phi_{\Lambda'}]$ preserves all infrared singularities.

Of course, (10.7) is only a formal expression. To make it useful, one has to assume that $|g| \ll 1$ and expand in powers of g. To the accuracy $0(g^3)$, one gets

$$\begin{aligned}S_{eff}[\Phi_{\Lambda'}] &= S_0[\Phi_{\Lambda'}]\\ &+ \langle S_1[\Phi_{\Lambda'}+h]\rangle_h\\ &- \frac{1}{2}\left(\langle S_1^2[\Phi_{\Lambda'}+h]\rangle_h - \langle S_1[\Phi_{\Lambda'}+h]\rangle_h^2\right) + \dots \qquad (10.8)\end{aligned}$$

Here we realize that the RG method is a *perturbative* one. Therefore, it is only applicable as long as renormalized couplings remain small.

In connection with this limitation, a question arises whether the RG method is applicable in the case when scale invariance is broken by a dynamical generation of a mass gap M. We already know that this is the case for the sine-Gordon model at $\beta^2 < 8\pi$. It is at this point where the requirement for the bare amplitude of the perturbation to be small is crucial. In the latter case, the mass gap is necessarily small compared to the ultraviolet cut-off, $M \ll \Lambda$, or equivalently, the correlation length is large, $\xi_c \gg \Lambda^{-1}$. Then there exists a wide interval of energies, $M \ll |E| \ll \Lambda$ (or distances $\Lambda^{-1} \ll |x| \ll \xi_c$) where the effective couplings remain small, and the RG equations allow to estimate the rate of their increase, and even the magnitude of the mass gap. The range of energies comparable to or less than M can be only treated by nonperturbative methods.

Let us now estimate the first order correction to $S_0[\phi_{\Lambda'}]$ in (10.8). We have:

$$\begin{aligned}\langle S_1[\Phi_{\Lambda'}+h]\rangle_h &= g\int d^2x\langle\cos\beta[\Phi_{\Lambda'}(x)+h(x)]\rangle_h\\ &= \frac{1}{2}g\sum_{\sigma=\pm1}\int d^2x\ \exp\{i\beta\sigma\Phi_{\Lambda'}(x)\}\langle\exp\{i\beta\sigma h(x)\}\rangle_h, \qquad (10.9)\end{aligned}$$

Since the field $h(x)$ contains only short-wavelength harmonics, $\langle h^2\rangle_h$ is finite:

$$\langle h^2\rangle_h = \int_{\Lambda'<k<\Lambda}\frac{d^2k}{(2\pi)^2}\frac{1}{k^2} = \frac{1}{2\pi}dl + 0(dl^2), \quad dl = \frac{d\Lambda}{\Lambda} = d\ln\Lambda,$$

and therefore

$$\langle\exp\{i\beta\sigma h(x)\}\rangle_h = 1 - \frac{\beta^2}{4\pi}dl + 0(dl^2)$$

So, in the first order approximation,

$$S_{eff}[\Phi_{\Lambda'}] = \frac{1}{2}\int d^2x\,(\nabla\Phi_{\Lambda'})^2 + z(1-ddl)\int\frac{d^2x}{a^2}\cos\beta\Phi_{\Lambda'}(\mathbf{x}) \qquad (10.10)$$

To recover the original cut-off Λ, we rescale the momentum, $\mathbf{k'} = (\Lambda/\Lambda')\mathbf{k} \simeq (1+dl)\mathbf{k}$. To keep the scalar product $\mathbf{k}\cdot\mathbf{x}$ intact, we have to rescale the coordinate in the opposite way, $\mathbf{x'} = (1-dl)\mathbf{x}$. Since the Gaussian action S_0 is scale invariant, we are only left to transform the second term in the r.h.s. of (10.10). Thus, in the first order in z, the renormalized action preserves its original form (8.1)

$$S_{eff}[\Phi_\Lambda] = \frac{1}{2}\int d^2x\,(\nabla\Phi_\Lambda)^2 + z'\int\frac{d^2x}{a^2}\cos\beta\Phi_\Lambda(\mathbf{x}) \qquad (10.11)$$

but with a different value of z:

$$z' = z[1+(2-d)dl] \qquad (10.12)$$

No renormalization of β occurs in this order.

For small dl the difference $z' - z$ is also small. So, Eq. (10.12) can be rewritten as a differential equation for effective coupling $z(l)$, continuously depending on the logarithmic variable l:

$$\frac{dz(l)}{dl} = (2-d)z(l) \qquad (10.13)$$

with the initial condition $z(l = 0) = z$. Increasing l means going in the infrared direction.

We see that, in the first order in g, scaling is solely determined by the free (Gaussian) dimension of the cosine term in the sine-Gordon model. The solution of Eq. (10.13)

$$z(l) = z_0\left(\frac{L}{a}\right)^{2-d} \qquad (10.14)$$

gives the already discussed criterion of relevancy for the cosine perturbation: on increasing the length scale L, $z(l) \to 0$, if $d > 2$ (irrelevant perturbation), or $z(l) \to \infty$, if $d < 2$ (relevant perturbation). The former case corresponds to a weak-coupling regime, and the semiaxis $z = 0$, $d > 2$ represents a line of infrared *stable* fixed points. In the latter case, a strong-coupling regime develops in the infrared limit, and the semiaxis $z = 0$, $d < 2$ is a line of infrared *unstable* fixed points.

In the strong-coupling case, the renormalized coupling $z(L)$ increases with L reaching values of the order of 1 at $L \sim \xi_c$. This gives an estimate for the correlation length and the related mass gap:

$$M \sim \xi_c^{-1} \sim \Lambda z^{1/(2-d)}$$

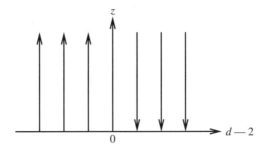

Fig. 10.1. The scaling 'portrait' of the sine-Gordon model obtained in the first loop approximation.

An equivalent way to obtain the scaling of the mass gap is to demand that

$$z \int^{\xi_c} \frac{d^2\mathbf{x}}{a^2} \langle \cos \beta \Phi \rangle \sim 1$$

assuming that integration in the two-dimensional space is limited by a sphere of radius $\xi_c \gg a$, within which one can substitute $\langle \cos \beta \Phi(\mathbf{x}) \rangle \sim |\mathbf{x}|^{-d}$.

The scaling 'portrait' in shown in Fig. 10.1

Notice that all scaling trajectories are vertical, reflecting the fact that β is not renormalized. We shall demonstrate how this oversimplified picture transforms to the beautiful Kosterlitz–Thouless phase diagram, when the second order corrections are taken into account.

Let us now turn to the second order terms in Eq. (10.8). One can easily verify that

$$\langle S_1^2[\Phi_{\Lambda'} + h] \rangle_h - \langle S_1[\Phi_{\Lambda'} + h] \rangle_h^2 = \frac{1}{2} g^2 (1 - 2d \, dl) \int d^2 x_1 \int d^2 x_2$$
$$\times \{ (\exp[-4\pi d \langle h(\mathbf{x}_1)h(\mathbf{x}_2) \rangle_h] - 1) \cos \beta [\Phi_{\Lambda'}(\mathbf{x}_1) + \Phi_{\Lambda'}(\mathbf{x}_2)]$$
$$+ (\exp[4\pi d \langle h(\mathbf{x}_1)h(\mathbf{x}_2) \rangle_h] - 1) \cos \beta [\Phi_{\Lambda'}(\mathbf{x}_1) - \Phi_{\Lambda'}(\mathbf{x}_2)] \} \quad (10.15)$$

Consider the correlation function of the fast components of the field, $\mathcal{G}(\mathbf{x}) = \langle h(\mathbf{x})h(0) \rangle_h$. It is clear that

$$\mathcal{G}(\mathbf{x}) = F(r) \, dl + 0(dl^2)$$

where $r = |\mathbf{x}|$. A straightforward calculation with the adopted sharp momentum cut-off prescription yields for $F(r)$ a Bessel function, $F(x) = (1/2\pi) J_0(\Lambda|\mathbf{x}|)$, which has a long oscillating tail and does not fall off rapidly on increasing its argument. However, it can be shown (Wiegmann (1978), Kogut (1979)) that implementing a smooth cut-off procedure makes $F(r)$ truly short-ranged, i.e. essentially nonzero at $r < \Lambda^{-1}$. As a result, in Eq. (10.15), the functions $\exp[\pm 4\pi d\mathscr{G}(\mathbf{x_1} - \mathbf{x_2})] - 1$ will be also short-ranged. This circumstance makes it possible to introduce the center-of-mass coordinate $\mathbf{R} = (\mathbf{x_1} + \mathbf{x_2})/2$ and relative coordinate $\mathbf{r} = \mathbf{x_1} - \mathbf{x_2}$ and expand the cosines in the right-hand side of Eq.(10.15) in $\mathbf{r}$. The first cosine generates the second harmonic, $\cos 2\beta\Phi_{\Lambda'}(\mathbf{R})$, which can be neglected as irrelevant, since in the region of interest ($\beta^2 \simeq 8\pi$) its critical dimension is $4d \simeq 8$. For the second cosine in (10.15) we get

$$\cos\beta[\phi_{\Lambda'}(\mathbf{x_1}) - \phi_{\Lambda'}(\mathbf{x_2})] \simeq \cos\beta(\mathbf{r} \cdot \nabla_{\mathbf{R}}\phi_{\Lambda'}(\mathbf{R})$$

$$\simeq 1 - \frac{1}{2}\beta^2(\mathbf{r} \cdot \nabla_{\mathbf{R}}\phi_{\Lambda'})^2$$

Here the first term contributes to renormalization of the free energy which we are not interested in, while the second (gradient) term is responsible for renormalization of the constant β. Then

$$\langle S_1^2\rangle_h - \langle S_1\rangle_h^2 = -Az^2 d^2 dl \int \frac{d^2x}{a^2}(\nabla_{\mathbf{R}}\Phi_{\Lambda'})^2(\mathbf{R}), \tag{10.16}$$

where A is a nonuniversal numerical constant determined by the convergent integral

$$A \simeq \int_0^\infty d\rho\, \rho^3 F(\rho).$$

Using Eqs.(10.8), (10.10) and (10.16) and rescaling the momenta and coordinates to map $\Phi_{\Lambda'}$ onto Φ_{Λ} we obtain:

$$S_{eff}[\Phi_{\Lambda}] = \frac{1}{2}(1 + Az^2 d^2 dl)\int d^2x\,(\nabla\Phi_{\Lambda})^2$$

$$+ z[1 + (2 - d)dl]\int \frac{d^2x}{\alpha^2}\cos\beta\Phi_{\Lambda}(\mathbf{x}) \tag{10.17}$$

The changed overall factor in front of the Gaussian part of the action requires a multiplicative renormalization of the field Φ and the constant β

$$\Phi_{\Lambda}(\mathbf{x}) \to Z^{1/2}\Phi_{\Lambda}(\mathbf{x}), \quad \beta \to Z^{-1/2}\beta, \tag{10.18}$$

such that the product $\beta\phi_{\Lambda}$ remains invariant. The constant Z is determined from the condition $Z(1 + Az^2 d^2 dl) = 1$, leading to the following renormalization of the scaling dimension:

$$d' = Z\Delta = d - Az^2 d^3 dl \tag{10.19}$$

This leads to the differential equation

$$\frac{dd(l)}{dl} = -Az^2 d^3(l) \tag{10.20}$$

with the initial condition $d(l = 0) = d \equiv \beta^2/4\pi$.

As already mentioned, it is our intention to clarify details of the weak-to-strong coupling transition in the vicinity of the critical point $d = 2$. Therefore, we set

$$d(l) = 2 + z_\parallel(l), \quad z(l) = \frac{z_\perp(l)}{\sqrt{8A}} \tag{10.21}$$

assuming that $z_\parallel$ and $z_\perp$ are small.* Thus, we arrive at the famous Kosterlitz–Thouless renormalization group equations:

$$\frac{dz_\parallel}{dl} = -z_\perp^2, \quad \frac{dz_\perp}{dl} = -z_\parallel z_\perp \tag{10.22}$$

with initial conditions

$$z_\parallel(0) = z_\parallel^0 \equiv d - 2, \quad z_\perp(0) = z_\perp^0 \equiv z\sqrt{8A}$$

Eqs. (10.22) are characterized by scaling invariant (first integral)

$$\mu^2 = z_\parallel^2 - z_\perp^2 = (z_\parallel^0)^2 - (z_\perp^0)^2 \tag{10.23}$$

which determines the phase diagram of the system on the plane $(z_\parallel, z_\perp)$ in the region where renormalized couplings are small: $|z_\parallel|, |z_\perp| \ll 1$. The phase diagram is shown in Fig. 10.2. Arrows indicate the infrared direction. This direction is easily established by observing that $z_\parallel(l)$ is a decreasing function of l, as follows from the first equation (10.22).

The essentially new feature of the second order scaling, seen in Fig. 10.2, is a strong renormalization of $z_\parallel$ (i.e. β) in the region where $|z_\parallel|$ is comparable to or less than $|z_\perp|$. There are two separatrices, $z_\parallel = \pm|z_\perp|$,[†] that divide the phase plane into three sectors:

(1) $z_\parallel \geq |z_\perp|$, – the *weak coupling (WC)* or *zero-charge* sector;

(2) $-|z_\perp| < z_\parallel < |z_\perp|$ – the *crossover (C)* sector;

(3) $z_\parallel \leq -|z_\perp|$ – the *strong coupling (SC)* or *asymptotic freedom* sector.

In the *WC* sector, $|z_\perp(l)| \to 0$ as $l \to \infty$, and the SG model scales to a Gaussian model. In this sector scale invariance of the system is preserved, and the spectrum is massless. In two other sectors, *C* and *SC*, the situation is qualitatively different. For any initial values of the coupling constants,

---

[*] The reason for introducing such a 'strange' notations, $z_{\parallel,\perp}$, is historical. The structure of the Kosterlitz–Thouless equations (10.22) coincides with that of analogous equations for the anisotropic Kondo problem (Anderson, Yuval and Hamann (1970)), as well as one-dimensional Fermi gas model with a U(1)-symmetric interaction (Dzyaloshinskii and Larkin (1972); Emery et al. (1974)). In these models, the coupling constants $z_\parallel$ and $z_\perp$ are the amplitudes for the spin-non-flip and spin-flip scattering processes, respectively.

[†] It can be shown that, along these two lines, the theory possesses a hidden SU(2) symmetry - see Chapters 17 and 18.

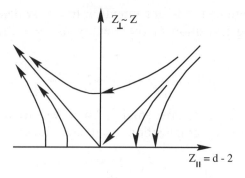

Fig. 10.2. The improved phase portrait of the sine-Gordon model.

the scaling trajectories eventually go to a strong coupling region, away from the Gaussian fixed line. In the SC-sector the increase of $z_\parallel$ and $z_\perp$ is immediate, while in the C-sector one observes a crossover from a weak-coupling regime ($z_\parallel = |z_\perp|$) to strong-coupling regime ($z_\parallel = -|z_\perp|$).

One important result of the second order scaling is that the location of the KT transition point is now modified. The new condition is $z_\parallel = |z_\perp|$, or equivalently

$$d = 2 + \sqrt{8Az_0} \tag{10.24}$$

This should be compared with the marginality condition $d = 2$ obtained in the first order approximation (i.e. by estimating the Gaussian dimension of the cosine perturbation).

The KT scaling equations allow to estimate the mass gap which is dynamically generated upon renormalization in sectors C and SC. The solution of Eqs. (10.22) is

$$z_\parallel(l) = \mu \coth(\mu l + f), \quad z_\perp(l) = \frac{\mu}{\sinh^2(\mu l + f)} \tag{10.25}$$

where

$$\tanh f = -\frac{\mu}{g_\parallel}$$

$z_\parallel$ can be transformed to an infinite sum

$$z_\parallel(\omega/M) = -\sum_{n=-\infty}^{\infty} \left(\ln \frac{|\omega|}{M} - \frac{i\pi n}{2\mu} \right)^{-1} \tag{10.26}$$

Here ω is an energy variable, and m is the mass gap. In the SC-sector μ is real, while in the C-sector it is purely imaginary, $\mu = -i\mu_1$,

$\mu_1 = \sqrt{(z_\perp^0)^2 - (z_\parallel^0)^2}$. The mass gap is given by

$$M = \Lambda \left(\frac{|z_\perp^0|}{z_\parallel^0} \right)^{1/z_\parallel^0}, \quad -z_\parallel^0 \gg |z_\perp^0|,$$

$$M = \Lambda \exp(-1/|z_\parallel^0|), \quad \mu \ll |z_\parallel^0|, \qquad (10.27)$$

$$M = \Lambda \exp(-\pi/2\mu_1), \quad \mu_1 \gg |z_\parallel^0|$$

Comparison with the exact (Bethe-ansatz) solution of the SG model shows that formulas (10.27) give a correct estimate of the quantum soliton mass. Of course, the exact solution tells us much more; for instance, it shows the existence of the so-called 'breather' modes (or soliton–antisoliton bound states) in the spectrum at $\beta^2 < 4\pi$, appearing inside the single-soliton gap m. These subtle details are not accessible by renormalization group approach whose range of applicability is restricted by the condition $|\omega| \gg m$.

Going back to the general definition of relevancy for perturbations to a critical theory, let us warn the reader that an attempt to obtain the criterion of the relevancy within a linear response theory would be misleading. By this we mean that one might deduce that the perturbation is relevant (irrelevant), if the corresponding susceptibility describing the *linear* response of the system to the perturbation is divergent (finite). To show that such a method would not work, consider the linear response to the cosine perturbation in the sine-Gordon model. The susceptibility as a function of two-dimensional momentum $\mathbf{q} = (q, -\omega/c)$ is given by

$$\mathcal{X}(\mathbf{q}) = \int d^2x \, e^{i\mathbf{q}\cdot\mathbf{x}} \langle \cos \beta \Phi(\mathbf{x}) \cos \beta \Phi(0) \rangle$$

$$= \int d^2x \, \frac{e^{i\mathbf{q}\cdot\mathbf{x}}}{(\mathbf{x}^2 + a^2)^d} \sim \int_0^\infty dr \, \frac{r J_0(qr)}{(r^2 + a^2)^d}$$

We find that, in the infrared limit $q \to 0$

$$\mathcal{X}(\mathbf{q}) \sim |q|^{-2(1-d)} \to \infty, \quad d < 1$$

$$\mathcal{X}(\mathbf{q}) \sim \ln(1/qa) \to \infty, \quad d = 1$$

$$\mathcal{X}(\mathbf{q}) \to \text{const}, \quad d > 1 \qquad (10.28)$$

Looking at these results, it would be tempting to conclude that $d = 1$ is the marginality point separating the region $d > 1$, where the gapless Gaussian model is stable against the perturbation, from the region $d < 1$ where an instability develops, and the perturbation drives the system to a presumably massive phase. Such a conclusion is in contradiction with the general definition of relevancy discussed in Chapter 8 and the RG analysis of the sine-Gordon model according to which the correct marginality condition is $d = 2$. In particular, it follows from the correct criterion that in the whole region $1 < d < 2$ where the linear response function is finite, the perturbation is in fact relevant, and a mass gap is generated in the spectrum.

The above results for the linear susceptibility allows one to estimate the lowest order correction to the temperature dependence of the free energy for the two-dimensional Gaussian model: $\Delta F(T, L) \sim -z^2 L T^{2(d-1)}$. As for the susceptibility, this expression is infrared divergent ($T \to 0$) at $d < 1$. But one should actually consider the *total* free energy, with the zero order contribution of free massless bosons included. Then one obtains

$$F(T, L) = -ALT^2 [\, 1 + Bz^2 T^{2(d-2)} + 0(z^4)\,]$$

where A and B are constants. Observe that the *relative* z^2-correction diverges as $T \to 0$ at $d < 2$, illustrating once again the validity of the general criterion.

II   Exact solution of the sine-Gordon model

The sine-Gordon model is one of the simplest and best studied exactly solvable non-critical models. The spectrum of the sine-Gordon model was obtained semiclassically (Dashen *et al.* (1975)) and from the exact solution (Takhtadjan and Faddeev (1975)). The S-matrix was calculated by Zamolodchikov (1977). A regularized integrable version of the sine-Gordon model was proposed and studied in detail by Japaridze *et al.* (1984). The formfactors which define the correlation functions were derived by Smirnov in the series of papers published between 1985 and 1992. The results are summarized in his book cited in the General bibliography. We shall use this well understood model to illustrate the problems one encounters outside the critical point.

The subject of exactly solvable models is very rich and complicated. We cannot possibly give it the space it deserves – a serious discussion would take another volume or two. At the same time we cannot resist a temptation to discuss the problem on some elementary level. Below we will give a review of the main results about exactly solvable problems in general and the sine-Gordon model in particular.

All exactly solvable $(1 + 1)$-dimensional models share certain common features. In what follows we shall discuss only the models whose low-energy spectrum displays Lorentz invariance. Since the behaviour of critical models is discussed at length throughout the rest of the book, we shall restrict the present consideration to the models with spectral gaps. In particular, the sine-Gordon model has a spectral gap in the entire interval $\beta^2 < 8\pi$. The spectrum of relativistic massive theories can be conveniently parametrized by the parameter called 'rapidity' θ:

$$\epsilon(\theta) = m \cosh \theta, \; p(\theta) = m \sinh \theta \qquad (10.29)$$

As a consequence of Lorentz invariance, the two-particle scattering matrix must be a function of $\theta_1 - \theta_2$, where $\theta_{1,2}$ are the rapidities of colliding particles.

There are different definitions of integrability; in particular, a model is integrable if all interactions can be taken into account by modification of commutation relations of the creation and annihilation operators. In such theories multiparticle scattering is factorizable, that is an N-particle scattering matrix is represented as a product of two-particle ones. Thus complete information about the interaction is contained in the two-particle S-matrix, i.e. in the commutation relations. Such a description of integrable theories was introduced by the Zamolodchikovs (see their review article published in 1979).

Let $Z_a^+(\theta)$ and $Z_a(\theta)$ be creation and annihilation operators for a particle state characterized by the isotopic index a and rapidity θ. These operators satisfy what is called the Zamolodchikov–Faddeev algebra:

$$Z_a^+(\theta_1)Z_b^+(\theta_2) = S_{a,b}^{\bar{a},\bar{b}}(\theta_1 - \theta_2)Z_{\bar{b}}^+(\theta_2)Z_{\bar{a}}^+(\theta_1)$$

$$Z^a(\theta_1)Z^b(\theta_2) = S_{\bar{a},\bar{b}}^{a,b}(\theta_1 - \theta_2)Z^{\bar{b}}(\theta_2)Z^{\bar{a}}(\theta_1)$$

$$Z^a(\theta_1)Z_b^+(\theta_2) = S_{b,\bar{a}}^{\bar{b},a}(\theta_2 - \theta_1)Z_{\bar{b}}^+(\theta_2)Z^{\bar{a}}(\theta_1) + \delta(\theta_1 - \theta_2)\delta_a^b \quad (10.30)$$

The commutation relations are self-consistent, that is the algebra has a property of associativity if the S-matrix satisfies the Yang–Baxter relations:

$$S_{\bar{a}_1,\bar{a}_2}^{a_1,a_2}(\theta_{12})S_{b_1,\bar{a}_3}^{\bar{a}_1,a_3}(\theta_{13})S_{b_2,b_3}^{\bar{a}_2,\bar{a}_3}(\theta_{23}) = S_{\bar{a}_2,\bar{a}_3}^{a_2,a_3}(\theta_{23})S_{\bar{a}_1,b_3}^{a_1,\bar{a}_3}(\theta_{13})S_{b_1,b_2}^{a_1,\bar{a}_2}(\theta_{12}) \quad (10.31)$$

One can understand these relations as equations whose solutions determine integrable theories.

The Zamolodchikov–Faddeev description is universal for most integrable models (there are those where a particle description is not possible, but we will not dwell on this at the moment). The only thing which changes from theory to theory is the spectrum content and the form of S-matrix. In the sine-Gordon model those change rather dramatically throughout the interval $8\pi > \beta^2 > 0$ where the cosine term is relevant. The most essential difference appears between the intervals $8\pi > \beta^2 > 4\pi$ and $4\pi > \beta^2$. Before going into details, however, we shall first consider the simplest point, namely the point $\beta^2 = 4\pi$ where the sine-Gordon model is equivalent to free massive Dirac fermions:

$$e^{i\sqrt{4\pi}\phi} \sim R^+L, \; e^{-i\sqrt{4\pi}\phi} \sim L^+R$$

Thus the correlation functions of the bosonic exponents with $\beta = \sqrt{4\pi}$ can be easily calculated. It is remarkable and highly nontrivial that at the free fermion point one can calculate two-point correlation functions of all bosonic exponents. This problem was solved by Bernard and LeClair (1994) who applied to this problem the methods developed earlier for the Ising model (see the reference on McCoy and Wu in the General bibliography). The corresponding expressions for the exponents with $\beta < \sqrt{4\pi}$ are

$$\langle \exp[i\sqrt{4\pi}\alpha\phi(\tau,x)] \exp[-i\sqrt{4\pi}\alpha'\phi(\tau,x)]\rangle \equiv \exp[\Sigma(\tau,x)] \quad (10.32)$$

$$\nabla^2\Sigma = -m^2\sinh^2\rho, \quad 2\nabla^2\rho = m^2\sinh 2\rho$$
$$+2(\alpha+\alpha')^2\sinh\rho/r^2\cosh^3\rho \quad (10.33)$$

Eqs. (10.33) allow us to add to Σ any analytic or antianalytic function. The solution depends on parameters α, α' via the boundary conditions which can be extracted from the long or short distance behaviour of the correlation function. In particular, for $\alpha = \alpha'$ we must have

$$\Sigma(r \to 0) = -2\alpha^2[\ln(mz) + \ln(m\bar{z})] + O(1)$$

Alternatively, the correlation function can be expressed as an infinite series:

$$G_{\alpha,\alpha'}(\tau,x) = m^{2\Delta+2\Delta_2}C_\alpha C_{\alpha'}\sum_{n=0}^{\infty}\frac{m^{2n\Delta_1+2n\Delta_2}C_\alpha^n C_{\alpha'}^n\sin^n(\pi\alpha)\sin^n(\pi\alpha')}{\pi^{2n}(n!)^2}$$

$$\times \int d\theta_1...\theta_{2n}\exp\left[-m\tau\sum_j\cosh\theta_j - imx\right.$$

$$\left.\times\sum_j\sinh\theta_j\right]F_\alpha(\theta_1,...\theta_{2n})F_{\alpha'}^*(\theta_1,...\theta_{2n}) \quad (10.34)$$

$$F_\alpha(\theta_1,...\theta_{2n}) = \left\{\prod_{j=1}^{n}\exp[(1/2-\alpha)\theta_j + (1/2+\alpha)\theta_{j+n}]\right\}$$

$$\times\frac{\prod_{i<j\leq n}(e^{\theta_i}-e^{\theta_j})\prod_{n+1\leq i<j}(e^{\theta_i}-e^{\theta_j})}{\prod_{r=1}^{n}\prod_{s=n+1}^{2n}(e^{\theta_r}+e^{\theta_s})} \quad (10.35)$$

where the coefficients C_α are determined from the one-point function

$$\langle \exp[i\alpha\sqrt{4\pi}\phi(\tau,x)]\rangle = C_\alpha m^{2\Delta} \quad (10.36)$$

Now we return to the problem of calculating the correlation functions with general β. In order to get an idea about the excitations, one may solve the corresponding classical equation, which is the Lagrange equation for the sine-Gordon action:

$$-(\partial_t^2 - \partial_x^2)\phi + \beta^{-1}m^2\sin\beta\phi = 0 \quad (10.37)$$

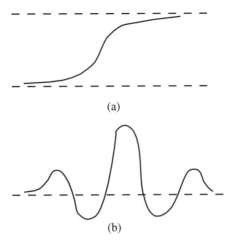

Fig. 10.3.   Classical solutions of the sine-Gordon equations corresponding to (a) a kink, (b) a breather.

Since β and m can be removed by rescaling of ϕ and the coordinates respectively, the classical equation is insensitive to these parameters. In the following discussion of the classical solution we shall put $\beta = 1, m = 1$.

One of the simplest solutions of Eq. (10.37) is the so-called 'kink' ('antikink')

$$\phi_{k,a}(x,t) = 4\tan^{-1}\left[\pm\frac{x - x_0 - vt}{\sqrt{1 - v^2}}\right] \tag{10.38}$$

(see Fig. 10.3). This solution has a nonzero 'topological charge'

$$Q = \frac{1}{2\pi}\int_{-\infty}^{\infty}\partial_x\phi\,dx \tag{10.39}$$

($Q = 1$ for a kink and -1 for an antikink).

In fact, kinks and antikinks look like moving domain walls. For this reason they are also called solitary waves, or 'solitons'. One can observe such waves on shallow water (and thus they were first observed and described) where they sometimes reach colossal size – in Japan such giant solitary waves are called 'tsunami'.

Another elementary solution of the sine-Gordon equation is called a 'breather'. Its analytical form is given by

$$\phi = 4\tan^{-1}\left\{\frac{\Im m\lambda}{|\Re e\lambda|}\frac{\sin\left[\frac{\Re e\lambda}{|\lambda|}\left(\frac{t - t_0 - x/v}{\sqrt{1 - v^2}}\right)\right]}{\cosh\left[\frac{\Im m\lambda}{|\lambda|}\left(\frac{x - x_0 - vt}{\sqrt{1 - v^2}}\right)\right]}\right\} \tag{10.40}$$

where

$$v = \frac{4|\lambda|^2 - 1}{4|\lambda|^2 + 1}$$

and x_0, t_0 are parameters. Breathers appear as soliton–antisoliton bound states and look more similar to ordinary waves, especially if $\Im m\lambda \ll |\lambda|$.

As soon as we return to the quantum world, the actual magnitude of β becomes important. As we have mentioned, the spectrum is different at $8\pi > \beta^2 > 4\pi$ and at $4\pi > \beta^2$. In the first sector breathers do not form and the spectrum consists solely of kinks and antikinks. In the latter sector breathers emerge and the number of their types depends on the magnitude of the coupling constant β. In general the coupling constant enters into S-matrices and the spectrum in a combination

$$\gamma = \frac{\pi\Delta}{1 - \Delta} = \frac{\pi\beta^2}{8\pi - \beta^2} \qquad (10.41)$$

We shall consider first the special points in the second sector, namely

$$\gamma = \pi/v, \quad (v = 1, 2, ...),$$

at which both the spectrum and the S-matrices are especially simple.

The simplicity stems from the fact that all scattering matrices are diagonal, that is upon scattering colliding particles acquire only phase shifts, but do not change their isotopic numbers:

$$S_{a,b}^{\bar{a},\bar{b}}(\theta) = \delta_a^{\bar{a}}\delta_b^{\bar{b}} S_{a,b}(\theta) \qquad (10.42)$$

so that Yang–Baxter equations (10.31) are trivially satisfied. The spectrum consists of kinks and antikinks with the mass M_s and their bound states. There are $v - 1$ breathers with different masses:

$$M_n = 2M_s \sin\left(\pi n/2v\right), \quad n = 1, 2, ...(v - 1) \qquad (10.43)$$

The particle masses are functions of β and g: $M_s \sim g^{1/(2-2\Delta)}$. The operators have the following transformation properties under parity conjugation:

$$C\Phi C^{-1} = -\Phi$$
$$CZ_s(\theta)C^{-1} = Z_{\bar{s}}(\theta)$$
$$CZ_n(\theta)C^{-1} = (-1)^n Z_n(\theta)\, (n = 1, ...v - 1) \qquad (10.44)$$

In this case the soliton–soliton, soliton–antisoliton and antisoliton–antisoliton S-matrices are all equal to each other:

$$S_{ss}(\theta) = S_{s\bar{s}}(\theta) = S_{\bar{s}\bar{s}}(\theta) \equiv S_0(\theta)$$

$$S_0(\theta) = -\exp\left\{-i\int_0^\infty dx \frac{\sin(x\theta)\sinh[(\pi - \gamma)x/2]}{x\cosh(\pi x/2)\sinh(\gamma x/2)}\right\}$$

$$= \prod_{m=0}^{\infty} \frac{\Gamma[v(2m - i\theta/\pi)]\Gamma[v(2m+1+i\theta/\pi)-1]\Gamma[v(2m+2-i\theta/\pi)-1]}{\Gamma[v(2m+i\theta/\pi)]\Gamma[v(2m+1-i\theta/\pi)-1]\Gamma[v(2m+2+i\theta/\pi)-1]}$$

$$\times \frac{\Gamma[v(2m+1+i\theta/\pi)]}{\Gamma[v(2m+1-i\theta/\pi)]} \qquad (10.45)$$

This S-matrix has a pole on the physical sheet $\theta = i(\pi - \gamma)$ which corresponds to the first bound state – the first breather with the mass $2M \sin(\gamma/2)$. The S-matrix of first breathers is

$$S_{11}(\theta) = \frac{\sinh\theta + i\sin\gamma}{\sinh\theta - i\sin\gamma} \qquad (10.46)$$

This S-matrix also has a pole at $\theta = i\gamma$ which means that two first breathers may create a bound state – a second breather etc.

Now recall that these S-matrices determine commutation relations of the creation and annihilation operators. We see that all S-matrices have the same asymptotic behaviour: $S(|\theta| \to \infty) = 1$ and $S(0) = -1$. This means that the excitations with rapidities far apart in rapidity space view each other as Bose particles, while those which are close become fermions! This explains why states with the same rapidity cannot be occupied twice: in a remarkable way the Pauli principle is generated by the interactions.

Taking this into account we can write an arbitrary multiparticle state in terms of the creation operators acting on the vacuum state $|0\rangle$:

$$|(\theta_1, a_1)...(\theta_N, a_N)\rangle = Z_{a_1}^+(\theta_1)...Z_{a_N}^+(\theta_N)|0\rangle \qquad (10.47)$$

where, in order to prevent overcounting, we assume that rapidities of particles of the same kind are ordered: $\theta_1 > \theta_2 > ...$

It is assumed that all physical operators can be expanded in terms of Z and Z^+. Let us consider some operator $A(\tau, x)$ and its matrix element between the vacuum and an N-particle eigenstate:

$$\langle 0|A(\tau, x)|(\theta_1, a_1)...(\theta_N, a_N)\rangle = e^{-\tau E - ixP}\langle 0|A(0,0)|(\theta_1, a_1)...(\theta_N, a_N)\rangle$$

$$\equiv e^{-\tau E - ixP} F_{a_1,...a_N}(A; \theta_1, ...\theta_N) \qquad (10.48)$$

The quantity $F_{a_1,...a_N}(A; \theta_1, ...\theta_N)$ is called the 'formfactor'.

The task of calculation of formfactors turns out to be not so daunting under close inspection. This is because formfactors satisfy certain general requirements greatly restricting their functional form. Below we shall confine our discussion to the one- and two-particle formfactors. The expressions for multiparticle formfactors with a detailed discussion of their derivation can be found in the book by Smirnov cited in the General bibliography, and also in Babelon *et al.* (1996).

Let us recall the requirements for formfactors.

- First of all, the Lorentz invariance dictates that the formfactors of scalar operators must be functions of θ_{ij}.

- The second requirement is for the asymptotics at $\theta_i \to \infty$. If at the critical point $m \to 0$ the operator A has a conformal dimension Δ_A, then its formfactor satisfies the following inequality:

$$F(A; \theta_1, ... \theta_N)_{\theta_i \to \infty} \leq \exp(\Delta_A \theta_i) \qquad (10.49)$$

- It is clear from the definition of the formfactor that

$$F_{a_1, a_2}(A; \theta_1, \theta_2) S_{\bar{a}_1, \bar{a}_2}^{a_1, a_2}(\theta_{12}) = F_{\bar{a}_1, \bar{a}_2}(A; \theta_2, \theta_1) \qquad (10.50)$$

$$F_{a_1, a_2}(A; \theta_1, \theta_2 + 2i\pi) = F_{a_2, a_1}(A; \theta_2, \theta_1) \qquad (10.51)$$

Combining the last two equations we obtain the Riemann problem for the two-particle formfactor:

$$F_{a_1, a_2}(A; \theta_1, \theta_2 + 2i\pi) = F_{\bar{a}_1, \bar{a}_2}(A; \theta_1, \theta_2) S_{a_1, a_2}^{\bar{a}_1, \bar{a}_2}(\theta_{12}) \qquad (10.52)$$

- If particles create bound states (as kinks and antikinks do at $\beta^2 < 4\pi$) their formfactors have poles in the complex plane of θ. One can use this property to derive formfactors of breathers from soliton formfactors. The following rule holds:

$$ResF_{a_1, a_2}(A; \theta_1, \theta_2) = \pm [ResS(\theta_{12})]^{1/2} F_{br}(A) \qquad (10.53)$$

For the sine-Gordon model the poles are formed at $\theta_{12} = i(\pi - m\gamma)$.

Let us find the simplest formfactors of the operators $A = \cos \beta \Phi$ and $\sin \beta \Phi$. The first operator is a Lorentz scalar and the second one is pseudoscalar, i.e. changes sign under charge conjugation $\Phi \to -\Phi$. Thus their formfactors are functions of θ_{ij}. At the critical point both of them have conformal dimensions (Δ, Δ). The fact that the operators behave differently under parity conjugation introduces a further restriction for their formfactors. According to Eq. (10.44), the expansion of $\cos \beta \Phi$ may contain only even breathers, and the expansion of $\sin \beta \Phi$ only odd ones. Since breathers are bound states of solitons and antisolitons, these requirements also give us information about analytic properties of the soliton formfactors.

The two-soliton sector is most representative. For two particles there are two invariant subspaces: C-even and C-odd. The S-matrices are the same in both cases, but the formfactors have different singularities in the θ-plane. From Eq. (10.44) we conclude that the odd state wave function changes its sign under parity conjugation and the even one does not. Since the wave function of the first breather has negative parity, it can emerge only as a bound state in the odd channel. Therefore the odd formfactor $F_o(\theta)$ has a pole at $\theta = i(\pi - \gamma)$ and the even formfactor $F_e(\theta)$ does not.

In the invariant subspaces the matrix Riemann problem (10.52) becomes a scalar one:

$$F_{e,o}(A; \theta - 2i\pi) = F_{e,o}(A; \theta)S_{e,o}(\theta) \qquad (10.54)$$

A general solution of this problem is given by

$$F_a(\theta) = \mathscr{R}_{A,a}(e^{\theta}) \exp \left\{ \int_{-\infty}^{\infty} \frac{dx}{x} \frac{1 - e^{ix\theta}}{e^{2\pi x} - 1} K_a(x) \right\}$$

$$K_a(x) = \frac{1}{2\pi i} \int d\theta e^{i\theta x} \frac{d}{d\theta} \ln S_a(\theta) \qquad (10.55)$$

where $a = $ o,e and $\mathscr{R}_{A,a}(y)$ is a rational function which is determined by the requirements specific for a given operator. For instance, as we have mentioned above, the singlet formfactor of $\sin \beta \Phi$ must have a pole at $\theta = i(\pi - \gamma)$. Therefore it is convenient to extract from $F_a(\theta)$ a function which does not have poles on the physical strip.

$$F_0(\theta) = -i \sinh(\theta/2) \exp \left\{ \int_0^{\infty} dx \frac{\sin^2[x(\theta + i\pi)/2] \sinh[(\pi - \gamma)x/2]}{x \sinh \pi x \cosh(\pi x/2) \sinh(\gamma x/2)} \right\}$$

$$(10.56)$$

At large θ we have $F_0(\theta \gg 1) \sim \exp(\theta)$.

Then the soliton–antisoliton formfactor of the $\cos \beta \Phi$-operator is

$$F_{s\bar{s}}^{\cos}(\theta) = Z^{1/2} \frac{i \cosh(\theta/2)}{\sinh[(\pi/2\gamma)(\theta - i\pi)]} F_0(\theta) \qquad (10.57)$$

where the normalization factor $Z \sim M$. This solution is 'minimal', that is contains the necessary pole and has the mildest asymptotic behaviour at infinity (Karowski *et al.* (1977, 1978)), Delfino and Mussardo (1996)).

For $\sin \beta \Phi$ we have

$$F_{s\bar{s}}^{\sin}(\theta) = Z^{1/2} \frac{\cosh(\theta/2)}{\cosh[(\pi/2\gamma)(\theta - i\pi)]} F_0(\theta) \qquad (10.58)$$

Using these formfactors and the rule (10.53) we can obtain formfactors for breathers. For the first breather we have

$$F_{b1}^{\sin} = \{ResS[\theta = i(\pi - \gamma)]\}^{-1/2} Res F_{s\bar{s}}^{\sin}[\theta = i(\pi - \gamma)] \equiv Z^{1/2} Z_1^{1/2}$$

$$Z_1^{1/2} = \frac{\gamma \sin \gamma}{\pi} \exp \left(- \int_0^{\infty} \frac{dx \sinh \gamma x/2 \sinh(\pi - \gamma)x/2}{x \sinh \pi x \cosh \pi x/2} \right) \qquad (10.59)$$

The formfactor (10.57) has a pole at the position of the second breather and so we have:

$$F_{b2}^{\cos} = \{ResS[\theta = i(\pi - 2\gamma)]\}^{-1/2} Res F_{s\bar{s}}^{\cos}[\theta = i(\pi - 2\gamma)] \equiv Z^{1/2} Z_2^{1/2}$$

$$Z_2^{1/2} = \frac{\gamma \sin 2\gamma}{\pi} \exp \left(-2 \int_0^{\infty} \frac{dx \sinh^2 \gamma x \sinh(\pi - \gamma)x/2}{x \sinh \pi x \cosh \pi x/2 \sinh \gamma x/2} \right) \qquad (10.60)$$

We are now in a position to describe single breather and kink–antikink contributions to the correlation functions of $\sin \beta \Phi$ and $\cos \beta \Phi$. For the sake of compactness we write down only the imaginary part of the corresponding retarded correlation functions:

$$\Im m D^{\sin}(\omega, q)$$

$$= Z \left\{ 2\pi \sum_{j=0} Z_{2j+1} \delta(s^2 - M_{2j+1}^2) + \Re e \frac{1}{s\sqrt{s^2 - 4M^2}} |F^{\sin}[\theta(s)]|^2 + ... \right\}$$

$$s^2 = \omega^2 - q^2, \quad \theta(s) = 2 \ln(s/2M + \sqrt{s^2/4M^2 - 1}) \tag{10.61}$$

and similarly

$$\Im m D^{\cos}(\omega, q)$$

$$= Z \left\{ 2\pi \sum_{j=1} Z_{2j} \delta(s^2 - M_{2j}^2) + \Re e \frac{1}{s\sqrt{s^2 - 4M^2}} |F^{\cos}[\theta(s)]|^2 + ... \right\}$$

$$\tag{10.62}$$

The dots here stand for multi-breather contributions which we have omitted.

In the sector $\gamma > \pi/2$ there are only kinks and antikinks, and the correlation functions have two-particle thresholds. We shall not discuss this sector in detail, referring the reader to Smirnov's book.

It is interesting to discuss how the shape of the spectral functions depends on β. According to Eqs. (10.61, 10.62) at low enough energies these spectral functions consist of series of delta functions originating from emissions of breathers, and two-particle thresholds. Let us imagine that the lowest threshold corresponds to emission of a kink–antikink pair (this imposes a certain constraint on β which we leave to the reader to derive). Let us discuss the shape of the spectral functions at the threshold where at $s - 2M \ll M$ we have

$$\theta(s) \approx \sqrt{8(s/2M - 1)} \tag{10.63}$$

At small θ we always have $F_0(\theta) \sim \theta$, but the trigonometric prefactors in Eqs.(10.57, 10.58) behave differently for different values of γ. Namely, the sine formfactor is finite at $\theta = 0$ when $\gamma = \pi/(1 + 2k)$ and the cosine formfactor is finite when $\gamma = \pi/2k$. Substituting (10.63) into Eqs. (10.57, 10.58) and then into Eqs. (10.61, 10.62) we obtain the following expressions for the spectral functions at $s - 2M \ll M$:

$$\Im m D^{\cos}(\omega, q) \sim \frac{\sqrt{s^2 - 4M^2}}{s^2/4M^2 - 1 + [(\gamma/\pi)\cos(\pi/2\gamma)]^2} \tag{10.64}$$

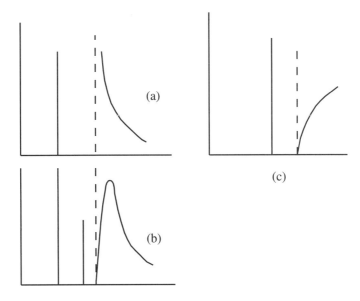

Fig. 10.4. The shape of the spectral function of sines at different values of γ.

$$\Im m D^{\sin}(\omega, q) \sim \frac{\sqrt{s^2 - 4M^2}}{s^2/4M^2 - 1 + [(\gamma/\pi)\sin(\pi/2\gamma)]^2} \qquad (10.65)$$

These expressions are applicable when $(\gamma/\pi)\cos(\pi/2\gamma)$ or $(\gamma/\pi)\sin(\pi/2\gamma)$ are small numbers.

Let us follow closely the change of the two-particle emission contribution upon changing the coupling constant γ. Without loss of generality we can choose the spectral function of sines. Let us start from $\gamma = \pi/(1+2k_1)$ where k_1 is some integer number. The spectral function has a $x^{-1/2}$-singularity at the threshold (see Fig. 10. 4(a)). At this point the spectrum includes $2k_1$ breathers and the last one, being even by number does not appear in the sin–sin correlation function. Instead its presence in the spectrum is signified by the square root singularity.

According to Eq. (10.65) this singularity is smeared out when γ decreases. However, as soon as $\gamma < \pi/(1 + 2k_1)$, a new breather emerges with mass $M_{(2k_1+1)} = 2M \sin[(2k_1 + 1)\gamma/2]$ and since its number is odd, it contributes to the sin–sin spectral function. Therefore the disappearance of the singularity is compensated by appearance of a new peak near the two-particle threshold (see Fig. 10.4(b)). When γ moves further towards $\pi/(2k_1 + 2)$ the splitting between this peak and the two-particle threshold grows and the spectral function at the threshold becomes featureless (Fig. 10.4(c)).

References

D. J. Amit, Y. Y. Goldschmidt and G. Grinstein, *J. Phys.* A**13**, 585 (1980).

P. W. Anderson, G. Yuval and D. R. Hamann, *Phys. Rev.* B**1**, 4464 (1970).

O. Babelon, D. Bernard and F. A. Smirnov, hep-th/9603010 (1996).

D. Bernard and A. Le Clair, hep-th/9402144 (1994).

R. F. Dashen, B. Hasslacher and A. Neveu, *Phys. Rev.* D**11**, 3424 (1975).

G. Delfino and G. Mussardo, *Nucl. Phys.* B**455**, 724 (1996).

I. E. Dzyaloshinskii and A. I. Larkin, *Soviet Phys. JETP* **34**, 422 (1972).

V. J. Emery, A. Luther and I. Peschel, *Phys. Rev.* B**13**, 1272 (1976).

G. E. Japaridze, A. A. Nersesyan and P. B. Wiegmann, *Nucl. Phys.* B**230**, 10 (1984).

J. V. Jose, L. P. Kadanoff, S. Kirkpatrick and D. R. Nelson, *Phys. Rev.* B**16**, 1217 (1977).

M. Karowski, H.-J. Thun, T. T. Truong, P. H. Weisz, *Phys. Lett.* B**67**, 321 (1977); M. Karowski and P. Weisz, *Nucl. Phys.* B**139**, 455 (1978).

J. Kogut, *Rev. Mod. Phys.* **51**, 659 (1979).

A. M. Polyakov, *ZhETF* **63**, 24 (1972).

N. Yu. Reshetikhin and F. A. Smirnov, *Comm. Math. Phys.* **131**, 157 (1990).

F. A. Smirnov, *J. Phys.* A**17**, L873 (1984); *ibid.*, A**19**, L575 (1985); *Nucl. Phys.* B**337**, 156 (1990);

L. A. Takhtadjan and L. D. Faddeev, *Sov. Teor. Math. Phys.* **25**, 147 (1975).

P. B. Wiegmann, *J. Phys.* C**11**, 1583 (1978); *Sov. Sci. Rev.* A**2**, 41 (1980).

A. B. Zamolodchikov, *Pisma ZhETP* **25**, 499 (1977).

A. B. Zamolodchikov and Al. B. Zamolodchikov, *Annals of Phys.(NY)* **120**, 253 (1979).

11

Spin $S = 1/2$ Heisenberg–Ising chain

In this chapter we are going to make a big step toward description of real physical systems. At the same time we are still not in applications and are continuing to develop technical aspects of the bosonization approach. The model we are going to discuss – the spin $S = 1/2$ Heisenberg chain – posseses a fortunate combination of properties: it is nontrivial, relatively simple and describes real physical systems. What is especially important for us in the present context is that the solution of this model by the bosonization approach has constituted a major development in perfecting the method. Therefore we have an opportunity to kill many birds with one stone – to learn about the method and to use the results to describe real quasi-one-dimensional magnets.

In fact, this model was the first one-dimensional quantum model solved exactly by a straightforward diagonalization of the Hamiltonian (Bethe (1931)). Bethe found the eigenvalues and eigenfunctions of the Hamiltonian which later on have been used to describe thermodynamics of the Heisenberg chain (Johnson *et al.* (1973)). However, the exact eigenfunctions are so complicated that they are almost useless for calculating the correlation functions. Therefore in the rest of the chapter we shall follow another route; namely, we shall use the bosonization approach which was applied for the first time to this model by Luther and Peschel (1975). In this approach one considers the low energy sector of the spin $S = 1/2$ Heisenberg–Ising chain, Eq. (1.1), where this model is equivalent to the Gaussian model (3.1).

We shall start our consideration of the XXZ spin chain with the case of *weak* anisotropy, $|\Delta| \ll 1$. Then the equivalent tight-binding model (1.4) describes *weakly interacting* Jordan–Wigner fermions. Under this assumption correlations at large distances are determined solely by the low-energy part of the spectrum. This means that only states close to the Fermi points $\pm k_F = \pm \pi/2a$ (a being the lattice constant) are important, and one can linearize the bare-particle spectrum $\epsilon(k) = J\cos ka$ in the vicinity of these points, as we have done in Chapter 2:

$$\epsilon(k) \simeq \mp v(k \mp k_F), \quad |k \mp k_F| \ll k_F \tag{11.1}$$

where $v = Ja$ is the Fermi velocity of the Jordan–Wigner fermions.

This procedure defines the continuum limit of the model in which the original lattice field ψ_j is replaced by a pair of slowly varying Fermi fields, $R(x)$ and $L(x)$, describing right-moving and left-moving particles:

$$\psi_j = \frac{1}{\sqrt{N}} \sum_{|k|<\pi/a} e^{ikx_j}\psi(k)$$

$$\rightarrow \sqrt{a}\,[(-i)^j R(x) + i^j L(x)] \tag{11.2}$$

where

$$R(x) = \frac{1}{\sqrt{L}} \sum_{|p|\ll\pi/a} e^{ipx}\psi(-k_F+p), \quad L(x) = \frac{1}{\sqrt{L}} \sum_{|p|\ll\pi/a} e^{ipx}\psi(k_F+p) \tag{11.3}$$

with $L = Na$ being the length of the system. The oscillatory factors $(\pm i)^j$ reflect the existence of two Fermi points in momentum space.

Although the decomposition of low-energy fermionic excitations into right- and left-movers is physically clear, one still may wonder how a *single* lattice field ψ_j gives rise to *two* continuous fields $R(x)$ and $L(x)$ when the limit $a \rightarrow 0$ is taken. This phenomenon is known as *fermion doubling*, and we refer the reader to the original publication by Fradkin and Susskind (1978) cited in the references for Chapter 12.

As follows from representation (1.3), the z-component of the spin-density operator of the XXZ chain coincides with the normal ordered density operator for the Jordan–Wigner fermions,

$$S_j^z =: \psi_j^\dagger\psi_j := \psi_j^\dagger\psi_j - 1/2.$$

Since in one-dimensional Fermi systems the Fermi 'surface' is reduced to two points $\pm k_F$, only Fourier components of the operator $\rho_q = \sum_k \psi_k^\dagger\psi_{k+q}$ with momenta q close to 0 and $\pm 2k_F$ describe low-energy density fluctuations. Consequently, in the continuum limit, the spin-density operator $S^z(x)$ is given by a sum of slow and rapidly oscillating (staggered) contributions:

$$S_j^z \rightarrow aS^z(x),$$
$$S^z(x) = \rho(x) + (-1)^j M(x) \tag{11.4}$$

Here

$$\rho(x) =\ : R^\dagger(x)R(x) :\ +\ : L^\dagger(x)L(x) :\ =\ J(x) + \bar{J}(x) \tag{11.5}$$

where $J, \bar{J}$ are the left and right Abelian currents satisfying the U(1) Kac–Moody algebra (see Chapter 2), and

$$M(x) =: R^\dagger(x)L(x) + L^\dagger(x)R(x) : \qquad (11.6)$$

Using decompositions (11.4) and (11.5), we obtain the continuuum version of the XXZ model

$$H = H_0 + H_{\text{int}},$$

$$H_0 = -iv \int dx(R^\dagger \partial_x R - L^\dagger \partial_x L), \qquad (11.7)$$

$$H_{\text{int}} = v\Delta \int dx[: \rho(x)\rho(x+a) : - M(x)M(x+a)] \qquad (11.8)$$

Let us now bosonize this Hamiltonian. As we already know from Chapter 2, a model of free massless fermions is equivalent to the model of free massless Bose field. Therefore

$$H_0 = \frac{v}{2} \int dx[\Pi^2 + (\partial_x \Phi)^2] \qquad (11.9)$$

where $\Pi(x)$ is the momentum conjugate to the field $\Phi(x)$, satisfying the canonical commutation relation

$$[\Phi(x,t), \Pi(x',t)] = i\delta(x - x') \qquad (11.10)$$

When bosonizing H_{int}, the limit $a \to 0$ can be safely taken in the first, $\rho\rho$-term of Eq. (11.8), and the smooth part of the density operator can be replaced by

$$\rho(x) = \frac{1}{\sqrt{\pi}} \partial_x \Phi(x) \qquad (11.11)$$

However, the second term in (11.8) should be treated with care. Representing $M(x)$ as

$$M(x) \simeq -\frac{1}{\pi a} : \sin \sqrt{4\pi}\Phi(x) : \qquad (11.12)$$

one can derive the following operator product expansion

$$\lim_{a\to 0} \left\{ \frac{1}{\pi a} : \sin \sqrt{4\pi}\Phi(x) : \right\} \left\{ \frac{1}{\pi a} : \sin \sqrt{4\pi}\Phi(x+a) : \right\}$$

$$= -\frac{1}{(\pi a)^2} \cos \sqrt{16\pi}\Phi(x) - \frac{1}{\pi}(\partial_x \Phi)^2 + \text{const} \qquad (11.13)$$

Using this relation, one finds the bosonized version of the XXZ model to be of the form (see Lukyanov (1997) for greater details)

$$H = \int dx \left\{ \frac{v}{2} \left[\Pi^2 + (1 + 4\Delta/\pi)(\partial_x \Phi)^2 \right] + \frac{v\Delta}{(\pi a)^2} : \cos \sqrt{16\pi}\Phi : \right\} \qquad (11.14)$$

The cosine term in (11.14) originates from Umklapp processes

$$R^\dagger(x)R^\dagger(x + a)L(x + a)L(x) + \text{H.c.}$$

'hidden' in the interaction of the staggered components of $S^z(x)$. These processes are allowed to appear because the band of the Jordan–Wigner fermions is $1/2$-filled, and $4k_F$ coincides with a reciprocal lattice vector (Haldane (1980); Black and Emery (1981); den Nijs (1981)). The Umklapp scattering breaks continuous chiral (γ^5) symmetry of the one-dimensional fermionic model (11.7) down to a discrete (Z_2) one. The latter can be broken spontaneously in two dimensions, and, in fact, at $\Delta = 1$ the system undergoes an Ising-like transition from the gapless disordered phase to a gapful, long-range ordered Neel state. However, in the whole range $|\Delta| \le 1$ the scaling dimension of the cosine perturbation is larger than 2. As was explained in Chapter 8, such a perturbation should be regarded as irrelevant and, therefore, can be taken into account perturbatively. This has been done by Eggert *et al.* (1994) who established an excellent agreement between the field theory and the exact solution.

The remaining Hamiltonian can be rewritten as follows

$$H = \frac{u}{2} \int dx \left[K\Pi^2 + \frac{1}{K}(\partial_x \Phi)^2 \right] \tag{11.15}$$

To transform it to the canonical form (11.9), one should rescale the field and momentum keeping the canonical commutator (11.10) preserved,

$$\Phi(x) \to \sqrt{K}\,\Phi(x), \quad \Pi(x) \to \frac{1}{\sqrt{K}}\Pi(x) \tag{11.16}$$

and renormalize the velocity

$$v \to u = \frac{v}{K} \tag{11.17}$$

At small Δ

$$K \simeq 1 - \frac{2\Delta}{\pi} + 0(\Delta^2)$$

Thus, the low-energy sector of the spin-$1/2$ XXZ chain is described by the Gaussian model with parameters u and K depending on the anisotropy Δ. The conformal dimensions of various operators are given by Eqs. (3.28). One should take into account, however, that operators local in terms of spins have integer conformal spins, and operators local in terms of fermions may have half-integer spins. In fact, the Gaussian model description, which we have derived in the limit $\Delta \ll 1$, holds throughout the entire critical region $-1 < \Delta \le 1$. The exact expressions for u and K can be extracted from the Bethe-ansatz solution of the spin-1/2 XXZ chain (Johnson *et al.* (1973)):

$$K = \frac{\pi}{2(\pi - \mu)}, \quad u = \frac{\pi \sin \mu}{2\mu} \tag{11.18}$$

According to the Abelian bosonization rules given in Table 3.1, with transformations (11.16) taken onto account, the spin-density operators are represented as

$$S^z(x) = \sqrt{\frac{K}{2\pi}} \partial_x \Phi(x) - \lambda_z (-1)^n \sin \sqrt{4\pi K} \, \Phi(x) \tag{11.19}$$

$$S^{\pm}(x) = \lambda_x (-1)^n \exp[\pm i \sqrt{\pi/K} \, \Theta(x)] \tag{11.20}$$

where $\lambda_z \lambda_x$ are given in Lukyanov (1997). In Eq. (11.20) we have retained the most singular, staggered part of the transverse magnetization.

Since the smooth part of $S^z(x)$ is proportional to the sum of the U(1) currents,

$$S^z_{\text{smooth}}(x) = \sqrt{K} [J(x) + \bar{J}(x)] \tag{11.21}$$

the decay of the corresponding correlation function is characterized by universal critical exponent 2:

$$\langle S^z_{\text{smooth}}(x, \tau) S^z_{\text{smooth}}(0, 0) \rangle$$

$$= K \left[\langle J(x, \tau) J(0, 0) \rangle + \langle \bar{J}(x, \tau) \bar{J}(0, 0) \rangle \right] = \frac{K}{4\pi^2} \left(\frac{1}{\bar{z}^2} + \frac{1}{z^2} \right) \tag{11.22}$$

where $z = u\tau + ix$, $\bar{z} = u\tau - ix$. Notice that under Wick's back rotation to (1 + 1)-dimensional space-time, this correlator is not Lorentz invariant, because $S^z_{\text{smooth}}(x)$ is proportional to a fixed (temporal) component j^0 of the 2-current $j^\mu \sim \epsilon^{\mu\nu} \partial_\nu \Phi$. On the other hand, the correlation functions of the staggered components of the magnetization are Lorentz invariant, with critical exponents continuously depending on the anisotropy parameter Δ (Luther and Peschel (1975)):

$$\langle S^z_{\text{stag}}(x, \tau) S^z_{\text{stag}}(0, 0) \rangle \sim \frac{1}{|z|^{\nu_z}} \tag{11.23}$$

$$\langle S^+_{\text{stag}}(x, \tau) S^-_{\text{stag}}(0, 0) \rangle \sim \frac{1}{|z|^{\nu_x}} \tag{11.24}$$

where

$$\nu_z = \frac{1}{\nu_x} = 2K \tag{11.25}$$

At the isotropic point, $\Delta = 1$, $K = 1/2$, the above correlation functions display manifestly SU(2)-invariant behavior, with $\nu_x = \nu_z = 1$. An alternative bosonization scheme for the Heisenberg (SU(2)-invariant) spin chain is discussed in Chapter 13.

I   Explicit expression for the dynamical magnetic susceptibility

The spin–spin correlation functions are directly measurable by inelastic neutron scattering. The neutron's differential cross section at energy transfer ω and wave vector q is proportional to the imaginary part of the Fourier image of the dynamical spin–spin correlation function. The latter is related to the retarded spin–spin correlation function $\chi^{(R)}(\omega, q)$:

$$\frac{d\sigma(\omega, k)}{d\Omega} \propto \frac{1}{1 - e^{-\omega/T}} \Im m \chi^{(R)}(\omega, k) \tag{11.26}$$

Since the bosonization approach usually provides us with correlation functions in space-time representation, it is convenient to use the following explicit expression where a retarded correlation function is written as a Fourier transformation of the space-time correlation function:

$$\Im m D^{(R)}(\omega) = \frac{1}{2} \int_{-\infty}^{\infty} dt [D_-(it + \epsilon) - D_+(it + \epsilon)] e^{i\omega t} \tag{11.27}$$

where the functions $D_\pm(\tau)$ are determined from the decomposition of the original thermodynamic Green's function:

$$D(\tau) = \theta(\tau)D_+(\tau) \pm \theta(-\tau)D_-(\tau) \tag{11.28}$$

For bosonic operators $D_-(\tau) = D_+(-\tau)$.

It is not really a problem for us to get an explicit expression for the thermodynamic spin–spin correlation function. As was explained at the beginning of Chapter 4, one should make a conformal transformation

$$z(\xi) = \exp(2\pi i T \xi)$$

Taking into account the operator relations (11.24) for the transverse spin components, we get the following expression for the thermodynamic correlation function of their staggered components:

$$\langle\langle S^+(\tau, x)S_-(0, 0)\rangle\rangle \propto \left\{ \frac{(\pi T)^2}{\sinh[\pi T(x - i\tau)]\sinh[\pi T(x + i\tau)]} \right\}^{2\Delta} (-1)^{x/a}$$

with $\Delta = 1/8K = \frac{1}{4}(1 - \mu/\pi)$.[*]

In order to find $D^+(\tau, q)$ and $D^-(\tau, q)$, it is convenient to calculate the Fourier transformation in x first ($q = \pi - k$, where k is a real wave vector):

$$D(\tau, q > 0) = \int_{-\infty}^{\infty} dx e^{iqx} \left\{ \frac{(\pi T)^2}{\sinh[\pi T(x - i\tau)]\sinh[\pi T(x + i\tau)]} \right\}^{2\Delta}$$

---

[*] Here and below the renormalized velocity of magnons is assumed to be $u = 1$.

Bending the contour of integration to the upper plane where the integrand has a cut we get

$$2\sin(2\pi\Delta) \int_{|\tau|}^{\infty} dy e^{-qy} \left\{ \frac{(\pi T)^2}{\sinh[\pi T(iy - i\tau)]\sinh[\pi T(iy + i\tau)]} \right\}^{2\Delta}$$

From this we find

$$D_+(\tau, q > 0) - D^-(\tau, q > 0)$$

$$= -2\sin(2\pi\Delta) \int_{-\tau}^{\tau} dy e^{-qy} \left\{ \frac{(\pi T)^2}{\sinh[\pi T(iy - i\tau)]\sinh[\pi T(iy + i\tau)]} \right\}^{2\Delta}$$

$$(11.29)$$

The latter expression allows the straightforward analytic continuation $\tau = it$. Finally we can write down the integral representation for $\Im m D(\omega, q)$:

$$-\Im m D(\omega, q > 0)$$

$$= \sin(2\pi\Delta) \int_{-\infty}^{\infty} dt e^{i\omega t} \int_{-|t|}^{|t|} dx e^{-iqx} \left\{ \frac{(\pi T)^2}{\sinh[\pi T(t - x)]\sinh[\pi T(x + t)]} \right\}^{2\Delta}$$

$$(11.30)$$

This integral is reduced to a stable one by the substitution $z = x + t$, $\bar{z} = t - x$:

$$-\Im m D(\omega, q > 0)$$

$$= \sin(2\pi\Delta)(\pi T)^{4\Delta} \Im m \int_0^{\infty} dz \frac{e^{iz(\omega - q)}}{[\sinh(\pi T z)]^{2\Delta}} \int_0^{\infty} d\bar{z} \frac{e^{i\bar{z}(\omega + q)}}{[\sinh(\pi T \bar{z})]^{2\Delta}}$$

$$\sim \sin(2\pi\Delta) \frac{1}{T^{2-4\Delta}} \Im m \left[\rho\left(\frac{\omega - q}{4\pi T}\right) \rho\left(\frac{\omega + q}{4\pi T}\right) \right] \qquad (11.31)$$

where $q = |\pi - k|$ and

$$\rho(x) = \frac{\Gamma(\Delta - ix)}{\Gamma(1 - \Delta - ix)}$$

The above result was first obtained by Schulz and Bourbonnais (1983) and Schulz (1986).

References

H. Bethe, *Z. für Phys.* **71**, 205 (1931).

J. L. Black and V. J. Emery, *Phys. Rev.* B**23**, 429 (1981).

M. den Nijs, *Phys. Rev.* B**23**, 6111 (1981).

S. Eggert, I. Affleck and M. Takahashi, *Phys. Rev. Lett.* **73**, 332 (1994).

F. D. M. Haldane, *Phys. Rev. Lett.* **45**, 1358 (1980); **47**, 1840 (1981).

J. D. Johnson, S. Krinsky and B. M. McCoy, *Phys. Rev.* A8, 2526 (1973).

A. Lukyanov, cond-mat/9712314 (1997).

A. Luther and I. Peschel, *Phys. Rev.* B12, 3908 (1975).

H. J. Schulz and C. Bourbonnais, *Phys. Rev.* B27, 5856 (1983); H. J. Schulz, *ibid.*, 34, 6372 (1986).

12
Ising model

The equivalence between two-dimensional models of classical statistics and quantum field theories in one space dimension has been known for a long time. In this chapter we shall consider in detail the relationship between the simplest statistical system, the two-dimensional Ising model, and a model of one-dimensional noninteracting fermions. Such a mapping established by Shultz, Mattis and Lieb (1964) by means of the transfer matrix method and the Jordan–Wigner transformation provides a simple and efficient tool to describe the universal properties of the Ising model near its critical point.

Let us first demonstrate how the transfer matrix method converts the two-dimensional Ising model to an effective quantum problem in one dimension. Consider a two-dimensional $N \times M$ lattice with Ising variables $\sigma_{nm} = \pm 1$, $n = 1, ..., N$; $m = 1, ..., M$ residing on its sites. Let K_x and K_τ be arbitrary dimensionless couplings between the nearest-neighbour Ising spins in the spatial (x) and temporal (τ) directions, the latter being actually the y-direction which, for later purposes, we identify as that of imaginary time. The lattice constants along the two directions are denoted by a and τ, and periodic boundary conditions are assumed. The partition function of the model is given by

$$Z = \sum_{\{\sigma_{nm}\}} \exp\left(-S[\sigma]\right) \tag{12.1}$$

where S is the Euclidean action (energy in units of the temperature),

$$S = -K_\tau \sum_{nm} \sigma_{nm}\sigma_{n,m+1} - K_x \sum_{nm} \sigma_{nm}\sigma_{n+1,m} \tag{12.2}$$

The two-constant Ising model (12.2) has a critical curve

$$\sinh 2K_\tau \sinh 2K_x = 1$$

along which the correlation length ξ_c diverges. In the isotropic case

101

$K_\tau = K_x$, spin correlations at large distances are rotationally invariant. In the limit of strong anisotropy, $K_\tau \gg K_x$, the correlations will be ellipsoidal, elongated in the τ-direction. To maintain the same long-distance physics in the two cases, one has to squeeze the lattice in the temporal direction by choosing $\tau \ll a$. Therefore, at $K_\tau \to \infty$, $K_x \to 0$ with $e^{2K_\tau}K_x$ kept finite, one can pass to the continuum limit in the τ-direction only, being sure that within such a description the critical point is accessible, and the universal scaling properties of the model remain unchanged. In this, the so-called 'τ-continuum' limit, the Ising model can be mapped onto an equivalent one-dimensional quantum problem in the most natural way (Ferrell (1973); Fradkin and Susskind (1978)).

Consider two neighbouring spatial rows, m and $m+1$, and relabel the Ising variables in these rows as $\sigma(n)$ and $s(n)$, respectively. Define a $2^N \times 2^N$ *transfer* matrix $T(\{s\},\{\sigma\})$ that couples the two rows as follows:

$$T(\{s\},\{\sigma\}) = \exp\left[-\frac{1}{2}K_\tau \sum_n (\sigma_n - s_n)^2 + \frac{1}{2}K_x \sum_n (\sigma_n \sigma_{n+1} + s_n s_{n+1})\right]$$

$$(12.3)$$

The partition function is then given by $Z = \mathrm{Tr}\hat{T}^M$. Define the τ-continuum limit by setting

$$K_x = \frac{1}{2}\tau, \quad e^{-2K_\tau} = \frac{1}{2}\lambda\tau, \quad \tau \to 0 \qquad (12.4)$$

with λ being an arbitrary dimensionless parameter. We expect that the quantum Hamiltonian we are looking for can be identified in this limit by representing the transfer matrix as

$$\hat{T} = 1 - \tau\hat{H} + 0(\tau^2) \simeq \exp(-\tau\hat{H}) \qquad (12.5)$$

One can easily check that, under conditions (12.4), keeping the accuracy $0(\tau^2)$, it is sufficient to take into account those matrix elements of $\hat{T}$ for which configurations of the two rows $\{\sigma\}$ and $\{s\}$ are either identical, or differ at most by one spin flip. Then the one-dimensional quantum Hamiltonian is given by

$$H = -\frac{J}{2}\sum_{n=1}^{N}(\sigma_n^z \sigma_{n+1}^z + \lambda\sigma_n^x) \qquad (12.6)$$

Here σ_n^α are the Pauli matrices, and J is an arbitrary constant introduced to fix the energy scale in the problem. The model described by the Hamiltonian (12.6) is called the *quantum Ising chain* (Pfeuty (1970)). Notice that it is the presence of the second, λ-proportional term in (12.6), having the meaning of a transverse magnetic field, that makes dynamics of the spins quantum.

The two-dimensional Ising model possesses the well-known Kramers–Wannier symmetry showing up in one-to-one correspondence between the high-temperature and low-temperature expansions of the partition function (see e.g. the book by Itsykson and Drouffe). At the level of the quantum Ising chain (12.6), this symmetry reveals itself in the following way. Introduce a set of sites $\{n + 1/2\}$ that defines a lattice dual to the original one, $\{n\}$. Define then dual variables $\mu^z_{n+1/2}$ and $\mu^x_{n+1/2}$ related to the original ones as follows

$$\mu^z_{n+1/2} = \prod_{j=1}^{n} \sigma^x_j, \quad \mu^z_{n-1/2}\mu^z_{n+1/2} = \sigma^x_n$$

$$\sigma^z_n = \prod_{j=0}^{n-1} \mu^x_{j+1/2}, \quad \sigma^z_n\sigma^z_{n+1} = \mu^x_{n+1/2} \tag{12.7}$$

In terms of the new variables the Hamiltonian preserves its original form

$$H = -\frac{J\lambda}{4} \sum_{n=1}^{N} (\mu^z_{n-1/2}\mu^z_{n+1/2} + \frac{1}{\lambda}\mu^x_{n+1/2}) \tag{12.8}$$

Comparing the Hamiltonians (12.6) and (12.8), one finds that each eigenvalue of H satisfies the relation $E(\lambda) = \lambda E(\lambda^{-1})$. The same relation must hold for the energy gap of the Hamiltonian H (the inverse correlation length of the two-dimensional Ising model). Therefore, since the critical point at which the gap vanishes exists and, for any fixed pair of couplings K_x and K_τ, is unique, it must coincide with the *self-duality* point $\lambda_c = 1$.

The case $\lambda < 1$ corresponds to the low-temperature ($T < T_c$) ordered phase of the Ising model, characterized by a nonzero expectation value of the *order* parameter, $\langle \sigma^z_n \rangle \neq 0$. The case $\lambda > 1$ corresponds to the high-temperature ($T > T_c$) disordered phase. By duality it then follows that, in this phase, $\langle \mu^z_{n+1/2} \rangle \neq 0$, so that the operator $\mu^z_{n+1/2}$ should be identified as the *disorder* parameter. As follows from the first relation of Eqs. (12.7), this operator is of topological nature, since when acting on the perfectly ordered vacuum state $|vac\rangle = \prod_n |\uparrow\rangle_n$ it creates a kink of the magnetization $\cdots \downarrow\downarrow\downarrow\downarrow\uparrow\uparrow\uparrow\uparrow \cdots$ at the point $x = n + 1/2$. The kinks introduce disorder; hence the nomenclature 'disorder variable' for $\mu^z_{n+1/2}$. The above, 'quantum' definition of the disorder operator is in correspondence with the general definition of the disorder parameter μ_{r^*} for the two-dimensional Ising model given by Kadanoff and Ceva (1971); in the latter case μ_{r^*} is defined as a magnetic dislocation line originating at the dual lattice site r^* and terminating at infinity.

The quantum Ising chain Hamiltonian (12.6) can be reduced to a tight-binding model of spinless fermions, using the Jordan–Wigner transformation:

$$\sigma_n^x = 2a^\dagger a_n - 1$$

$$\sigma_n^z = (-1)^{n-1} \exp[\pm i\pi \sum_{j=1}^{n-1} a_j^\dagger a_j] \, (a_n^\dagger + a_n) \tag{12.9}$$

(This representation differs from the conventional one, Eq. (1. 3), by the interchange of the $\hat{z}$ and $\hat{x}$ axes in spin space.) The Fermi operators a_n, $a_n^\dagger$ obey the standard anticommutation relations

$$\{a_n, a_m^\dagger\} = \delta_{nm}, \quad \{a_n, a_m\} = 0$$

Then the Hamiltonian H becomes quadratic:

$$H = \frac{J}{2} \sum_n [-(a_n^\dagger - a_n)(a_{n+1}^\dagger + a_{n+1}) + \lambda(a_n^\dagger - a_n)(a_n^\dagger + a_n)] \tag{12.10}$$

The structure of the Hamiltonian (12.10) immediately suggests introduction of two *lattice* Majorana (i.e. real) fields, $\zeta_1(n)$ and $\zeta_2(n)$, which are defined as the real and imaginary parts of the Jordan–Wigner fermion operators a_n

$$\zeta_1(n) = \frac{a_n^\dagger + a_n}{\sqrt{2}}, \quad \zeta_2(n) = \frac{a_n^\dagger - a_n}{\sqrt{2}i} \tag{12.11}$$

$$\zeta_j^\dagger(n) = \zeta_j(n) \quad (j = 1, 2)$$

and satisfy the anticommutation relations

$$\{\zeta_j(n), \zeta_{j'}(n')\} = \delta_{jj'}\delta_{nn'}, \quad \zeta_j^2(n) = \frac{1}{2} \tag{12.12}$$

Thus, the quantum Ising chain is equivalent to the following lattice model of free Majorana fermions:

$$H = -iJ \sum_n \{\zeta_2(n)[\zeta_1(n+1) - \zeta_1(n)] - (\lambda - 1)\zeta_2(n)\zeta_1(n)\} \tag{12.13}$$

At small deviations from criticality, $|\lambda - 1| \ll 1$, one can pass to the continuum limit which is defined as $a_0 \to 0$, $J \to \infty$, $\lambda \to 1$ with the parameters

$$c = Ja, \quad m = c(\lambda - 1)/a_0 \sim (T - T_c)/T_c \ll 1 \tag{12.14}$$

kept fixed. The lattice Majorana operators transform to slowly varying fields $\zeta_j(x)$,

$$\zeta_{1,2}(n) \to \sqrt{a_0} \, \zeta_{1,2}(x), \quad \{\zeta_j(x), \zeta_{j'}^\dagger(x')\} = \delta(x - x') \tag{12.15}$$

and the Hamiltonian acquires a formally relativistic form

$$H = \int dx \, [-ic\zeta_2(x)\partial_x\zeta_1(x) + im\zeta_2(x)\zeta_1(x)] \tag{12.16}$$

with c and m being the 'speed of light' and 'relativistic mass', respectively. The two-component spinor

$$\zeta = \begin{pmatrix} \zeta_1 \\ \zeta_2 \end{pmatrix}$$

satisfies the relativistic equation

$$(i\gamma^\mu \partial_\mu - m)\zeta(x) = 0 \qquad (12.17)$$

with the γ matrices

$$\gamma^0 = \hat{\tau}_2, \quad \gamma^1 = -i\hat{\tau}_3, \quad \gamma^5 = \gamma^0\gamma^1 = \hat{\tau}_1$$

Sometimes it is more convenient to work with the representation of the γ-matrices in which γ^5 is diagonal:

$$\gamma^0 = \hat{\tau}_2, \quad \gamma^1 = i\hat{\tau}_1, \quad \gamma^5 = \hat{\tau}_3$$

In this representation

$$H = \int dx \; \bar{\xi}(x)\left(-\frac{ic}{2}\hat{\tau}_3\partial_x + m\hat{\tau}_2 \right)\xi(x)$$

$$= \int dx \left[-\frac{ic}{2}(\xi_R\partial_x\xi_R - \xi_L\partial_x\xi_L) - im\xi_R\xi_L \right] \qquad (12.18)$$

where the spinors ξ and ζ are related by a chiral rotation

$$\xi_R = \frac{\zeta_1 + \zeta_2}{\sqrt{2}}, \quad \xi_L = \frac{-\zeta_1 + \zeta_2}{\sqrt{2}} \qquad (12.19)$$

There are two immediate consequences following from the very reduction of the two-dimensional Ising model to one-dimensional Majorana fermions and thus demonstrating the efficiency of this equivalence. First of all, one readily observes that the fermion mass $m \sim (T - T_c)/T_c$ sets a unique energy scale in the problem. By virtue of dimensionality, the correlation length ξ_c of the Ising model should be proportional to m^{-1}. Therefore near the critical point

$$\xi_c \sim \frac{T_c}{|T - T_c|} \qquad (12.20)$$

The second consequence is related to the logarithmic singularity of the specific heat of the Ising model near the critical point which can also be deduced from the fermionic Hamiltonian (12.18). The reader familiar with basic properties of massive fermions in one dimension will immediately associate this singularity with the $m^2 \ln |m|$ contribution to the free energy of the system generated by the mass term. To derive this result, let us elaborate more on the Majorana model (12.18).

Since the fields $\xi_{R,L}(x)$ are real, their Fourier expansion

$$\xi_j(x) = \frac{1}{L}\sum_{k>0}[\ \xi_j(k)e^{ikx} + \xi_j^\dagger(k)e^{-ikx}\ ], \quad j = R, L \tag{12.21}$$

involves complex operators $\xi_j(k)$ and $\xi_j^\dagger(k)$ which are only independent on the semiaxis $k > 0$. Such a restriction follows from the relation $\xi_j^\dagger(k) = \xi_j(-k)$ and guarantees that each independent mode is counted exactly once. On the semiaxis $k > 0$ $\xi_j(k)$ and $\xi_j^\dagger(k)$ satisfy the standard anticommutation relations

$$\{\xi_j(k), \xi_{j'}^\dagger(k')\} = \delta_{jj'}\delta_{kk'} \qquad (k, k' > 0)$$

The Hamiltonian (12.18) can be rewritten in a compact form

$$H = \sum_{k>0} \xi^\dagger(k)\ (k\hat{\tau}_3 + m\hat{\tau}_2)\ \xi(k) \tag{12.22}$$

where

$$\xi(k) = \begin{pmatrix} \xi_R(k) \\ \xi_L(k) \end{pmatrix}, \qquad \xi^\dagger(k) = (\xi_R^\dagger(k), \xi_L^\dagger(k))$$

The partition function can be represented as a path integral over Grassmann fields

$$Z = \int D\xi_R\ D\xi_L\ \exp(-S_M[\xi]) \tag{12.23}$$

with the Euclidean action of two-dimensional Majorana fermions (we have set the velocity $c = 1$)

$$S_M[\xi] = \sum_\varepsilon \sum_{k>0} \xi^\dagger(k, \varepsilon)\ (-i\varepsilon + k\hat{\tau}_3 + m\hat{\tau}_2)\ \xi(k, \varepsilon) \tag{12.24}$$

and the integration measure

$$D\xi_R\ D\xi_L = \prod_\varepsilon \prod_{k>0} d\xi_R^*(k, \varepsilon)d\xi_R(k, \varepsilon)d\xi_L^*(k, \varepsilon)d\xi_L(k, \varepsilon)$$

where ε is the Matsubara frequency. Close to the critical point the specific heat of the Ising model is proportional to

$$C \sim \frac{\partial^2}{\partial m^2}\ln Z = \frac{1}{Z}\frac{\partial^2 Z}{\partial m^2} - \left(\frac{1}{Z}\frac{\partial Z}{\partial m}\right)^2$$

which, as follows from (12.23) and (12.24), reduces to

$$C \sim \langle[\sum_{k>0}\sum_\varepsilon \xi^\dagger(k, \varepsilon)\hat{\tau}_2\xi(k, \varepsilon)]^2\rangle - \langle\sum_{k>0}\sum_\varepsilon \xi^\dagger(k, \varepsilon)\hat{\tau}_2\xi(k, \varepsilon)\rangle^2$$

$$= -\sum_{k>0}\sum_\varepsilon \text{Tr}\ \hat{G}(k, \varepsilon)\hat{\tau}_2\hat{G}(k, \varepsilon)\hat{\tau}_2 \tag{12.25}$$

Here $\hat{G}(k,\varepsilon)$ is the Majorana Green's function defined as a 2×2 matrix

$$G_{jj'}(k,\varepsilon) = -\langle \xi_j(k,\varepsilon)\xi_{j'}^*(k,\varepsilon)\rangle$$
$$= -Z^{-1}\int D\xi_R\, D\xi_L\, \xi_j(k,\varepsilon)\xi_{j'}^*(k,\varepsilon)\exp(-S[\xi]) \quad (12.26)$$

It is easily found to be equal to

$$\hat{G}(k,\varepsilon) = -\frac{i\varepsilon + k\hat{\tau}_3 + m\hat{\tau}_2}{\varepsilon^2 + k^2 + m^2}. \quad (12.27)$$

Using (12.25) and (12.27), one then finds that, in the massless limit $m \to 0$ (i.e. approaching the critical point), the specific heat per unit area is given by a two-dimensional logarithmically divergent integral:

$$C \sim \int \frac{d^2q}{q^2 + m^2} \sim \ln\frac{\Lambda}{|m|} \sim \ln\frac{T_c}{|T - T_c|} \quad (12.28)$$

(12.20) and (12.28) are the well-known results for the Ising model.

Turning back to the lattice definitions (12.7) and (12.9), we see that the order and disorder operators are nonlocal in Majorana fields $\zeta_{1,2}$:

$$\sigma_n^z = (-i)^{n-1}\sqrt{2}\exp\left(\pi\sum_{m=1}^{n-1}\zeta_1(m)\zeta_2(m)\right)\zeta_1(n) \quad (12.29)$$

$$\mu_{n+1/2}^z = (-i)^n\sqrt{2}\exp\left(\pi\sum_{m=1}^{n}\zeta_1(m)\zeta_2(m)\right) \quad (12.30)$$

On the other hand, the Majorana operators are realized as local products of the order and disorder parameters:

$$\sigma_n^z\mu_{n-1/2}^z = \mu_{n-1/2}^z\sigma_n^z = \sqrt{2}\,\zeta_1(n)$$
$$\sigma_n^z\mu_{n+1/2}^z = -\mu_{n+1/2}^z\sigma_n^z = -i\sqrt{2}\,\zeta_2(n) \quad (12.31)$$

From (12.31) one can extract more 'fusion' rules on the lattice:

$$\sigma_n^z\zeta_1(n) = \zeta_1(n)\sigma_n^z = \frac{1}{\sqrt{2}}\mu_{n-1/2}^z$$

$$\sigma_n^z\zeta_2(n) = -\zeta_2(n)\sigma_n^z = \frac{i}{\sqrt{2}}\mu_{n+1/2}^z \quad (12.32)$$

$$\mu_{n-1/2}^z\zeta_1(n) = \zeta_1(n)\mu_{n-1/2}^z = \frac{1}{\sqrt{2}}\sigma_n^z$$

$$\mu_{n+1/2}^z\zeta_2(n) = -\zeta_2(n)\mu_{n+1/2}^z = -\frac{i}{\sqrt{2}}\mu_{n+1/2}^z \quad (12.33)$$

The above formulas can be extended to account for arbitrary space separations between different operators. Then one obtains the following set of commutation relations:

$$\sigma^z(x)\mu^z(y) = \mu^z(y)\sigma^z(x)\text{sign}\ (x-y)$$
$$\sigma^z(x)\zeta(y) = \zeta(y)\sigma^z(x)\text{sign}\ (x-y)$$
$$\mu^z(x)\zeta(y) = -\zeta(y)\mu^z(x)\text{sign}\ (x-y) \tag{12.34}$$

There is a striking resemblance between the relations (12.34) and those appearing in the theory of the free Dirac field treated by Abelian bosonization method. According to the basic bosonization formula, the Dirac field operator is a local product of two phase exponentials depending on the scalar field $\Phi(x)$ and its dual counterpart $\Theta(x)$:

$$\psi_{R,L} \sim \exp\left(\pm i\sqrt{\pi}\Phi(x) - i\sqrt{\pi}\Theta(x)\right) \tag{12.35}$$

Using the commutation relations

$$[\Phi(x), \Phi(y)] = [\Theta(x), \Theta(y)] = 0, \quad [\Phi(x), \Theta(y)] = i\theta(y-x)$$

one easily finds that the algebra of operators $\exp[\pm i\sqrt{\pi}\Phi]$, $\exp[\pm i\sqrt{\pi}\Theta]$ and ψ is identical to the Ising model algebra (12.34) for σ, μ and ζ, respectively:

$$\exp[is\sqrt{\pi}\Phi(x)]\exp[is'\sqrt{\pi}\Theta(y)] = \exp[is'\sqrt{\pi}\Theta(y)]\exp[is\sqrt{\pi}\Phi(x)]\ \text{sign}(x-y)$$
$$\exp[is\sqrt{\pi}\Phi(x)]\psi(y) = \psi(y)\exp[is\sqrt{\pi}\Phi(x)]\ \text{sign}(x-y) \tag{12.36}$$
$$\exp[is'\sqrt{\pi}\Theta(x)]\psi(y) = -\psi(y)\exp[is'\sqrt{\pi}\Theta(x)]\ \text{sign}(x-y)$$

(s and s' take values ± 1). Comparing Eqs. (12.34) and (12.36) we observe that there should be an intimate relationship between the two-dimensional Ising model and the theory of free massive Dirac fermions, the latter, as we already know, being equivalent to the sine-Gordon model at the special value of the coupling constant $\beta^2 = 4\pi$. Such a relationship was explicitly constructed by Zuber and Itsykson (1977) (see also Schroer and Truong (1978)) who considered two noninteracting copies of the Ising model and thus built up a complex (Dirac) fermion out of two decoupled real (Majorana) fermions. Such a trick makes possible straightforward application of Abelian bosonization to the calculation of correlation functions in the Ising model. It then turns out that squared Ising correlators reduce to correlation functions of phase exponentials of the sine-Gordon model.

Following Zuber and Itsykson (1977), let us introduce two copies of the Majorana field, $\zeta^a(x)$ (or $\xi^a(x)$), $a = 1, 2$, and construct a Dirac field as follows:

$$\psi_j = \frac{\xi_j^1 + i\xi_j^2}{\sqrt{2}}, \quad j = R, L \tag{12.37}$$

Then the Hamiltonian of the doubled Majorana field takes the form of a

Hamiltonian of the single massive Dirac field:

$$H_D[\psi] = H_M[\xi^1] + H_M[\xi^2] = \int dx \; [-ic\psi^\dagger \hat{\tau}_3 \partial_x \psi + m\psi^\dagger \hat{\tau}_2 \psi] \quad (12.38)$$

This is the free massive Thirring model equivalent to the sine-Gordon model at $\beta^2 = 4\pi$:

$$H_D \to H_{SG} = \int dx \; \left\{ \frac{c}{2}[\Pi^2 + (\partial_x \Phi)^2] - \frac{m}{\pi a_0} \cos\sqrt{4\pi}\Phi \right\} \quad (12.39)$$

This correspondence provides a possibility to relate correlation functions of the Ising model to those of phase exponentials in the sine-Gordon model.

Let $\sigma_a^z(n)$, $a = 1, 2$, be the order parameters of the two identical Ising models. Since the two Ising systems are decoupled, the spin–spin correlation function of the original system can be squared and represented as

$$\mathscr{K}^2(n) \equiv \langle \sigma^z(n)\sigma^z(0) \rangle = \langle \sigma_1^z(n)\sigma_2^z(n)\sigma_1^z(0)\sigma_2^z(0) \rangle \quad (12.40)$$

Representing

$$\sigma_a^z(n)\sigma_a^z(0) = \prod_{m=0}^{n-1} \sigma_a^z(m+1)\sigma_a^z(m) = \prod_{m=0}^{n-1} 2i\zeta_1^a(m+1)\zeta_2^a(m) \quad (12.41)$$

we get

$$\mathscr{K}^2(n) = \langle F_2(0)\mathscr{F}(n-1)F_1(n) \rangle \quad (12.42)$$

where

$$F_j(m) = 2\zeta_j^1(m)\zeta_j^2(m) = \exp[\pi\zeta_j^1(m)\zeta_j^2(m)] \quad (12.43)$$

$$\mathscr{F}(n-1) = \prod_{m=1}^{n-1} F_1(m)F_2(m) = \exp[\pi \sum_{m=1}^{n-1}\sum_{j=1,2} \zeta_j^1(m)\zeta_j^2(m)] \quad (12.44)$$

Making the chiral rotation (12.19) from ζ^a to ξ^a and passing to the continuum limit, we obtain

$$2\zeta_{1,2}^1(n)\zeta_{1,2}^2(n) \to 2a_0 \; \zeta_{1,2}^1(x)\zeta_{1,2}^2(x)$$

$$= a_0[\xi_R^1(x)\xi_R^2(x) + \xi_L^1(x)\xi_L^2(x)] \mp a_0[\xi_R^1(x)\xi_L^2(x) + \xi_L^1(x)\xi_R^2(x)] \quad (12.45)$$

According to (12.37), the right-hand side of (12.45) can be expressed in terms of the Dirac fermion operators and then bosonized, using the standard rules (see Chapter 3). As a result

$$F_{1,2}(n) \to -ia_0[\; J(x) + \bar{J}(x) \;] \mp i[\; R^\dagger(x)L(x) + L^\dagger(x)R(x) \;]$$

$$= -\frac{ia_0}{\sqrt{\pi}}\partial_x\Phi(x) \mp \frac{i}{\pi}\sin\sqrt{4\pi}\Phi(x) \quad (x = na_0) \quad (12.46)$$

where $R(x)$ and $L(x)$ are the right and left chiral components of ψ, and J and $\bar{J}$ are the corresponding $U(1)$ currents. In the continuum limit, the operator $\mathcal{F}(n-1)$ in Eq. (12.44) transforms to a phase exponential of Φ:

$$\mathcal{F}(n-1) \rightarrow \exp\left\{-i\pi \int dy\, [J(y) + \bar{J}(y)]\right\}$$

$$= \exp\left\{-i\sqrt{\pi}[\Phi(x-a_0) - \Phi(a_0)]\right\} \qquad (12.47)$$

When expression (12.47) is substituted in (12.42) together with the end-points $F_2(0)$ and $F_1(n)$ given by Eqs. (12.43), one should carefully take the limit $a_0 \rightarrow 0$ by means of the following operator product expansions:

$$\partial_x \Phi(x)\, e^{\pm i\sqrt{\pi}\Phi(x+a)} = \pm i\sqrt{\pi}\frac{(ma)^{1/4}}{2\pi a}\; :e^{\pm i\sqrt{\pi}\Phi(x)}: \; [1+0(a)] \quad (12.48)$$

$$\sin\sqrt{4\pi}\Phi(x)\, e^{\pm i\sqrt{\pi}\Phi(x+a)} = \mp\frac{(ma)^{1/4}}{2i}\; :e^{\mp i\sqrt{\pi}\Phi(x)}: \; [1+0(a)] \quad (12.49)$$

Here we have used the well-known formula for normal ordering of phase exponentials in the sine-Gordon model (Coleman (1975)):

$$e^{i\lambda\Phi(x)} = (ma_0)^{\lambda^2/4\pi}\; :e^{i\lambda\Phi(x)}:$$

As a result of all above transformations, the correlation function being nonlocal in terms of fermionic operators (see Eqs. (12.42) – (12.44)) acquires a simple structure of a local two-point function in terms of the bosonic field Φ:

$$\mathcal{K}^2(x) = \frac{(ma_0)^{1/2}}{\pi}\langle : \sin\sqrt{\pi}\Phi(x) : \; : \sin\sqrt{\pi}\Phi(0) :\rangle \qquad (12.50)$$

Consequently, the following correspondence holds between the local product of two order parameters of the doubled Ising model and the scalar field of the related sine-Gordon model:

$$\sigma_1^z(x)\sigma_2^z(x) = \frac{(ma_0)^{1/4}}{\sqrt{\pi}}\; : \sin\sqrt{\pi}\Phi(x) : \qquad (12.51)$$

This representation is valid not only at the critical point ($m=0$), but at small deviations from criticality as well. At the critical point $\sin\sqrt{\pi}\Phi(x)$ is a primary field of the Gaussian field theory, with scaling dimension $1/4$. Therefore, $\mathcal{K}^2(x) \sim |x|^{-1/2}$, and the two-spin correlation function of the critical Ising model falls out as

$$\mathcal{K}(x) \sim |x|^{-1/4} \qquad (12.52)$$

Clearly, representation (12.51) allows us to estimate long-distance asymptotics for any multi-point correlation function of the order parameter.

Given representation (12.51) for the local product $\sigma_1^z\sigma_2^z$, it is easy to figure out similar representations for $\mu_1^z\mu_2^z$, $\sigma_1^z\mu_2^z$ and $\mu_1^z\sigma_2^z$, using duality arguments. Consider the Hamiltonian (12.18) for a single Majorana field ξ. Since $m \sim (T - T_c)/T_c$, the Kramers–Wannier duality transformation that interchanges the order and disorder parameters is related to the change of the sign of the mass, $m \to -m$. In the continuum description, this is achieved by changing the sign of one of the two chiral components of the spinor ξ, e.g.

$$\xi_R \to \xi_R, \quad \xi_L \to -\xi_L \tag{12.53}$$

Suppose first that the duality transformation (12.53) is applied to both copies of the Ising model, so that $\sigma_{1,2}^z \leftrightarrow \mu_{1,2}^z$. As a result, the Dirac spinor (12.37) undergoes a discrete γ^5-transformation $\psi \to \gamma^5 \psi$, i.e.

$$R \to R, \quad L \to -L \tag{12.54}$$

and the amplitude of the cosine term of the sine-Gordon model (12.39) changes its sign. According to formula (12.35), transformation (12.54) is equivalent to shifts of the Bose fields Φ and Θ

$$\Phi \to \Phi \pm \frac{\sqrt{\pi}}{2}, \quad \Theta \to \Theta \pm \frac{\sqrt{\pi}}{2} \tag{12.55}$$

Under (12.55), $\sin \sqrt{\pi}\Phi$ is changed by $\cos \sqrt{\pi}\Phi$, and we thus arrive at the bosonic representation for the local product of two disorder operators

$$\mu_1(x)\mu_2(x) = \frac{(ma_0)^{1/4}}{\sqrt{\pi}} : \cos \sqrt{\pi}\Phi(x) : \tag{12.56}$$

It is clear from (12.56) that at the critical point the two-point correlation function of the disorder operators, $\langle \mu^z(x)\mu^z(0) \rangle$, displays the same power-law decay as that in Eq. (12.52).

Suppose now that the duality transformation (12.53) is applied only to the first Ising copy. Under such transformation the two Ising systems occur on the opposite sides of the transition point, so that $\sigma_1^z \leftrightarrow \mu_1^z$ with σ_2^z and μ_2^z remaining unchanged. In particular,

$$\sigma_1^z \sigma_2^z \to \mu_1^z \sigma_2^z, \quad \mu_1^z \mu_2^z \to \sigma_1^z \mu_2^z \tag{12.57}$$

The right chiral component of ψ is unchanged, while the left component undergoes a particle-hole transformation

$$R \to R, \quad L \to L^\dagger \tag{12.58}$$

implying that the Bose fields Φ and Θ are interchanged:

$$\Phi \to -\Theta, \quad \Theta \to -\Phi \tag{12.59}$$

Notice that the sine-Gordon model (12.39) for the field Φ transforms to that for the dual field Θ.

Thus, from (12.51), (12.56) and (12.57) it follows that local products of
the order and disorder operators belonging to different Ising systems are
given by:

$$\mu_1(x)\sigma_2(x) = -\frac{(ma_0)^{1/4}}{\sqrt{\pi}} : \sin\sqrt{\pi}\Theta(x) : \qquad (12.60)$$

$$\sigma_1(x)\mu_2(x) = \frac{(ma_0)^{1/4}}{\sqrt{\pi}} : \cos\sqrt{\pi}\Theta(x) : \qquad (12.61)$$

Relations (12.60) and (12.61) can be alternatively derived from (12.51) and
(12.56) by making use of fusion rules (12.33), (12.33) and operator product
expansions similar to those in (12.49) (see e.g. Boyanovsky (1988)).

The established relations (12.51), (12.56), (12.60) and (12.61) which bring
to correspondence four products of the order and disorder operators of two
different Ising models, $\sigma_1\sigma_2, \mu_1\mu_2, \sigma_1\mu_2, \mu_1\sigma_2$, and four phase exponentials

$$\exp(\pm i\sqrt{\pi}\Phi), \quad \exp(\pm i\sqrt{\pi}\Theta)$$

of the related sine-Gordon model (12.39) can be easily understood, using
the following nonrigorous but simple arguments (Shelton *et al.* (1996)). As
we already demonstrated, the sine-Gordon model (12.39) is equivalent to
a model of two degenerate massive Majorana fermions, i.e. two decoupled
Ising models, with the fermionic mass $m \sim (T - T_c)/T_c$. At criticality (zero
fermionic mass) the above bilinears of the Ising variables have the same
conformal weight $(1/8, 1/8)$ as that of the phase exponentials for a free
bosonic field. Therefore there must be some correspondence between the
two groups of four operators which should also hold at small deviations
from criticality. To find this correspondence, notice that, as follows from
(12.39), at $m > 0$ $\langle\cos\sqrt{\pi}\Phi\rangle \neq 0$, while $\langle\sin\sqrt{\pi}\Phi\rangle = 0$. Since the case
$m > 0$ corresponds to the disordered phase of the Ising systems ($T > T_c$),
$\langle\sigma_1\rangle = \langle\sigma_2\rangle = 0$, while $\langle\mu_1\rangle = \langle\mu_2\rangle \neq 0$. At $m < 0$ (ordered Ising
systems, $T < T_c$) the situation is inverted: $\langle\cos\sqrt{\pi}\Phi\rangle = 0$, $\langle\sin\sqrt{\pi}\Phi\rangle \neq 0$,
$\langle\sigma_1\rangle = \langle\sigma_2\rangle \neq 0, \langle\mu_1\rangle = \langle\mu_2\rangle = 0$. This leads to formulas (12.51) and
(12.56).

Clearly, the exponentials of the dual field Θ must be expressed in terms
of $\sigma_1\mu_2$ and $\mu_1\sigma_2$. To find the correct correspondence, one has to take into
account the fact that a local product of the order and disorder operators
of a single Ising model results in the Majorana fermion operator (see Eqs.
(12.31)):

$$\xi^1 \sim \cos\sqrt{\pi}(\phi_+ + \theta_+) \sim \sigma_1\mu_1, \quad \xi^2 \sim \sin\sqrt{\pi}(\phi_+ + \theta_+) \sim \sigma_2\mu_2$$

Using relations (12.51) and (12.56), we find

$$\mu_1\mu_2 \cos\sqrt{\pi}\tilde{\phi} \pm \sigma_1\sigma_2 \sin\sqrt{\pi}\tilde{\phi} \sim \sigma_1\mu_1$$
$$\mu_1\mu_2 \cos\sqrt{\pi}\tilde{\phi} \pm \sigma_1\sigma_2 \sin\sqrt{\pi}\tilde{\phi} \sim \sigma_2\mu_2$$

The latter two equations lead unambiguously to relations (12.60) and (12.61).

It should be pointed out that the case of two identical Ising systems weakly coupled by a four-spin interaction corresponds to the Baxter model (Baxter (1972)) which in turn is related to the spin-1/2 quantum chain with fully broken spin-rotational symmetry – XYZ model. A field-theoretical approach to this model based on its mapping onto one-dimensional *interacting* fermions with subsequent bosonization has been developed by Luther and Peschel (1975) and Ogilvie (1981). This approach has proven to be very efficient in calculating critical exponents for the Baxter model from the correlation functions of the Tomonaga–Luttinger model. Thus it provides an alternative possibility to find a bosonic representation for the operators of the Ising model by considering the decoupling limit of the Baxter model when the latter reduces to two noninteracting Ising systems.

We conclude this section by writing down (without derivation) asymptotic expansions of two-point correlation functions for the order and disorder parameters of the *off-critical* two-dimensional Ising model in the limits $\tilde{r} \ll 1$ and $\tilde{r} \gg 1$, where $\tilde{r} = mr$. (Details of calculations can be found in Wu *et al.* (1976).) Assuming that $m > 0$ (i.e. $T > T_c$), the long-distance asymptotics ($\tilde{r} \gg 1$) are given by

$$\langle \sigma(r)\sigma(0) \rangle \equiv G_\sigma(\tilde{r}) = \frac{A_1}{\pi} K_0(\tilde{r}) + O(e^{-3\tilde{r}}) \tag{12.62}$$

$$\langle \mu(r)\mu(0) \rangle \equiv G_\mu(\tilde{r})$$

$$= A_1 \left\{ 1 + \frac{1}{\pi^2} \left[\tilde{r}^2 \left(K_1^2(\tilde{r}) - K_0^2(\tilde{r}) \right) - \tilde{r} K_0(\tilde{r}) K_1(\tilde{r}) + \frac{1}{2} K_0^2(\tilde{r}) \right] \right\} + O(e^{-4\tilde{r}}) \tag{12.63}$$

where A_1 is a nonuniversal parameter, and $K_n(\tilde{r})$ are the Bessel functions of imaginary argument.

In the short-distance limit, $\tilde{r} \ll 1$, the leading asymptotics of the correlation functions are of power law form:

$$G_\sigma(\tilde{r}) = G_\mu(\tilde{r}) = A_2 \tilde{r}^{-1/4} \tag{12.64}$$

The ratio of the constants A_1 and A_2 is a universal quantity involving the Glaisher's constant A:

$$\frac{A_2}{A_1} = 2^{-1/6} A^{-3} e^{1/4}$$
$$A = 1.282\,427\,129... \tag{12.65}$$

One can find information about Ising model correlation functions at finite temperatures in the paper by Leclair *et al.* (1996). Expressions for *n*-point correlation functions were derived by Abraham (1978).

Exercise. According to Eqs. (12.51) and (12.56) N-point correlation function of σs (μs) can be represented by a square root of the N-point correlation function of $\sin\sqrt{\pi}\Phi$ ($\cos\sqrt{\pi}\Phi$) fields. Use this property to calculate the four-point correlation function. Make sure you reproduce the result obtained from the conformal theory (6.30).

Exercise. Consider two Ising models with the same temperature T coupled together by a weak exchange interaction. Obtain the expression for the effective action as the Gaussian models perturbed by two cosine terms. Consider the case $T = T_c$ when one of the cosines vanishes. Using the results of Chapter 10, determine the excitation spectrum of the resulting sine-Gordon model.

References

D. Abraham, *Comm. Math. Phys.* **60**, 205 (1978).

R. J. Baxter, *Ann. Phys.* (N.Y.) **70**, 193 (1972).

D. Boyanovsky, *Nucl. Phys.* B (Proc. Suppl.) 5A, 20 (1988).

S. Coleman, *Phys. Rev.* **D11**, 2088 (1975).

R. A. Ferrell, *J. Stat. Phys.* **8**, 265 (1973).

E. Fradkin and L. Susskind, *Phys. Rev.* **D17**, 2637 (1978).

C. Itsykson and J.-M. Drouffe, *Statistical Field Theory*, Cambridge University Press, 1992, vol. 1, p. 61.

L. P. Kadanoff and H. Ceva, *Phys. Rev.* **B3**, 3918 (1971).

A. Leclair, F. Lesage, S. Sachdev and H. Saleur, cond-mat/9606104 (1996).

A. Luther and I. Peschel, *Phys. Rev.* **B12**, 3908 (1975).

M. C. Ogilvie, *Ann. Phys.* (N.Y.) **136**, 273 (1981).

P. Pfeuty, *Ann. Phys.* (N.Y.) **57**, 79 (1970).

B. Schroer and T. T. Truong, *Nucl. Phys.* **B144**, 80 (1978).

T. D. Schultz, D. C. Mattis and E. H. Lieb, *Rev. Mod. Phys.* **36**, 859 (1964).

D. G. Shelton, A. A. Nersesyan and A. M. Tsvelik, *Phys. Rev.* **B53**, 8521 (1996).

T. T. Wu, B. McCoy, C. A. Tracy and E. Barouch, *Phys. Rev.* **B13**, 316 (1976).

J. B. Zuber and C. Itsykson, *Phys. Rev.* **D15**, 2875 (1977).

13
More about the WZNW model

I Special cases

I.1 $SU_1(2)$ WZNW model as a Gaussian model

The possibility of an Abelian bosonization of the $SU_1(2)$ WZNW model stems from the fact that its conformal charge is equal to one: $C_{su(2),k=1}^{WZNW} = C_{boson} = 1$. Using the relations $(1/3)\, \mathbf{J}_{R(L)} \cdot \mathbf{J}_{R(L)} = J_{R(L)}^z J_{R(L)}^z$, $H_{su(2)}^0$ can be expressed in terms of J^z-currents only; introducing then a pair of canonical variables, ϕ_s and π_s, via

$$J^z + \bar{J}^z = \frac{1}{\sqrt{2\pi}}\partial_x\Phi_s, \quad J^z - \bar{J}^z = \frac{1}{\sqrt{2\pi}}\pi_s, \tag{13.1}$$

one finds

$$H_{su(2)}^0 \to H_B = \frac{v_s}{2}\int dx\ [\pi_s^2(x) + (\partial_x\phi_s(x))^2] \tag{13.2}$$

The price we pay for this simplification is the loss of spin rotational invariance in the bosonized structure of the spin currents: the J^x and J^y cannot be represented as simply as J^z, and require bosonization of the Fermi fields:

$$\{R, L\}_\alpha(x) \simeq (2\pi a_0)^{-1/2} \exp\left(\pm i\sqrt{4\pi}\ \varphi_{R,L;\alpha}(x)\right) \tag{13.3}$$

Linear combinations

$$\Phi_\alpha = \varphi_{R\alpha} + \varphi_{L\alpha}, \quad \Theta_\alpha = -\varphi_{R\alpha} + \varphi_{L\alpha}$$

constitute scalar fields Φ_α and their dual counterparts Θ_α introduced for each spin component. The fields describing the charge and spin degrees of freedom are defined as follows:

$$\phi_c = \frac{\Phi_\uparrow + \Phi_\downarrow}{\sqrt{2}}, \qquad \theta_c = \frac{\Theta_\uparrow + \Theta_\downarrow}{\sqrt{2}}$$

$$\phi_s = \frac{\Phi_\uparrow - \Phi_\downarrow}{\sqrt{2}}, \qquad \theta_s = \frac{\Theta_\uparrow - \Theta_\downarrow}{\sqrt{2}} \tag{13.4}$$

where $\partial_x \theta_{c,s} = \Pi_{c,s}$.

To bosonize $J^\pm, \bar{J}^\pm$ we use Table 3.1 to obtain:

$$\bar{J}^+ = R_\uparrow^\dagger R_\downarrow = \frac{1}{2\pi a_0} \exp[-i\sqrt{2\pi}(\phi_s - \theta_s)]$$

$$J^+ = L_\uparrow^\dagger L_\downarrow = \frac{1}{2\pi a_0} \exp[i\sqrt{2\pi}(\phi_s + \theta_s)] \tag{13.5}$$

Note that, as expected, the charge field ϕ_c does not contribute to the spin SU(2) currents. Moreover, despite the fact that definitions (13.3) contain the cut-off a_0 explicitly, the current–current correlation functions are cut-off independent and reveal the underlying SU(2) symmetry:

$$\langle J^a(x) J^b(x') \rangle = -\frac{\delta^{ab}}{4\pi^2} \frac{1}{(x - x')^2} \tag{13.6}$$

The SU(2) currents $\mathbf{J}(x), \bar{\mathbf{J}}(x)$ determine the smooth parts of the spin operators in the continuum limit. Namely, at $a_0 \to 0$

$$\mathbf{S}_n \to a_0 \mathbf{S}(x), \quad \mathbf{S}(x) = \mathbf{J}(x) + \bar{\mathbf{J}}(x) + (-1)^n \mathbf{n}(x) \tag{13.7}$$

where

$$\mathbf{n}(x) = R_\alpha^\dagger(x) \frac{\sigma_{\alpha\beta}}{2} L_\beta(x) + \text{H.c.} \tag{13.8}$$

is the staggered part of the local spin density.

When bosonizing the staggered magnetization, the (redundant) charge excitations emerge, since off-diagonal bilinears like $R^\dagger L$ and $L^\dagger R$ describe particle–hole *charge* excitations with momentum transfer $\pm 2k_F$. We find:

$$n^z = -\frac{1}{\pi a_0} \cos \sqrt{2\pi}\phi_c \ \sin \sqrt{2\pi}\phi_s$$

$$n^\pm = \frac{1}{\pi a_0} \cos(\sqrt{2\pi}\phi_c) \ \exp(\pm i\sqrt{2\pi}\theta_s)$$

Assuming that charge excitations have a finite energy gap m_c and being interested in the energy range $|E| \ll m_c$, we can replace the charge operator $\cos(\sqrt{2\pi}\phi_c)$ by its nonzero vacuum expectation value; we denote this (nonuniversal) value by $\lambda = \langle \cos(\sqrt{2\pi}\phi_c) \rangle$ and arrive at the bosonization formulas for $\mathbf{n}(x)$:

$$n^z(x) = -\frac{\lambda}{\pi a_0} : \sin\sqrt{2\pi}\phi_s(x) :$$

$$n^{\pm}(x) = \frac{\lambda}{\pi a_0} : \exp[\pm i\sqrt{2\pi}\theta_s(x)] : \qquad (13.9)$$

Since the $SU_1(2)$ WZNW model describes the continuous limit of the isotropic spin-1/2 Heisenberg chain, the presented procedure completes the bosonization of the spin operators for the isotropic Heisenberg chain. Notice that the critical dimensions of the smooth and staggered parts of the spin densities are different:

$$\dim J^a = 1, \quad \dim n^a = 1/2 \qquad (13.10)$$

Exercise. Starting from the definitions of the right and left fermionic currents

$$\bar{J}(x^-) =: R^{\dagger}(x^-)R(x^-) :, \quad J(x^+) =: L^{\dagger}(x^+)L(x^+) :, \quad (x^{\pm} = x^0 \pm x^1)$$

and using bosonization formulas for the fermion operators

$$R \simeq \frac{1}{\sqrt{2\pi a}}e^{i\sqrt{4\pi}\phi_R}, \quad L \simeq \frac{1}{\sqrt{2\pi a}}e^{-i\sqrt{4\pi}\phi_L}$$

apply operator product expansion to reproduce the standard bosonization formulas for the currents in the limit $a \to 0$:

$$\bar{J}(x^-) = -\frac{1}{\sqrt{\pi}}\partial_-\phi_R, \quad J(x^+) = -\frac{1}{\sqrt{\pi}}\partial_+\phi_L$$

I.2 $SU_2(2)$ WZNW model and the Ising model

The $SU_2(2)$ Kac–Moody algebra is defined by the following commutation relations:

$$[J^a(x), J^b(y)] = i\epsilon^{abc}\delta(x-y)J^c(y) + \frac{i\delta_{ab}}{2\pi}\delta'(x-y) \qquad (13.11)$$

This case is very special because, as we are going to show, it can be represented as a current algebra of free Majorana fermions. This can be conjectured from the fact that, according to Eq. (7.14) the WZNW central charge in this case ($N = 2, k = 2$) is equal to 3/2. Since a single Majorana mode carries central charge $C = 1/2$, we conclude that the $SU_2(2)$ WZNW model is equivalent to the theory of three species of massless Majorana fermions. It is straightforward to check that the currents being expressed in terms of a triplet of Majorana fermionic fields χ^a,

$$J^a_{R,L} = \frac{1}{2}i\epsilon^{abc}\chi^b_{R,L}\chi^c_{R,L} \qquad (13.12)$$

satisfy the algebra (13.11). Here we assume that the Majorana fields have the Hamiltonian (12.18) with $m = 0$, so that

$$\langle\langle \chi(z)\chi(0)\rangle\rangle = 1/2\pi z \tag{13.13}$$

As we know from Chapter 7, however, the current algebra (13.11) can be realized using four species of Dirac fermions with spin and flavour (or chain index) $R_{\sigma,n}$ ($n = 1,2$; $\sigma = \pm 1/2$). Naturally, it is interesting to know how these 'normal' fermions are related to these Majorana modes.

Let us recall some basic facts about our problem. The central charge of the model of U(2)×SU(2)-symmetric free massless fermions is equal to four. So the problem is equivalent to the Gaussian model of four bosonic fields. From the available fermions one can construct three sorts of currents: the U(1) charge current and $SU_2(2)$-symmetric spin and flavour currents respectively. Both spin and flavour sectors can be described by Majorana fermions which we denote χ^a ($a = 1,2,3$) and η^j ($j = 1,2,3$) respectively.

Let us show that one can express all possible chiral fermion bilinears in terms of χ and η. Consider right- (or left-, it does not matter) moving Dirac fermions transforming according to U(2)×SU(2) group. For each flavour we can write the spin current:

$$\bar{J}_n^a = R_{\sigma,n}^+ S_{\sigma,\sigma'}^a R_{\sigma',n} \tag{13.14}$$

The combination obeying the $SU_2(2)$ Kac–Moody algebra is $\mathbf{J} = \mathbf{J}_1 + \mathbf{J}_2$. The expression for $\mathbf{J}$ in terms of Majoranas is already available (13.12). The remaining problem is to express the 'wrong' combination $\mathbf{K} = \mathbf{J}_1 - \mathbf{J}_2$. Using the fact that $\mathbf{J}_1$, $\mathbf{J}_2$ mutually commute and separately satisfy the $SU_1(2)$ Kac–Moody algebra, we derive the following commutation relations:

$$[K^a(x), K^b(y)] = i\epsilon^{abc}\delta(x-y)J^c(y) + \frac{i\delta_{ab}}{2\pi}\delta'(x-y) \tag{13.15}$$

$$[J^a(x), K^b(y)] = i\epsilon^{abc}\delta(x-y)K^c(y) \tag{13.16}$$

Now taking into account Eq. (13.13), it is easy to check that these relations are satisfied if we assume

$$K^a = i\chi^a\eta^3 \tag{13.17}$$

The choice of the index of the flavour Majorana fermion corresponds to the fact that $K^a = L^+S^a\tau^3L$. The symmetry considerations dictate

$$L^+S^a\tau^jL = i\chi^a\eta^j \tag{13.18}$$

The sector with zero conformal spin is represented by the two multiplets of primary fields: $\Phi_{\alpha\beta}^{(1/2)}$ and $\Phi_{ab}^{(1)}$ with the conformal dimensions (3/16,

3/16) and (1/2, 1/2) respectively. These fields are 2×2 and 3×3 matrices – tensors realizing the representations of SU(2) group with isospins $S = 1/2$ and $S = 1$ respectively. The fact that these fields can be expressed in terms of the Ising model operators gives a key to construct the deformed theory. According to Zamolodchikov and Fateev (1986) the following identities hold:

$$\Phi^{(1/2)}(z, \bar{z}) = \sum_{j=0}^{3} \hat{\tau}_j G_j(z, \bar{z}) \tag{13.19}$$

$$G_0 = \sigma_1 \sigma_2 \sigma_3 + i\mu_1\mu_2\mu_3, \quad G_1 = \sigma_1\mu_2\mu_3 + i\mu_1\sigma_2\sigma_3$$
$$G_2 = \mu_1\sigma_2\mu_3 + i\sigma_1\mu_2\sigma_3, \quad G_3 = \mu_1\mu_2\sigma_3 + i\sigma_1\sigma_2\mu_3 \tag{13.20}$$

where $\hat{\tau}_j$ ($j = 1, 2, 3$) are Pauli matrices, σ_0 is the identity matrix and σ_a, μ_a ($a = 1, 2, 3$) are the order and disorder parameter fields of the Ising models. The $\Phi^{(1)}$-operator is just a tensor composed of fermion bilinears:

$$\Phi^{(1)}_{ab} = \chi_a \bar{\chi}_b \tag{13.21}$$

I.3 $SU_4(2)$ as a theory of two bosonic fields

The $SU_4(2)$ WZNW model has the central charge $C = 2$ which means that it can be represented as the Gaussian model with two species of bosonic fields. Such representation was suggested by Zamolodchikov and Fateev (1986) and by Fabrizio and Gogolin (1994):

$$J^x = \frac{\sqrt{2}}{\pi a} \cos(\sqrt{2\pi}\phi_s) \cos(\sqrt{6\pi}\phi_f), \quad J^y = -\frac{\sqrt{2}}{\pi a} \sin(\sqrt{2\pi}\phi_s) \cos(\sqrt{6\pi}\phi_f),$$

$$J^z = \sqrt{\frac{2}{\pi}} \partial_x \phi_s \tag{13.22}$$

where ϕ_i are chiral components of the corresponding bosonic fields.

I.4 $SU_{10}(2)$ as a theory of three bosonic fields

The $SU_{10}(2)$ WZNW model has the central charge $C = 5/2$. Obviously it can be represented as a theory of five Majorana fermions, but the current algebra can be represented even more economically. The corresponding representation was found by Sengupta and Kim (1996) and requires three free bosonic fields.

$$J^3 = \sqrt{20\pi}\partial\phi_1$$
$$J^\pm = \frac{1}{\pi a_0} \left[\sqrt{3} \cos \sqrt{4\pi}\phi_3 e^{\pm i\sqrt{4\pi/5}(\phi_1 + 2\phi_2)} + e^{i\sqrt{4\pi/5}(\phi_1 - 3\phi_2)} \right] \tag{13.23}$$

where ϕ_i are chiral components of the corresponding bosonic fields. Note that the currents involve the bosonic fields $\phi_{1,2}$ and a chiral Majorana fermion field $\cos \sqrt{4\pi}\phi_3$. Since the second Majorana fermion represented by the chiral field $\sin \sqrt{4\pi}\phi_3$ is not included, the overall central charge is not 3, but 5/2.

II   Deformation of the WZNW model and coset constructions

To illustrate the concept of a deformed WZNW model let us consider the $SU_2(2)$ model. As we know, it can be represented as a model of three massless Majorana fermions. These fermions realize the adjoint representation of the SU(2) group. Let us take two Majoranas and combine them into a Dirac fermion; the latter can be bosonized. As a result we obtain the following action:

$$S = \int d\tau dx \left[\frac{1}{2}\chi(\partial_\tau - i\partial_x)\chi + \frac{1}{2}\bar{\chi}(\partial_\tau + i\partial_x)\bar{\chi} + \frac{1}{8\pi}(\partial_\mu\phi)^2 \right] \qquad (13.24)$$

Here we introduce the notation $\chi \equiv \chi_3, \bar{\chi} \equiv \bar{\chi}_3$.

In this formulation the original SU(2) symmetry is hidden, as in the Abelian representation of the $SU_1(2)$ model. What is manifest is the fact that the SU(2) group contains the Abelian subgroup U(1). This can be expressed as follows:

$$H[SU_2(2)] = H[U(1); K = 1] + H_{\text{Ising}} \qquad (13.25)$$

Now we can deform the theory changing the constant K (do not confuse the capital K with the level of the WZNW model k). Naturally, the SU(2) symmetry is now lost, the remaining symmetry being U(1)$\times Z_2$. Then the deformed action is

$$S_{\text{def}} = \int d\tau dx \left[\frac{1}{2}\chi(\partial_\tau - i\partial_x)\chi + \frac{1}{2}\bar{\chi}(\partial_\tau + i\partial_x)\bar{\chi} + \frac{1}{8\pi K}(\partial_\mu\phi)^2 \right] \qquad (13.26)$$

where K is a parameter of deformation. At $K \neq 1$ the SU(2) symmetry is broken and the deformed theory has only combined $Z_2 \times$U(1) symmetry. There is a one-to-one correspondence between the primary fields of the SU(2) symmetric and the deformed theories. Using the correspondence between the Ising model operators and bosonic exponents (12.60, 12.61) we establish that the multiplet of four complex fields (13.19) splits into two multiplets with two complex fields in each:

$$(\sigma, \mu)\exp\left(\pm\frac{i}{2}\Phi\right), \quad \Delta_{(1/2,1)} = \frac{1}{16} + \frac{K}{8} \qquad (13.27)$$

$$(\sigma, \mu)\exp\left(\pm\frac{i}{2}\Theta\right), \quad \Delta_{(1/2,2)} = \frac{1}{16} + \frac{1}{8K} \qquad (13.28)$$

and the multiplet of nine real fields of $\Phi^{(1)}$ splits into four multiplets:

$$\chi\bar{\chi}, \quad \Delta_{(1,1)} = 1/2 \qquad (13.29)$$

$$\chi \exp\left[\pm\frac{i}{2}(\Phi - \Theta)\right] \quad \text{and} \quad \bar{\chi}\exp\left[\pm\frac{i}{2}(\Phi + \Theta)\right],$$

$$\Delta_{(1,2)} = \frac{1}{8}(\sqrt{K} + 1/\sqrt{K})^2 \tag{13.30}$$

$$\exp(+i\Theta), \quad \Delta_{(1,3)} = \frac{1}{2K} \tag{13.31}$$

$$\exp(\pm i\Phi), \quad \Delta_{(1,4)} = \frac{K}{2} \tag{13.32}$$

Here, as usual, Θ is the field dual to Φ.

With χ and Φ-fields being conveniently separated one can consider a curious problem where the U(1) current is set to be identically zero (gauged away):

$$J_{u(1)} = 0 \tag{13.33}$$

It is clear that the theory (13.26) with this constraint remains well defined and reduces to the Ising model. Of course, the reader may say that this is trivial. However, both deformation and gauging procedures (the latter being the limiting case of the former) can be generalized for the WZNW model on SU(N) group of an arbitrary level k, which is already nontrivial. Furthermore, the gauging procedure can be introduced for the WZNW model on any Lie group G which has an Abelian subgroup H. The gauged model called the *coset* model on G/H is defined by setting to zero all H-invariant currents of the original WZNW model. The reader can learn more about the coset construction from the book by Itsykson and Drouffe; the procedure for calculation of correlation functions of primary fields has been developed by Kogan *et al.* (1997). Recently interesting physical realizations of coset models were suggested by Andrei *et al.* (1995).

References

N. Andrei, M. R. Douglas and A. Jerez, cond-mat 9502082 (1995).

M. Fabrizio and A. O. Gogolin, *Phys. Rev.* B**52**, (1994).

I. I. Kogan, A. Lewis and O. A. Soloviev, *Int. J. Mod. Phys.* A, to appear (1997); hep-th 9703028.

A. M. Sengupta and Y. B. Kim, *Phys. Rev.* B**54**, 14 918 (1996).

A. B. Zamolodchikov and V. A. Fateev, *Yad. Fiz. [Sov. J. Nucl. Phys.]***43**, 1043 (1986).

14

Non-Abelian bosonization

I   WZNW model in the Lagrangian formulation

Despite the fact that we have already learned a lot about WZNW model, we still lack a procedure which would be as powerful for non-Abelian models, as Abelian bosonization for more simple ones. Such a procedure exists; to formulate it we need to have the WZNW model in its Lagrangian form. The Lagrangian form of the WZNW model was derived by Polyakov and Wiegmann (1983). We shall reproduce their derivation in the next Section.

Theorem.   *The Euclidean action for the Sugawara Hamiltonian*

$$\hat{H} = \frac{2\pi}{k + c_v} \sum_{a=1}^{D} \int dx [: J^a(x)J^a(x) : + : \bar{J}^a(x)\bar{J}^a(x) :]$$

where the currents J^a, $\bar{J}^a$ satisfy the Kac–Moody algebras for the group G with the central extension k defined by Eq. (7.4), is given by $S = kW(U)$, where U is a matrix from the fundamental representation of the group G and $W(U)$ is

$$W(U) = \frac{1}{16\pi} \int d^2x \text{Tr}(\partial_\mu U^{-1} \partial_\mu U) + \Gamma[U] \quad (14.1)$$

$$\Gamma[U] = -\frac{i}{24\pi} \int_0^\infty d\xi \int d^2x \epsilon^{\alpha\beta\gamma} \text{Tr}(U^{-1}\partial_\alpha U U^{-1}\partial_\beta U U^{-1}\partial_\gamma U) \quad (14.2)$$

It is supposed that the field $U(\xi, \mathbf{x})$ in Eq. (14.2) is defined on a three-dimensional hemisphere whose boundary coincides with the two-dimensional plane where the original theory is defined, so that $U(\xi = 0, \mathbf{x}) = \mathbf{U}(\mathbf{x})$. The appearance of the third coordinate may look mysterious, but close inspection reveals that the integral is, in fact, ξ-independent (or rather *almost* independent). The reason for this is that the integrand in (14.2) is a total derivative – a Jacobian of transformation from three dimensional plane coordinates to the coordinates of the group. The first remarkable fact about Lagrangian (14.1) is that it has local equations of motion:

$$\bar{\partial}\bar{J} = 0, \; \bar{J} = \frac{k}{2\pi} U^{-1} \partial U,$$

$$\partial J = 0, \; J = -\frac{k}{2\pi} U \partial U^{-1} \tag{14.3}$$

In fact, as we shall show later, the currents $J, \bar{J}$ are the WZNW currents introduced in Chapter 7, i.e. they have the same correlation functions as the currents obeying the Sugawara Hamiltonian of the Theorem.

The easiest way to obtain the equations of motion (14.3) is to use the following identity:

$$W(gU) = W(U) + W(g) + \frac{1}{2\pi} \int d^2x \mathrm{Tr}(g^{-1}\bar{\partial}gU\partial U^{-1}) \tag{14.4}$$

We suggest to the reader to check the validity of this identity by the direct substitution. The meaning of Eq. (14.4) is easy to grasp considering its Abelian limit. Let $U = \exp(\mathrm{i}\phi)$, $g = \exp(\mathrm{i}\eta)$, where ϕ, η are scalar fields. In this case the daunting Γ-term vanishes identically and the WZNW action (14.1) becomes

$$W[\exp(\mathrm{i}\phi)] = \frac{1}{16\pi} \int d^2x(\partial_\mu\phi\partial_\mu\phi)$$

Substituting this into the identity (14.4) we get

$$\frac{1}{16\pi} \int d^2x[\partial_\mu(\phi + \eta)\partial_\mu(\phi + \eta)] =$$

$$\frac{1}{16\pi} \int d^2x(\partial_\mu\phi\partial_\mu\phi) + \frac{1}{16\pi} \int d^2x(\partial_\mu\eta\partial_\mu\eta) + \frac{1}{8\pi} \int d^2x(\partial_\mu\eta\partial_\mu\phi)$$

In order to appreciate the necessity of introduction of the auxiliary variable ξ in Eq. (14.2), let us consider the case when $G = SU(2)$. There is a simple parametrization for the fundamental representation of the $SU(2)$ which allows us to express the action explicitly in terms of the group coordinates. We mean the Euler parametrization:

$$U = \exp\left(\frac{\mathrm{i}}{2}\phi\sigma^z\right) \exp\left(\frac{\mathrm{i}}{2}\theta\sigma^x\right) \exp\left(\frac{\mathrm{i}}{2}\psi\sigma^z\right)$$

$$= \begin{pmatrix} \cos(\theta/2)\exp[\mathrm{i}(\phi + \psi)/2] & \mathrm{i}\sin(\theta/2)\exp[\mathrm{i}(\phi - \psi)/2] \\ \mathrm{i}\sin(\theta/2)\exp[\mathrm{i}(\psi - \phi)/2] & \cos(\theta/2)\exp[-\mathrm{i}(\phi + \psi)/2] \end{pmatrix} \tag{14.5}$$

where σ^a are the Pauli matrices. The group coordinates ψ, θ, ϕ belong to the manifold which is a sphere of radius 2π whose boundary is equivalent to its central point. The matrix U is invariant with respect to the transformations of coordinates $\phi \to \phi + 2\pi$, $\psi \to \psi + 2\pi$, $\theta \to \theta$.

We advise the reader to prove the following valuable identities:

$$\Omega_\mu^3 \equiv -i\text{Tr}[\sigma^z U^{-1}\partial_\mu U] = \partial_\mu\phi + \cos\theta\partial_\mu\psi$$

$$\Omega_\mu^\pm \equiv -i\text{Tr}[\sigma^\pm U^{-1}\partial_\mu U] = e^{\pm i\phi}[\partial_\mu\theta \pm i\sin\theta\partial_\mu\psi] \qquad (14.6)$$

With these identities we can express $iU^{-1}\partial_\mu U$ as follows:

$$-iU^{-1}\partial_\mu U = \frac{1}{2}\sigma_a\Omega_\mu^a \qquad (14.7)$$

Substituting the latter expression into the Eq. (14.2) we find that the integrand is equal to the Jacobian of transformation from ψ, θ, ϕ to ξ, x:

$$\Gamma_{\text{su}(2)} = -\frac{i}{96\pi}\int_0^\infty d\xi \int d^2 x \epsilon^{\alpha\beta\gamma}\epsilon_{abc}\Omega_\alpha^a\Omega_\beta^b\Omega_\gamma^c$$

$$= \frac{i}{4\pi}\int d\xi d^2 x \frac{\partial(\phi,\theta,\psi)}{\partial(\xi,x_1,x_2)} = \frac{i}{4\pi}\int d^2 x \epsilon_{\mu\nu}\phi\sin\theta\partial_\mu\theta\partial_\nu\psi \qquad (14.8)$$

As we see, the result of integration cannot be expressed unambiguously in terms of $U(x)$; U is periodic in ϕ, but $\Gamma_{\text{su}(2)}$ is not: when ϕ changes by $2\pi q$ it changes by

$$\delta\Gamma_{\text{su}(2)} = \frac{iq}{2}\int d^2 x \epsilon_{\mu\nu}\sin\theta\partial_\mu\theta\partial_\nu\psi \qquad (14.9)$$

The latter integral is a topological invariant; it is an integer counting the number of times the vector field $\mathbf{n} = (\cos\theta, \sin\theta\cos\psi, \sin\theta\sin\psi)$ covers the sphere.

Thus the WZNW action (14.1) is what is called *multivalued* functional. One may wonder how we can use such an ambiguously defined thing as an action of a field theory. However, as it often happens in quantum theory, we will turn this ambiguity to our advantage. Indeed, the fact that the action (14.1) is defined modulo $2\pi i$ does not prevent the equations of motion (14.3) from being local. Neither does it prevent the partition function from being well defined, *provided k is an integer*. This means that the Γ-functional can become a part of an action of a physical theory only if it has an integer coefficient k. In other words, the multivaluedness of the action yields the quantization condition for k! Needless to say, k does not renormalize.

From the previous chapters we are already familiar with primary fields of the WZNW model. In the Lagrangian formulation we identify them with the following matrix fields.

(i) U-field – a primary field transforming according to the fundamental representation of the group G. If $G = SU(N)$ ($c_v = N$) its scaling dimension is equal to (see Eq. (7.22))

$$\Delta_U = \bar{\Delta}_U = \frac{N^2 - 1}{2N(N+k)} \tag{14.10}$$

(ii) $\phi_1^{ab} = \text{Tr}(U^{-1}t^a U t^b)$ where t^a are generators of G. This field is also a primary one and belongs to the adjoint representation; its scaling dimensions are equal to

$$\Delta_1 = \bar{\Delta}_1 = \frac{c_v}{c_v + k} \tag{14.11}$$

(iii) The 'wrong currents' $K^a \sim \text{Tr}(t^a U^{-1}\partial U)$, $\bar{K}^a \sim \text{Tr}(t^a \bar{\partial} U U^{-1})$ whose scaling dimensions are equal to $(\Delta_1 + 1, \Delta_1)$ and $(\Delta_1, \Delta_1 + 1)$, respectively. These fields are not primary.

(iv) The Lagrangian density $\text{Tr}(\partial_\mu U^{-1}\partial_\mu U)$ is also not a primary field and has scaling dimensions equal to $(\Delta_1 + 1, \Delta_1 + 1)$. As we see, the Lagrangian density is an irrelevant operator. Therefore the general theory with the action

$$W(\lambda; U) = \frac{1}{2\lambda} \int d^2x \text{Tr}(\partial_\mu U^{-1}\partial_\mu U) + k\Gamma[U] \tag{14.12}$$

scales to the critical point $\lambda = 8\pi/k$. The crossover was described exactly by Polyakov and Wiegmann who solved this model by the Bethe ansatz (1983). In fact, it would be proper to call this model the Wess–Zumino–Novikov–Witten model; then the theory described by the action (14.1) is 'the WZNW model at critical point' or *critical WZNW model*. Using the above results we can compile the following table (Table 14.1).

The presence of the topological term in the action of WZNW model (14.12) is strictly necessary for existence of the critical point. Once this term is removed, the model becomes an asymptotically free theory.

II   Derivation of the Lagrangian

Let us consider relativistic fermions described by creation and annihilation operators $\psi_{n\alpha}^+$, $\psi_{n\alpha}$ ($n = 1, ..., k$; $\alpha = 1, ..., N$) interacting with a non-Abelian gauge field

$$A_\mu^{\alpha\beta} = A_\mu^a \tau_a^{\alpha\beta}$$

where τ_a are matrices-generators of some Lie group G. These matrices act only on the Greek indices; the interaction is diagonal with respect to the

Table 14.1.

Critical U(N) WZNW-model with $k = 1$	Massless Dirac Fermions
Action	Action
$\frac{1}{2}\int d^2x(\partial_\mu\Phi)^2 + W[g]; g \in SU(N)$	$2\int d^2x(R_\alpha^+\partial_{\bar{z}}R_\alpha + L_\alpha^+\partial_z L_\alpha)$
Operators	Operators
$1/(2\pi a)\exp\left[i(N/4\pi)^{1/2}\Phi\right]g_{\alpha\beta}$	$R_\alpha^+ L_\beta$
$i(N/4\pi)^{1/2}\partial\Phi$	$\frac{1}{2} : L_\alpha^+ L_\alpha :$
$-i(N/4\pi)^{1/2}\bar{\partial}\Phi$	$\frac{1}{2} : R_\alpha^+ R_\alpha :$
$1/2\pi\mathrm{Tr}(\tau^a g^{-1}\bar{\partial}g)$	$: R_\alpha^+\tau_{\alpha\beta}^a R_\beta :$
$-1/2\pi\mathrm{Tr}[\tau^a g\partial g^{-1}]$	$: L_\alpha^+\tau_{\alpha\beta}^a L_\beta :$
Critical O(N) WZNW-model with $k = 1$	Massless Real Fermions
Action	Action
$W[g]; g \in O(N)$	$2\int d^2x\sum_{\alpha=1}^N$ $(\chi_{R,\alpha}\partial_{\bar{z}}\chi_{R,\alpha} + \chi_{L,\alpha}\partial_z\chi_{L,\alpha})$
Operators	Operators
$1/(2\pi a)g_{\alpha\beta}$	$\chi_{R,\alpha}\chi_{L,\beta}$
$1/2\pi\left(g^{-1}\bar{\partial}g\right)_{\alpha\beta}$	$: \chi_{R,\alpha}\chi_{R,\beta} :$
$-1/2\pi\left(g\partial g^{-1}\right)_{\alpha\beta}$	$: \chi_{L,\alpha}\chi_{L,\beta} :$

English indices. The fermionic part of the action has the standard form:

$$S_f = \int d^2x\sum_{n=1}^k \bar{\psi}_{n\alpha}\gamma_\mu(\partial_\mu\delta_{\alpha\beta} + iA_\mu^{\alpha\beta})\psi_{n\beta} \tag{14.13}$$

Let us integrate over the fermions and consider the effective action for the gauge field:

$$\exp\{-S_{eff}[A]\} = Z^{-1}\int D\bar{\psi}D\psi e^{-S_f[\bar{\psi},\psi,A]} \equiv e^{kW[A]} \tag{14.14}$$

$$W[A] = \ln\det[\gamma_\mu(\partial_\mu\delta_{\alpha\beta} + iA_\mu^{\alpha\beta})] - \ln\det(\gamma_\mu\partial_\mu) \tag{14.15}$$

The field interacts only with the SU(N) currents; using the results of the previous chapter on the non-Abelian bosonization, we can separate the relevant degrees of freedom and rewrite the above effective action as an average over the WZNW action:

$$e^{-S_{eff}[A]} = \langle \exp\left(-i \int d^2 x A_\mu^a J^{a,\mu}\right)\rangle_W \tag{14.16}$$

At the present stage we are not yet in a position to write down an explicit expression for this average in path integral form. The reason is that we do not know the expression for the WZNW-Lagrangian. In order to find it, we need to do some preparatory work.

Our first step is to calculate $W[A]$. Let us define the current:

$$j_\mu(x) = ik \frac{\delta W[A]}{\delta A_\mu(x)}$$

The functional $W[A]$ is completely determined by the two equations satisfied by its derivatives $j_\mu(x)$:

$$\partial_\mu j_\mu + i[A_\mu, j_\mu] = 0 \tag{14.17}$$

$$\epsilon_{\mu\nu}(\partial_\mu j_\nu + i[A_\mu, j_\nu]) = -\frac{k}{2\pi}\epsilon_{\mu\nu}F_{\mu\nu} \tag{14.18}$$

where

$$F_{\mu\nu} = \partial_\mu A_\nu - \partial_\nu A_\mu + i[A_\mu, A_\nu]$$

In the operator language these equations should be understood as equations for motion for the current operators. In the path integral approach we understand them as identities for the correlation functions. Eqs. (14.17) and (14.18) were derived by Johnson (1963) and solved by Polyakov and Wiegmann (1983). The first one, Eq. (14.17), follows from the gauge invariance and expresses the fact that the non-Abelian charge

$$Q^a = \int dx_1 J_0^a(x)$$

is a conserved quantity. The total charge is the sum of charges created by left- and right-moving particles. Naively one can expect that for a massless theory, where there are no transitions between fermionic states with different chirality, right and left charges conserve independently. However, as we have already seen in Chapter 2, this expectation is false because the measure is invariant only with respect to simultaneous gauge transformations of right and left particles. The phenomenon of nonconservation of chiral components of the charge is called *chiral anomaly*. As was shown by Johnson, the difference between right and left charges is proportional to the total flux of the gauge field and Eq. (14.18) expresses this fact. Thus we have two equations (14.17) and (14.18) for two unknown functions j_μ and thus can attempt to solve this system (see also Polyakov (1989)). In order to do it we rewrite these equations in a more convenient form, namely, in the light cone coordinates. The corresponding notations are:

$$\bar{j} = j_0 + ij_x, \; j = j_0 - ij_x$$

$$\bar{A} = A_0 + iA_x, \; A = A_0 - iA_x;$$

$$\partial_\tau = \partial + \bar{\partial}, \partial_x = i(\partial - \bar{\partial})$$

The equations (14.17) and (14.18) are gauge invariant; we shall solve them in a particular gauge:

$$\bar{A} = 0, \; A = -2ig\partial g^{-1} \tag{14.19}$$

where g is a matrix from the G group. In these notations Eqs. (14.17) and (14.18) acquire the following form:

$$\partial \bar{j} + \frac{i}{2}[A, \bar{j}] + \bar{\partial} j = 0 \tag{14.20}$$

$$\partial \bar{j} + \frac{i}{2}[A, \bar{j}] - \bar{\partial} j = -\frac{ik}{\pi} \bar{\partial} A \tag{14.21}$$

Substituting $\partial \bar{j} + \frac{1}{2}[A, \bar{j}]$ from the first equation into the second one we find

$$j = \frac{ik}{2\pi} A = \frac{ik}{\pi} g\partial g^{-1} \tag{14.22}$$

Then the first equation transforms into

$$\partial \bar{j} + [g^{-1}\partial g, \bar{j}] + \frac{ik}{\pi} \bar{\partial} \left(g\partial g^{-1} \right) = 0 \tag{14.23}$$

whose solution is given by

$$\bar{j} = -\frac{ik}{\pi} g\bar{\partial} g^{-1} \tag{14.24}$$

In order to find the effective action $W[A]$ one has to integrate the equation

$$k\delta W = \frac{i}{2} \int d^2 x \mathrm{Tr}(\bar{j}\delta A) = -\frac{k}{2\pi} \int d^2 x \mathrm{Tr}[\delta gg^{-1}\bar{\partial}(g\partial g^{-1})] \tag{14.25}$$

(in the derivation of this expression we have used the invariance of trace under cyclic permutations and the identity $\delta g^{-1} = -g^{-1}\delta gg^{-1}$).

Exercise. Check that the solution of this problem is given by the following functional described by Novikov (1982) and Witten (1984).[*]

---

[*] Do not forget that in our notations $4\partial\bar{\partial} = \partial_\mu^2$ and $\mathrm{Tr}\tau^a\tau^b = 1/2\delta_{ab}$.

To prove this theorem we use identity (14.4). Let us show that the integral (14.16) can be written as the path integral with the WZNW-action:

$$e^{-S_{eff}[A]} = \langle \exp(-i \int d^2x A_\mu^a J^{a,\mu}) \rangle_w =$$

$$\frac{\int [U^{-1}DU] \exp[-kW(U) - (k/2\pi) \int d^2x \mathrm{Tr}(U^{-1}\bar{\partial}Ug\partial g^{-1})]}{\int [U^{-1}DU] \exp[-kW(U)]} \quad (14.26)$$

where $[U^{-1}DU]$ means the measure of integration on the group G. Now using the identity (14.4) and invariance of the measure with respect to the transformations

$$U \to Ug^{-1}$$

we can rewrite the integral as follows:

$$e^{-S_{eff}[A]} = \frac{\int [U^{-1}DU] \exp[kW(g) - kW(Ug)]}{\int [U^{-1}DU] \exp[-kW(U)]} = \exp[kW(g)] \quad (14.27)$$

and get the answer obtained before. The theorem is proven.

III   Calculation of a nontrivial determinant

As we have seen from the previous discussion certain combinations of fermionic fields governed by the Dirac Hamiltonian can be written in terms of fields of the WZNW model. In particular, we have the explicit equivalences for currents. At the end of the previous chapter we have also shown that the slow components of the R^+L-field in the $U(N) \times SU(k)$-model with the current–current interaction are proportional to the Wess–Zumino matrix U. The latter property holds also for noninteracting Dirac fermions with the symmetry $U(N)$ (see Knizhnik and Zamolodchikov (1984)) and for noninteracting Majorana fermions with the symmetry $O(N)$ (Witten (1984)). Thus we can compose the bosonization table for non-Abelian theories.

Using Table 14.1 and the properties of the WZNW action we can calculate the following fermionic determinant (Tsvelik (1994)):

$$D[U] = \mathrm{Tr} \ln[\gamma_\mu \partial_\mu + (1 + \gamma_S)mU/2 + (1 - \gamma_S)mU^+/2] \quad (14.28)$$

where U is an external matrix field from the $SU(N)$ group and m is a parameter of the dimension of mass. We assume that the field $U(\tau, x)$ varies slowly on the scale m^{-1} and will be interested in the expansion of $D[U]$ in powers of m^{-1}. Such problems appear frequently in applications. Using Table 14.1 we can rewrite the determinant as path integral over the matrix field g with the WZNW action:

$$e^{D[U]} = \int DgD\Phi e^{-S}$$

$$S = \int d^2x \left[\frac{m}{2\pi a} \mathrm{Tr}(Ug^+e^{-i\gamma\Phi} + gU^+e^{i\gamma\Phi}) \right] + W(g)$$

$$+ \frac{1}{2} \int d^2x (\partial_\mu\Phi)^2 \qquad\qquad (14.29)$$

where $\gamma = (4\pi/N)^{1/2}$. Now we make a shift of variables in the path integral introducing a new variable G:

$$g = GU$$

This shift leaves the measure of integration unchanged. Using the identity (14.4) we get the following result:

$$e^{D[U]} = \int DGD\Phi \exp\{-S[G,\Phi] - S_{int}[U,G] - W[U]\} \quad (14.30)$$

$$S[G,\Phi] = \int d^2x \left[\frac{m}{2\pi a} \mathrm{Tr}(e^{i\gamma\Phi}G + G^+e^{-i\gamma\Phi}) \right] + W(G)$$

$$+ \frac{1}{2} \int d^2x (\partial_\mu\Phi)^2 \qquad\qquad (14.31)$$

$$S_{int} = \frac{1}{2\pi} \int d^2x \mathrm{Tr}\left(U\partial U^{-1}G^{-1}\bar{\partial}G \right) \qquad (14.32)$$

The only term in the action connecting G and U is the interaction term S_{int}. Let us show that at small energies $\ll m$ this term is irrelevant. In order to make it completely obvious we fermionize the new action using Table 14.1. Then we have

$$e^{D[U]} = e^{-W(U)} \int D\bar{\psi}D\psi \exp\{-\int d^2x[\bar{\psi}\gamma_\mu\partial_\mu\psi + m\bar{\psi}\psi + \mathrm{Tr}(U\partial U^{-1}\bar{J})]\}$$

$$(14.33)$$

where $\bar{J}$ is the right current of free *massive* fermions.

Calculating the first diagram in the expansion in $U\partial U^{-1}$ we get the term of order of

$$\frac{1}{m^2}\mathrm{Tr}[\bar{\partial}(U\partial U^{-1})]^2$$

which is irrelevant at low energies. Therefore from Eq. (14.30) we find

$$D[U]/D[I] = -W(U) + O(m^{-2}) \qquad (14.34)$$

The latter result is nontrivial since it includes the topological term which is difficult to get by more conventional methods.

References

K. Johnson, *Phys. Lett.* **5**, 253 (1963).

V. G. Knizhnik and A. B. Zamolodchikov, *Nucl. Phys.* **B247**, 83 (1984).

S. Novikov, *Usp. Math. Nauk* **37**, 3 (1982).

A. M. Polyakov in *Les Houches, Session XLIX, Fields, Strings and Critical Phenomena, 1988*, ed. by E. Brezin and J. Zinn-Justin, Elsevier Science Publ. 1989.

A. M. Polyakov and P. B. Wiegmann, *Phys. Lett.* **B131**, 121 (1983); **141**, 223 (1983).

A. M. Tsvelik, *Phys. Rev. Lett.* **72**, 1048 (1994).

E. Witten, *Commun. Math. Phys.* **92**, 455 (1984).

Part II

Application of the bosonization technique to physical models in (1 + 1)-dimensions

All theory, my friend, is grey, but the golden tree of life is green.
Faust (*Göthe*).[†]

The area of quasi-one-dimensional materials has been enjoying an increasing attention of theorists and experimentalists for more than 25 years. During these years an enormous amount of experimental data have been collected and a lot of interesting theoretical papers published. The accumulated knowledge suggests that the riches of quasi-one-dimensional world are almost inexaustible. To give a comprehensive review of what is known would require many volumes. Fortunately, there are many review articles on the subject to which we refer the reader (see the references at the end of this Section). On our part we shall concentrate on those properties which are common to many quasi-one-dimensional systems and can be explained using the ideas developed in Part I. Even this exposition will require a considerable amount of space.

We begin with a brief survey of experimental systems which are traditionally regarded as quasi-one-dimensional. This list is incomplete and probably inaccurate. The right of some materials to be there may be disputed. This is because it is not always clear how one-dimensional a given system is and whether a more traditional approach based on the Fermi liquid theory description would work better. So we ask the reader to be indulgent.

There are several classes of quasi-one-dimensional systems, organic and inorganic.

- Conjugated polymers. The best known example is polyacetylene, $(CH)_x$. It is a linear polymer consisting of weakly coupled chains of CH units forming a quasi-one-dimensional lattice. Three of four carbon valence electrons are in sp^2 (σ-type) hybridized orbitals; two of these bonds connect the neighbouring C-atoms and one holds an H-atom. The energies of these orbitals are far below the chemical potential. The remaining π-orbital has $2p^z$ symmetry; overlaping π-orbitals create a half-filled tight binding one-dimensional band (see Fig. II.1).

- Pure (undoped) polyacetylene undergoes a spontaneous dimerization and becomes an insulator with the gap $E_g \approx 1.8$ eV. This is considered as an illustration of the Peierls idea about instability of one-dimensional metals with respect to lattice dimerization (Peierls (1955)). For further details about conjugated polymers we refer the reader to the review article by Heeger *et al.* (1988).

[†] This translation belongs to Raymond Lucas.

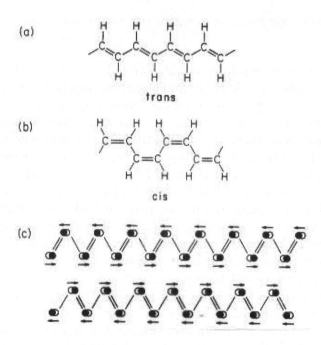

Fig. II.1.   Schematic picture of polyacetylene.

• Doped polyacetylene becomes a conductor remaining magnetically inert. The fact that spin excitations have a spectral gap, makes it certain that electric current is transported by collective modes – charge density waves (CDW). Other CDW systems are so-called blue bronzes of which $K_{0.3}MoO_3$ is the best representative. In these materials the MoO_6 octahedra form chains separated by chains of alkali-metal atoms leading to a quasi-two-dimensional crystal structure, but to nearly one-dimensional band structure. One can estimate the value of spin gap in $K_{0.3}MoO_3$ from the photoemission experiments (Dardel *et al.* (1992)) as $M_s \approx 0.25$ eV. The material undergoes a transition to an insulating (three dimensional) phase at 180 K. There is an excellent review article by Grüner about CDW (1988); the most recent measurements of the optical conductivity of $K_{0.3}MoO_3$ were done by Gorshunov *et al.* (1994).

• Bechgaard salts. These are organic materials described by a chemical formula $(TMTSF)_2X$ ($X = PF_6$ and ClO_4) or $(TMTTF)_2X$ ($X = PF_6$ and Br). In these materials big and flat organic molecules TMTSF (TMTTF) are arranged in stacks separated by anions (see Fig. II.2). Bechgaard salts have an extremely rich phase diagram part of which is represented on Fig. II.3. It is almost certain that the materials on

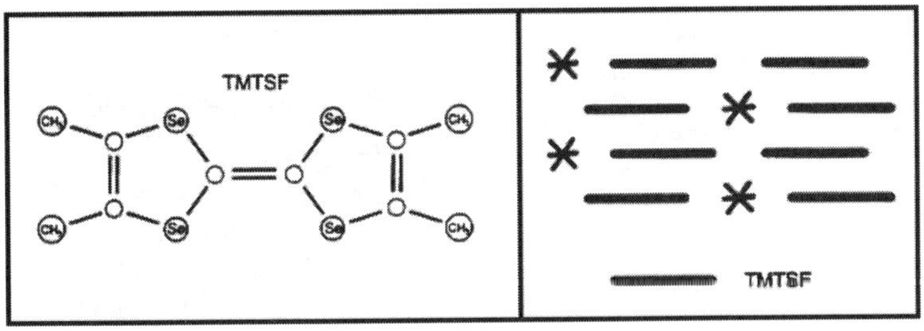

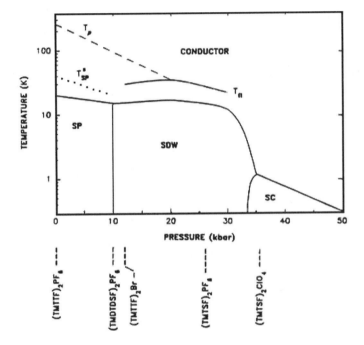

Fig. II.2.   Chemical formula and schematic crystal structure of $(TMTSF)_2PF_6$.

Fig. II.3.   Unified phase diagram and energy scales of Bechgaard salts as a function of hydrostatic pressure and anion X substitution. The abbreviations SP, SDW and SC stand for *spin-Peierls*, *Spin Density Wave* and *superconductor*. From Wzietek *et al.* (1993).

the right hand side of this figure are not one-dimensional and are better described within a more traditional framework (see Gor'kov and Lebed' (1984) and Yakovenko (1987), Yakovenko and Goan (1996)). Magnetic field, however, may increase quantum fluctuations and reduce the effective dimensionality (Danner and Chaikin (1995)). Further information can be obtained from the articles by Jerome and Schulz (1982), Brazovskii and Yakovenko (1985), Wzietek *et al.* (1993).

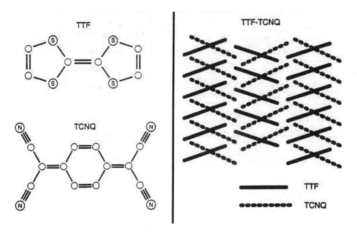

Fig. II.4.   Chemical formula and schematic crystal structure of TTF-TCNQ.

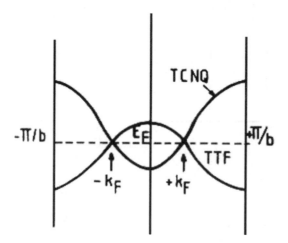

Fig. II.5.   Electron energy dispersion for TTF-TCNQ assuming no tunneling interchain coupling. From Jerome and Schulz (1982).

- TTF-TCNQ materials are composed of stacks of two types of large flat organic molecules (see Fig. II.4). TTF-TCNQ has two electrons per unit cell, but since the conduction bands are formed by both electrons and holes, it is not a trivial insulator. The suggested band structure is shown in Fig. II.5. The estimates of the transverse hopping integral derived from the NMR experiments and the optical conductivity give a very low value $t_\perp \sim 5$ meV (see Section 2.2.3 in the review article by Jerome and Schulz (1982)). Many valuable details about these materials can also be found in the review article by Skov Pedersen and Carneiro (1987).

Fig. II.6. Victor Yakovenko

- Quasi-one-dimensional antiferromagnets and spin liquids. These materials are magnetic insulators and some of them are almost ideally one-dimensional. Those include the spin-1/2 Heisenberg antiferromagnet Sr_2CuO_3 with the intrachain exchange integral $J \approx 1800$ K and the three-dimensional Neel temperature $T_N \approx 5$ K, and the material called NENP – spin $S = 1$ spin liquid with no magnetic transition. Another extremely one-dimensional $S = 1$ antiferromagnet is $YBaNiO_5$ (Di Tusa *et al.* (1994), Yokoo *et al.* (1996)). Spin liquids will be discussed further in Chapters 21–23.

- Edge states in quantum Hall effect. These systems are ideally one-dimensional and will be briefly discussed in Chapter 24.

Developments in chemistry constantly add materials to this list. In order not to get lost in this labyrinth one needs the Ariadne thread of a unified theory. In three dimensions the role of the parent theory of metals is played by the Landau Fermi liquid theory. In one dimension the concept of Fermi liquid is replaced by the one of Tomonaga–Luttinger liquid. The reader may remember, that we have already discussed this concept in Chapter 4 in application to spinless fermions. As will be shown in the next chapter, for fermions with spin the Tomonaga–Luttiger liquid Hamiltonian is a direct sum of the Gaussian model describing charge excitations and the level $k = 1$ SU(2)-invariant WZNW model describing the spin sector. The Tomonaga–Luttinger liquid model is a critical theory and its relevant operators are known. Thus one can classify possible states of quasi-one-dimensional materials as instabilities of the Tomonaga–Luttinger liquid. This program will be carried out in this part of the book.

References

S. A. Brazovskii and V. M. Yakovenko, *Sov. Phys. –JETP* **62**, 1340 (1985).

G. M. Danner and P. M. Chaikin, *Phy. Rev. Lett.* **75**, 4690 (1995).

B. Dardel, D. Malterre, M. Grioni, P. Weibel, Y. Baer, C. Schlenker and Y. Petroff, *Europhys. Lett.* **19**, 525 (1992).

J. F. Di Tusa, S.-W. Cheong, C. Broholm, G. Aeppli, L. W. Rupp, Jr. and B. Batlogg, *Physica* **B194-196**, 181 (1994).

L. P. Gor'kov and A. G. Lebed', *J. Phys. Lett.* (Paris) **45**, L433 (1984).

B. P. Gorshunov, A. A. Volkov, G. V. Kozlov, L. Degiorgi, A. Blank, T. Csiba, M. Dressel, Y. Kim, A. Schwartz and G. Grüner, *Phys. Rev. Lett.* **73**, 308 (1994).

G. Grüner, *Rev. Mod. Phys.* **60**, 1129 (1988).

A. J. Heeger, S. Kivelson, J. R. Schrieffer and W. P. Su, *Rev. Mod. Phys.* **60**, 781 (1988).

D. Jerome and H. J. Schulz, *Adv. Phys.* **31**, 299 (1982).

R. E. Peierls, *Quantum Theory of Solids*, Clarendon Press, Oxford (1955).

J. Skov Pedersen and K. Carneiro, *Rep. Prog. Phys.* **50**, 995 (1987).

P. Wzietek, F. Creuzet, C. Bourbonnais, D. Jerome and A. Moradpour, *J. Phys.*(Paris) I3, 171 (1993).

V. M. Yakovenko, *Sov. Phys. JETP* **66**, 355 (1987) and *Europhys. Lett.* **3**, 1041 (1987).

V. M. Yakovenko and H. S. Goan, *J. de Physique* **6**, 1917 (1996).

T. Yokoo, T. Sakaguchi, K. Kakurai and J. Akimitsu, *J. Phys. Soc. Jpn.* **64**, 3651 (1996).

15

Interacting fermions with spin

In this chapter we consider a realistic model of one-dimensional electrons with a short range interaction. The Hamiltonian is given by

$$H = H_0 + H_{int} \tag{15.1}$$

Here we represent the noninteracting Hamiltonian H_0 in the bosonized Sugawara form

$$H_0 = \frac{\pi v_F}{2} \int dx \, (: JJ : + : \bar{J}\bar{J} :)$$

$$+ \frac{2\pi v_F}{3} \int dx \, (: \mathbf{J} \cdot \mathbf{J} : + : \bar{\mathbf{J}} \cdot \bar{\mathbf{J}} :) \tag{15.2}$$

where

$$\bar{J} = \sum_\alpha R_\alpha^+ R_\alpha, \quad J = \sum_\alpha L_\alpha^+ L_\alpha \tag{15.3}$$

and the vector currents are defined by Eq. (7.1). The latter's currents satisfy the $SU_1(2)$ Kac–Moody algebra. The commutator of the Abelian currents is

$$[J(x), J(y)] = \frac{i}{\pi} \partial_x \delta(x - y) \tag{15.4}$$

The most general translationally invariant interaction which preserves spin rotational SU(2) symmetry is*

$$H_{int} = \frac{1}{2} \int dx \, dx' \psi_\alpha^\dagger(x) \psi_\beta^\dagger(x') Q_{\alpha\beta\alpha'\beta'}(x - x') \psi_{\beta'}(x') \psi_{\alpha'}(x), \tag{15.5}$$

where

$$Q_{\alpha\beta\alpha'\beta'}(x - x') = U_1(x - x')\delta_{\alpha\alpha'}\delta_{\beta\beta'} - U_2(x - x')\delta_{\alpha\beta'}\delta_{\beta\alpha'} \tag{15.6}$$

---

* Here we assume that the interaction is time independent. The effects of retardation which appear if the interaction is mediated by some low-energy excitations (phonons) will be considered later.

Interacting fermions with spin

The continuum representation of the Fermi field operators

$$\psi_\sigma(x) \to \exp(ik_Fx)R_\sigma + \exp(-ik_Fx)L_\sigma \tag{15.7}$$

yeilds the following expression for their bilinears:

$$\psi_\sigma^+(x)\psi_\mu(x) \to [R_\sigma^+(x)R_\mu(x) + L_\sigma^+(x)L_\mu(x)]$$
$$+[\exp(-2ik_Fx)R_\sigma^+(x)L_\mu(x) + H.c.] \tag{15.8}$$

First we shall assume that $4k_F \neq 2\pi$ and drop the oscillatory terms containing $\exp(\pm4ik_Fx)$. For a one-dimensional Fermi system on a lattice, this is the case when the band is not half filled. Substituting (15.8) into (15.5) and using the identities

$$R_\sigma^+ L_\sigma L_{\sigma'}^+ R_{\sigma'} = -\frac{1}{2}J\bar{J} - 2\mathbf{J}\cdot\bar{\mathbf{J}} \tag{15.9}$$

$$\lim_{x\to x'} R_\sigma^+(x)R_{\sigma'}(x)R_{\sigma'}^+(x')R_\sigma(x') = \frac{1}{2}J\bar{J} + 2\bar{\mathbf{J}}\cdot\bar{\mathbf{J}} \tag{15.10}$$

$$\lim_{x\to x'} L_\sigma^+(x)L_{\sigma'}(x)L_{\sigma'}^+(x')L_\sigma(x') = \frac{1}{2}JJ + 2\mathbf{J}\cdot\mathbf{J} \tag{15.11}$$

we find:

$$H_{int} = \int dx \{g_4(J^2 + \bar{J}^2) + g_cJ\bar{J} + g_4'(\mathbf{J}\cdot\mathbf{J} + \bar{\mathbf{J}}\cdot\bar{\mathbf{J}}) + g_s\mathbf{J}\cdot\bar{\mathbf{J}}\} \tag{15.12}$$

where

$$g_4 = \frac{1}{4}[2V_1(0) - V_2(0)], \quad g_4' = -V_2(0), \quad g_s = -2[V_2(0) + V_1(2k_F)]$$

$$g_c = [V_1(0) + V_2(2k_F)] - \frac{1}{2}[V_2(0) + V_1(2k_F)]$$

$V_i(k)$ being the Fourier transforms of $U_i(x)$.[†] In particular, for the Hubbard model with on-site interaction U

$$V_1(0) = V_1(2k_F) = \frac{2}{3}U, \quad V_2(0) = V_2(2k_F) = \frac{1}{3}U$$

leading to

$$g_4 = \frac{U}{4}, \quad g_4' = -\frac{U}{3}, \quad g_s = -2U, \quad g_c = \frac{U}{2}$$

The resulting Hamiltonian splits naturally into two mutually commuting parts describing charge and spin degrees of freedom, respectively:

$$H = H_c + H_s \equiv H_{U(1)} + H_{SU_1(2)} \tag{15.13}$$

where

$$H_c = \frac{\pi v_c}{2}\int dx \,(:JJ: + :\bar{J}\bar{J}:) + g_c\int dx \,J\bar{J} \tag{15.14}$$

[†] We suppose that these matrix elements are nonsingular.

$$H_s = \frac{2\pi v_s}{3} \int dx \; (: \mathbf{J} \cdot \mathbf{J} : + : \bar{\mathbf{J}} \cdot \bar{\mathbf{J}} :) + g_s \int dx \; \mathbf{J} \cdot \bar{\mathbf{J}} \qquad (15.15)$$

with renormalized velocities

$$v_c = v_F + \frac{2g_4}{\pi}, \quad v_s = v_F + \frac{3g_4'}{2\pi} \qquad (15.16)$$

Using the Abelian bosonization we reduce the Hamiltonian H_c to the canonical Gaussian form (11.15):

$$H_c = \frac{v_c}{2} \int dx \; [K_c(\partial_x \Theta_c)^2 + K_c^{-1}(\partial_x \Phi_c)^2] \qquad (15.17)$$

with v_c and K_c dependent on the coupling constants g_c and g_4.

In fact, the form of the effective Hamiltonian (15.13), (15.15), (15.17) for the low-energy sector is dictated by the symmetry of the theory due to which any product of four fermion operators can be written as a product of currents. The suggested derivation serves pedagogical purposes and is necessary only to relate parameters of the effective theory to the bare interaction. These relations hold only for small values of the bare coupling constants and therefore are not universal. Meanwhile, the general form of the low-energy Hamiltonian, being determined by the symmetry requirements, survives even if the interactions are strong.

In what follows we shall consider K_c, v_c, v_s and g_s as phenomenological parameters. For the Hubbard model

$$\hat{H} = -t \sum_{j,\sigma}(c_{j+1,\sigma}^+ c_{j,\sigma} + c_{j,\sigma}^+ c_{j+1,\sigma}) + U \sum_j n_{j\uparrow} n_{j\downarrow} \qquad (15.18)$$

these parameters are known exactly from the Bethe ansatz solution (Kawakami and Yang (1991), Frahm and Korepin (1990)).

In particular, for $U/t \gg 1$ we have

$$v_s/v_c \sim (t/U) \sin \pi v$$

$$K_c = \frac{1}{2}[1 - (8t \ln 2/\pi U) \sin \pi v] + O(t^2/U^2) \qquad (15.19)$$

where v is the band filling, that is the average number of electrons per site divided by 2.

The example of the Hubbard model (Fig. 15.1) demonstrates a general trend: the ratio v_s/v_c may change widely with the interaction contrary to K_c which for models with local interaction is always greater than $1/2$. Below we shall demonstrate that retardation effects may lift this limitation.

The fact that the charge and spin sectors have different velocities, $v_c \neq v_s$, is general for one-dimensional systems. This phenomenon has a general name of *spin–charge separation*. In fact, the difference between the spectra of excitations with different symmetry may become even more dramatic when the interaction between spin currents is antiferromagnetic.

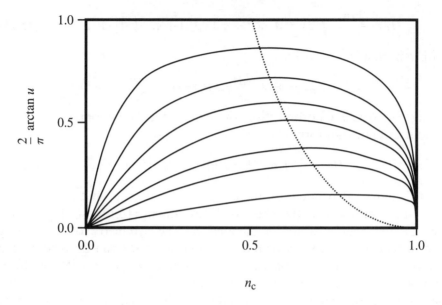

Fig. 15.1. Lines of constant K_c for the Hubbard model as a function of band filling n_c for various $u = U/t$. The dotted line denotes the value of n_c where K_c takes its minimal value for given u (from Frahm and Korepin (1990)).

This interaction has a scaling dimension $d = 2$ and therefore is marginal. This means that depending on the sign of g_s, it either dies out or scales to strong coupling. The latter happens when $g_s > 0$ and, as we shall demonstrate later, the spin sector acquires a spectral gap. Meanwhile the charge sector always remains gapless away from half-filling. This leads to a peculiar situation when at low temperatures the material has an exponentially small Pauli susceptibility, remaining at the same time an excellent conductor.

When the band is half-filled ($2k_F = \pi$), we have $\exp(4ik_F na) = 1$, and Umklapp process

$$g_u(R_\sigma^\dagger L_\sigma R_{\sigma'}^\dagger L_{\sigma'} + h.c.), \qquad g_u = V_1(4k_F) + V_2(4k_F) \qquad (15.20)$$

should be taken into account in the derivation of the continuum limit of the Hamiltonian. This extra term contributes to the charge part of the Hamiltonian, modifying its simple Gaussian form (15.17). In fact, with the Umklapp term included, the charge Hamiltonian H_c can be written entirely in terms of new, the so-called *pseudospin*, vector currents $\mathbf{I}$ and $\bar{\mathbf{I}}$. The latter can be obtained from the usual spin vector currents $\mathbf{J}, \bar{\mathbf{J}}$ by a particle–hole transformation in one spin component of the Fermi field (Lieb and Wu (1968), Yang (1989), Yang and Zhang (1990)):

$$c_{j\downarrow} \to (-1)^j c_{j\downarrow}^\dagger : \qquad R_\downarrow \to R_\downarrow^\dagger, \quad L_\downarrow \to L_\downarrow^\dagger. \qquad (15.21)$$

which implies that

$$I^3 = \sum_\sigma \; :R_\sigma^\dagger R_\sigma: \; = \frac{1}{2}\bar{J}, \quad I^+ = R_\uparrow^\dagger R_\downarrow^\dagger, \quad I^- = (\bar{I}^+)^\dagger \qquad (15.22)$$

with similar expressions for the left chiral components I^a. One can easily check that the pseudospin currents satisfy the $SU_1(2)$ Kac–Moody algebra.

Written in terms of $\mathbf{I}, \bar{\mathbf{I}}$, the Hamiltonian H_c takes the form

$$H_c = \frac{2\pi v_c}{3} \int dx \; (:\mathbf{I}\cdot\mathbf{I}: + :\bar{\mathbf{I}}\cdot\bar{\mathbf{I}}:)$$
$$+ \int dx \; [4g_c I^3 \bar{I}^3 + 2g_u(I^+\bar{I}^- + I^-\bar{I}^+)] \qquad (15.23)$$

This Hamiltonian will be discussed in detail in Section 17.IV, where the formation of the charge gap at half-filling is considered. Here we would like to note that, if $g_u = g_c$, which is the case for the half-filled Hubbard model, the U(1) symmetry of the charge sector is extended to the (pseudospin) SU(2), and H_c acquires the form identical to that of H_s in the spin sector (see Eq. (15.15)). The total Hamiltonian is then characterized by SU(2)×SU(2) ≈O(4) symmetry.

In general, a system of electrons with SU(2) spin symmetry in one dimension may be in the following states:[‡]

1. *Not half-filled band*

   - $g_s < 0$. Both spin and charge excitations are gapless with different velocities; it is a metal with ill-defined single electron excitations.

   - $g_s > 0$. Charge excitations are gapless and spin excitations have a spectral gap.

2. *Half-filled band*

   - If $g_c < 0$ (more exactly, if $g_c < -|g_u|$; see Chapter 17) and $g_s < 0$, both sectors remain gapless.

   - If $g_c > 0$ ($g_c > -|g_u|$) and $g_s < 0$, the charge excitations have a spectral gap and the spin excitations are gapless (Mott–Hubbard insulator or Spin Density Wave state (SDW)). Recently spin–charge separation was directly observed in this state by the Angle Resolved Photoemission method (see Kim *et al.* (1996)). We shall discuss this state later.

[‡] We do not consider here the effects of disorder.

- $g_c > 0$ ($g_c > -|g_u|$) and $g_s > 0$. Spectral gaps are present both for the charge and spin excitations. Such a situation can arise only in a system with a half-filled band *and* non-point-like interaction (e.g. in the extended, so-called UV Hubbard model which includes next-nearest-neighbour interactions). The ground state is a commensurate charge density wave (CDW) with true long-range order.

References

H. Frahm and V. V. Korepin, *Phys. Rev.* **B42**, 10 553 (1990).

N. Kawakami and S. K. Yang, *J. Phys: Condens. Matter* **3**, 5983 (1991).

C. Kim, A. Y. Matsuura, Z.-X. Shen, N. Motoyama, H. Eisaki, S. Uchida, T. Tohyama and S. Maekawa, *Phys. Rev. Lett.* **77**, 4054 (1996).

E. H. Lieb and F. Y. Wu, *Phys. Rev. Lett.* **20**, 1445 (1968).

C. N. Yang, *Phys. Rev. Lett.* **63**, 2144 (1989).

C. N. Yang and S. C. Zhang, *Mod. Phys. Lett.* **B4**, 759 (1990).

16

Spin-1/2 Tomonaga–Luttinger liquid

Let us consider the easiest case when the current–current interaction in Eq. (15.15) is ferromagnetic ($g_s < 0$). In this case the renormalization group analysis indicates that this interaction is irrelevant and in the first approximation we can just forget about it.[*] Thus we have shown that a one-dimensional electronic system may be in a critical state. Since in this state the system has a sharp Drude peak in the optical conductivity, finite $T = 0$ paramagnetic susceptibility and linear specific heat, this state is a metal. However, one feature distinguishes this metal from the conventional ones: it does not support coherent single electron excitations. In other words, this metallic state is not a Fermi liquid. Haldane suggested calling it *Luttinger liquid*; throughout this book we adopt another name – *Tomonaga–Luttinger liquid*, the concept we came across in Chapter 4.

The effective Hamiltonian describing the spin sector of the spin-1/2 Tomonaga–Luttinger liquid is the critical $SU_1(2)$ WZNW model which has been discussed earlier in Chapter 13. As we know, such a model is equivalent to the Gaussian model with $K = 1$. Thus the complete description of the low-energy sector of spin-1/2 fermions is given by two Gaussian models, the one for the charge sector being represented by Eq. (15.17). This description is highly universal being dependent only on two dimensionless parameters v_s/v_c and K_c ($K_s = 1$ is fixed by the SU(2) symmetry).

As we have mentioned in Chapter 4, these velocities and K_c are related to the static susceptibilities. For spinless TL liquid we had identity (4.25); in the presence of spin this identity is modified due to the summation over spin indices in definition of the charge current (15.3). In general we have

$$\chi(\omega, q) = \frac{2K}{\pi} \frac{q^2 v}{\omega^2 - v^2 q^2} + \chi_{vV}(\omega, q) \qquad (16.1)$$

---

[*] If it is necessary to take into account corrections to the critical point, one can replace the bare coupling $g_s^{(0)}$ with its renormalized value $g_s(T) = g_s^{(0)}/[1 - g_s^{(0)}/2\pi v_s \ln(\Lambda/T)]$.

where $\chi_{vV}(\omega, q)$ is the van Vleck contribution coming from valence bands. Since these bands do not cross the chemical potential the latter term does not contribute the static susceptibility: $\chi_{vV}(\omega, 0) = 0$ and we have

$$K_c = \frac{1}{2}\pi v_c \chi_c(0, 0) \tag{16.2}$$

When the SU(2) is unbroken, the smooth part of the spin susceptibility is the same as for free spin-1/2 fermions. In the general case we have

$$K_s = \frac{1}{2}\pi v_s \chi_s(0, 0)(2/g_L)^2$$

where g_L is the Landee factor.

In what follows we denote the corresponding free bosonic fields in charge and spin sector Φ_c and Φ_s respectively and write down expressions for the most interesting operators.

charge density
$$\rho(x) = J + \bar{J} + a_c \cos(\sqrt{2\pi}\Phi_s)\sin(2k_F x + \sqrt{2\pi}\Phi_c)$$
$$+a_u \cos(4k_F x + \sqrt{8\pi}\Phi_c) \tag{16.3}$$

spin density
$$S(x) = J + \bar{J} + a_s \cos(2k_F x + \sqrt{2\pi}\Phi_c)n(x) \tag{16.4}$$

singlet pairing
$$\Delta_0(x) = i\psi(x)\sigma^y\psi(x) \approx \sum \sigma R_\sigma L_{-\sigma}$$
$$= d_u e^{-i\sqrt{2\pi}\Theta_c}\sin(\sqrt{2\pi}\Phi_s) \tag{16.5}$$

triplet pairing
$$\Delta^a(x) = k_F^{-1}i\psi(x)\sigma^y\sigma^a\partial_x\psi(x) \approx R_\sigma\sigma^a_{\sigma\sigma'}\sigma^y L_{\sigma'}$$
$$= p_u e^{-i\sqrt{2\pi}\Theta_c}[\sin(\sqrt{2\pi}\Theta_s), \cos(\sqrt{2\pi}\Theta_s), -i\cos(\sqrt{2\pi}\Phi_s)] \tag{16.6}$$

where the staggered magnetization n is given by Eq. (13.9) and a_i, d_u, p_u are nonuniversal constants.

The $4k_F$-term in the charge density appears only in the presence of interaction (see Fig. 16. 1); for small magnitudes of the interaction $a_u \sim V/\epsilon_F$.

From these expressions we can find the corresponding two-point correlation functions:

$$\langle S(1)S(2)\rangle = \frac{\chi_s}{8\pi v_s}\left[\frac{1}{(\tau + ix/v_s)^2} + \frac{1}{(\tau - ix/v_s)^2}\right] + \frac{a_s^2\cos(2k_F x)}{|\tau + ix/v_c|^{K_c}|\tau + ix/v_s|} \tag{16.7}$$

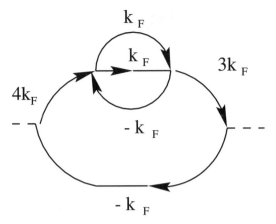

Fig. 16.1. The first diagram for the density–density correlation function containing $4k_F$-singularity. The diagram contains two Green's functions with wave vectors far from the Fermi surface.

$$\langle \rho(1)\rho(2) \rangle = \frac{\chi_c}{8\pi v_c} \left[\frac{1}{(\tau + ix/v_c)^2} + \frac{1}{(\tau - ix/v_c)^2} \right]$$
$$+ \frac{a_c^2 \cos(2k_F x)}{|\tau + ix/v_c|^{K_c}|\tau + ix/v_s|} + \frac{a_u^2 \cos(4k_F x)}{|\tau + ix/v_c|^{4K_c}} \qquad (16.8)$$

$$\langle \Delta_0(1)\Delta_0(2) \rangle = \frac{d_u^2}{|\tau + ix/v_c|^{1/K_c}|\tau + ix/v_s|} \qquad (16.9)$$

$$\langle \Delta(1)\Delta(2) \rangle = \frac{d_p^2}{|\tau + ix/v_c|^{1/K_c}|\tau + ix/v_s|} \qquad (16.10)$$

One can use these equations to calculate important physical quantities – static susceptibilities. These susceptibilities determine which ordered low temperature state will be stabilized by weak interchain hopping in a three-dimensional array of coupled chains. Then the most divergent susceptibility will determine which symmetry will be broken and what order parameter will emerge.

To give an example let us calculate the pairing susceptibility χ_{sc}. From Eq. (16.9) we deduce the following expression at $\omega, q = 0$:

$$\chi_{sc} = d_u^2 \int_{\tau_0}^{1/T} d\tau \int_{|x|>a_0} dx \frac{\pi T}{|\sin \pi T(\tau + ix/v_s)|} \left[\frac{\pi T}{|\sin \pi T(\tau + ix/v_c)|} \right]^{1/K_c} \qquad (16.11)$$

where $\tau_0 \sim \epsilon_F^{-1}$. If $K_c > 1$ the integrals converge at small τ and x and we

Fig. 16.2.   Vladimir Korepin.

can replace the lower limits by 0 and rescale the variables:

$$\tau = t/T, \; x = x'/T\sqrt{v_c v_s}$$

As a result we get

$$\chi_{sc} = \frac{d_u^2 I(v_c/v_s)}{\sqrt{v_c v_s}}(T\tau_0)^{-1+1/K_c} \tag{16.12}$$

$$I(v_c/v_s) = \int_0^1 d\tau \int_{-\infty}^{\infty} dx \frac{\pi}{|\sin \pi(\tau + ix\sqrt{v_c/v_s})|}\left[\frac{\pi}{|\sin \pi(\tau + ix\sqrt{v_s/v_c})|}\right]^{1/K_c}$$

We see that at $K_c > 1$ the pairing susceptibility diverges at low temperatures which means that the system has a superconducting instability at $T = 0$. As we have said, in real quasi-one-dimensional systems interchain interactions will shift this instability to finite T where a real three-dimensional superconducting transition will take place.

Another example of an experimentally observable quantity is the imaginary part of the local magnetic susceptibility. This quantity is related to the nuclear relaxation rate $1/T_1$ measurable by Nuclear Magnetic Resonance

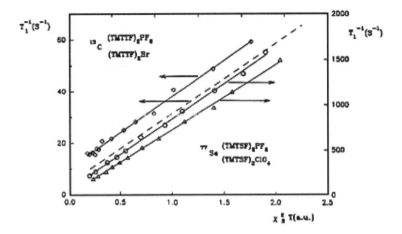

Fig. 16.3. Plots of T_1^{-1} vs. measured $T[\chi_s(T)]^2$ for Bechgaard salts. From Wzietek *et al.* (1993).

(NMR):

$$T_1^{-1} \sim T \lim_{\omega \to 0} \int d^D q \frac{\Im m \chi^{(R)}(\omega, q)}{\omega} \qquad (16.13)$$

For one-dimensional systems the main contributions to NMR relaxation rate come from the vicinity of $q = 0$ and $q = 2k_F$. In the first case we have

$$\Im m \chi^{(R)}(\omega, q \ll k_F) = -\chi_s \Im m \frac{q^2 v_s^2}{(\omega + i\delta)^2 - v_s^2 q^2}$$

$$= \frac{\pi}{2} \chi_s \omega [\delta(q v_s - \omega) + \delta(q v_s + \omega)] \qquad (16.14)$$

This gives the following contribution to T_1^{-1}:

$$T_1^{-1}(q \sim 0) \sim T \pi \chi_s / 2 v_s = \pi^2 T [\chi_s(T)]^2 \qquad (16.15)$$

In the last equation we have used the relation (16.2) and the fact that $K_s = 1$. The replacement of χ_s by its finite temperature value is a conjecture which finds strong experimental support (see Fig. 16.3).

From Eq. (16.4) we also find the $2k_F$-contribution:

$$T_1^{-1}(q \sim 2k_F) = A(T/\epsilon_F)^{K_c} \qquad (16.16)$$

In case there is a charge gap M_c, like in the materials represented on Fig. 16.3, the temperature in this equation must be replaced by M_c such that we have

$$T_1^{-1} = A(M_c/\epsilon_F)^{K_c} + B T [\chi_s(T)]^2 \qquad (16.17)$$

Fig. 16.3 clearly shows the constant part of T_1^{-1} (note, however, that its magnitude decreases for less one-dimensional materials from the right-hand side part of the phase diagram Fig. II.3). The prediction that the $2k_F$ contribution must decrease with M_c is also supported by the experiments (see the reference to Wzietek *et al.* (1993) given in the introductory Section).

Depending on the value of K_c a leading singularity appears in different correlation functions. There are three possibilities.

- $K_c > 1$. Only the pairing operators have scaling dimensions less than 1. Therefore only the superconducting susceptibilities are singular at $T \to 0$.

- $1/3 < K_c < 1$. The most relevant operators are staggered components of charge and spin densities with $q = 2k_F$.

- $K_c < 1/3$. The most relevant operator is the $4k_F$-component of charge density.

Perhaps, the most dramatic manifestation of the spin–charge separation is a loss of coherence of the single particle excitations. The right and left components of the annihilation operator are given by

$$R_\sigma, L_\sigma = \exp\{i\pi \pm [\Phi_c - \Theta_c + \sigma \pm (\Phi_s - \Theta_s)]\} \qquad (16.18)$$

This leads to the following expression for the single electron Green's function:

$$G(x, \tau) = \left[\frac{a^2}{v_c^2\tau^2 + x^2}\right]^{\theta/2}$$

$$\times \left[\frac{\exp(ik_Fx)}{\sqrt{(v_c\tau - ix)(v_s\tau - ix)}} + \frac{\exp(-ik_Fx)}{\sqrt{(v_c\tau + ix)(v_s\tau + ix)}}\right] \qquad (16.19)$$

where

$$\theta = 1/4(1/K_c + K_c - 2) \qquad (16.20)$$

We shall discuss the Fourier transformation of this Green's function later in Chapter 19 and here restrict ourselves to a much simpler task of calculating the single particle density of states

$$\rho(\omega) = -\frac{1}{\pi}\Im m G^{(R)}(\omega, x = 0)$$

$$= \frac{2}{\pi}\int_{\sim a/v_c}^{\infty} dt \sin(|\omega|t)\left(\frac{\pi T}{\sinh \pi T t}\right)^{1+\theta}(a/v_c)^\theta \qquad (16.21)$$

If $\theta < 1$ (which corresponds to $K_c > 3 - 2\sqrt{2} \approx 0.17$) the integral converges at small t and the lower limit may be put to zero. In this case we get a

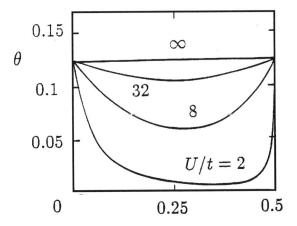

Fig. 16.4. The exponent θ for the Hubbard model as a function of band filling for various U/t (from Kawakami and Yang (1990)).

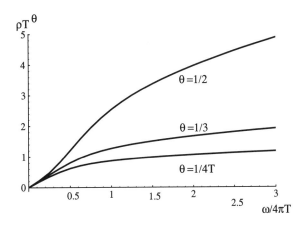

Fig. 16.5. The single particle density of states as a function of $\omega/4\pi T$ for $\theta = 1/3$ (after Eq. (16.22)).

universal result:

$$\rho(\omega) \sim (Ta/v_c)^\theta \, \Im m \frac{\Gamma(1/2 + \theta/2 - i\omega/4\pi T)}{\Gamma(1/2 - \theta/2 - i\omega/4\pi T)}, \quad (\theta < 1) \qquad (16.22)$$

Otherwise the value of the integral is determined by short times and we get a temperature independent (at $T \ll v_c/a$) result

$$\rho(\omega) \sim |\omega|, \quad (\theta > 1) \qquad (16.23)$$

Gradually improving resolution of photoemission experiments will probably make it possible to compare these theoretical expressions with data for real materials.

References

N. Kawakami and S.-K. Yang, *Phys. Lett.* A**148**, 359 (1990).

V. Meden and K. Schönhammer, *Phys. Rev.* B**46** 15 753 (1992).

Y. Ren and P. W. Anderson, *Phys. Rev.* B**48**, 16 662 (1993).

J. Voit, *Phys. Rev.* B**45**, 4027 (1992).

P. Wzietek, F. Creuzet, C. Bourbonnais, D. Jerome and A. Moradpour, *J. Phys.* (Paris) I**3**, 171 (1993).

17

Instabilities of a
Tomonaga–Luttinger liquid

I   Electron–phonon interaction

The effects of electron–phonon interaction in one-dimensional systems have been extensively studied by different authors (see the review articles by Heeger *et al.* (1988) and Grüner (1988)), but mostly for noninteracting electrons. A combination of interactions and phonon effects was studied by Voit and Schulz (1986) and Voit (1990) using the renormalization group approach. Here we shall show how these results can be put into the scheme of weakly perturbed Tomonaga–Luttinger liquid.

The most natural way to include lattice effects into the Hubbard model is to make the hopping integral t dependent on the intersite distance. For small anharmonisms we can expand in lattice deformations u_j:

$$t_{ij} \approx t + \frac{1}{2a}\kappa(u_i - u_j)$$

Thus the term which should be added to the Hubbard Hamiltonian is

$$H_{\text{el–ph}} = -\frac{1}{2a}\kappa \sum_{j,\sigma}(u_j - u_{j+1})(c^+_{j+1,\sigma}c_{j,\sigma} + c^+_{j,\sigma}c_{j+1,\sigma}) \qquad (17.1)$$

where u_j is dimensionless and κ has a dimension of energy.

I.1   Incommensurate band filling, the effect on K_c

In the continuous approximation Hamiltonian (17.1) will generate a coupling between the lattice deformations and the $2k_F$ and $4k_F$ components of the charge density. In order to get a continuous description we separate fast and slow components of the deformation field:

$$u_j = u_0(x) + \sum_{l=1,2}[\exp(2ilk_F aj)u_l(x) + \exp(-2ilk_F aj)u_l^*(x)] \qquad (17.2)$$

155

It is assumed that the fields $u_l(x)$ vary much more slowly than the corresponding exponents (as usual, such assumptions must be checked for self-consistency). The field $u_0(x)$ is real and the other fields are complex provided the wave vectors $2k_F$ and $4k_F$ are incommensurate with the lattice. Substituting (17.2) together with the similar decomposition for the fermions (2.5) into Eq. (17.1) we obtain

$$H_{el-ph} = i \sum_{l=1,2} \int dx \, [\Delta_l(x)\rho(2lk_F;x) - \text{H.c}] \qquad (17.3)$$

where

$$\Delta_l(x) = \kappa \frac{\sin(lk_F a)}{a} u_l(x)$$

Here we have kept only the most relevant terms. The term with $\rho(4k_F;x)$ appears only for interacting electrons (see the explanations around Fig. 16.1).

The action for the rescaled deformation fields $\Delta_{1,2}$ has a standard quadratic form:

$$S_{ph} = \frac{1}{2\pi v_c} \sum_{l=1,2} \lambda_l^{-1} \int d\tau dx \left[\omega_l^{-2} \partial_\tau \Delta_l^* \partial_\tau \Delta_l + \Delta_l^* \Delta_l \right] \qquad (17.4)$$

where $\omega_l = \omega(2lk_F)$ and the dimensionless coupling constants are

$$\lambda_l = [\kappa \sin(lk_F a)]^2 / \pi v_c \rho_l \omega_l^2 a^2$$

with ρ_l being the effective mass density for the l-th mode. In the first approximation we have neglected the fact that phonons disperse around these wave vectors. It can be shown that the dispersion along the chain leads to insignificant corrections. To the contrary, a dependence of $\omega(q)$ on transverse components of the wave vector is potentially significant since it generates interchain coupling which may lead to a three-dimensional phase transition at low temperatures. As always we shall assume that the bare values of three-dimensional couplings are small and therefore there is a temperature interval where the correlations are essentially one-dimensional.

Since the action is quadratic in the displacement fields, we can integrate them out to obtain the following retarded interaction between the electronic densities:

$$S_{int} = - \int d\tau d\tau' dx \sum_l \rho(2lk_F, \tau, x) D_l(\tau - \tau') \rho(-2lk_F, \tau', x) \qquad (17.5)$$

where $D_l(\tau)$ is the phonon Green's function determined by the action (17.4). As above here we neglect the phonon dispersion and treat the phonons as local. Writing the kernel in frequency representation, we get

$$D_l(\omega) = \frac{\pi v_c \lambda_l \omega_l^2}{\omega^2 + \omega_l^2} = \pi v_c \lambda_l \left[1 - \frac{\omega^2}{\omega^2 + \omega_l^2}\right] \tag{17.6}$$

In real time representation this corresponds to

$$D_l(\tau) = \pi v_c \lambda_l \left[\delta(\iota) + \frac{1}{2\omega_l}\partial_\tau^2 e^{-\omega_l|\tau|}\right] \tag{17.7}$$

Let us first consider the contribution coming from the $4k_F$ terms. Since $\rho(4k_F, x)$ is proportional to the complex exponent, its convolution with the delta-functional part of the kernel (17.7) gives a zero contribution to the action. The convolution with the second term yields the following to the action density:

$$4\pi^2 a_u^2 v_c \frac{\lambda_2}{\omega_2} \partial_\tau \Phi_c(\tau) \int d\tau' e^{-\omega_2|\tau'|} \exp\{i\sqrt{8\pi}[\Phi_c(\tau) - \Phi_c(\tau - \tau')]\}\partial_\tau \Phi_c(\tau - \tau') \tag{17.8}$$

In this expression we shall replace the complex exponent by its average value:

$$\int d\tau' e^{-\omega_2|\tau'|} \langle\exp\{i\sqrt{8\pi}[\Phi_c(\tau) - \Phi_c(\tau - \tau')]\}\rangle$$
$$= \int_{-\infty}^{\infty} d\tau' e^{-\omega_2|\tau'|} \left(\frac{\tau_0}{|\tau'|}\right)^{2d} \approx \frac{2\tau_0}{2d - 1} \tag{17.9}$$

where d is the bare scaling dimension of the corresponding operator (in the given case it is $4K_c^{(0)}$). Here we assume that $2d > 1$; then the integral converges at small times $|\tau'| \sim \tau_0 \sim \epsilon_F^{-1}$. This is an important result because it shows that in dealings with Eq. (17.5) one expands in $\omega_l \tau_0$. Substituting (17.9) into Eq. (17.8) we get the following contribution to the Lagrangian density:

$$\mathcal{L}(4k_F) = 8\pi^2 a_u^2 v_c \frac{\tau_0 \lambda_2}{(4K_c^{(0)} - 1)\omega_2}(\partial_\tau \Phi_c)^2 \tag{17.10}$$

Now let us consider the $2k_F$-terms. The local term in Eq. (17.7) gives the following contribution:

$$\rho(2k_F, \tau, x)\rho(-2k_F, \tau, x) = R_\sigma^+ L_\sigma L_{\sigma'}^+ R_{\sigma'} = -(\frac{1}{2}J\bar{J} + 2\mathbf{J}\bar{\mathbf{J}}) \tag{17.11}$$

From this equation we conclude that the $2k_F$ phonons give a positive contribution to the current–current coupling constant in the spin sector:

$$g_s = g_s^{(0)} + 2\pi \lambda_1 v_c \tag{17.12}$$

This is the effect which has been known for a long time: the electron–phonon interaction competes with repulsive forces responsible for a negative $g_s^{(0)}$ tending to produce a spin gap. Since the gap will appear only when $g_s > 0$, this requires the coupling constant to be larger than

$$\lambda_1 > -g_s^{(0)}/2\pi v_c \sim v_s/v_c \qquad (17.13)$$

For small $g_s/2\pi v_s$ the magnitude of the spin gap is exponential:

$$M_s \sim a^{-1}\sqrt{v_s g_s}\exp[-2\pi v_s/g_s] \qquad (17.14)$$

For large values of this parameter it becomes linear:

$$M_s \sim g_s/a \qquad (17.15)$$

The spin gap is observed in many quasi-one-dimensional materials and associated with creation of Charge Density Waves (CDW).

In addition to producing a spin gap and *quite independently of it*, phonons may have a strong influence on the dynamics and scaling dimensions in the charge sector renormalizing the charge velocity v_c and K_c. These effects come primarily from the quantum fluctuations, i.e. from the second term in the expansion of the phonon propagator (17.7). We have already calculated the contribution to the Lagrangian density coming from the $4k_F$ terms (17.10). Now let us calculate the full $2k_F$-contribution assuming that the spin sector is gapless. In this case the interactions in the spin channel die out and the other parameters of the effective action does not renormalize until we reach ω_l.

Substituting the expression (16.3) into Eq. (17.5) and using (17.7) we get

$$2\frac{\pi^2 a_c^2 \lambda_1 v_c \tau_0}{\omega_1 K_c^{(0)}}\left\{(\partial_\tau \Phi_c)^2 + [: \partial_\tau \cos(\sqrt{2\pi}\Phi_s) :]^2\right\} \qquad (17.16)$$

Then we can show (see the Appendix) that with the accuracy $\sim \omega_1 \tau_0$ the second term is a total time derivative:

$$[: \partial_\tau \cos(\sqrt{2\pi}\Phi_s) :]^2 = \frac{1}{4}\partial_\tau^2 \cos(\sqrt{8\pi}\Phi_s) + O(\omega_l^{-1}) \qquad (17.17)$$

and therefore may be discarded. This is consistent with the absence of renormalization of K_s (the SU(2) symmetry!).

Adding (17.16) and (17.10) we get the following addition to the Lagrangian density:

$$\delta\mathscr{L} = \frac{1}{2v_c K_c^{(0)}}\left(\frac{m^*}{m} - 1\right)(\partial_\tau \Phi_c)^2$$

$$\frac{m^*}{m} = 1 + (2\pi v_c/a)^2 \tau_0 \left\{a_c^2(\lambda_1/\omega_1) + 4a_u^2[\lambda_2 K_c^{(0)}/\omega_2(4K_c^{(0)} - 1)]\right\} \quad (17.18)$$

Here we follow the notations accepted in the literature on CDW where m^* is interpreted as a renormalized electron mass with m being its bare mass. Combining Eq. (17.18) with the bare Gaussian action we obtain the following renormalized values of K_c and the charge velocity $\tilde{v}_c$:

$$K_c = \left(\frac{m}{m^*}\right)^{1/2} K_c^{(0)}, \ \tilde{v}_c = \left(\frac{m}{m^*}\right)^{1/2} v_c \quad (17.19)$$

The strong renormalization of these parameters by the electron–phonon interaction was first pointed out by Brazovskii and Dzyaloshinskii (1976), Fukuyama (1976) and Fukuyama and Lee (1978) in the contex of CDW. Eqs. (17.19) give renormalized parameters in the limit of small frequencies $|\omega| \ll \omega_l$. In the intermediate frequency regime one may use the modified Gaussian action with nonlocal in time kinetic energy:

$$S = \frac{1}{2} \sum_{\omega,q} \Phi_c(-\omega,-q)[\frac{1}{v_c}\omega^2 f(i\omega) + v_c q^2]\Phi_c(\omega,q) \quad (17.20)$$

where the function $f(\omega)$ interpolates between $f(0) = m^*/m$ and $f(\infty) = 1$.

Experimental data on CDW materials like $K_{0.3}MoO_3$ give the ratio m^*/m of order of several hundred (see Grüner (1988)). Strong renormalizations are to be expected even in those materials where the spin gap is not formed, provided $K_c^{(0)} < 1$ (see Eq. (17.18)). According to Eq. (17.13) an absence of spin gap is not equivalent to the absence of electron–phonon interaction, it just means that the electron–phonon coupling constant is smaller than v_s/v_c. Since the ratios $\pi v_c/a\omega_l$ may be quite large there still may be a very substantial change of K_c.

According to Eqs. (16.19) and (16.20) small K_c would correspond to irrelevance of single particle hopping.

Thus phonon effects may lead not only to opening of the spectral gap in the spin spectrum, but also to *substantial decrease of K_c*.

Since the phonon correlation function is directly observable by X-rays, it is good to calculate it. From Eq. (17.5) we have

$$D(\omega_q) = D_0(\omega_q) + D_0(\omega_q)\langle\langle\rho(-\omega,-q)\rho(\omega,q)\rangle\rangle D_0(\omega_q) \quad (17.21)$$

where the density–density correlation function is the exact one, i.e. calculated with the electron–phonon interaction taken into account. If there is no spin gap both the $2k_F$ and the $4k_F$ contributions are singular at small frequencies. In this case the corresponding expressions for the density–density correlation functions are given by Eq. (16.3). When the spin gap M_s is present, the $2k_F$ singularity appears only at high frequencies $\omega > M_s$.

I.2 Commensurate band filling

Let us now assume that the band is half-filled, i.e. $4k_F = 2\pi/a$. Let us first consider the case when there is no interaction in the band except the electron–phonon one. In the commensurate case we have to replace the expansion (17.2) by

$$u_j = u_0(x) + (-1)^j \kappa^{-1} \Delta(x) \tag{17.22}$$

where now both $u_0(x)$ and $\Delta(x)$ are real fields. Substituting it into Eqs. (17.1) and (17.4) we obtain the following Lagrangian density:

$$\mathscr{L} = R_\sigma^+(\partial_\tau - iv\partial_x)R_\sigma + L_\sigma^+(\partial_\tau + iv\partial_x)L_\sigma + i\Delta(R_\sigma^+ L_\sigma - L_\sigma^+ R_\sigma)$$
$$+ \frac{1}{2\pi v\lambda}[(\partial_\tau\Delta)^2 + \omega_0^2\Delta^2] \tag{17.23}$$

This model was introduced by Takayama, Lin-Liu and Maki (1980) to describe an undoped polyacetylene. We shall postpone the further discussion of this model until the next Section.

I.3   Appendix

Here we give a proof of identity (17.17). The square of the operator in the left-hand side of Eq.(17.17) is really the time-split product ($\epsilon \sim \omega_l^{-1}$):

$$[: \partial_\tau \cos(\sqrt{2\pi}\Phi) :]^2 = \partial_\tau\partial_\epsilon \left\{ \cos[\sqrt{2\pi}\Phi(\tau)] \cos[\sqrt{2\pi}\Phi(\tau + \epsilon)] \right\}$$

$$= \frac{1}{2}\partial_\tau\partial_\epsilon \left\{ \cos[\sqrt{2\pi}\Phi(\tau) + \sqrt{2\pi}\Phi(\tau + \epsilon)] + \cos[\sqrt{2\pi}\Phi(\tau) - \sqrt{2\pi}\Phi(\tau + \epsilon)] \right\}$$

$$= \frac{1}{2}\partial_\tau\partial_\epsilon \left\{ \cos[\sqrt{8\pi}\Phi(\tau) - \epsilon\sqrt{2\pi}\partial_\tau\Phi(\tau)] \sin[\sqrt{8\pi}\Phi(\tau) + O(\epsilon^2)] \right\} \tag{17.24}$$

which gives the required result (17.17).

To double check this important result we calculate the pair correlation function of the operator

$$O = [: \partial_\tau \cos(\sqrt{2\pi}\Phi_s) :]^2 \tag{17.25}$$

we again treat the square as a product of two operators taken at slightly different points:

$$O = \partial_{\tau_1} A(z_1)\partial_{\tau_1} A(z_1 + \epsilon), \quad A = \cos(\sqrt{2\pi}\Phi_s) \tag{17.26}$$

Below we use the notations $z_2 = z_1 + \epsilon$, $z_3 = z_4 - \epsilon$.

Taking into account the normal ordering, we get the following expression for the pair correlation function:

$$\langle\langle O(1,2)O(3,4)\rangle\rangle \sim \partial_1\partial_2\partial_3\partial_4 \frac{1}{|z_{12}||z_{34}|}\left[\frac{|z_{13}||z_{24}|}{|z_{14}||z_{23}|} + \frac{|z_{14}||z_{23}|}{|z_{13}||z_{24}|} - 2\right]$$

$$= \partial_1\partial_2\partial_3\partial_4 \frac{|z_{12}||z_{34}|}{|z_{14}||z_{23}||z_{13}||z_{24}|} \approx \partial_1\partial_4 \frac{1}{|z_{14}|^4} \tag{17.27}$$

Thus the operator (17.25) has scaling dimension 3 which is consistent with Eq. (17.17).

II Spectral gap in the spin sector

Let us consider the case when the current–current interaction in Eq. (15.15) is antiferromagnetic ($g_s > 0$). Then the renormalization group analysis shows that the interaction is marginally relevant. Such a situation is described by the model where the electron–electron interaction is mediated by phonons. This model has the following Lagrangian density:

$$\mathscr{L}_f = R_\sigma^+(\partial_\tau - iv\partial_x)R_\sigma + L_\sigma^+(\partial_\tau + iv\partial_x)L_\sigma + \Delta^* R_\sigma^+ L_\sigma + \Delta L_\sigma^+ R_\sigma$$
$$+\frac{1}{2\pi v\lambda}\left(|\Delta|^2 + \omega_0^{-2}|\partial_\tau\Delta|^2\right) \tag{17.28}$$

For simplicity we use a single velocity v having in mind that for the spin sector we are interested in $v = v_s$.

Apart from the case where $\Delta(x)$ is complex, one can also consider the case of real Δ describing a half-filled band. Then the model (17.28) coincides with the model (17.23) introduced in the previous Section. We shall call these two models chiral and nonchiral TLM models respectively, after Takayama, Lin-Liu and Maki. It is instructive to study the two cases in parallel.

Let us start with the nonchiral model. Observe that, due to the particle–hole symmetry of the half-filled band, the U(1) symmetry of the charge sector is extended to SU(2). Thus the symmetry group of the nonchiral TLM model is SU(2)×SU(2) = O(4), as opposed to U(1)×SU(2) of the chiral model. For this reason it is natural to choose a basis of real fermions like in Eq. (12. 11):

$$R_\sigma = \frac{1}{\sqrt{2}}\left(\xi_R^{1,\sigma} + i\xi_R^{2,\sigma}\right) \tag{17.29}$$

(the same for left-moving particles). In this representation the Lagrangian of the nonchiral TLM model becomes manifestly O(4) invariant:

$$\mathscr{L} = \frac{1}{2}\xi_R^a(\partial_\tau - iv\partial_x)\xi_R^a + \frac{1}{2}\xi_L^a(\partial_\tau + iv\partial_x)\xi_L^a + i\Delta\xi_R^a\xi_L^a$$
$$+\frac{1}{2\pi v\lambda}\left[\Delta^2 + \omega_0^{-2}(\partial_\tau\Delta)^2\right] \tag{17.30}$$

where $a = 1,...4$.

There are two limiting cases where exact results are available. Namely, when $\omega_0 \to \infty$, the TLM models reduce to fermionic Gross–Neveu models well known in field theory. The Gross–Neveu models are exactly solvable by Bethe ansatz (Zamolodchikov and Zamolodchikov (1978, 1979), Andrei and Lowenstein (1980), see also the book by Smirnov in the General bibliography). For the opposite case $\omega_0 \to 0$ when the field $\Delta(x)$ is static, very interesting exact results were obtained by Brazovskii *et al.*

(1982) and Dzyaloshinskii and Krichever (1983). The spin gap persists in both limits. As we shall see below, it is exponentially small in $1/\lambda$: $M_s \sim (v_s/a)P(\lambda)\exp(-1/\lambda)$ with a power law prefactor $P(\lambda)$.

In our discussion we shall first consider the adiabatic description where the Δ-field is treated as a slow variable. The adiabatic approach is self-consistent when $\omega_0/M_s \ll 1$ which is the case for polyacetylene. The limit of the Gross–Neveu model corresponds to the opposite case $M_s/\omega_0 \ll 1$. Here the adiabatic approximation becomes self-consistent only for a modified model where the symmetry is extended from SU(2) and O(4) to SU(N) and O($2N$) respectively, with $N \gg 1$.

Since we expect Δ to be slow, we may obtain its effective action integrating over the fast variables (that is over fermions). The result for the chiral model is

$$S_{\text{eff}}[\Delta, \Delta^*] = \frac{1}{2\pi v\lambda}\int d\tau dx \left(|\Delta|^2 + \omega_0^{-2}|\partial_\tau\Delta|^2\right)$$
$$-N \ln\det\left[i\gamma_\mu\partial_\mu + \left(\frac{1}{2}-\sigma^3\right)\Delta + \left(\frac{1}{2}+\sigma^3\right)\Delta^*\right] \qquad (17.31)$$

where N is the number of fermionic species and $\gamma_0 = \sigma^x, \gamma_1 = \sigma^y$ are (1 + 1)-dimensional gamma matrices.

For the O($2N$) model we get

$$S_{\text{eff}}[\Delta] = \frac{1}{2\pi v\lambda}\int d\tau dx \left[\Delta^2 + \omega_0^{-2}(\partial_\tau\Delta)^2\right] - N\ln\det\left(i\gamma_\mu\partial_\mu + \Delta\right) \qquad (17.32)$$

The similarity between the models increases when we separate the U(1) charge density mode implicitly present in the action (17.31). To this end, we change the variables in the path integral over the complex Δ-field:

$$\Delta(x) = M(x)e^{i\sqrt{4\pi}\Phi_c(x)} \qquad (17.33)$$

and use the results derived in Section 14.III to extract the action for Φ_c.[*] Then the partition function of the chiral model becomes

$$Z = \int D\Phi_c(x)M(x)DM(x)\exp\{-S[\Phi_c] - S[M]\}$$

$$S[M] = \frac{1}{2\pi v\lambda}\int d\tau dx \left[M^2 + \omega_0^{-2}(\partial_\tau M)^2\right] - N\ln\det\left(i\gamma_\mu\partial_\mu + M\right) \qquad (17.34)$$

and $S[\Phi_c]$ for the charge field which we have discussed at length in Section II. As we see the resulting partition function for the radial component looks very similar to the partition function for the O($2N$) model (17.32). There are differences, however: (i) the field $M(x)$ is strictly positive and $\Delta(x)$ is not; and (ii) the $M(x)$-field has a different measure of integration.

---

[*] The field Φ_c is, of course, our familiar charge density field.

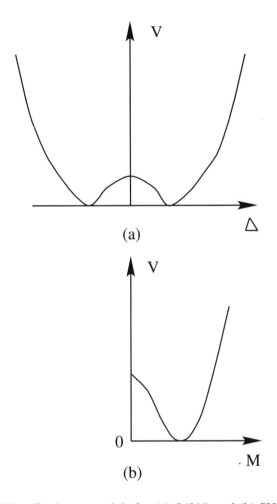

Fig. 17.1.   The effective potentials for (a) O($2N$) and (b) SU(N) models.

Now we have got rid of the massless field Φ_c, the ground is ready for the next step. The essence of the adiabatic approximation is that the determinants are calculated as if Δ is a constant field $\Delta = M_s$, but in the final result Δ is treated as a function of coordinates. The corrections to adiabatic approximation contain gradients of Δ, each gradient being multiplied by M_s^{-1}. A simple calculation gives the following effective potential (that is a part of the Lagrangian density not containing gradients)

$$V_{\text{eff}}(\Delta) = \frac{\Delta^2}{2\pi v \lambda} - \frac{N\Delta^2}{4\pi v} \ln(v/a|\Delta|) \qquad (17.35)$$

for the O($2N$) model with Δ replaced by M for the SU(N) model (Fig. 17.1).

The potential for $M(x)$ contains one minimum and the potential for Δ contains two minima at symmetric points $\Delta = \pm M_s$. From Eq. (17.35) we obtain

$$M_s = (v/a)\exp(-2/N\lambda) \tag{17.36}$$

This gives us (in the first approximation) the value of mass gap for fermions (the spin gap). Corrections to this expression will come from fluctuations of the Δ-field.

It is obvious that if the potential minimum is deep enough the radial component of the Δ-field is frozen. Therefore the asymptotic of its correlation function is determined by the gapless phase mode:

$$\langle \Delta(1)\Delta^*(2)\rangle \approx M_s^2[\omega_0^2(\tau^2 + x^2/\tilde{v}_c^2)]^{-K_c} \tag{17.37}$$

where $\tilde{v}_c$ and K_c are given by Eqs. (17.19). We emphasize that the presence of the gapless phase mode guarantees that the average displacement vanishes, $\langle \Delta(x)\rangle = 0$ even at zero temperature – a continuous symmetry (in the given case it is the U(1)-symmetry) cannot be spontaneously broken in two dimensions.

The form of the potential well suggests that the potential supports excitations – massive particles described by the deviations of $M(x)$ from its minimal value M_s. However, the existence of such a coherent mode depends crucially on the adiabaticity parameter ω_0/M_s. The exact solution for $\omega_0 \to \infty$ shows that, besides the gapless phason mode the only coherent excitations are fermionic.

Exercise. Derive the effective action for the field $u(x) = M(x) - M_s$ and find the dispersion law of this mode in the limit $\omega_0/M_s \ll 1$.

The situation with average displacement changes drastically in the commensurate case where the order parameter has a discrete Z_2 symmetry which can be broken at $T = 0$. Using the result of Rim and Weisberger (1984) we get the following expression for the two-point correlation function of the displacement fields:

$$\langle\langle \Delta(-\omega,-q)\Delta(\omega,q)\rangle\rangle \equiv D(\omega,q)$$

$$[D(\omega_n,q)]^{-1} = \frac{\omega^2}{\pi\lambda\omega_0^2} + \frac{N}{\pi}\sqrt{1 + 4M_s^2/s^2}\ln\left[s/2M_s + \sqrt{s^2/4M_s^2 + 1}\right] \tag{17.38}$$

where $s^2 = (\omega_n^2 + q^2v^2)$. (See also Fig. 17.2.)

We see that this function behaves very differently at $\omega_0 \ll M_s$ and

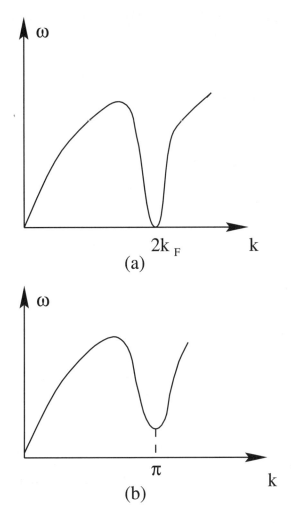

Fig. 17.2. The phonon spectrum for (a) incommensurate and (b) commensurate dimerized lattices.

$\omega_0 \gg M_s$. In the first case it has a pole in the complex plane at

$$i\omega_n = \omega \approx \sqrt{N\lambda}\omega_0 \left(1 + \frac{q^2 v^2}{24 M_s^2} + ... \right) \qquad (17.39)$$

corresponding to the optical phonon mode. In the second case there are no poles, just a cut at $|\omega^2 - v^2 q^2| > 4M_s^2$ signifying a threshold of the two-particle continuum.

Let us return to the physical case $N = 2$. As we have seen, the symmetry changes at half-filling from $U(1) \times SU(2)$ to $O(4)$. The latter symmetry, however is equivalent to SU(2)×SU(2), that is at half-filling the charge symmetry extends to SU(2).

Exercises.

(a) Demonstrate that the operators

$$R_\sigma^+ R_{-\sigma}^+, \quad R_{-\sigma} R_\sigma, \quad R_\sigma^+ R_\sigma$$

and their left moving counterparts are currents satisfying the $k = 1$ SU(2) Kac–Moody algebra.

(b) Show that bosonized expressions for these current operators are given by Eqs. (13.1) and (13.5) with ϕ_s, θ_s replaced by ϕ_c, θ_c.

(c) Show that the O(4) Gross–Neveu Hamiltonian can be represented as a sum of two identical $SU_1(2)$ WZNW Hamiltonians (15.15).

The latter fact enables us to derive a curious identity for the single electron Green's function. As we know, due to spin–charge separation asymptotics of this function always factorize like in Eq. (16.19):

$$G(\tau, x) = \exp(ik_F x)G_c^{(11)}(\tau, x)G_s^{(11)}(\tau, x) + \exp(-ik_F x)G_c^{(22)}(\tau, x)G_s^{(22)}(\tau, x)$$

$$(17.40)$$

where G_c and G_s are two-point correlation functions of operators with conformal spin $1/4$ belonging to the charge and the spin sectors respectively. Thus one may say that a real electron consists of two 'particles' with fractional spin. We take the word 'particle' in quotes because the corresponding Green's functions never have poles which means that objects with fractional spin do not propagate coherently. Since the charge sector of the chiral Gross–Neveu model is described by the Gaussian Hamiltonian (15.17), the functions G_c^{aa} for this model are the same as in Eq.(16.19):

$$G_c^{(11)}(\tau, x) = \frac{1}{\sqrt{(v_c\tau - ix)}} \left[\frac{a^2}{v_c^2\tau^2 + x^2} \right]^{\theta/2}$$

$$G_c^{(22)}(\tau, x) = \frac{1}{\sqrt{(v_c\tau + ix)}} \left[\frac{a^2}{v_c^2\tau^2 + x^2} \right]^{\theta/2} \qquad (17.41)$$

The expression for $G_s^{(aa)}$, however, is modified when the spin sector has a gap. We can find the asymptotic form of this expression using the fact that at half-filling, where the charge sector becomes identical to the spin one, we get $G_c^{(aa)} = G_s^{(aa)}$, which means that

$$\langle\langle R_\sigma(\tau, x)R_\sigma^+(0, 0)\rangle\rangle_{GN} = [G_s^{(11)}(\tau, x)]^2$$
$$\langle\langle L_\sigma(\tau, x)L_\sigma^+(0, 0)\rangle\rangle_{GN} = [G_s^{(22)}(\tau, x)]^2 \qquad (17.42)$$

where the functions on the left-hand side are correlation functions of the Gross–Neveu model. This identity is rigorous. In order to exploit it, we have to make some assumptions about the behaviour of the single electron Green's function in the Gross–Neveu model. It seems plausible that in the absence of soft modes this Green's function must contain a single particle pole, that is the asymptotics of these Green's functions must be proportional to that of free massive Dirac fermions (2.11). Then from (17.42) we get the following expression for the asymptotics:

$$G_s^{(11)}(\tau, x) = Z\left[rK_1(2M_s r)/(v_s\tau - ix)\right]^{1/2} \tag{17.43}$$

where $r^2 = \tau^2 + x^2/v_s^2$, and the same expression with $x \to -x$ for $G^{(22)}$. Since the asymptotics of G_s must be $\sim \exp[-M_s r]$, we suggest that the value of the mass gap of the Gross–Neveu fermions in the given regularization scheme is twice the spin gap in the chiral model with the same regularization.

Now we can calculate the single particle density of states. Combining Eqs. (17.40), (17.41) and (17.43) we get the following expression for the asymptotics at $M_s|\tau| > 1$:

$$G(\tau, x) \sim \frac{1}{\tau}|\tau|^{1/2-\theta}[K_1(2M_s|\tau|)]^{1/2} \tag{17.44}$$

from where it is not difficult to get that at zero temperature the single particle density of states has the following singularity at $0 < \omega - M_s \ll M_s$:

$$\rho(\omega) \sim (\omega - M_s)^{-1/4+\theta} \tag{17.45}$$

Recall that free massive fermions has exponent $(-1/2)$ which is never achieved here even for $K_c = 1$ when $\theta = 0$.

III Optical conductivity

In the absence of charge gap, electric current along the chains is carried by the charge field Φ_c. Therefore measurements of frequency dependent conductivity may provide valuable information about the dynamics of this mode. In what follows we shall describe the charge density waves with the modified Gaussian model (17.20) which takes into account the retardation effects. A more detailed analysis can be found in papers by Giamarchi (1991, 1992, 1997). The charge–charge and current–current correlation functions at small q in this model are given by

$$\langle\langle\rho(-\omega_n, -q)\rho(\omega_n, q)\rangle\rangle = -\frac{2e^2}{\pi}\frac{q^2}{\omega_n^2 f(i\omega_n)/v_c + q^2 v_c}$$

$$\langle\langle j(-\omega_n, -q)j(\omega_n, q)\rangle\rangle = -\frac{2e^2}{\pi}\frac{\omega_n^2}{\omega_n^2 f(i\omega_n)/v_c + q^2 v_c} \tag{17.46}$$

We can use the first expression to find the plasma frequency. For wave vectors along the chain direction the equation for the plasma wave dispersion is

$$1 - b^2 \frac{4\pi e^2}{q^2} \langle\langle \rho(-\omega_n, -q)\rho(\omega_n, q) \rangle\rangle = 0 \qquad (17.47)$$

where b is the lattice constant in the direction perpendicular to the chains, which gives us

$$\omega_p^2 = 8e^2 b^2 v_c / f(\omega_p) \rightarrow 8e^2 b^2 v_c \qquad (17.48)$$

Here we assume that the plasma frequency exceeds the scale where the retardation is effective.

The optical conductivity is related to the current–current correlation function at $q = 0$:

$$\sigma_\parallel(\omega) = i \left[\frac{b^2}{\omega + i\delta} \langle\langle j(-\omega_n, 0)j(\omega_n, 0) \rangle\rangle |_{i\omega_n = \omega + i\delta} \right] = \frac{\omega_p^2}{4\pi(\delta - i\omega)f(\omega)} \qquad (17.49)$$

Thus we see that the real part of the conductivity is pinned to zero frequency:

$$\Re e\sigma_\parallel(\omega) = \frac{m}{4m^*}\omega_p^2 \delta(0) \qquad (17.50)$$

In real systems this delta-functional peak is broadened and moved to finite frequencies due to interactions with impurities and imperfections of the lattice (see Fig. 17.3).[†]

Such low-frequency features are observed in real quasi-one-dimensional materials. Calculating the weight concentrated at low frequencies one can find K_c. Alternatively, one can calculate the weight at higher frequencies and compare it with the exact sum rule for the optical conductivity:

$$K_c^2 = \frac{m}{m^*} = \frac{8}{\omega_p^2} \int_0^\epsilon \Re e\sigma_\parallel(\omega)d\omega = \frac{8}{\omega_p^2}[1 - \int_\epsilon^\infty \Re e\sigma_\parallel(\omega)d\omega] \qquad (17.51)$$

where ϵ is some suitable low-frequency cut-off.

In contrast to $\sigma_\parallel$ the conductivity in the transverse direction $\sigma_\perp$ in a really one-dimensional system is always incoherent. In fact, this incoherence can be taken as a criterion of one-dimensionality. Let us suppose that the interchain hopping is very small and calculate the first nonvanishing contribution to $\sigma_\perp$. Since current in transverse direction may be carried by processes of different kinds (single- and multi-particle ones) it will be convenient to derive a general formula for $\sigma_\perp(\omega)$ assuming the following

---

[†] One could not fail to notice that Fig. 17.3 (a) looks exactly like the optical conductivity of a superconductor.

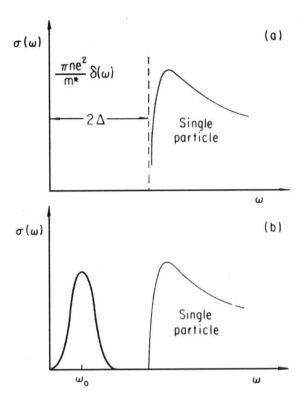

Fig. 17.3. Optical conductivity (a) without pinning, and (b) with pinning and damping. The response at frequencies $\omega > 2\Delta$ is due to single-particle excitations (from Grüner (1988)).

expression for the current density in the **b**-direction:

$$J_\mathbf{b}(\mathbf{n}, x) = iet_\perp [O_\mathbf{n}(x)O^+_{\mathbf{n}+\mathbf{b}}(x) - O_{\mathbf{n}+\mathbf{b}}(x)O^+_\mathbf{n}(x)] \tag{17.52}$$

where $O_\mathbf{n}(x)$ is an electron or pair annihilation operator on a chain with a coordinate $\mathbf{n}$ and $t_\perp$ is a hopping integral connecting neighbouring chains. In the leading order in $t_\perp$ we get the following general expression for the real part of the transverse conductivity:

$$\sigma_\perp(\omega) = \frac{e^2 t_\perp^2}{2\pi^2\omega} \int dq\, dx [n(x) - n(\omega + x)]\Im m D^{(R)}(x, q)\Im m D^{(R)}(\omega + x, q) \tag{17.53}$$

where $n(x)$ is either the Fermi or the Bose distribution function depending on the nature of the operator $O(x)$ and $D(\omega, q)$ is the pair correlation function of these operators. A simple power counting gives the following scaling form for $\sigma_\perp(\omega, T)$:

$$\sigma_\perp(\omega, T) = T^{-3+4d}\mathscr{F}_d(\omega/T) \tag{17.54}$$

where $d = \Delta + \bar{\Delta}$ and $\mathscr{F}_d$ is a crossover function. At $d = 1/2$ this crossover function simplifies for bosons:

$$\mathscr{F}_{1/2}(x) = \frac{\tanh(x/4)}{x}$$

If $d < 3/4$ the transverse conductivity has a maximum at $\omega = 0$ which can be mistaken for a Drude peak. The width of this peak, however, scales with the first power of T which betrays its incoherent nature.

For the Tomonaga–Luttinger liquid we have $d = (\theta + 1)/2$ for electrons and $d = 1 + K_c^{-1}$ for pairs. The former dimension is smaller for all physical values of K_c and therefore single particle processes always give dominant contribution to the conductivity in the transverse direction unless there is a spin gap. From Eq. (17.54) we see that when $d > 3/4$ the interchain conductivity of a Tomonaga–Luttinger liquid does not have a maximum at $\omega = 0$. In this state any semblance of the Drude peak in transverse direction disappears.

Single particle transport is suppressed when a spin gap is opened. Then electric current in the transverse direction is transported by pairs. In the presence of spin gap the $\cos\sqrt{2\pi}\Phi_s$-part of the pair operator (16.5) is frozen which leads to the change of the dimensionality of the pair operator: $d = K_c^{-1}$. In general we have for the pair conductivity

$$\sigma(T)_{\text{pair}} \sim T[\chi_{\text{sc}}(T)]^2 \tag{17.55}$$

Therefore we can have a situation where the susceptibility diverges and the conductivity (in the leading order in $t_\perp$, of course) increases.

IV   Gap in the charge sector at half-filling and the case of small doping

We have shown in Chapter 15 that at half-filling ($2k_F = \pi$), when Umklapp processes become important, the Hamiltonian in the charge sector, Eq. (15.23), can be expressed entirely in terms of the pseudospin currents (15.22). In the condensed matter community this model is known as the Luther–Emery model (Luther and Emery (1974))[‡] and is called the U(1) Thirring model by field theorists. In the Appendix to this chapter we derive renormalization group equations for the model (15.23), and show

---

[‡] In their original paper, Luther and Emery considered the model (15.23) in the spin sector (see Chapter 18).

that they have the form of the Kosterlitz–Thouless equations (10.22)

$$\frac{dz_c}{dl} = z_u^2, \quad \frac{dz_u}{dl} = z_c z_u, \tag{17.56}$$

where

$$z_i = \frac{2g_i}{\pi v_c}, \quad i = c, u$$

are dimensionless coupling constants. These equations directly indicate that there should be a connection between the Hamiltonian (15.23) and the SG model (see Chapter 10). To make this relationship transparent, let us apply Abelian bosonization to H_c, using the rules of Chapter 13. We obtain:

$$H_c = \frac{u_c}{2} \int dx \left[\frac{1}{K_c}(\partial_x \Phi_c)^2 + K_c(\partial_x \Theta_c)^2 \right] - \frac{m}{\pi a_0} \int dx \cos \sqrt{8\pi} \Phi_c \tag{17.57}$$

$$v_c + \frac{g_c}{\pi} = \frac{u_c}{K_c}, v_c \frac{g_c}{\pi} = u_c K_c, m = \frac{g_u}{\pi a} \tag{17.58}$$

As before, relations (17.58) are universal only at small coupling constants where

$$K_c = 1 - \frac{g_c}{\pi v_c} + O(g_c^2), \quad u_c = v_c[1 + O(g_c^2)]$$

We see from (17.57) that the perturbation generated by Umklapp processes is proportional to $\rho(4k_F) \sim \cos(\sqrt{8\pi}\Phi_c)$. Thus one can say that at half-filling the $4k_F$-component of charge density couples to the lattice. This operator is relevant at $K_c < 1$ and creates a spectral gap in the charge sector. The spin sector remains massless and is described by the $SU_1(2)$ WZNW model, as has already been discussed in Chapter 13.

Rescaling the fields,

$$\Phi_c \to \sqrt{K_c}\Phi, \quad \Theta_c \to \frac{1}{\sqrt{K_c}}\Theta,$$

brings the Hamiltonian (17.57) to the canonical form with $\beta^2 = 8\pi K_c$:

$$H_c = \int dx \left[\frac{u_c}{2} \left(\Pi^2 + (\partial_x \Phi)^2 \right) - \frac{m}{\pi a_0} \cos \beta \Phi \right] \tag{17.59}$$

The spectral gap disappears when the band filling deviates from $1/2$. We shall call this deviation 'doping'. If the amount of doping δ is small, there is an additional energy scale in the system $\sim \delta(u_c/a_0)$. One may suggest that at energies much smaller than this scale the standard Tomonaga–Luttinger liquid description can be employed, while for energies much larger than $\delta v_c a_0^{-1}$ the effects of commensurability are important, and the adequate description should be given by the sine-Gordon Hamiltonian

(17.59). The most challenging problem is therefore to describe a crossover between these two regimes.

At the bosonization level, the problem of quantum 'commensurate–incommensurate' transition was studied by Japaridze and Nersesyan (1978) in the spin context, i.e. as a problem of the phase transition driven by a magnetic field in a one-dimensional Fermi system with a spin gap (see Chapter 18), and by Pokrovsky and Talapov (1979) who considered this problem in the charge sector. The exact solution based on the Bethe-ansatz equations for the U(1) Thirring model was given by Japaridze *et al.* (1984). Finite doping can be introduced in the Hamiltonian (17.59) by adding a term

$$H_{\text{dop}} = -\mu \int dx [I^z(x) + \bar{I}^z(x)] = -\mu \sqrt{\frac{K_c}{2\pi}} \int dx \partial_x \Phi \qquad (17.60)$$

where μ is the chemical potential. The presence of this term makes it necessary to consider the ground state of the sine-Gordon model in the sector with nonzero topological charge. As follows from (17.60), the chemical potential tends to create a finite and uniform density of the topological charge along the chain, whereas the potential energy $-m \cos \beta \Phi$ tends to fix the field Φ in one of its minima ($\Phi = 0 \pmod{2\pi/\beta}$). This competition is resolved as a continuous phase transition from the gapful ground state with zero concentration of holes, $\langle \partial_x \Phi \rangle = 0$, to a gapless state with a finite concentration of holes, $\langle \partial_x \Phi \rangle \neq 0$, taking place at some critical value μ_c (due to the particle–hole symmetry, there are actually two symmetric critical points, $\mu = \pm \mu_c$). The transition is related to the instability of the system with respect to the spontaneous creation of a finite density of quantum solitons at $|\mu| > \mu_c$.

As explained in detail in the next chapter, this transition is most easily described at the Luther–Emery point $\beta^2 = 4\pi$, where an exact mapping onto free massive spinless fermions becomes available. The fermionized form of the chemical potential term is

$$H_{\text{dop}} = -\mu \int dx (: R^\dagger R : + : L^\dagger L :) \qquad (17.61)$$

At $\mu = 0$ the ground state of the system is a two-band insulator, with the chemical potential located at the center of the gap ($\mu = 0$) (see Fig. 17.4.).

As long as $|\mu| < m$, no changes in the ground state occur. The transition takes place at the critical value of μ equal to the quantum soliton mass, $\mu_c = m$. At $|\mu| > \mu_c$, the upper band gets partially filled (or equivalently, the lower band gets partially emptied), a finite soliton density appears in the ground state, and the Fermi surface opens with Fermi points $\pm k_0 = \pm \sqrt{\mu^2 - m^2}/u_c$. The value of doping, proportional to the particle (hole) density in the upper (lower) band, and the compressibility display square-

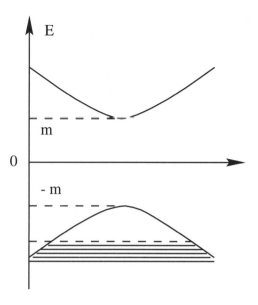

Fig. 17.4. The spectrum in the charge sector in the presense of small doping.

root singularities at the right-hand (left-hand) vicinity of the transition point (being zero everywhere below the transition):

$$\delta \sim (|\mu| - \mu_c)^{1/2}, \quad \chi_c \sim (|\mu| - \mu_c)^{-1/2}, \quad (|\mu| \to \mu_c + 0) \qquad (17.62)$$

These singularities are *universal*, i.e. valid for any finite β, since interaction between Thirring fermions (quantum sine-Gordon solitons), which is present at $\beta^2 \neq 4\pi$, does not lead to infrared divergences in the corresponding perturbative expansions because of a finite mass of the particles.

At $K_c < 1/2$ breathers emerge (see the discussion in Chapter 11). In the framework of the TLS model discussed in Section II, these particles correspond to the optical phonons. The quasiclassical limit $\beta \to 0$ is especially interesting because solitons in this case are much heavier than the breathers. It was shown by Dzyaloshinskii (1965) that in this limit

$$\delta \sim \left[\ln \left(\frac{\mu_c}{|\mu| - \mu_c} \right) \right]^{-1}$$

At small but finite β, there exists a narrow vicinity of the critical point

$$(|\mu| - \mu_c)/\mu_c \ll \beta^2/8\pi$$

where quantum fluctuations are important, and the universal square-root singularities are recovered. The quasiclassical limit has been thoroughly studied by Haldane (1982) and recently by Caux and Tsvelik (1996).

V   Appendix. RG equations for the model of one-dimensional electrons from the SU(2) current algebra

In this Appendix we derive renormalization group equations for the model (15.12) and its anisotropic generalization (17.49). As was shown in Chapter 15, the Hamiltonian of interacting fermions with spin-1/2 separates into two commuting parts describing the U(1)-symmetric charge and SU(2)-symmetric spin excitations, $H = H_c + H_s$. Since the charge sector is described by an Abelian Gaussian theory, with the marginal interaction $g_1 J \bar{J}$ easily eliminated by a Bogoliubov rotation of the U(1) currents, the coupling constant g_1 is not renormalized. On the other hand, the symmetry group in the spin sector is non-Abelian; therefore the coupling constant g_s is subject to renormalization. The corresponding RG equations can be easily derived, using OPE for the SU(2) spin currents (see e.g. Affleck and Ludwig (1991)). Below we shall consider the most general case when interaction in the spin sector is characterized by three independent couplings

$$g \mathbf{J} \cdot \bar{\mathbf{J}} \to \sum_{a=1,2,3} g_a J^a \bar{J}^a \qquad (17.63)$$

Apart from the SU(2)-symmetric case ($g_a = g$, $a = 1, 2, 3$) corresponding to the SU(2)-Thirring (or chiral Gross–Neveu) model, this will also allow us to obtain RG equations for the U(1)-symmetric Thirring (or Luther–Emery) model ($g_1 = g_2 \neq g_3$). We shall also consider the case of fully broken SU(2) symmetry ($g_1 \neq g_2 \neq g_3$) which becomes realistic when spin–orbit interaction, giving rise to spin-nonconserving processes, is taken into account (Giamarchi and Schulz (1988)).

Consider S-matrix operator for the Hamiltonian H_σ

$$S = T \exp\left(-i \int_{-\infty}^{\infty} dt H_{\text{int}}(t)\right) \qquad (17.64)$$

and write down perturbation theory expansion

$$
\begin{aligned}
S &= T \exp\left(-i \sum_a g_a \int d^2x \, \bar{J}^a(x^-) J^a(x^+)\right) \\
&= 1 - i \sum_a g_a \int d^2x \, \bar{J}^a(x^-) J^a(x^+) \\
&\quad - \frac{1}{2} \sum_{ab} g_a g_b \int d^2x_1 \int d^2x_2 T \, [\bar{J}^a(x_1^-) J^a(x_1^+) \bar{J}^b(x_2^-) J^b(x_2^+)]
\end{aligned}
$$

$$+ \text{ higher-order terms.} \qquad (17.65)$$

The idea is to use OPE for products of two right and left SU(2) currents in the second-order term of (17.65) and retain those singular terms which, after integrating over the relative 2-coordinate $x_1 - x_2$, recover the form of the first-order term of the expansion (17.65). The result will be the lowest-order (one-loop) renormalization of the g_a.

Rewrite the time-ordered product of four currents in (17.65) as

$$\theta(t_1 - t_2)J^a(1)J^b(2)\bar{J}^a(1)\bar{J}^b(2) + \theta(t_2 - t_1)J^b(2)J^a(1)\bar{J}^b(2)\bar{J}^a(1) \quad (17.66)$$

and use the OPE:

$$J^a_{R,L}(x^{\mp}_1)J^b_{R,L}(x^{\mp}_2) = \frac{\epsilon^{abc}J_{R,L}(x^{\mp}_1)}{2\pi(x^{\mp}_1 - x^{\mp}_2 - i\delta)} - \frac{\delta^{ab}}{8\pi^2(x^{\mp}_1 - x^{\mp}_2 - i\delta)^2} + \cdots \quad (17.67)$$

where dots stand for nonsingular terms. Then

$$T\,[\bar{J}^a(x^-_1)J^a(x^+_1)\bar{J}^b(x^-_2)J^b(x^+_2)]$$

$$= \theta(t_1 - t_2)\frac{\epsilon^{abc}\epsilon^{abd}\bar{J}^c(x^-_1)J^d(x^+_1)}{4\pi^2(x^-_1 - x^-_2 - i\delta)(x^+_1 - x^+_2 - i\delta)}$$

$$+\theta(t_2 - t_1)\frac{\epsilon^{bac}\epsilon^{bad}\bar{J}^c(x^-_1)J^d(x^+_1)}{4\pi^2(x^-_2 - x^-_1 - i\delta)(x^+_2 - x^+_1 - i\delta)}$$

$$= -\frac{1}{4\pi^2}\epsilon^{abc}\epsilon^{abd}\bar{J}^c(X)J^d(X)\left(\frac{\theta(\tau)}{(\rho - \tau + i\delta)(\rho + \tau - i\delta)}\right.$$

$$\left.+\frac{\theta(-\tau)}{(\rho - \tau - i\delta)(\rho + \tau + i\delta)}\right)$$

where $\tau = t_1 - t_2$ and $\rho = x_1 - x_2$, while X is the center-of-mass 2-coordinate.

Now we integrate this over τ and ρ:

$$\int d\tau \int d\rho \left(\frac{\theta(\tau)}{(\rho - \tau + i\delta)(\rho + \tau - i\delta)} + \frac{\theta(-\tau)}{(\rho - \tau - i\delta)(\rho + \tau + i\delta)}\right)$$

$$= 2\pi i \int d\tau \left(-\frac{\theta(\tau)}{2\tau} + \frac{\theta(-\tau)}{2\tau}\right)$$

$$= -i\pi \int \frac{d\tau}{|\tau|} = -2\pi i \ln \frac{L}{a}$$

As a result

$$\delta^{(2)}S = -\frac{i}{2\pi} \ln \frac{L}{a} \int d^2x\,(g_1g_2J^3\bar{J}^3 + g_2g_3J^1\bar{J}^1 + g_3g_1J^2\bar{J}^2) \quad (17.68)$$

Combining this with $\delta^{(1)}S$ in (17.65), we get

$$\delta^{(1)}S + \delta^{(2)}S = -i\sum_a g'_a \int d^2x J^a(x^-)\bar{J}^a(x^+) \quad (17.69)$$

with renormalized coupling constants

$$g_a' = g_a + \frac{1}{2\pi} g_b g_c \ln \frac{L}{a} \tag{17.70}$$

where (a,b,c) form a cyclic permutation of $(1,2,3)$.

From (17.70) we deduce differential RG equations which have a very symmetric form:

$$\frac{dg_1}{dl} = \frac{1}{2\pi} g_2 g_3$$
$$\frac{dg_2}{dl} = \frac{1}{2\pi} g_3 g_1 \tag{17.71}$$
$$\frac{dg_3}{dl} = \frac{1}{2\pi} g_1 g_2$$

where $l = \ln L/a$.

In the SU(2)-symmetric case, the three equations (17.71) reduce to a single one

$$\frac{dg}{dl} = \frac{1}{2\pi} g^2 \tag{17.72}$$

indicating the increase of the effective amplitude $g(l)$ upon renormalization for positive values of the bare constant g. In the Hubbard model this corresponds to the case of *attractive* on-site interaction, $U < 0$. The development of a strong-coupling regime leads to dynamical generation of a mass gap in the spin sector whose magnitude can be determined with exponential accuracy by the position of the (unphysical) pole in the solution of Eq. (17.72):

$$|\Delta| \simeq \Lambda \exp(-2\pi/g) \tag{17.73}$$

At $g < 0$, which is the case of a *repulsive*, non-half-filled Hubbard model, $g(l)$ scales down to zero (zero-charge regime),

$$g(l) \simeq \frac{1}{\ln(\Lambda/a)}$$

Interaction in the spin sector is therefore irrelevant, and spin degrees of freedom remain gapless, described by the critical $SU_1(2)$ WZW model.

In the U(1)-symmetric case, the action (15.12) must be replaced by its anisotropic generalization (15.23). Then we recover a pair of the Kosterlitz–Thouless equations already discussed in detail in Section 10.I:

$$\frac{dg_\parallel}{dl} = -g_\perp^2, \tag{17.74}$$

$$\frac{dg_\perp}{dl} = -g_\parallel g_\perp, \tag{17.75}$$

where we have relabeled the coupling constants as follows

$$g_1 = g_2 = -2\pi g_\perp, \quad g_3 = -2\pi g_\parallel$$

If the band is half-filled, the action (15.12) must be supplemented by Umklapp processes describing hopping of two right (left) fermions across the Fermi surface

$$g_U[R^\dagger_\uparrow R^\dagger_\downarrow L_\downarrow L_\uparrow + \text{H.c.}] \tag{17.76}$$

These processes contribute to the charge sector and at $g_c > 0$ are marginally relevant. There is no need to start deriving RG equations in the charge sector independently. These can be obtained from Eqs. (17.75) by applying to the action (15.23) the particle–hole transformation only in the spin-up component of the Fermi field:

$$R_\uparrow \leftrightarrow R^\dagger_\uparrow, \quad L_\uparrow \leftrightarrow L^\dagger_\uparrow, \quad R_\downarrow \to R_\downarrow, \quad L_\downarrow \to L_\downarrow \tag{17.77}$$

Under this transformation, the free part of the fermionic action is invariant, while the charge and spin degrees of freedom are interchanged. In particular, the $g_\perp$–proportional, spin-flip backscattering term in the U(1)-symmetric model (15.23) maps onto Umklapp term (17.76), while the $g_\parallel$-term maps onto the g_c-one. As a result, we get one more pair of the Kosterlitz–Thouless equations for effective couplings $g_c(l)$ and $g_U(l)$:

$$\frac{dg_c}{dl} = -g_U^2 \tag{17.78}$$

$$\frac{dg_U}{dl} = -g_c g_U \tag{17.79}$$

References

I. Affleck and A. W. W. Ludwig, *Nucl. Phys.* **B360**, 641 (1991).

N. Andrei and J. Lowenstein, *Phys. Lett.* **B90**, 106 (1980).

S. A. Brazovskii and I. E. Dzyaloshinskii, *Sov. Phys. JETP* **44**, 1233 (1976).

S. A. Brazovskii, I. E. Dzyaloshinskii and I. M. Krichever, *Sov. Phys. - JETP* **56**, 212 (1982).

J.-S. Caux and A. M. Tsvelik, *Nucl. Phys.* **B474**, 715 (1996).

I. E. Dzyaloshinskii, *Sov. Phys. JETP* **20**, 665 (1965).

I. E. Dzyaloshinskii and I. M. Krichever, *Sov. Phys. - JETP* **58**, 1031 (1983).

K. B. Efetov and A. I. Larkin, *Sov. Phys. JETP* **42**, 390 (1976).

H. Fukuyama, *J. Phys. Soc. Jap.* **41**, 513 (1976).

H. Fukuyama and P. A. Lee, *Phys. Rev.* **B17**, 535 (1978).

T. Giamarchi, *Phys. Rev.* **B44**, 2905 (1991); *ibid.* **46**, 343 (1992); *Physica* **B230**, 975 (1997).

T. Giamarchi and H. J. Schulz, *J. de Physique* (Paris) **49**, 819 (1988).

G. Grüner, *Rev. Mod. Phys.* **60**, 1129 (1988).

F. D. M. Haldane, *J. Phys.* A**15**, 507 (1982).

A. J. Heeger, S. Kivelson, J. R. Schrieffer and W. P. Su, *Rev. Mod. Phys.* **60**, 781 (1988).

G. I. Japaridze and A. A. Nersesyan, *Pis'ma Zh. Eksp. Teor. Fiz.* **27**, 356 (1978) [*Sov. Phys. JETP Lett.* **27**, 334 (1978)].

G. I. Japaridze, A. A. Nersesyan and P. B. Wiegmann, *Nucl. Phys.* B**230**, 511 (1984).

A. Luther and V. J. Emery, *Phys. Rev. Lett.* **33**, 589 (1974).

V. Meden and K. Schönhammer, *Phys. Rev.* B**46**, 15 753 (1992).

V. L. Pokrovsky and A. Talapov, *Phys. Rev. Lett.* **42**, 65 (1979).

C. Rim and W. I. Weisberger, *Phys. Rev.* D**30**, 1763 (1984).

M. Takayama, Y. R. Lin-Liu and K. Maki, *Phys. Rev.* B **21**, 2388 (1980).

J. Voit, *Phys. Rev. Lett.* **63**, 324 (1990); *Phys. Rev.* B**47**, 6740 (1993); *J. Phys.: Condens. Matt.* **5**, 8305 (1993).

J. Voit and H. J. Schulz, *Phys. Rev.* B**34**, 7429 (1986).

A. B. Zamolodchikov and Al. B. Zamolodchikov, *Nucl. Phys.* B **133**, 525 (1978); *Phys. Lett.* B**72**, 481 (1978); *Ann. Phys.* (N.Y.) **120**, 253 (1979).

18
Interacting fermions with broken spin rotational symmetry

In previous chapters we have been concerned with one-dimensional interacting Fermi systems possessing SU(2) spin rotational symmetry. In this chapter we generalize the discussion to the case of partially ($Z_2 \times U(1)$) or fully ($Z_2 \times Z_2 \times Z_2$) broken SU(2) symmetry. There are several reasons why the spin-anisotropic case is interesting.

First of all, it is important from the theoretical point of view. As we have seen in previous Chapters 16 and 17, systems with spin-isotropic short-range interaction can display two qualitatively different types of infrared behaviour. For repulsive interactions, $g_s < 0$, a Tomonaga–Luttinger liquid phase in realized, with gapless collective excitation spectrum, whereas in the case of attraction $g_s > 0$ a strong-coupling regime develops, accompanied by dynamical generation of a spin gap. Extending exchange interaction to an anisotropic, two-constant one,

$$g_s \mathbf{J} \cdot \bar{\mathbf{J}} \rightarrow g_\parallel J^z \bar{J}^z + \frac{1}{2} g_\perp (J^+ \bar{J}^- + \text{H.c.})$$

one expects to observe a richer phase diagram and, on changing the anisotropy parameter, study in detail the crossover between the weak-coupling ($g_s = -|g_\perp|$) and strong-coupling ($g_s = |g_\perp|$) isotropic regimes.

The resulting field-theoretical fermionic model is known as the isotopic U(1)-symmetric Thirring model

$$\mathcal{L} = i\bar{\Psi}_\alpha \gamma^\mu \partial_\mu \Psi_\alpha - \frac{1}{4} \sum_{a=1,2,3} g_a j^a_\mu j^{a\mu}, \quad g_1 = g_2 = g_\perp, \ g_3 = g_\parallel, \quad (18.1)$$

where $j^a_\mu = (1/2)\bar{\Psi}_\alpha \gamma^\mu \tau^a_{\alpha\beta} \Psi_\beta$ are the (iso)spin currents. We have already been dealing with the 'charge' version of this model in the previous chapter, with the spin currents replaced by the pseudospin ones (see Eqs. (15.22), (15.23)). In this chapter we shall discuss the U(1) Thirring model (and its XYZ extension) in much more detail, and focus on the magnetic properties of one-dimensional Fermi systems with spin-exchange anisotropy.

Together with the Hubbard and SU(2)-symmetric Thirring models, the U(1) model (18.1) is exactly integrable (Dutyshev (1980); Truong and Schotte (1981)); its (iso)spin properties have been studied in detail on the basis of the Bethe-ansatz solution (Japaridze *et al.* (1984)). Probably the most remarkable fact about the U(1) Thirring model is that it serves as a perfect example of 'Fermi–Bose democracy'. Indeed, the infrared properties of the fermionic model (18.1) in the (iso)spin sector are equivalently described by a nonlinear bosonic model, – the quantum sine-Gordon model (Luther and Emery (1974)). (This equivalence has already been exploited in Chapter 17 to describe generation of a charge gap in the case of a half-filled band.) This equivalence reveals the fact that, in the strong-coupling, massive phase, spin dynamics of one-dimensional Fermi systems is essentially nonlinear: elementary spin excitations are identified as quantum solitons of the SG model, with the spin gap coinciding with the soliton rest mass. This point of view is confirmed by one more observation that, in the strong-coupling regime ($\beta^2 < 8\pi$), the SG model can be mapped onto a system of massive spinless interacting fermions, the massive Thirring (MT) model (Coleman (1975)). Thus, quantum solitons of the SG model show up as fundamental fermions in the MT theory. It is very fortunate that there exists a special (though nonuniversal) value of $g_\parallel$, for which the SG coupling constant $\beta^2 = 4\pi$, and the massive Thirring fermions become free! It is the sequence of mappings, U(1) Thirring model $\to$ SG model $\to$ MT model, and the existence of an exactly solvable point that allowed Luther and Emery to obtain a full nonperturbative description of one-dimensional interacting spin-1/2 fermions in the strong-coupling regime. This equivalence triad will be described below at the level of bosonization.

On the other hand, the spin-anisotropic case is not purely academic. There are quasi-one-dimensional systems whose magnetic properties are caused by the spin-exchange anisotropy. Typical examples are compounds with tetramethyltetraselenafulvalene and tetramethyltetrathiafulvalene, (TMTSF)$_2$X and (TMTTF)$_2$X, undergoing magnetic phase transitions at a temperature $T \sim 10$ K. A three-dimensional antiferromagnetically ordered state below this temperature is characterized by an easy anisotropy axis in a direction perpendicular to the conducting chains, and one more anisotropy in the plane perpendicular to the easy axis (Torrance (1983)). Under a magnetic field parallel to the easy axis, the so-called *spin-flop* transition takes place: at $H \sim 4.5$ kG the antiferromagnetically ordered spins flop to the other axis (Torrance *et al.* (1982)). To account for the two-axis anisotropy, one has to incorporate spin-orbit and dipole–dipole magnetic interactions between the electrons into the 'g-ology' approach. This was done by Giamarchi and Schulz (1988) on the basis of a model of interacting one-dimensional fermions with *fully* broken spin rotational symmetry – the XYZ version of the model (18.1) with three independent coupling constants g_a. Notice that such a model possesses an Ising-like, discrete symmetry and does not conserve the total spin.[*]

At the theoretical level, the case of fully broken SU(2) symmetry is also quite interesting. As we shall discuss below, it turns out to be related to a nontrivial extension of the SG model in which, apart from the cosine of the original scalar field, a cosine of the dual field also appears in the Hamiltonian due to spin-nonconserving scattering processes of the original fermions. Such a model, in turn, has its counterpart in two-dimensional classical statistics where it is known as the Z_4-model of planar spins (Jose *et al.* (1978); Wiegmann (1978)), and can be reformulated in terms of a 2D Coulomb gas of charges and monopoles. Curiously enough, magnetic properties of the Thirring model (18.1) with broken SU(2)-symmetry find an unexpected application in the description of an orbital antiferromagnetic state in a toy model of two coupled chains (Nersesyan (1991)).

I U(1)-symmetric Thirring model: relation to sine-Gordon and massive Thirring models

We start from the charge–spin separated Hamiltonian density, $H = H_c + H_s$, corresponding to the U(1) Thirring model (18.1), with H_c describing gapless charge excitations (no Umklapp processes). In what follows, we shall focus solely on the spin degrees of freedom. The spin Hamiltonian density H_s expressed in terms of the chiral vector currents takes the form identical to the charge Hamiltonian H_c, Eq. (15.23), with the following replacements:

$$\mathbf{I} \to \mathbf{J}, \quad \bar{\mathbf{I}} \to \bar{\mathbf{J}} \tag{18.2}$$

$$g_c \to \frac{1}{4}g_{\parallel}, \quad g_u \to \frac{1}{4}g_{\perp}, \quad v_c \to v_s \tag{18.3}$$

The Abelian bosonization of the spin currents transforms H_s to the quantum SG model

$$H_s = \frac{u_s}{2}[\Pi_s^2 + (\partial_x \Phi_s)^2] - \frac{m}{\pi a_0} \cos \beta_s \Phi_s \tag{18.4}$$

with

$$\beta_s^2 = 8\pi K_s, \quad m = \frac{g_{\perp}}{4\pi a_0} \tag{18.5}$$

---

[*] In spite of this fact, in zero magnetic field the XYZ Thirring model is also exactly solvable (Truong and Schotte (1981)).

where

$$K_s = 1 - \frac{g_\parallel}{4\pi v_s} + 0(g_\parallel^2) \qquad (18.6)$$

The relations (18.5) and (18.6) are valid at small coupling constants, i.e. in the vicinity of $\beta_s^2 = 8\pi$. It is this region where the infrared behaviour of the U(1) Thirring model can be attributed to the Kosterlitz–Thouless universality class. As follows from the comparison of their Bethe-ansatz solutions (see e.g. Japaridze *et al.* (1984)), the equivalence between the U(1) Thirring and SG models holds beyond this region, including the whole asymptotic freedom massive phase, $\beta_s^2 < 8\pi$ (at least when $g_\perp$ is kept small).

Consider now the SG model in the strong-coupling phase ($\beta_s^2 < 8\pi$). Let us map this model onto massive spinless fermions. This can be achieved by the following canonical transformation of the field and its conjugate momentum:

$$\Phi_s \to \sqrt{\frac{4\pi}{\beta_s^2}}\Phi_s, \quad \Pi_s \to \sqrt{\frac{\beta_s^2}{4\pi}}\Pi_s \qquad (18.7)$$

The Hamiltonian density (18.4) then takes the form

$$H_s = \frac{u_s}{2}\left[\frac{\beta_s^2}{4\pi}\Pi^2(x) + \frac{4\pi}{\beta_s^2}(\partial_x\Phi_s(x))^2\right] - \frac{m}{\pi a_0}\cos\sqrt{4\pi}\Phi_s$$

Suppose that β_s^2 is close to 4π:

$$\frac{4\pi}{\beta_s^2} = 1 + \frac{g}{\pi u_s}, \quad |g|/\pi u_s \ll 1 \qquad (18.8)$$

Then, to the accuracy of $0(g^2)$

$$H_s = \frac{u_s}{2}[\Pi^2(x) + (\partial_x\Phi_s(x))^2] + \frac{g}{2\pi}[(\partial_x\Phi_s(x))^2 - \Pi^2(x)]$$
$$- \frac{m}{\pi a_0}\cos\sqrt{4\pi}\Phi_s \qquad (18.9)$$

Notice that, as a result of transformation (18.7), the critical dimension of the cosine term in (18.9) becomes equal to 1. This is the dimension of a mass bilinear $\bar{\psi}\psi$ for a Dirac field. This fact immediately suggests 'fermion-ization' of the Hamiltonian (18.9). Introduce a single (spinless) Fermi field, $\psi = (R, L)$. According to the standard Bose–Fermi correspondence

$$\partial_x\Phi_s = \sqrt{\pi}(J + \bar{J}), \quad \Pi_s = \sqrt{\pi}(J - \bar{J})$$
$$\cos\sqrt{4\pi}\Phi_s = i\pi a_0(R^\dagger L - \text{H. c.}) \qquad (18.10)$$

where $\bar{J} =: R^\dagger R :$ and $J =: L^\dagger L :$ are the Abelian chiral currents of the new fermions. Then the first and third terms in (18.9) describe free massive fermions, while the second term represents a marginal four-fermion interaction expressed in terms of chiral currents.

Thus, we arrive at the massive Thirring model, with the Lagrangian density

$$\mathscr{L}_{\mathrm{MT}} = \bar{\psi}(i\gamma^\mu \partial_\mu - m)\psi - \frac{1}{2}g j^\mu j_\mu \qquad (18.11)$$

In the region $\beta_s^2 < 8\pi$ its correspondence to the SG model is given by formulas

$$j^\mu =: \bar{\psi}\gamma^\mu\psi :\leftrightarrow \frac{\beta_s}{2\pi}\epsilon_{\mu\nu}\partial_\nu\Phi_s \qquad (18.12)$$

$$\bar{\psi}\psi \leftrightarrow \frac{1}{\pi a_0}\cos\beta_s\Phi_s \qquad (18.13)$$

The relation (18.8) is universal only at small g. This equivalence has been originally established by Coleman (1975) as a one-to-one correspondence between the perturbation series expansions for the two models in powers of m.

The established sequence of mappings between different models provides us with a nonperturbative approach to one-dimensional Fermi systems in the regime of *strong* effective interaction. Consider, for example, the region β^2 close to 4π, not accessible by perturbation theory for the U(1) Thirring or SG models. An equivalent description in terms of MT model is amazingly simple: the massive fermions interact *weakly*, and the corresponding perturbation theory is infrared convergent. Exactly at $\beta^2 = 4\pi$ the massive Thirring fermions become free ($g = 0$). This is the well-known Luther–Emery point (Luther and Emery (1974)) at which the low-temperature thermodynamic properties of the original U(1) Thirring model (18.1), as well as asymptotics of all correlation functions can be exactly estimated. The emerging physical picture is representative for the whole region $4\pi \le \beta_s^2 < 8\pi$ where the spin excitation spectrum consists only of massive particles (the SG quantum solitons). At $\beta_s^2 < 4\pi$ the soliton–antisoliton bound states (breathers) also appear in the spectrum, influencing both thermodynamics and correlation functions (see Chapter 11). Notice that the appearance of these bound states is easily understood from the equivalence between the SG and MT models. At $\beta_s^2 < 4\pi$, the massive fermions repel ($g > 0$) implying that particles (solitons) and holes (antisolitons) attract; the latter circumstance leads to the formation of particle–hole excitonic states below the single-particle mass gap. In the limit $\beta_s \to 0$, when the breather states fill the one-soliton gap, the SG model can be adequately treated by the quasiclassical approximation.

Let us make the following remark. Quite often one finds in the literature a claim that the SG model possesses U(1) symmetry. At first sight, this sounds like a puzzle, because looking at the Hamiltonian (18.4) one observes only a discrete Z_∞ symmetry, i.e. the symmetry under shifts $\Phi_s \to \Phi_s + (2\pi/\beta_s)n$, $(n \in Z)$, originating from periodicity of the cosine potential. However, one should remember that, apart from the original scalar field $\Phi = \phi + \bar{\phi}$, there also exists the dual field $\Theta = -\bar{\phi} + \phi$. The latter does not explicitly enter the SG Hamiltonian; therefore there is a continuous symmetry with respect to *arbitrary* translations of Θ,

$$\Theta \to \Theta + \alpha$$

at a fixed Φ. The right and left fields transform in an opposite way

$$\phi \to \phi - \alpha/2, \quad \bar{\phi} \to \bar{\phi} + \alpha/2 \tag{18.14}$$

Using the correspondence between the SG and MT models, one finds that, under (18.14), the right and left Fermi fields transform as

$$R \sim e^{i\sqrt{4\pi}\phi} \to e^{-i\sqrt{\pi}\alpha}R$$
$$L \sim e^{-i\sqrt{4\pi}\bar{\phi}} \to e^{-i\sqrt{\pi}\alpha}L$$

Thus, the hidden U(1) symmetry of the SG model is nothing but the U(1) invariance of the MT model under phase transformations of the field ψ, associated with conservation of the fermionic current $j_\mu = \bar{\psi}\gamma_\mu\psi$.

II   XYZ Thirring model

Let us now turn to the massless Thirring model (18.1) with interaction characterized by fully broken SU(2) symmetry: $g_1 \neq g_2 \neq g_3(\neq g_1)$. It is useful to redefine the coupling constants

$$g_\| = g_3, \quad g_\perp = \frac{1}{2}(g_1 + g_2), \quad g_f = \frac{1}{2}(g_1 - g_2) \tag{18.15}$$

and rewrite the Hamiltonian density as follows:

$$H_s = \frac{2\pi v_F}{3}[: \mathbf{J} \cdot \mathbf{J} : + : \bar{\mathbf{J}} \cdot \bar{\mathbf{J}} :]$$
$$+ g_\| J^z \bar{J}^z + \frac{g_\perp}{2}(J^+\bar{J}^- + \text{H. c.}) + \frac{g_f}{2}(J^+\bar{J}^+ + \text{H. c.}) \tag{18.16}$$

The new ingredient of this model is the appearance of spin-nonconserving processes

$$\sim g_f(R_\uparrow^\dagger L_\uparrow^\dagger L_\downarrow R_\downarrow + \text{H. c.})$$

which upon bosonization

$$J^+(x)\bar{J}^+(x) + \text{H. c.} = -\frac{1}{2(\pi a_0)^2} \cos \sqrt{8\pi}\Theta_s(x)$$

modify the SG model, giving rise to a cosine of the dual field in the bosonic Hamiltonian density:

$$H_s = \frac{u_s}{2}[\Pi_s^2 + (\partial_x \Phi_s)^2] - \frac{m}{\pi a_0}\cos\beta_s\Phi_s - \frac{\tilde{m}}{\pi a_0}\cos\tilde{\beta}_s\Theta_s \qquad (18.17)$$

Here, as above, β_s is given by relations (18.6), (18.5),

$$\beta_s\tilde{\beta}_s = 8\pi, \quad m = \frac{g_\perp}{4\pi a_0}, \quad \tilde{m} = \frac{g_f}{4\pi a_0} \qquad (18.18)$$

Notice the duality property of the model (18.17): it remains invariant under the change

$$\Phi_s \leftrightarrow \Theta_s, \quad m \leftrightarrow \tilde{m} \ (g_\perp \leftrightarrow g_f), \quad \beta_s \leftrightarrow \tilde{\beta}_s \ (g_\parallel \leftrightarrow -g_\parallel) \qquad (18.19)$$

This property is also seen from the RG equations for the XYZ-symmetric case derived in the Appendix to Chapter 17. Being rewritten in terms of the dimensionless couplings $z_i = g_i/2\pi v_s$ ($i = 1, 2, 3$), the equations read:

$$\frac{dz_1}{dl} = z_2 z_3 \quad \frac{dz_2}{dl} = z_3 z_1 \quad \frac{dz_3}{dl} = z_1 z_2 \qquad (18.20)$$

or in terms of $z_\parallel, z_\perp, z_f$

$$\frac{dz_\parallel}{dl} = z_\perp^2 - z_f^2 \quad \frac{dz_\perp}{dl} = z_\parallel z_\perp \quad \frac{dz_f}{dl} = -z_\parallel z_f \qquad (18.21)$$

The symmetry of the model is discrete. Its breakdown is associated with the ordering of the Φ-field or Θ-field, and indicates the existence of two Ising-like massive phases. This is easily seen from the structure of H_s in (18.17) and the RG equations (18.21). Indeed, let us consider the case $|z_\parallel| \gg |z_\perp|, |z_f|$ when renormalization of $z_\parallel$ can be neglected. At $z_\parallel^0 > 0$, i.e. $\beta_s^2 < 8\pi$, $\cos\beta_s\Phi_s$ is a relevant perturbation, while $\cos\tilde{\beta}_s\Theta_s$ is irrelevant. From the RG equations it follows that, in the infrared limit ($l \to \infty$), $|z_\perp| \to \infty$, while $z_f \to 0$. By duality, at $z_\parallel^0 < 0$, ($\beta_s^2 > 8\pi$), the picture is inverted: now $\cos\tilde{\beta}_s\Theta_s$ is relevant, $\cos\beta_s\Phi_s$ irrelevant, and $|z_f| \to \infty, z_\perp \to 0$. So, in each of these two cases, the model (18.17) scales towards a SG model in the strong-coupling regime. This allows one to estimate the magnitude of the mass gaps (see Chapter 10):

$$M \simeq \Lambda |z_\perp^0|^{\frac{1}{2(1-K_s)}} \quad at \ z_\parallel^0 > 0 \qquad (18.22)$$

$$M \simeq \Lambda |z_f^0|^{\frac{1}{2(1-1/K_s)}} \quad at \ z_\parallel^0 < 0 \qquad (18.23)$$

At $\beta_s^2 = \tilde{\beta}_s^2 = 8\pi$ both cosines in (18.17) become marginal. Therefore the scaling portrait of the model in the region where $|z_\parallel| \sim |z_\perp|, |z_f|$ depends on the relation between the coupling constants. This follows from the RG equations (18.21). These equations have two scaling integrals, $z_\perp z_f$ and $(z_\parallel - |z_\perp| + |z_f|)(z_\parallel + |z_\perp| - |z_f|)$. When one of these integrals is zero, the symmetry increases up to U(1), and the RG equations (18.21) reduce to the Kosterlitz–Thouless ones. In particular, equation

$$z_\parallel + |z_\perp| - |z_f| = 0 \qquad (18.24)$$

defines a critical plane of infrared stable Gaussian fixed points. In the three-dimensional parameter space of the model, this plane separates two, mutually dual, strong-coupling phases:

$$z_\parallel + |z_\perp| - |z_f| > 0 : \quad z_\parallel \to \infty, \;\; |z_\perp| \to \infty, \;\; z_f \to 0 \qquad (18.25)$$

and

$$z_\parallel + |z_\perp| - |z_f| < 0 : \quad z_\parallel \to -\infty, \;\; |z_f| \to \infty, \;\; z_\perp \to 0 \qquad (18.26)$$

Exercise. Show that near the critical plane (18.24) the mass gap is given by (Giamarchi and Schulz (1988))

$$M \simeq \Lambda \delta^{-\nu} \qquad (18.27)$$

where

$$\delta = [z_\parallel^2 - (|z_\perp| - |z_f|)^2]/|z_\perp z_f|, \quad \nu = 2|z_\perp z_f|^{1/2} \qquad (18.28)$$

III Spin correlation functions

In this section we shall consider the behaviour of the correlation functions both in the U(1) and XYZ symmetric Thirring models. Reflecting charge–spin separation at the level of the total bosonized Hamiltonian, various operators which represent different order parameters acquire a multiplicative structure exhibiting independent contributions from the charge and spin sectors. The corresponding expressions for the staggered particle density (CDW), pairing operators and staggered magnetization (SDW) have been given in Chapter 16:

$$\rho_{st}(x) \sim \sin\left[\frac{\beta_c \Phi_c(x)}{2} + 2k_F x\right] \cos\frac{\beta_s \Phi_s(x)}{2} \qquad (18.29)$$

$$\Delta_0(x) \sim \sin\frac{\tilde{\beta}_c \Theta_c(x)}{2} \cos\frac{\beta_s \Phi_s(x)}{2} \qquad (18.30)$$

$$\mathbf{\Delta}(x) \sim \sin\frac{\tilde{\beta}_c \Theta_c(x)}{2} \mathbf{n}(x) \qquad (18.31)$$

$$n^z(x) \sim \cos\left[\frac{\beta_c \Phi_c(x)}{2} + 2k_F x\right] \sin\frac{\beta_s \Phi_s(x)}{2} \qquad (18.32)$$

$$n^\pm \sim \cos\left[\frac{\beta_c \Phi_c(x)}{2} + 2k_F x\right] \exp\left[\pm i\tilde{\beta}_s \Theta_s(x)/2\right] \qquad (18.33)$$

U(1) case

Being mostly interested in effects caused by the spin anisotropy, we shall focus on the behaviour of SDW-correlations. We shall therefore assume the interaction in both sectors to be repulsive,

$$g_c > 0, \quad g_\|, g_\perp < 0 \tag{18.34}$$

The first condition (18.34) implies that $K_c < 1$; so the correlation function of exponents of the dual charge field Θ_c

$$\langle e^{i\tilde{\beta}_c \Theta_c(1)/2} e^{-i\tilde{\beta}_c \Theta_c(2)/2} \rangle \sim |x + iv_c\tau|^{-1/K_c}$$

decays faster than that of exponents of the original field Φ_c

$$\langle e^{i\beta_c \Phi_c(1)/2} e^{-i\beta_c \Phi_c(2)/2} \rangle \sim |x + iv_c\tau|^{-K_c}$$

For this reason pairing correlations will not be considered.

Let us analyse the changes in the behaviour of the correlations upon the change of the sign of the exchange anisotropy parameter

$$\delta_A = g_\| - g_\perp \tag{18.35}$$

Notice that, since both $g_\|$ and $g_\perp$ are negative, the isotropic case $\delta_A = 0$ corresponds to the weak-coupling separatrix on the Kosterlitz–Thouless phase diagram. Therefore, at $\delta_A < 0$ the effective couplings flow to infrared stable Gaussian fixed points:

$$g_\| \to g_\|^* = -\sqrt{g_\|^2 - g_\perp^2}, \quad g_\perp \to g_\perp^* = 0$$

implying that $K_s \to K_s^* > 1$, $\tilde{K}_s \to \tilde{K}_s^* = 1/K_s^* < 1$. As follows from the above definitions of order parameters

$$\langle \rho_{\text{stag}}(1)\rho_{\text{stag}}(2) \rangle \sim \langle n^z(1)n^z(2) \rangle \sim |x + iv_c\tau|^{-K_c}|x + iv_s\tau|^{-K_s^*} \tag{18.36}$$

$$\langle n^+(1)n^-(2) \rangle \sim |x + iv_c\tau|^{-K_c}|x + iv_s\tau|^{-1/K_s^*} \tag{18.37}$$

We see that the transverse spin correlations, SDW^x, SDW^y show the slowest algebraic decay and, reflecting the underlying U(1) symmetry, are degenerate in the xy-plane of spin space. Therefore, at $\delta_A < 0$ the case of easy plane anisotropy is realized.

At $\delta_A = 0$ the SU(2) symmetry is recovered, and CDW and isotropic SDW correlations have the same critical exponents.

On changing the sign of the anisotropy parameter, $\delta_A > 0$, the weak-coupling regime changes by a strong-coupling one in which couplings $|g_\||$ and $|g_\perp|$ increase upon renormalization. A spin gap opens in the spectrum, and the spin field Φ_s gets ordered. Since $g_\perp < 0$, the amplitude of the cosine term in the Hamiltonian (18.4) is positive, and vacuum states of the field Φ_s, determined as degenerate minima of the potential

$-g_\perp \cos(\beta_s \Phi_s/2)$, are given by

$$(\Phi_s)_{vac} = \pi/\beta_s \quad \mathrm{mod}(2\pi/\beta_s) \tag{18.38}$$

As a result, the operator $\sin \beta_s \Phi_s/2$ acquires a nonzero vacuum expectation value, and, hence, its correlation function has a finite limit at $|x|, |\tau| \to \infty$. On the other hand, by the symmetry of the degenerate vacua (18.38), the operator $\cos \beta_s \Phi_s/2$ has zero expectation value; moreover, its correlations are short-ranged, i.e. decay exponentially over the scale $\xi_s \sim v_s/M_s$, where M_s is the mass gap (see also below). Therefore, CDW correlations are suppressed. Since Θ_s is a disorder field with respect to Φ_s, correlations of the transverse magnetization are also exponentially suppressed. The only remaining longitudinal spin correlation function shows a power law decay, with a critical exponent which at $|x|, v_s\tau \gg \xi_s$ is entirely contributed by the gapless charge sector

$$\langle n^z(1)n^z(2)\rangle \sim |x + iv_s\tau|^{-K_c} \tag{18.39}$$

Thus, changing the sign of the exchange anisotropy parameter δ_A, we observe a transition from the easy plane anisotropy to an Ising-like, easy axis anisotropy.

We can confirm the picture emerging in the massive phase ($\delta_A > 0$) by exact calculations at the Luther–Emery point, $\beta_s^2 = 8\pi/\tilde\beta_s = 4\pi$. First we notice that, at this point, the exponential of the dual spin field in the definition (18.33) of the transverse spin density is expressed as a pairing bilinear operator in terms of the Thirring fermions

$$e^{i\sqrt{4\pi}\Theta_s} \to -2\pi i a_0 R^\dagger L^\dagger \tag{18.40}$$

Since the mass term in the free ($g = 0$) Thirring model (18.11) is not of the BCS type but instead of the CDW nature, it is clear that pairing correlations in such a Fermi system will fall out exponentially. Indeed,

$$\langle \exp[i\sqrt{4\pi}\Theta_s(x,\tau)] \exp[-i\sqrt{4\pi}\Theta_s(0,0)]\rangle$$
$$= (2\pi a_0)^2 \left[G_{RR}(x,\tau)G_{LL}(x,\tau) - G_{RL}(x,\tau)G_{LR}(x,\tau)\right] \tag{18.41}$$

where

$$\hat{G}(x,\tau) = \frac{T}{L}\sum_{k,\varepsilon_n} e^{ikx - i\varepsilon_n\tau}\hat{G}(k,\varepsilon_n), \quad \hat{G}(k,\varepsilon_n) = (i\varepsilon_n - ku_s\hat\tau_3 - m\hat\tau_2)^{-1} \tag{18.42}$$

is a 2×2 matrix for single-fermion Green's functions (ε_n is the Matsubara frequency, and Pauli matrices τ^a operate in the two-dimensional space of right and left components). We shall do calculations in the zero-temperature limit. Using two-dimensional Euclidean notations,

$$\mathbf{k} = (k, -\varepsilon/u_s), \quad \mathbf{x} = (x, u_s\tau)$$

we find that

$$G_{R(L),L(R)}(x,\tau) = \pm\frac{i}{\xi_s}J(\mathbf{x}), \quad G_{R(L),R(R)}(x,\tau) = (\partial_2 \pm i\partial_1)J(\mathbf{x})$$

where

$$J(\mathbf{x}) - \int \frac{d\mathbf{k}}{(2\pi)^2}\frac{e^{i\mathbf{k}\cdot\mathbf{x}}}{\mathbf{k}^2 + \xi_c^{-2}} - \frac{1}{2\pi}K_0(r/\xi_c)$$

Here $K_0(x)$ is the Bessel function, and $r = \sqrt{x^2 + u_s^2\tau^2}$. The correlator in (18.41) equals

$$\langle\exp[i\sqrt{4\pi}\Theta_s(x,\tau)]\exp[-i\sqrt{4\pi}\Theta_s(0,0)]\rangle = \left(\frac{a_0}{\xi_s}\right)^2 [K_1^2(r/\xi_s) - K_0^2(r/\xi_s)]$$

$$\simeq \pi\frac{a_0^2}{x^2 + u_s^2\tau^2}\exp[-2\sqrt{x^2 + u_s^2\tau^2}/\xi_s] \qquad (18.43)$$

On the other hand, at $\beta_s^2 = \sqrt{4\pi}$, the spin parts of ρ_{st} and n^z in (18.32) are given by $\cos\sqrt{\pi}\Phi_s(x)$ and $\sin\sqrt{\pi}\Phi_s(x)$, respectively. These operators cannot be expressed in a local way in terms of the massive Fermi fields. However, we have been already dealing with such operators when discussing the two-dimensional Ising model in Chapter 12. We have shown there that the SG model at $\beta_s^2 = \sqrt{4\pi}$, or the free massive Thirring model, are equivalent to two identical copies of a noncritical Ising system. Using this equivalence, the phase exponentials of the SG field with dimension $1/4$ were represented as products of order (σ_1, σ_2) and disorder (μ_1, μ_2) operators corresponding to different Ising models. According to (12. 51), (12. 56)

$$\cos\sqrt{\pi}\Phi_s(\mathbf{x}) \sim \mu_1(\mathbf{x})\mu_2(\mathbf{x}), \quad \sin\sqrt{\pi}\Phi_s(\mathbf{x}) \sim \sigma_1(\mathbf{x})\sigma_2(\mathbf{x}) \qquad (18.44)$$

The correlation functions factorize:

$$\langle\cos\sqrt{\pi}\Phi_s(\mathbf{x})\cos\sqrt{\pi}\Phi_s(\mathbf{0})\rangle \sim \langle\mu_1(\mathbf{x})\mu_1(\mathbf{0})\rangle_1\langle\mu_2(\mathbf{x})\mu_2(\mathbf{0})\rangle_2$$

$$= \langle\mu_1(\mathbf{x})\mu_1(\mathbf{0})\rangle_1^2 \qquad (18.45)$$

$$\langle\sin\sqrt{\pi}\Phi_s(\mathbf{x})\sin\sqrt{\pi}\Phi_s(\mathbf{0})\rangle \sim \langle\sigma_1(\mathbf{x})\sigma_1(\mathbf{0})\rangle\langle\sigma_2(\mathbf{x})\sigma_2(\mathbf{0})\rangle$$

$$= \langle\sigma_1(\mathbf{x})\sigma_1(\mathbf{0})\rangle_1^2 \qquad (18.46)$$

As explained in Chapter 12, the fermionic mass $m \sim (T - T_c)/T_c$, where T_c is the critical temperature for the Ising models. Recall now that $m \sim g_\perp < 0$. Therefore, the two Ising systems are in the ordered, low-temperature phase with $\langle\sigma_1\rangle = \langle\sigma_2\rangle \equiv \langle\sigma\rangle \neq 0$, $\langle\mu_1\rangle = \langle\mu_2\rangle = 0$. Therefore, in the limit $r/\xi_s \to \infty$

$$\langle\cos\sqrt{\pi}\Phi_s(\mathbf{x})\cos\sqrt{\pi}\Phi_s(\mathbf{0})\rangle \sim K_0^2(r/\xi_s) \sim (\xi_s/r)\exp(-2r/\xi_s) \qquad (18.47)$$

$$\langle\sin\sqrt{\pi}\Phi_s(\mathbf{x})\sin\sqrt{\pi}\Phi_s(\mathbf{0})\rangle \sim \langle\sigma\rangle^4 [1 + 0(\exp(-2r/\xi_s)] \qquad (18.48)$$

This result is in full agreement with the previous analysis for weak exchange anisotropy, confirming the picture in the strong-coupling phase.

Notice that, due to the presence of the gapless charge mode, spin correlations decay algebraically for both signs of δ_A. If the original fermions were put on a lattice and the energy band were half-filled, a charge gap would open due to Umklapp processes, leading to generation of a Mott–Hubbard gap in the charge sector. This would lead to ordering of the charge field Φ_c, implying a nonzero expectation value of $\cos \beta_c \Phi_c / 2$. At energies well below the charge gap, the fermion spin dynamics would transform to that of the anisotropic (XXZ) spin $S = 1/2$ chain, Eq. (1.1), and the above described 'easy plane – easy axis' transition in the U(1) Thirring model would be nothing but the well-known transition from the disordered, gapless, XY-like phase to the long-range ordered Ising-like phase, taking place at $\Delta = J_{\parallel}/J_{\perp} = 1$ (see Chapter 11).

XYZ case

Let us now turn to the XYZ-symmetric case, characterized by two anisotropy axes. We shall study dominant correlations in both massive phases of the system. Assuming as before that $g_{\parallel} < 0$, the initial conditions for each of the two phases are

$$(a) \quad |g_{\parallel}| > |g_{\perp}| - |g_f|, \quad (b) \quad |g_{\parallel}| < |g_{\perp}| - |g_f| \qquad (18.49)$$

These conditions determine two different strong-coupling regimes given by Eqs. (18.26) and (18.25), respectively. They reduce to single-axis, U(1) cases $\delta_A < 0$ and $\delta_A > 0$ when $g_{\perp} = 0$.

In the case (a) the effective coupling constants $|g_{\parallel}|$ and $|g_f|$ increase upon renormalization, while $|g_{\perp}|$ scales down to zero. The field Θ_s gets ordered, while Φ_s is disordered, being characterized by a finite correlation length ξ_s. The latter circumstance leads to the suppression of the CDW and SDW^z correlations. The type of dominant SDW correlations in the xy-plane depends on the sign of g_f. At $g_f > 0$ the degenerate vacuum states of Θ_s are defined as $(\Theta_s)_{vac} = 0 \mod(2\pi/\beta_s)$, while at $g_f < 0$ $(\Theta_s)_{vac} = \pi/\beta_s \mod(2\pi/\beta_s)$. Therefore, at $g_f > 0$ the SDW^x correlation will be dominant, whereas at $g_f < 0$ it will be SDW^y. Here we observe the role of spin-nonconserving (g_f)-processes removing the xy spin degeneracy.

Since in the case (b) the g_f-processes are irrelevant ($|g_{\parallel}|, |g_{\perp}| \to \infty$, $g_f \to 0$), the resulting picture of correlations is qualitatively the same as the case $\delta_a > 0$ in the U(1)-symmetric model. At $g_{\perp} < 0$ SDW^z correlations are dominant, while at $g_{\perp} > 0$ it will be a CDW. Since the spin gap is nonzero in both phases, the asymptotics of the dominant correlation functions in all cases are determined by the gapless charge sector, the critical exponent being K_c.

IV   The role of magnetic field

Phase transition in the U(1) Thirring model

The U(1) generalization of the SU(2) Thirring model emphasizes the crucial role of spin-flop backscattering ($g_\perp$)-processes responsible for the development of the strong-coupling regime and dynamical generation of a spin gap at $|g_\parallel| > -|g_\perp|$. It is worth repeating once again that at $g_\perp = 0$ the system would represent a Tomonaga–Luttinger liquid at any sign of $g_\parallel$. An external magnetic field, leading to Zeeman splitting $2\mu_B H$ (μ_B being the Bohr magneton) of the bare-particle spectrum, shifts the Fermi momenta making the latter dependent on the spin projection of the particles,

$$k_F \to k_{F,\sigma} = k_F + \sigma\mu_B H, \quad (\sigma = \pm 1) \qquad (18.50)$$

Considering the states of right- and left-moving fermions in the vicinity of the shifted Fermi points, one finds out that in spin-flip backscattering processes, $g_\perp (R_\uparrow^\dagger L_\downarrow^\dagger L_\uparrow R_\downarrow + \text{H. c.})$, the total momentum is no longer conserved. As a result, these processes will be frozen out at energies below $\mu_B H$. Thus the role of the magnetic field is expected to be most pronounced in the strong-coupling phase of the U(1) Thirring model. In zero field, the ground state of the system is nonmagnetic due to the existence of a finite spin gap. When the magnetic field is applied, the resulting picture in the infrared limit will be determined by the interplay of two competing tendencies: the increase of the effective coupling $z_\perp(l)$ upon renormalization in the energy range $\Lambda v_s \gg |E| \gg \mu_B H$, and its suppression at $|E| < \mu_B H$. We shall show below that this competition is resolved as a continuous transition from the nonmagnetic ground state to a magnetic gapless phase, taking place at some critical value of the magnetic field (Japaridze and Nersesyan (1978)).

Let us first look at the problem from a perturbative point of view. It is apparent from the above discussion that a *two-cut-off* RG procedure should be implemented to account for the already mentioned suppression of $g_\perp$-processes. In such a procedure, one deals with two cut-offs – the ultraviolet real-space cut-off (or the lattice constant) a_0 and an infrared one defined as the magnetic correlation length $\xi_H \sim v_s/\mu_B H$ ($\xi_H \gg a_0$). Introducing the logarithmic variable $l = \ln(a/a_0)$, one has to distinguish between two, qualitatively different cases, $l < h$ and $l > h$, where $h = \ln(\xi_H/a_0)$. In the region $l < h$, the effective couplings $z_\parallel$ and $z_\perp$ do not depend on h (within logarithmic accuracy) and are determined by the

Kosterlitz–Thouless scaling equations (10.22). In the region $l > h$ the magnetic field suppresses the $g_\perp$-processes, and further renormalization of $z_\perp$ stops. So, in the whole range $l > h$, $z_\parallel(l)$ and $z_\perp(l)$ coincide with new 'bare' couplings $z_\parallel(h)$ and $z_\perp(h)$. However, since the RG method is a perturbative approach, such a conclusion will only make sense, if $z_\parallel(h)$ and $z_\perp(h)$ are small. The last requirement is always satisfied in the weak-coupling phase occurring at $g_\parallel \le -|g_\perp|$. But this may not be the case at $g_\parallel > -|g_\perp|$ when a strong-coupling regime develops in zero magnetic field, because, if h is large enough, the effective interaction increasing in the range $l < h$ can become strong at $l \sim h$. Therefore, for the two-cut-off scaling to be applicable, one has to assume that the magnetic field significantly exceeds the spin gap M_s (or equivalently, $|g_{\parallel,\perp}|h \ll 1$).

Thus, from the RG analysis one concludes that the originally ($H = 0$) gapful phase of the U(1) Thirring model transforms to a gapless Tomonaga–Luttinger phase in a strong magnetic field, $\mu_B H \gg M_s$. To describe the most interesting, strong-coupling region $\mu_B H \sim M_s$, one has to apply nonperturbative methods. We shall now obtain such a description using bosonization which, as we have seen, establishes correspondence between the U(1) Thirring model and the SG model.

The magnetic field adds to the Hamiltonian density a term $H_{\mathrm{mag}} = -2\mu_B H J^z$. Correspondingly, the SG model (18.4) is supplemented with a term proportional to the density of the topological charge:

$$H_s = \frac{u_s}{2}[\Pi_s^2 + (\partial_x \Phi_s)^2] - \frac{\beta_s \mu_B H}{2\pi}\partial_x \Phi_s - \frac{m}{\pi a_0}\cos\beta_s\Phi_s \qquad (18.51)$$

The cosine term in (18.51) is irrelevant in two cases: at $\beta_s^2 > 8\pi$ for arbitrary magnetic field, and at $\beta_s^2 < 8\pi$ in the strong field limit, $\mu_B H \gg M_s$. In these cases this term can be dropped out, with β_s and u_s replaced by their effective values $\beta_s(h)$ and $u_s(h)$. The magnetic field is then eliminated by a shift

$$\partial_x \Phi_s \to \partial_x \Phi_s + \frac{\beta_s(h)H}{2\pi u_s(h)}$$

and H_S transforms to

$$H_S = \frac{u_s(h)}{2}[\Pi_s^2 + (\partial_x \Phi_s)^2] - \frac{K_s(h)\mu_B^2 H^2}{\pi u_s(h)} \qquad (18.52)$$

displaying Pauli paramagnetism in a Tomonaga–Luttinger liquid.

Turning now to the model (18.51) at arbitrary magnetic field, we realize that we have already been dealing with such a problem in Chapter 17 where effects of small doping were considered in one-dimensional Fermi systems with a Mott–Hubbard charge gap. As we have already mentioned, there exists a mapping between charge and spin sectors generated by a particle–hole transformation (15.21). In particular, Umklapp processes in the charge sector map onto spin-flip backscattering in the spin sector, while the chemical potential transforms to a magnetic field. Using the correspondence

$$\Phi_c \to \Phi_s, \quad \beta_c \to \beta_s, \quad \mu \to 2\mu_B H$$

the commensurate–incommensurate transition decribed in Chapter 17 can be straightforwardly translated to our case.

The emerging picture is extremely simple at the Luther–Emery point $\beta_s^2 = 4\pi$ where it can be recast as a metal–insulator transition in a one-dimensional system of free massive fermions. The phase transition takes place at the critical field $H_c = M_s/\mu_B$. As long as $H < H_c$, no changes occur in the nonmagnetic ground state. At $H > H_c$ the ground state becomes paramagnetic, with a finite magnetization proportional to the induced density of topological charge. At the transition, the magnetization and magnetic susceptibility display square-root singularities

$$M(H) \sim (H - H_c)^{1/2}, \quad \chi_s(H) \sim (H - H_c)^{-1/2}, \quad (H \to H_c + 0) \quad (18.53)$$

universal at all finite values of β_s. In this region of fields corresponds to a strong-coupling regime, characterized by low density of solitons. On increasing the magnetic field, the system gradually crosses over to the regime of high soliton density, representing a Tomonaga–Luttinger liquid.

It is interesting to figure out how the SDW^z correlations, dominant at $H = 0$, are affected by the transition in the magnetic field.

Exercise.   Calculate asymptotics of subdominant $SDW^{x,y}$ correlations in a magnetic field $H > H_c$ at the Luther–Emery point.

Suppose that β_s^2 is close to 4π, so that one can use an equivalent formulation in terms of weakly interacting massive Thirring fermions. The corresponding Hamiltonian density now includes a chemical potential term:

$$H_{MT} = H_0 + H_{int}$$
$$= \sum_k \psi^\dagger(k)(ku_s\hat{\tau}_3 - m\hat{\tau}_2 - \mu)\psi(k) + 2g\sum_k J(k)\bar{J}(-k) \quad (18.54)$$

Here $\psi(k)$ is a Dirac 2-spinor, $\mu = \mu_B H$ and $m = M_s$. The free part of the Hamiltonian, H_0, is easily diagonalized by a Bogoliubov transformation

$$\psi(k) = \exp(i\gamma(k)\hat{\tau}^1/2)\,\Psi(k), \quad \Psi(k) = \begin{pmatrix} a_+(k) \\ a_-(k) \end{pmatrix} \quad (18.55)$$

where $\tan[\gamma(k)/2] = m/ku_s$. We obtain;

$$H_0 = \sum_k [\varepsilon_+(k)a_+^\dagger(k)a_+(k) + \varepsilon_-(k)a_-^\dagger(k)a_-(k)] \qquad (18.56)$$

with $\varepsilon_\pm(k) = \pm\sqrt{k^2 u_s^2 + m^2} - \mu$

Above the transition ($\mu > m$) the upper band $\varepsilon_+(k)$ gets partially filled, and a Fermi surface opens at $k = \pm k_0$, where $k_0 = \sqrt{\mu^2 - m^2}/u_s$. As long as one is interested in long-distance asymptotics of the correlation functions ($k_0|x| \gg 1$), it is sufficient to linearize the spectrum $\varepsilon_+(k)$ and introduce right and left fermionic states near $\pm k_0$:

$$r(k) = a_+(k_0 + k), \quad l(k) = a_+(-k_0 + k) \quad (|k| \ll k_0)$$

The lower-band states separated by the energy gap $2|m|$ can be completely ignored. Since renormalization of the mass and velocity is not important for our purposes, such an approximation can be considered as legitimate at energies $|E - \mu| \ll m$.

Exercise.   Using the Hamiltonian (18.54), prove universality of the square-root singularities (18.53) in the lowest order in g.

In this approximation, the Hamiltonian (18.54) reduces to a Tomonaga–Luttinger model

$$H_{\text{eff}} = \pi u_s(k_0) \{N_\mu[j^2(x)] + N_\mu[\bar{j}^2(x)]\} + \lambda(k_0)j(x)\bar{j}(x), \qquad (18.57)$$

where the effective coupling and velocity

$$\lambda(k_0) = 4g\frac{k_0^2 u_s^2 m^2}{\mu^4}, \quad v_s(k_0) = \frac{k_0 u_s^2}{\mu}\left(1 + \frac{2g}{\pi}\frac{k_0 m^2}{\mu^3} + 0(g^2)\right) \qquad (18.58)$$

vanish on approaching the transition point. The new chiral currents are defined in the standard way

$$\bar{j}(x) = N_\mu[r^\dagger(x)r(x)], \quad j(x) = N_\mu[l^\dagger(x)l(x)]$$

However, the normal ordering prescription (N_μ) is now defined with respect to the *new* vacuum at $\mu > m$ which differs from that at $\mu = 0$ by the presence of filled states of the upper band in the momentum range $-k_0 < k < k_0$:

$$: \psi^\dagger \psi := N_\mu[\psi^\dagger \psi] + k_0/\pi \qquad (18.59)$$

It follows from (18.59) that the Gaussian scalar field $\phi(x)$, appearing upon bosonization of H_{eff} in (18.57), differs from the original (SG) field $\Phi_s(x)$ by a shift:

$$\Phi_s(x) = \phi(x) + k_0 x/\sqrt{\pi} \qquad (18.60)$$

This relation reflects the fact that the ground state of the system above the transition is characterized by a nonzero density of topological charge, $\langle \partial_x \Phi_s(x) \rangle_{\mu > m} = k_0/\sqrt{\pi}$. Diagonalization of the Hamiltonian (18.57) will cause a simple rescaling of the field ϕ, $\phi \to \sqrt{Q(k_0)}\,\phi$, where (at small g)

$$Q(k_0) = 1 - \frac{\lambda(k_0)}{2\pi v_s(k_0)} = 1 - \frac{2g k_0 m^2}{\pi \mu^3} + 0(g^2) \qquad (18.61)$$

Then, using the definition (18.32), the SDW^z correlation function is easily estimated as (Schulz (1996)):

$$\langle n^z(x,\tau) n^z(0,0) \rangle \sim \cos[2i(k_F \pm k_0)x]|x + iv_C\tau|^{-K_C}|x + iu_S(k_0)\tau|^{-Q(k_0)/2}$$
$$(18.62)$$

From (18.62) we observe that at $H > H_C$ there are two, incommensurate with respect to the $H = 0$ case, spin-density waves, with periods $2\pi/2(k_F \pm k_0)$ changing with the field. In the limit of strong field ($H \gg H_C$, $k_0 = \mu_B H/v_F$), these periods are inversely proportional to the Fermi diameters of the spin-up and spin-down branches of the bare-particle spectrum, in agreement with (18.50). We also see that the critical exponent of the SDW^z-correlator, $v(k_0) = K_S + (1/2)Q(k_0)$, being a nonuniversal function of the magnetic field at $H > H_C$, displays a *universal jump* at the transition point:

$$v(H_c + 0) - v(H_c - 0) = \frac{1}{2} \qquad (18.63)$$

Universality of this result stems from the fact that, just above the transition where the induced vacuum density n_{sol} of quantum SG solitons (massive fermions) is low, $n_{sol}\xi_s \ll 1$ ($\xi_s = u_s/M_s$), interaction between them is weak and for any $\beta_s < 8\pi$ vanishes in the zero-density limit (Haldane (1982)).

IV.1   Spin-flop transition in the XYZ model

Further lowering the U(1) symmetry down to XYZ turns out to be quite interesting: here one finds a spin-flop transition induced by the magnetic field (Giamarchi and Schulz (1988)).

Suppose that $g_{\parallel}, g_{\perp} < 0$ and $|g_{\parallel}| < |g_{\perp}| - |g_f|$. As we have seen in the previous section, under these conditions the spin-nonconserving processes are irrelevant, and the system scales towards the strong-coupling phase of the U(1) model, with dominant SDW^z correlations. Let us understand how this picture is modified in the magnetic field.

The asymptotic U(1) behaviour of the system will persist below the critical field H_c and also in some range of fields above the magnetic transition, where the effective coupling $z_{\perp}$ remains large and the g_f-processes are still irrelevant. Therefore, as we have seen above, at $H = H_c$ a 'commensurate-incommensurate' transition from a SDW^z phase with wave vector $2k_F$ to a SDW^z phase with a field dependent wave vector $2(k_F \pm k_0)$ will take place. However, such a picture cannot survive further increase of the magnetic field because spin-flip backscattering gets suppressed, and the neglect of the g_f-processes may no longer be correct. Let us look therefore at the RG equations (18.21), using the two-cut-off scaling method. In the region $l < h$ these equations are not influenced by the field. In the region $l > h$ renormalization of $z_\perp$ is stopped, and one is left with a pair of Kosterlitz–Thouless equations for $z_\parallel$ and z_f:

$$\frac{dz_\parallel}{dl} = -z_f^2, \qquad \frac{dz_f}{dl} = -z_\parallel z_f \qquad (18.64)$$

We shall assume that the field is large enough, so that the new 'bare' charges $z_\parallel(h)$ and $z_f(h)$ are small. This makes Eqs. (18.64) applicable.

Two cases are to be distinguished: (i) $z_\parallel(h) > |z_f(h)|$ and (ii) $z_\parallel(h) < |z_f(h)|$. On increasing the field one expects the case (i) to be realized first. This is a weak-coupling regime in which z_f still scales down to zero. Therefore in case (i) we are still in the 'incommensurate' SDW^z phase. However, on further increasing the field $z_\parallel(h)$ decreases, and the inequality (ii) should inevitably be realized. This means that the system crosses over to a new strong-coupling regime in which g_f-processes become relevant. As we have seen in the previous section, in this regime the field Θ_s gets ordered, implying the suppression of the SDW^z correlations and, depending on the sign of $z_f(h)$, the onset of SDW^x or SDW^y.

Thus, the presence of spin-nonconserving processes, originating from spin–orbit or magnetic dipole–dipole interactions, leads to a sequence of transitions in an external magnetic field. At the first transition taking place at the critical field H_c, the 'commensurate' SDW^z state with a field independent period transforms to an 'incommensurate' SDW^z state with a magnetization dependent wave vector. The critical exponent, characterizing a power-law decay of the SDW^z correlation function, exhibits a universal jump at this transition. On further increasing the magnetic field, the SDW^z state gradually crosses over into a SDW^x or SDW^y state – the spin-flop transition. The transverse SDW states are 'commensurate', i.e. again with wavevector $2k_F$.

IV.2 Toy model for an orbital antiferromagnet

We conclude this chapter with an example illustrating application of the magnetic properties of the Thirring model with broken SU(2) symmetry to a two-chain model of an orbital antiferromagnet (OAF).

A long time ago Halperin and Rice (1968), discussing the excitonic insulator problem, predicted the existence of ordered phases characterized by nonzero local charge or spin currents. More recently, after the discovery of high-T_c superconductivity, the so-called flux phases have been studied in two-dimensional strongly correlated electron systems (for a review see Fradkin's book). Apart from the usual CDW and SDW phases, the OAF and spin nematic (SN) states with alternately circulating charge and spin currents were also conjectured to exist in weakly interacting electron systems, with a Fermi surface having the perfect nesting property. It was expected that, extending the repulsive, half-filled two-dimensional Hubbard model to include finite-range interactions, one would find conditions under which the OAF or SN instability would dominate over the CDW and SDW. However, since the general problem of instabilities of two-dimensional interacting electrons still remains unresolved, it is natural to find the simplest model that preserves the possibility for circulating currents and, at the sime time, makes the problem solvable.

Such a toy model describes weakly interacting spinless fermions on a strip of plaquettes (ladder) formed by two coupled chains. The Hamiltonian is given by

$$H = -t_\parallel \sum_{i,\sigma} (c_{i,\sigma}^\dagger c_{i+1,\sigma} + \text{H.c.}) - t_\perp \sum_{i,\sigma} c_{i,\sigma}^\dagger c_{i,-\sigma}$$
$$+ \sum_{i,\sigma} (\frac{1}{2} U n_{i,\sigma} n_{i,-\sigma} + V_1 n_{i,\sigma} n_{i+1,\sigma} + V_2 n_{i,\sigma} n_{i+1,-\sigma}) \qquad (18.65)$$

Here $\sigma = \pm 1$ is the chain index, $\langle n_{i,\sigma} \rangle = 1/2$, $t_\parallel$ and $t_\perp$ are the intrachain and interchain hopping amplitudes respectively. The last three terms in (18.65) include *repulsive* interactions between particles on the neighbouring sites belonging either to the same chain (V_1), or to different ones (U), and the interaction along diagonals of the plaquettes (V_2).

Although the spinless case is, of course, an oversimplification, it allows one to decrease the number of competing instabilities: for repulsive interactions, the only possible types of orderings are different kinds of CDW (both site-diagonal and off-diagonal) and the OAF. We shall see that the 'frustrating' interaction V_2, together with the interchain hopping ($t_\perp$), is crucial in suppressing the CDW correlations in favour of the OAF.

Notice that the chain index σ can be formally treated as a spin-1/2 variable. Then the Hamiltonian (18.65) is recognized as a single-chain, repulsive, half-filled Hubbard model extended to include U(1)-symmetric exchange anisotropy term $2(V_1 - V_2)S_i^z S_{i+1}^z$, where $S_i^z = (n_{i+} - n_{i-})/2$. The interchain hopping reduces to a magnetic field $\mathbf{H} = 2t_\perp \hat{\mathbf{x}}$ perpendicular to the spin anisotropy axis. Such a system can exhibit either CDW, or SDW instabilities. The corresponding site-diagonal order parameters are given by the operators:

$$\mathcal{O}_{CDW}(n) = (-1)^n \sum_\sigma c_{n\sigma}^\dagger c_{n\sigma}, \tag{18.66}$$

$$\mathcal{O}_{SDW^z}(n) = (-1)^n \sum_\sigma \sigma c_{n\sigma}^\dagger c_{n\sigma}, \tag{18.67}$$

$$\mathcal{O}_{SDW^x}(n) = (-1)^n \sum_\sigma c_{n\sigma}^\dagger c_{n,-\sigma}, \tag{18.68}$$

$$\mathcal{O}_{SDW^y}(n) = -i\,(-1)^n \sum_\sigma \sigma c_{n\sigma}^\dagger c_{n,-\sigma}, \tag{18.69}$$

In terms of the original two-chain model, the CDW and SDW^z correspond to charge-density waves on the two chains with relative phases 0 and π, respectively (CDW$^{0,\pi}$), while SDW^x and SDW^y describe, respectively, a modulation of the vertical bonds (rungs) of the two-chain lattice (bond wave), BW$_\perp$, and a modulation of the transverse currents, OAF$_\perp$. We will also need two more operators describing off-diagonal SDW^x and SDW^y ordering:

$$\mathcal{O}_{SDW^x}(n.n+1) = (-1)^n \sum_\sigma c_{n\sigma}^\dagger c_{n+1,-\sigma}, \tag{18.70}$$

$$\mathcal{O}_{SDW^y}(n.n+1) = -i\,(-1)^n \sum_\sigma \sigma c_{n\sigma}^\dagger c_{n+1,-\sigma}, \tag{18.71}$$

In terms of the two-chain model, these operators correspond to modulation of effective bonds and currents along the diagonals of the plaquettes, BW$_d$ and OAF$_d$. The origin of the OAF ordering can be understood as follows.

Exercise. Consider the special case $t_\perp = 0$, $V_1 = V_2 \equiv V$ corresponding to the so-called UV model. Show that, for $U, V > 0$, the phase diagram of this model has a critical line $U = 2V$ separating the long-range ordered CDW state ($U < 2V$) and a SDW state ($U > 2V$) with a power-law decay of spin correlation functions.

Clearly, the SDW phase exists in the anisotropic case, $V_1 \neq V_2$, too; however the degeneracy between the longitudinal (SDW^z) and transverse (SDW^{x,y}) spin-density waves is removed by the exhange anisotropy. As we already know, changing the sign of the anisotropy parameter leads to a transition from an ordered Ising-like phase SDW^z(CDW$^\pi$) at $V_2 < V_1$ to a quasi- ('algebraically') ordered, XY-like phase with degenerate SDW^x(BW$_\perp$) and SDW^y(OAF$_\perp$) correlations at $V_2 > V_1$. Applying in the region $V_2 > V_1$ a uniform magnetic field $2t_\perp$ along the x-axis in spin space suppresses the staggered SDW^x(BW$_\perp$) correlations, thus leaving the SDW^y(OAF$_\perp$) as the only instability. Since the field $2t_\perp$ breaks the U(1) symmetry down to Z_2, the OAF$_\perp$ ordering will be *long ranged*. The corresponding alternative alignment of the currents along the horizontal links simply follows from the conservation of the total current at each site of the two-chain lattice. In what follows, we shall advance scaling arguments to support this conclusion.

Assuming that interaction and interchain coupling are weak ($U, V_{1,2}, t_\perp \ll t_\parallel$), we pass to the continuum limit of the Hamiltonian (18.65) and arrive at a charge–spin separated model which includes Umklapp (g_u)and backscattering ($g_\perp$) processes. The charge sector is described by a SG model:

$$H_c = \frac{u_c}{2}[\Pi_c^2 + (\partial_x \Phi_c)^2] - \frac{m_c}{\pi a_0} \cos \beta_c \Phi_c \qquad (18.72)$$

where

$$\beta_c^2/8\pi = 1 - g_c/2\pi v_F + 0(g_c^2), \quad m_c = g_u/2\pi a_0 \qquad (18.73)$$
$$g_c = U + 4V_1 + 2V_2, \quad g_u = -U + 2V_2 \qquad (18.74)$$

Since all interactions are repulsive, the condition $g_c > -|g_u|$ is always satisfied, and the SG model (18.72) occurs in its strong-coupling phase with a finite commensurability (charge) gap. The charge field Φ_c is ordered with vacuum values $\langle \Phi_c \rangle = 0$ at $m_c > 0$ ($U < 2V_2$), and $\langle \Phi_c \rangle = \pi/\beta_c$ at $m_c < 0$ ($U > 2V_2$).

The spin sector is represented by the U(1)-symmetric Thirring model (18.1), with an extra ($t_\perp$) term

$$\mathscr{L}_\perp = -t_\perp (R_\sigma^\dagger R_{-\sigma} + L_\sigma^\dagger L_{-\sigma}) \qquad (18.75)$$

appearing as a *transverse* magnetic field. The coupling constants $g_\parallel$ and $g_\perp$ are parametrized as follows

$$g_\parallel = -U + 4V_1 - 2V_2, \quad g_\perp = -U + 2V_2 \qquad (18.76)$$

Let us first consider the case $t_\perp = 0$. As folows from the above discussion, we have to single out the 'easy plane' phase with degenerate SDW^x and SDW^y correlations. To this end, we must get to the weak-coupling regime, $g_\parallel < -|g_\perp|$, of the Kosterlitz–Thouless phase diagram in the spin sector. This gives us the region

$$V_2 > V_1, \quad U > 2V_1 \qquad (18.77)$$

In the continuum limit, the site-diagonal and off-diagonal transverse SDW operators have the following factorized structure:

$$\mathcal{O}_{\text{SDW}^x} = (-1)^n \cos \frac{\beta_c \Phi_c}{2} \cos \frac{\tilde{\beta}_s \Theta_s}{2} \qquad (18.78)$$

$$\mathcal{O}_{\text{SDW}^y} = (-1)^n \cos \frac{\beta_c \Phi_c}{2} \sin \frac{\tilde{\beta}_s \Theta_s}{2} \qquad (18.79)$$

$$\overline{\mathcal{O}}_{\text{SDW}^x} = (-1)^n \sin \frac{\beta_c \Phi_c}{2} \cos \frac{\tilde{\beta}_s \Theta_s}{2} \qquad (18.80)$$

$$\overline{\mathcal{O}}_{\mathrm{SDW}^y} = (-1)^n \sin \frac{\beta_c \Phi_c}{2} \sin \frac{\tilde{\beta}_s \Theta_s} {2} \tag{18.81}$$

where $\overline{\mathcal{O}}$ stand for off-diagonal operators (18.70), (18.71). Then we conclude that, at $t_\perp = 0$, the system is characterized by coexistence of the SDW^x (BW$_\perp$) and SDW^y (OAF$_\perp$) correlations at $2V_1 < U < 2V_2$, and $\overline{\mathrm{SDW}^x}$ (BW$_d$) and $\overline{\mathrm{SDW}^y}$ (OAF$_d$) at $U > 2V_2$.

Consider now the role of a finite 'magnetic field' $2t_\perp$. It is convenient to make a $\pi/2$-rotation in spin space around the y-axis to have new z-axis along the magnetic field. In terms of the two-chain model, this is equivalent to choosing a new basis built up from the symmetric and antisymmetric states, $\psi_{s,a} = (\psi_+ - \psi_-)/\sqrt{2}$. The spin rotation does not affect the charge sector, but gives rise to spin-nonconserving processes in the spin channel:

$$g_\parallel J^z \bar{J}^z + \frac{1}{2} g_\perp (J^+ \bar{J}^- + \text{H. c.})$$

$$\rightarrow G_\parallel J^z \bar{J}^z + \frac{1}{2} G_\perp (J^+ \bar{J}^- + \text{H. c.}) + \frac{1}{2} G_f (J^+ \bar{J}^+ + \text{H. c.}) \tag{18.82}$$

where

$$G_\parallel = g_\perp = -U + 2V_2, \quad G_\perp = \frac{1}{2}(g_\parallel + g_\perp) = -U + 2V_1,$$

$$G_f = \frac{1}{2}(g_\parallel - g_\perp) = 2(V_1 - V_2) \tag{18.83}$$

The bosonized Hamiltonian in the spin sector acquires the two-cosine structure (18.17) of the XYZ model with a topological term:

$$H_s = \frac{u_s}{2}[\Pi_s^2 + (\partial_x \Phi_s)^2] - \frac{m}{\pi a_0} \cos \beta_s \Phi_s - \frac{\tilde{m}}{\pi a_0} \cos \tilde{\beta}_s \Theta_s$$

$$- \frac{\beta_s t_\perp}{2\pi} \partial_x \Phi_s \tag{18.84}$$

Under the spin rotation, the form of the SDW^y and $\overline{\mathrm{SDW}^y}$ operators remains intact, while the SDW^x and $\overline{\mathrm{SDW}^x}$ operators are replaced by

$$\mathcal{O}_{\mathrm{SDW}^x} \rightarrow (-1)^n \cos \frac{\beta_s \Phi_s}{2} \sin \frac{\beta_s \Phi_s}{2} \tag{18.85}$$

$$\overline{\mathcal{O}}_{\mathrm{SDW}^x} \rightarrow (-1)^n \sin \frac{\beta_c \Phi_c}{2} \sin \frac{\beta_s \Phi_s}{2} \tag{18.86}$$

Applying the two-cut-off scaling method, we consider first the region $l < h$, $h = \ln(u_s/|t_\perp| a_0)$, where the effective couplings $Z_\parallel(l), Z_\perp(l)$ and $Z_f(l)$ do not depend on h. Notice that, in this region, the XYZ model occurs on its critical plane, $G_\parallel + |G_\perp| - |G_f| = 0$. This is a manifestation of the 'hidden' U(1) symmetry of the model (18.84) at $t_\perp = 0$, and the fact that the system flows to a weak-coupling fixed point (in agreement with our discussion of the 'unrotated' model). In the region $l > h$, the magnetic field suppresses the G_f-processes and thus drives the system away from the critical plane. Here one is left with Kosterlitz–Thouless equations (18.64) for $Z_\parallel$ and $Z_\perp(l)$, which are applicable in the whole infrared region since the new 'bare' couplings $Z_\parallel(h)$ and $Z_f(h)$ are small due to conditions (18.77). Moreover, one can easily check that, at any h, the condition $Z_\parallel(h) < |Z_f(h)|$ is satisfied, and Eqs. (18.64) describe the development of a strong-coupling regime, with $Z_\parallel \to -\infty$, $|Z_f| \to \infty$. This means that H_s in (18.84) crosses over to a SG model

$$H_s \to \frac{u_s}{2}[\tilde{\Pi}_s^2 + (\partial_x\Theta_s)^2] - \frac{\beta_s t_\perp}{2\pi}\,\tilde{\Pi}_s - \frac{\tilde{m}}{\pi a_0}\cos\tilde{\beta}_s\Theta_s \qquad (18.87)$$

written in terms of the dual field Θ_s, using the relationship $\tilde{\Pi}_s = \partial_x\Phi_s$. Notice that $t_\perp$ is coupled to the momentum of the SG field Θ. Then, by Galilean invariance, the term linear in $\tilde{\Pi}$ is eliminated by an appropriate shift, and $\partial_x\Phi_s$ acquires a vacuum expectation value, $\langle\partial_x\Phi_s\rangle = \beta_s t_\perp/2\pi u_s$, implying that the particles are coherently delocalized between the two chains. Moreover, the field Θ_s gets ordered, with vacuum values $\langle\Theta_s\rangle_{\text{vac}} = \pi/\tilde{\beta}_s$. Using this result and the vacuum values of the charge field Φ_c, we deduce from the structure of operators (18.79), (18.81), (18.85), (18.86) that the BW, BW_d correlations are suppressed in the whole region (18.77), while the $OAF_\perp$ and OAF_d order parameters have nonzero average values at $2V_1 < U < 2V_2$ and $U > 2V_2$, respectively.

Thus, in the range $2V_1 < U < 2V_2$, one finds that the OAF state, with currents circulating alternately around the plaquettes, is indeed the stable ground state of the model with true long-range order. Such a state represents a two-chain analog of two-dimensional OAF ordering on a square lattice.

Exercise. Construct the order parameter describing alternation of the in-chain currents and find its continuum bosonized form. Prove that the onset of long-range order in the interchain (transverse) currents is indeed accompanied by ordering of the in-chain currents on the horizontal links of the two-chain lattice.

At $U > 2V_2$ another OAF phase with $\langle OAF_d\rangle \neq 0$ is realized. Interestingly, in contrast with the previous OAF state, the current conservation law does not require long-range ordering of the currents flowing through horizontal and vertical links of the lattice, although they contribute to formation of effective currents flowing across diagonals of the plaquettes.

More exotic phases, such as spin nematic and a two-chain analogue of d-wave superconductivity, the latter occurring upon hole doping, have been found in the case of two spinful chains (Khveshchenko (1993); Schulz (1996)).

References

S. Coleman, *Phys. Rev.* D**11**, 2088 (1975).

V. N. Dutyshev, *Zh. Eksp. Teor. Fiz.* **78**, 1698 (1980).

T. Giamarchi and H. J. Schulz, *J. de Physique* (Paris) **49**, 819 (1988).

F. D. M. Haldane, *J. Phys.* A**15**, 507 (1982).

B. I. Halperin and T. M. Rice, in *Solid State Physics*, vol. 21, p. 116, ed. by F. Seitz, D. Turnbull and H. Ehrenreich (New York, Academic), (1968).

G. I. Japaridze and A. A. Nersesyan, *Pis'ma Zh. Eksp. Teor. Fiz.* **27**, 356 (1978) [*Sov. Phys. JETP Lett.* **27**, 334 (1978)].

G. I. Japaridze, A. A. Nersesyan, and P. B. Wiegmann, *Nucl. Phys.* B**230**, 511 (1984).

J. V. Jose, L. P. Kadanoff, S. Kirkpatrick and D. R. Nelson, *Phys. Rev.* B **18**, 2318 (1978).

D. V. Khveshchenko, *Phys. Rev.* B**50**, 380 (1993).

A. Luther and V. J. Emery, *Phys. Rev. Lett.* **33**, 589 (1974).

A. A. Nersesyan, *Phys. Lett.* A**153**, 49 (1991).

H. J. Schulz, *Phys. Rev.* B**53**, R2959 (1996).

J. B. Torrance, *J. de Physique Colloq.* (Paris) **44**, C3 799 (1983).

J. B. Torrance, H. B. Pedersen and K. Bechgaard, *Phys. Rev. Lett.* **49**, 882 (1982).

T. T. Truong and K. D. Schotte, *Phys. Rev. Lett.* **47**, 285 (1981).

P. B. Wiegmann, *J. Phys.* C**11**, 1583 (1978); see also P. B. Wiegmann, in *Sov. Sci. Rev.*, vol. 2, p. 41, (1980), Harwood Acad. Publ., ed. by I. M. Khalatnikov.

19

What may happen with a Tomonaga–Luttinger liquid in three dimensions

Needless to say, real materials are usually three dimensional (with exceptions of specially prepared quantum wires or quantum Hall effect edge states, which will be discussed later). This is made manifest by the fact that almost all of them undergo phase transitions at low temperatures which may happen only in three dimensions. Thus one has to consider boundaries of applicability of the approach which takes the Tomonaga–Luttinger liquid as a reference point. This problem is not at all simple and is far from being resolved. Necessary conditions for a Tomonaga–Luttinger liquid are probably model dependent. Sufficient conditions, however, are much easier to formulate. In our further presentation we follow the paper by Boies et al. (1995).

The main reason why the problem of stability of Tomonaga–Luttinger liquids is so complicated is that the interchain hopping simultaneously generates effects of different nature. Thus there are processes leading to the creation of a three-dimensional Fermi surface and processes of multi-particle exchange leading to three-dimensional phase transitions (we shall discuss such mechanisms in greater detail in the subsequent Chapters). In principle, one can imagine the following possibilities.

- Single-particle interchain tunneling leads to formation of a three-dimensional Fermi surface at some characteristic temperature T_{FS}.[*] The remaining interaction may destabilize this Fermi surface at some lower temperature T_{c} where the system undergoes a real phase transition into some symmetry-broken three-dimensional state.[†]

---

[*] The reader should bear in mind that there is no phase transition associated with this and T_{FS} is a crossover temperature.

[†] In principle, the conventional Fermi liquid is also unstable. Whatever realistic interaction we consider, it always has negative harmonics on the Fermi surface which leads to superconducting pairing in this channel. Of course, for higher channels the transition temperature may be quite small.

- Interchain interactions are stronger than the single-particle hopping ($T_c > T_{FS}$) and a transition to a symmetry-broken state occurs from the Tomonaga–Luttinger state.

We see that whether a Fermi surface is formed or not, the interchain exchange will always destabilize the TL fixed point and create a three-dimensional ground state with a broken symmetry. In this and only this sense the interchain coupling is always relevant. There is another question, however, and it is whether the state above the transition temperature is a normal metal or something else. It was rightly pointed out by Anderson that the earmark of a Fermi liquid is a coherent transport in all directions. Contrary to that, in a one-dimensional Tomonaga–Luttinger liquid state the Drude peak exists only in optical conductivity in the direction along the chains.

In this chapter we shall discuss only the one scenario, namely what happens if the single-particle tunneling is stronger than the interchain interactions. To isolate effects of this tunneling we need to have a situation where $T_{FS} \gg T_c$. This can be achieved in a model describing a multi-dimensional array of chains coupled only by a weak interchain hopping $t_\perp$, by considering a limit $D \gg 1$, where D is the coordination number of the array. The large value of D will allow us to ignore all feedback effects coming from electrons returning to the same chain and generating effective interchain many-body exchange interactions which may (and usually do) lead to phase transitions at low temperatures. In order to minimize similarity between the Tomonaga–Luttinger liquids existing on single chains and a Fermi liquid we shall consider a strong spin–charge separation $v_s/v_c \ll 1$. The quantity K_c will remain arbitrary.

Taking into account only the diagrams which can be cut along a single hopping integral as leading ones in $1/D$, we get the following expression for the single electron Green's function:

$$G(\omega, k, \mathbf{k}) = [G_0^{-1}(\omega, k) - t_\perp(\mathbf{k})]^{-1} \tag{19.1}$$

Here $G_0(\omega, k)$ is the Fourier transformation of the single chain Green's function (16.19), $\mathbf{k}$ represents transverse components of the wave vector and

$$t_\perp(\mathbf{k}) = t_\perp \sum_{i=1}^{D} \cos(\mathbf{k}\mathbf{b}_i)$$

with $\mathbf{b}_i$ being lattice vectors.

Eq. (19.1) was studied by Wen (1990) who used various approximate expressions for $G_0(\omega, k)$. Wen demonstrated that at $T = 0$ the pole appears if $\theta < 1$ where

$$\theta = (K_c + K_c^{-1} - 2)/4$$

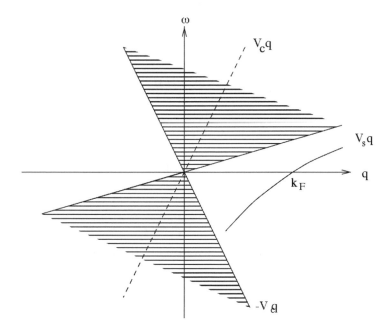

Fig. 19.1.   The shaded region is the region of the (ω, q) plane where $\Im m G_0^{(R)}$ for the single chain does not vanish. The solid line shows the dispersion law of the coherent fermion excitation (19.6) near the right Fermi point.

Below we will partially repeat Wen's arguments and derive some new results. The retarded Green's function has singularities near the right and left Fermi points $k = \pm k_F$. In what follows we shall concentrate on the vicinity of the right Fermi point and introduce the wave vector $q = k - k_F$. The calculations of the single particle Green's function of the Tomonaga–Luttinger liquid were carried out by Meden and Schönhammer (1992), Ren and Anderson (1993) and Voit (1992) (see also references in the previous chapter) whose results are rederived in the Appendix to this chapter. In particular, it was established that at $T = 0$ its imaginary part (the spectral function $\rho(\omega, q)$) is finite in the area represented on Fig. 19.1.

$$G_0^{(R)}[\omega, k = \pm(q + k_F)] = \frac{|q/k_F|^\theta}{q v_c} \left[\mathscr{F}_1(\omega/v_s q) + \mathscr{F}_2(\omega/v_c q) \right] \qquad (19.2)$$

where the functions $\mathscr{F}_a$ can be obtained from Eq. (19.27). For $q > 0$ we have (see the Appendix):

$$\mathscr{F}_1(\omega/v_s q) = (v_c/v_s)^{1+\theta} \times \int_0^{v_c/v_s - 1} \frac{dx}{x^{1/2}} (v_c/v_s - 1 - x)^{-1/2 - \theta/2}$$

$$(x + v_c/v_s + 1)^{-\theta/2} (x + 1 + \omega/v_s q)^{-1+\theta} \quad (19.3)$$

$$\mathscr{F}_2(\omega/v_c q) = \int_0^\infty dx x^{-\theta/2}(x+2)^{-1/2-\theta/2}(x+1+v_s/v_c)^{-1/2}(x+1-\omega/v_c q)^{-1+\theta}$$

$$+ \int_0^\infty dx x^{-1/2-\theta/2}(x+2)^{-\theta/2}(x+1-v_s/v_c)^{-1/2}(x+1-\omega/v_c q)^{-1+\theta} \quad (19.4)$$

For $\theta < 1$ these integrals converge at small x and the answers are universal.

To study the Fermi surface formation, we need to know $G_0(\omega, q)$ at $\omega \to 0$. As follows from Eqs. (19.3, 19.4), in the area $-v_c q < \omega < v_s q$ ($q > 0$) the Green's function is purely real and can be expanded in powers of ω. At $v_s/v_c \ll 1$ the leading contribution comes from the $\mathscr{F}_1$-term:

$$\mathscr{F}_1(\omega/v_s q) = a - b\frac{\omega}{qv_s} + \dots \quad (19.5)$$

Substituting (19.5) into Eq.(19.1) we obtain the following expression for the dispersion law:

$$\epsilon(q, \mathbf{k}) = \frac{a}{b}v_s\left[-q + av_c^{-1}t_\perp(\mathbf{k})(|q|/k_F)^\theta\right] \quad (19.6)$$

So we see, that as it was originally found by Wen, the Fermi surface is formed at $\theta < 1$ which corresponds to $3 - 2\sqrt{2} < K_c < 3 + 2\sqrt{2}$. A possibility of large K_c is open to question, but as it has been explained in Section 17.I.1, small K_c can be realized.

From Eqs. (19.1) and (19.6) we find the expressions for the Fermi vector

$$\delta k_F = k_F\left[at_\perp(\mathbf{k})/\epsilon_F\right]^{1/1-\theta} \quad (19.7)$$

and the quasiparticle residue:

$$Z \approx \frac{v_s}{v_c}\frac{a^2}{b}\left[at_\perp(\mathbf{k})/\epsilon_F\right]^{\theta/1-\theta} \quad (19.8)$$

As we shall see in a moment, the spin–charge separation plays an important role in the behaviour of Z providing an additional small parameter in the region $\theta < 1/2$.

From Eq. (19.3) we find

$$a \approx (v_c/v_s)^{1+\theta}\int_0^{v_c/v_s} dx x^{-1/2}(1+x)^{-1+\theta}(v_c/v_s - x)^{-1/2-\theta/2}(x + v_c/v_s)^{-\theta/2}$$
$$(19.9)$$

The integral for b is determined by $x \sim 1$ and can be calculated analytically:

$$b \approx (1 - \theta)(v_c/v_s)^{1/2}B(1/2, 3/2 - \theta) \quad (19.10)$$

where $B(x, y)$ is the digamma function.

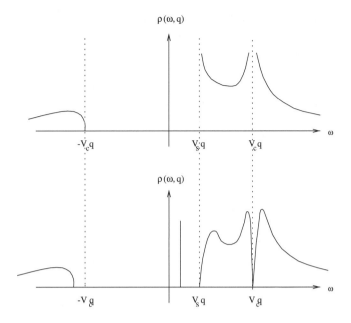

Fig. 19.2. (Top) The spectral function for the spin-1/2 Luttinger liquid at $\theta < 1/2$, $q > 0$; (bottom) the imaginary part of the single electron Green's function G in the presence of interchain hopping.

The integral (19.9) has different areas of convergence depending on whether $\theta < 1/2$ or $1/2 < \theta < 1$. In the former case it converges at $x \sim 1$:

$$a \approx (v_{\rm c}/v_{\rm s})^{1/2} B(1/2, 1/2 - \theta) \tag{19.11}$$

In the latter case the integral converges at large x, so that one can neglect the difference between x and $x + 1$ to obtain

$$a \approx (v_{\rm c}/v_{\rm s})^{\theta} B(\theta - 1/2, 1 - \theta/2) F(\theta/2, \theta - 1/2, 1/2 + \theta/2; -1) \tag{19.12}$$

Substituting Eqs. (19.11, 19.10, 19.12) into (19.8) we get

$$Z \sim \left(\frac{v_{\rm s}}{v_{\rm c}}\right)^{\gamma} \left[t_{\perp}(\mathbf{k})/\epsilon_{\rm F}\right]^{\theta/1-\theta} \tag{19.13}$$

where

$$\gamma = (1 - 2\theta)/2(1 - \theta) > 0, \quad (\theta < 1/2)$$
$$\gamma = (3 - \theta)(1 - 2\theta)/2(1 - \theta) < 0 \quad (1 > \theta > 1/2) \tag{19.14}$$

Therefore we see that the exponent γ changes its sign at $\theta = 1/2$ which leads to a rather peculiar situation. Namely, at $\theta < 1/2$ the single particle pole is formed, but its residue may be small even at moderately small $t_{\perp}/\epsilon_{\rm F}$. This means that the Fermi liquid phase will retain many features of the Tomonaga–Luttinger liquid (see Fig. 19.2). So for $\theta < 1/2$ the spin–charge separation works in favour of the Luttinger liquid. For $\theta > 1/2$ the situation is reversed: one needs a very high degree of one-dimensionality to get small Z. This apparently paradoxical behaviour has a simple explanation. The calculations show that at $\theta < 1/2$ the single particle spectral function is singular at $\omega = v_s q$ and $\omega = v_c q$ (see Fig. 19.2 (top)). In the presence of hopping these singularities are smeared out, but maxima still remain (Fig. 19.2 (bottom)) and therefore, as we have said, many features of the system are still Tomonaga–Luttinger-liquid-like. At $\theta < 1/2$ these singularities disappear and the spectral function becomes featureless. Then when the pole is formed we will have a situation very much like that in a conventional Fermi liquid: a featureless incoherent background and a sharp coherent pole.

Exercise.   Consider the influence of the coherent branch on the collective modes.

As we have mentioned above, at $\theta > 1/2$ ($K_c < 0.28$) the coherent branch disappears altogether. Interestingly, according to Eq. (17.54) this is accompanied by disappearance of the Drude-like maximum in the transverse conductivity. The system remains in a spin-charge separated state dominated by collective modes. Those are usually taken into account in RPA and lead to all sorts of three-dimensional phase transitions (see for example, the reference for Brazovskii and Yakovenko (1985) in the introduction to this Part).

The above calculation was made at zero temperature which makes sense only if the interchain hopping is negligible ($D \gg 1$). We have established that in this case the single pole may appear only when $K_c < 2 - \sqrt{3}$. When the interchain coupling is significant, one can imagine another possibility, namely, that the three-dimensional phase occurs sooner then the single particle pole is formed:

$$T_c > T_{FS} = \epsilon_F \left(t_\perp / \epsilon_F \right)^{1/(1-\theta)} \tag{19.15}$$

Then the normal phase is dominated by Tomonaga–Luttinger liquid effects even if $\theta < 1/2$. We shall consider such cases in subsequent Chapters.

I   Appendix. Fermionic Green's function

This Appendix was written by A. G. Green.

I.1 Coordinate space Green's function

The thermodynamic single electron Green's function for spin-1/2 TL liquid is given by Eq. (16.19). Later we will need to know the retarded Green's function. This is defined in real time, ($\tau = $ it), as

$$G_{r\sigma}^R(x,t) = -i\theta(t)\langle\{\psi_{r\sigma}(x,t),\psi_{r\sigma}^\dagger(0,0)\}\rangle$$
$$= -i\theta(t)e^{ik_F r x}\left\{G_r^0(x,t,\alpha) + G_r^0(-x,-t,\alpha)\right\} \qquad (19.16)$$

where $r = \pm 1$. The Fourier transform of this Green's function is

$$G_{r\sigma}^R[r(k_F + q),\omega] = \int_{-\infty}^{\infty} dx \int_{-\infty}^{\infty} dt\, e^{-i(k_F+q)rx+i\omega t} G_{r\sigma}^R(x,t) \qquad (19.17)$$

The evaluation of this Fourier transform is a complicated problem in contour integration. First of all, we will calculate its imaginary part, the spectral function.

I.2 The spectral function ($v_c > v_s$)

The spectral function is defined as

$$\rho_{r\sigma}(q,\omega) = -\frac{1}{\pi}\Im m G_{r\sigma}^R(k_F + q,\omega). \qquad (19.18)$$

Since $G^{0*}(x,t,\alpha) = G^0(-x,-t,\alpha)$ we may substitute it into Eq. (19.16) and obtain

$$\rho_{r\sigma}(q,\omega) = -\frac{1}{2\pi i}\left[G_{r\sigma}^R(k,\omega) - G_{r\sigma}^{*R}(k,\omega)\right]$$
$$= -\frac{1}{2\pi i}\int_{-\infty}^{\infty} dxdt\left[e^{-i(kx-\omega t)}G_{r\sigma}^R(x,t) - e^{i(kx-\omega t)}G_{r\sigma}^{R*}(x,t)\right]$$
$$= -\frac{1}{2\pi i}\int_{-\infty}^{\infty} dxdt\left\{\begin{array}{l}-i\theta(t)e^{-i(qx-\omega t)}\left[G^0(x,t,\alpha) + G^0(-x,-t,\alpha)\right]\\ -i\theta(t)e^{i(qx-\omega t)}\left[G^0(x,t,\alpha) + G^0(-x,-t,\alpha)\right]\end{array}\right\}$$
$$= \frac{1}{2\pi}\int_{-\infty}^{\infty} dxdt e^{-i(qx-\omega t)}\left[G^0(x,t,\alpha) + G^0(-x,-t,\alpha)\right] \qquad (19.19)$$

$$= \frac{a^{\theta_c+\theta_s}}{2\pi}\int_{-\infty}^{\infty} dxdt e^{-i(qx-\omega t)}\left\{\begin{array}{l}[\alpha + i(v_c t - rx)]^{-1/2}[\alpha + i(v_s t - rx)]^{-1/2}\\ \times[(a + iv_c t)^2 + x^2]^{-\theta_c/2}[(a + iv_s t)^2 + x^2]^{-\theta_s/2}\\ +x \rightarrow -x, t \rightarrow -t\end{array}\right\}$$

In the above we have allowed for the possibility that the spin sector has an anomalous dimension, as well as the charge sector. In fact, if the SU(2) symmetry is preserved, $\theta_s = 0$. In this case one may write the spectral function in terms of a single integration. We will show this calculation later, but first we show how one may extract the singular behaviour of the spectral function from Eq. (19.19) by simple power counting arguments. First, we make the change of variables, $s = (v_s t - rx)/(v_s + v_c)$ and $s' = (v_c t + rx)/(v_s + v_c)$ in Eq.(19.19);

$$\rho_{r\sigma}(q, \omega) = \frac{a^{\theta_c}}{2\pi} \int_{-\infty}^{\infty} ds ds' e^{-i(\Omega s + \Omega' s')}$$

$$\times \left\{ \begin{array}{c} [\alpha + i(v_s + v_c)s]^{-1/2} [a + i(v_s + v_c)s']^{-\theta_c/2} \\ \times [\alpha + 2iv_c s + i(v_c - v_s)s']^{-1/2} [a + 2iv_c s + i(v_c - v_s)s']^{-\theta_c/2} \\ +s \rightarrow -s, s' \rightarrow -s' \end{array} \right\}$$

$$\Omega = q v_c + \omega$$
$$\Omega' = -q v_s + \omega \qquad (19.20)$$

For the regions in which the spectral function is nonzero, it may be deduced by considering the nonanalyticities of the integrand in Eq. (19.20). For the $+s$, $+s'$ part of the integrand, nonanalyticities for both s and s' occur only in the upper half plane. From Jordan's lemma it follows that the contribution of this part of the integral is nonzero only when $\Omega < 0$ and $\Omega' < 0$. For the $-s$, $-s'$ part of the integrand, nonanalyticities occur only in the lower half plane. The contribution to the spectral function from this part of the integral is nonzero only when $\Omega > 0$ and $\Omega' > 0$. The spectral function is therefore nonzero in the regions $\omega > q v_s$ and $\omega < -q v_c$.

We expect singularities in the spectral function near $\omega = \pm v_s q, \pm v_c q$, (for the SU(2) invariant case, the singularity at $-v_s q$ is suppressed). Near to $\omega = q v_s$, $\Omega' \approx 0$ and $\Omega \approx q(v_c - v_s)$. The integral in s is dominated by $s < 1/\Omega$ whereas that in s' is dominated by very large values. Power counting implies that near to $\omega = q v_s$,[‡]

$$\rho(q, \omega \approx q v_s) \sim \theta(\omega - q v_s)(\omega - q v_s)^{\theta - 1/2} \qquad (19.21)$$

Near to $\omega = -q v_c$, $\Omega \approx 0$ and $\Omega' \approx q(v_c - v_s)$. The integral in s' is dominated by $s' < 1/\Omega'$ and that in s by very large values. By power counting we obtain the singular form of the spectral function;

$$\rho(q, \omega \approx -q v_c) \sim \theta(-\omega - q v_c)(-\omega - q v_c)^{\theta} \qquad (19.22)$$

To find the behaviour near $\omega = v_c q$, we rewrite Eq. (19.20) using the substitution $s = v_c t - rx$, $s' = v_c t + rx$. Again, we may find the singular behaviour by power counting. The result is

$$\rho(q, \omega \approx q v_c) \sim (\omega - q v_c)^{(\theta - 1)/2} \qquad (19.23)$$

[‡] The reader should be careful not to confuse the theta function $\theta(x)$ with the exponent θ.

The spectral function may be written in terms of a single integration as follows:

$$\rho_{+\sigma}(q,\omega) = 2\pi \left(\frac{a}{v_c}\right)^\theta \frac{(1+v_s/v_c)^{1/2}}{(1-v_s/v_c)^{\theta/2}} \left[\Gamma(1/2)\Gamma(\theta/2)\Gamma((\theta+1)/2)\right]^{-1}$$

$$\times \exp\left\{-\frac{a}{v_c+v_s}[2\omega + q(v_c - v_s)]\right\} \theta(\omega - v_s q)(\omega - v_s q)^{\theta-1/2}$$

$$\times \int_0^1 (1-s)^{\theta/2-1} s^{(\theta-1)/2} \left[(\omega + v_c q)(1 - v_s/v_c) - (\omega - v_s q)2s\right]^{-1/2}$$

$$\times \exp\left(2as\frac{\omega - v_s q}{v_s + v_c}\right) \theta\left[(\omega + v_c q)(1 - v_s/v_c) - (\omega - v_s q)2s\right]$$

$$+q \to -q, \omega \to -\omega. \tag{19.24}$$

The rigorous proof of this is rather complicated; here we will just simply outline the derivation. The starting point is Eq.(19.24). We make the change of variables, $x, t \to x' = x - v_s t, t$. The integral over t may be written in terms of confluent hypergeometric functions using the expression 3.384.8 given in the book by Gradstein and Ryzhik (further referred as GR):

$$\int_{-\infty}^{\infty} dx \, e^{-ipx}(\beta+ix)^{-\mu}(\gamma+ix)^{-\nu} = \frac{2\pi e^{\gamma p}(-p)^{\mu+\nu+1}}{\Gamma(\mu+\nu)} \Phi[\mu, \mu+\nu, (\beta-\gamma)p]\theta(-p),$$

$$[\Re\beta > 0, \Re\gamma > 0, \Re(\mu + \nu) > 1] \tag{19.25}$$

with $p = v_s q - \omega$, $\mu = (\theta_c + 1)/2$, $\nu = \theta_c/2$, $\beta = (\alpha - ix')/(v_c - v_s)$ and $\gamma = (\alpha + ix')/(v_c + v_s)$. Next, we use the integral representation of the confluent hypergeometric function given in GR 9.211;

$$\Phi(a, b, z) = \frac{\Gamma(b)}{\Gamma(b-a)\Gamma(a)} \int_0^1 ds \, e^{zs} s^{a-1}(1-s)^{b-a-1} \tag{19.26}$$

(We have also substituted for GR 8.384.) The resulting integral over x' may be written in terms of gamma functions using GR 3.382.7.[§] Substituting this completes the derivation.[¶]

Finally, one may use GR3.197.3 to recast the representation of the spectral function given in Eq. (19.24) in terms of hypergeometric functions.[‖] The regions $\omega < -v_c q$, $v_s q < \omega < v_c q$ and $\omega > v_c q$ are considered

---

[§] $\int_{-\infty}^{\infty}(\beta - ix)^{-\nu}e^{-ipx}dx = \frac{2\pi}{\Gamma(\nu)}\theta(p)p^{\nu-1}e^{-\beta p}$, $[\Re\nu > 0, \Re\beta > 0]$.

[¶] Note that there is a factor of 2π difference between our normalization of the Green's function and Voit's. Also note that the expression for the spectral function is the same as that for the retarded Green's function, with integrals in time over the full line instead of the half line.

[‖] $\int_0^1 x^{\lambda-1}(1-x)^{\mu-1}(1-\beta x)^{-\nu}dx = B(\lambda, \mu)F(\nu, \lambda; \lambda+\mu; \beta)$, $[\Re\lambda > 0, \Re\mu > 0, |\beta| < 1]$.

separately and the result is

$$\rho(q, \omega < -v_c q) = \frac{2\pi a^\theta}{\Gamma(\theta/2)\Gamma(\theta/2+1)}$$

$$\times (2v_c)^{-1/2-\theta/2}(v_c + v_s)^{1/2-\theta/2}(-\omega - v_c q)^{\theta/2}(v_s q - \omega)^{\theta/2-1}$$

$$\times F\left(1 - \theta/2, 1/2 + \theta/2; 1 + \theta/2; \frac{v_c - v_s}{2v_c} \frac{v_c q + \omega}{\omega - v_s q}\right)$$

$$\rho(q, v_s q < \omega < v_c q) = \frac{2\pi a^\theta}{\Gamma(1/2)\Gamma(\theta + 1/2)}$$

$$\times (v_c + v_s)^{1/2-\theta/2}(v_c - v_s)^{-1/2-\theta/2}(\omega - v_s q)^{\theta/2-1/2}(\omega + v_c q)^{-1/2}$$

$$\times F\left(1/2, 1/2 + \theta/2; 1/2 + \theta; \frac{2v_c}{v_c - v_s} \frac{\omega - v_s q}{\omega + v_c q}\right)$$

$$\rho(q, \omega > v_c q) = \frac{2\pi a^\theta}{\Gamma(\theta/2)\Gamma(\theta/2+1)}$$

$$\times (2v_c)^{-\theta/2-1/2}(v_c + v_s)^{1/2-\theta/2}(\omega + v_c q)^{\theta/2}(\omega - v_s q)^{\theta/2-1}$$

$$\times F\left(1 - \theta/2, 1/2 + \theta/2; 1 + \theta/2; \frac{v_c - v_s}{2v_c} \frac{v_c q + \omega}{\omega - v_s q}\right) \qquad (19.27)$$

I.3 Fourier transform of the Green's function $(v_c > v_s)$

In this section we derive an integral representation for the Fourier transform of the Green's function, similar to Eq. (19.24) for the spectral function.

From Eq. (19.16) this Fourier transform may be written

$$G_{r\sigma}^R(x, t) = -i\theta(t)e^{ik_F r x} \lim_{\alpha \to 0} \left[G_{r\sigma}^0(x, t, \alpha) + G_{r\sigma}^0(-x, -t, \alpha) \right]$$

$$G_{r\sigma}^R[\omega, r(k_F + q)] = -i a^\theta \int_0^\infty dt \int_{-\infty}^\infty e^{-i(qrx - \omega t)}$$

$$\times \left\{ \begin{array}{c} [\alpha + i(v_c t - rx)]^{-1/2-\theta/2} [\alpha + i(v_s t - rx)]^{-1/2} [\alpha + i(v_c t + rx)]^{-\theta/2} \\ +x, t \to -x, -t \end{array} \right\}$$

$$\qquad (19.28)$$

It should be noted that in order to make the integration tractable, we make the replacement $a \to \alpha$. This will affect the high frequency part of the spectrum, but will have no effect on the low frequency part. The same assumption is implicit in our derivation of the integral form of the spectral function. We make the change of variables, $x, t \to X = x/t, t$ to obtain

$$G^R_{r\sigma}[\omega, r(k_F + q)]$$

$$= -ia^\theta \int_{-\infty}^\infty dX (v_c - rX)^{-1/2-\theta/2} (v_s - rX)^{-1/2} (v_c + rX)^{-\theta/2} \int_0^\infty tdt e^{-i(qrX-\omega)t}$$

$$\times \left\{ [it + \alpha/(v_c - rX)]^{-1/2-\theta/2} [it + \alpha/(v_s - rX)]^{-1/2} [it + \alpha/(v_c + rX)]^{-\theta/2} \atop +t \to -t \right\}$$

$$(19.29)$$

The integral in X is now split into the sum of four parts, corresponding to the ranges $X < v_c$, $-v_c < X < v_s$, $v_s < X < v_c$ and $X > v_c$ respectively;

$$G^1(\omega, q) =$$

$$-ia^\theta \int_0^\infty dX (2v_c + X)^{-1/2-\theta/2} (v_s + v_c + X)^{-1/2} X^{-\theta/2} \int_0^\infty tdt e^{i(qX+qv_c+\omega)t}$$

$$\times \left\{ [it + \alpha/(2v_c + X)]^{-1/2-\theta/2} [it + \alpha/(v_s + v_c + X)]^{-1/2} (-it + \alpha/X)^{-\theta/2} \atop +t \to -t \right\}$$

$$G^2(\omega, q) =$$

$$-ia^\theta \int_0^{v_c+v_s} dX (2v_c - X)^{-1/2-\theta/2} (v_s + v_c - X)^{-1/2} X^{-\theta/2} \int_0^\infty tdt e^{-i(qX-qv_c-\omega)t}$$

$$\times \left\{ [it + \alpha/(2v_c - X)]^{-1/2-\theta/2} [it + \alpha/(v_s + v_c - X)]^{-1/2} (it + \alpha/X)^{-\theta/2} \atop +t \to -t \right\}$$

$$G^3(\omega, q) =$$

$$-ia^\theta \int_0^{v_c-v_s} dX (v_c - v_s - X)^{-1/2-\theta/2} (X)^{-1/2} (v_c + v_s + X)^{-\theta/2}$$

$$\int_0^\infty tdt e^{-i(qX+qv_s-\omega)t}$$

$$\times \left\{ [it + \alpha/(v_c - v_s - X)]^{-1/2-\theta/2} (-it + \alpha/X)^{-1/2} [it + \alpha/(v_c + v_s + X)]^{-\theta/2} \atop +t \to -t \right\}$$

$$G^4(\omega, q) =$$

$$-ia^\theta \int_0^\infty dX X^{-1/2-\theta/2}(v_c - v_s + X)^{-1/2}(2v_c + X)^{-\theta/2} \int_0^\infty t dt e^{-i(qX + qv_s - \omega)t}$$

$$\times \left\{ [it + \alpha/(v_c - v_s - X)]^{-1/2-\theta/2} (-it + \alpha/X)^{-1/2} [it + \alpha/(v_c + v_s + X)]^{-\theta/2} \atop +t \to -t \right\}$$

$$G^4(\omega, q) =$$

$$-ia^\theta \int_0^\infty dX X^{-1/2-\theta/2}(v_c - v_s + X)^{-1/2}(2v_c + X)^{-\theta/2} \int_0^\infty t dt e^{-i(qX + qv_c - \omega)t}$$

$$\times \left\{ (-it + \alpha/X)^{-1/2-\theta/2} [-it + \alpha/(v_c - v_s + X)]^{-1/2} [it + \alpha/(2v_c + X)]^{-\theta/2} \atop +t \to -t \right\}$$

$$(19.30)$$

The integrals in t may be calculated in the limit $\alpha \to 0$ in terms of gamma functions with prefactors determined by the positioning of the branch cuts in the t plane. The difficulty is to choose the positioning of the branch cuts appropriately. The phases of each factor ($\pm it + \alpha$...) in the left-hand part of the integrals are chosen in the range $\pi/2$ to $-\pi/2$. The form of the Green's function used in Eq. (19.16) determines the topology of the way in which branch cuts must be positioned. This must be such that (i) $G_r^0(-x, -t, \alpha) = -G_r^0(x, t, -\alpha)$ and that (ii) making the change $t \to -t$ in the first and third factors simply gives a factor of $-i$. The phases and branch cuts in the $-t$ parts of the integrand are chosen to be consistent with the $+t$ part. A suitable choice of branch cuts is shown in Fig. 19.3.

Applying this procedure to the t-integral for G^1, the t-integral has the form

$$\int_0^\infty t dt e^{iQt}(it + \alpha_a)^{-1/2-\theta/2}(it + \alpha_b)^{-1/2}(-it + \alpha_c)^{-\theta/2}$$

with α_a, α_b and α_c positive constants which tend to zero in the limit $\alpha \to 0$. As we take the limit $\alpha \to 0$, the only effect of the branch cuts is to introduce prefactors to the various integrals. In principle there may be contributions from contours between α_a and α_c for example; however, in the limit as $\alpha \to 0$ the support for these is zero. Using the branch cuts as shown in Fig. 19.3., we find

$$\int_0^\infty t dt e^{iQt}(it + \alpha_a)^{-1/2-\theta/2}(it + \alpha_b)^{-1/2}(-it + \alpha_c)^{-\theta/2}$$

$$= \left(e^{i\frac{\pi}{2}(-1/2-\theta/2)}e^{i\frac{\pi}{2}(-1/2)}e^{-i\frac{\pi}{2}(-\theta/2)} + e^{-i\frac{\pi}{2}(-1/2-\theta/2)}e^{-i\frac{\pi}{2}(-1/2)}e^{-i\frac{3\pi}{2}(-\theta/2)} \right)$$

$$\times \int_0^\infty t^{-\theta} e^{iQt} dt = 2ie^{-i\pi\theta} \sin(\pi\theta/2)\Gamma(1 - \theta)(-Q)^{\theta-1} \qquad (19.31)$$

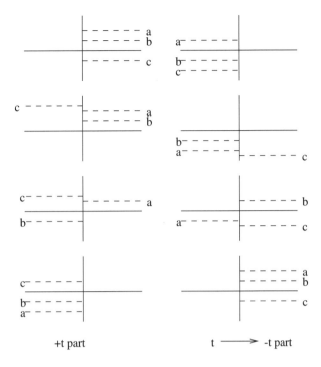

Fig. 19.3. Branch cuts for evaluation of the *t*-integrals.

The only subtlety to note here, is that the positioning of the branch cuts and consistency with the $+t$ part of the integral means that the factor $(-it + \alpha_c)$ has a negative argument.

The t integrals for G^2, G^3 and G^4 are calculated similarly.[**] Substituting these results into Eq. (19.30), we obtain

$$G^1(\omega, q) = 2a^\theta e^{-i\pi\theta} \sin(\pi\theta/2)\Gamma(1-\theta)$$

$$\times \int_0^\infty dX(2v_c + X)^{-1/2-\theta/2}(v_s + v_c + X)^{-1/2}X^{-\theta/2}[q(X+v_c)+\omega]^{\theta-1}$$

[**] The cancelation of G^2 may also be seen by considering the contour of integration in the x-plane. We note that $G_r^0(-x, -t, \alpha) = -G_r^0(x, t, -\alpha)$. One may then express the retarded Green's function in terms of a contour integral in the x plane;

$$G_{r\sigma}^R(x, t) = -i\theta(t)e^{ik_F rx} \lim_{\alpha \to 0} \left[G_{r\sigma}^0(x, t, \alpha) + G_{r\sigma}^0(-x, -t, \alpha) \right]$$

$$= -i\theta(t)e^{ik_F rx} \lim_{\alpha \to 0} \left[G_{r\sigma}^0(x, t, \alpha) - G_{r\sigma}^0(x, t, -\alpha) \right]$$

$$G_{r\sigma}^R[\omega, r(k_F + q)] = -i \int_0^\infty dt \int_C dx e^{-i(qrx - \omega t)} G_{r\sigma}^0(x, t, \alpha = 0)$$

The contour of integration, C, follows from a consideration of the positions of the nonanalyticities of $G_{r\sigma}^0(x, t, \alpha)$ and $G_{r\sigma}^0(x, t, -\alpha)$. The major consequence is that the portion of the x integral between $-v_c t$ and $v_s t$ cancels.

$$G^2(\omega, q) = 0$$

$$G^3(\omega, q) = 2a^\theta e^{-\pi\theta} \Gamma(1 - \theta)$$

$$\times \int_0^{v_c - v_s} dX (v_c - v_s - X)^{-1/2 - \theta/2} (X)^{-1/2} (v_c + v_s + X)^{-\theta/2} [\omega - q(X + v_s)]^{\theta - 1}$$

$$G^4(\omega, q) = -2a^\theta e^{-\pi\theta} \sin(\pi\theta/2) \Gamma(1 - \theta)$$

$$\int_0^\infty dX (X)^{-1/2 - \theta/2} (v_c - v_s + X)^{-1/2} (2v_c + X)^{-\theta/2} [\omega - q(X + v_c)]^{\theta - 1}$$

$$\tag{19.32}$$

I.4   The spectral function, $v_s > v_c$

The expression for the spectral function, prior to completing one of the integrals, is as before. The integrand now has a different analytic structure to that which it had previously and the integral is altered accordingly. The singular behaviour of the spectral function may be extracted in the same way as before, with the same result. In order to find the form of the spectral function in terms of a single integration we follow a similar procedure to before. Firstly, we make the change of variables $x, t \to X = x - v_c t, t$ and follow by using GR 3.384.8, GR 9.211 and GR 3.382.7 to obtain

$$\rho_{+\sigma}(q, \omega) =$$

$$\frac{2\pi a^\theta}{\Gamma(1/2)\Gamma(\theta/2)\Gamma[(\theta+1)/2]} (2v_c)^{1/2 - \theta} (v_s - v_c)^{-\theta/2} (\omega - qv_c)^{\theta/2 - 1/2} \theta(\omega - qv_c)$$

$$\times \exp\left\{ -\frac{a}{v_c + v_s} [2\omega + q(v_c - v_s)] \right\} \theta(\omega - v_s q)(w - v_s q)^{\theta - 1/2}$$

$$\times \int_0^1 s^{-1/2} (1 - s)^{\theta/2 - 1} [(\omega + v_c q)(v_s - v_c) + s(v_s q + \omega)(v_s + v_c)]^{\theta/2 - 1/2}$$

$$\times \theta [(\omega + v_c q)(v_s - v_c) + s(v_s q + \omega)(v_s + v_c)] \exp\left(\frac{\alpha s}{2v_c} \frac{3v_c - v_s}{v_s - v_c} \right)$$

$$\times \exp\left(\alpha \frac{\omega + v_s q}{2v_c} \right) + q \to -q, \omega \to -\omega \tag{19.33}$$

The spectral function in terms of hypergeometric functions is

$$\rho(q, \omega < -v_c q) =$$

$$\frac{2\pi a^\theta}{\Gamma(\theta/2)\Gamma(\theta/2 + 1)} (2v_c)^{1/2 - \theta/2} (v_s + v_c)^{-1/2} (-\omega - v_c q)^{\theta/2} (-\omega + v_c q)^{\theta/2 - 1}$$

$$\times F\left(1 - \theta/2, 1/2; 1 + \theta/2; \frac{v_s - v_c}{v_s + v_c} \frac{\omega + v_c q}{\omega - v_c q} \right) \tag{19.34}$$

$$\rho(q, v_c q < \omega < v_s q) =$$

$$\frac{2\pi a^\theta}{\Gamma^2(\theta/2 + 1/2)}(2v_c)^{1/2-\theta}(v_s - v_c)^{1/2}(\omega - v_c q)^{\theta/2-1/2}(\omega + v_c q)^{\theta/2-1/2}$$

$$\times F\left(1/2 - \theta/2, 1/2; 1/2 + \theta/2; \frac{v_s + v_c}{v_s - v_c}\frac{\omega - v_c q}{\omega + v_c q}\right) \tag{19.35}$$

$$\rho(q, \omega > v_s q) =$$

$$\frac{2\pi a^\theta}{\Gamma(\theta/2)\Gamma(\theta/2 + 1)}(2v_c)^{1/2-\theta}(v_s + v_c)^{-1/2}(\omega + v_c q)^{\theta/2}(\omega - v_c q)^{\theta/2-1}$$

$$\times F\left(1 - \theta/2, 1/2; 1 + \theta/2; \frac{v_s - v_c}{v_s + v_c}\frac{\omega + v_c q}{\omega - v_c q}\right) \tag{19.36}$$

I.5 Fourier transform of Green's function, $v_s > v_c$

The expression for the Fourier transform of the Green's function in this case is identical to Eq. (19.29). The reversal in relative size of the spin and charge velocities simply alters the analytic structure of the integrand. Taking this into acount and making a simple change of variables, the integral may be split into three nonzero parts as in the case $v_c > v_s$;

$$G^1(\omega, q) =$$

$$-ia^\theta \int_0^\infty dX(2v_c + X)^{-1/2-\theta/2}(v_s + v_c + X)^{-1/2}X^{-\theta/2}\int_0^\infty tdt e^{i(qX + qv_c + \omega)t}$$

$$\left\{ \begin{array}{c} [it + \alpha/(2v_c + X)]^{-1/2-\theta/2}\,[it + \alpha/(v_s + v_c + X)]^{-1/2}\,(-it + \alpha/X)^{-\theta/2} \\ +t \to -t \end{array} \right\}$$

$$G^2(\omega, q) =$$

$$-ia^\theta \int_0^{v_s - v_c} dX X^{-1/2-\theta/2}(v_s - v_c - X)^{-1/2}(2v_c + X)^{-\theta/2}\int_0^\infty tdt e^{-i(qX + qv_c - \omega)t}$$

$$\times \left\{ \begin{array}{c} (-it + \alpha/X)^{-1/2-\theta/2}\,[it + \alpha/(v_s - v_c - X)]^{-1/2}\,[it + \alpha/(2v_c + X)]^{-\theta/2} \\ +t \to -t \end{array} \right\}$$

$$G^3(\omega, q) =$$

$$-ia^\theta \int_0^\infty dX (X+v_s-v_c)^{-1/2-\theta/2}(X)^{-1/2}(v_c+v_s+X)^{-\theta/2} \int_0^\infty tdt e^{-i(qX+qv_s-\omega)t}$$

$$\times \left\{ \begin{array}{l} [-it+\alpha/(X+v_s-v_c)]^{-1/2-\theta/2} \, (-it+\alpha/X)^{-1/2} \, [it+\alpha/(v_c+v_s+X)]^{-\theta/2} \\ +t \to -t \end{array} \right\}$$

$$(19.37)$$

The t integrals may be evaluated in the limit $\alpha \to 0$ as before. Parts 1 and 3 have t integrals indentical to the case $v_c > v_s$. Part 2 has an additional factor of $\cos(\pi\theta/2)$.

$$G^1(q, \omega) = 2a^\theta e^{-i\pi\theta} \sin(\pi\theta/2)\Gamma(1-\theta)$$

$$\times \int_0^\infty dX (2v_c+X)^{-1/2-\theta/2}(v_s+v_c+X)^{-1/2}X^{-\theta/2}[q(X+v_c)+\omega]^{\theta-1}$$

$$G^2(q, \omega) = 2a^\theta e^{-i\pi\theta} \cos(\pi\theta/2)\Gamma(1-\theta)$$

$$\times \int_0^{v_s-v_c} dX X^{-1/2-\theta/2}(v_s-v_c-X)^{-1/2}(2v_c+X)^{-\theta/2}[\omega-q(X+v_c)]^{\theta-1}$$

$$G^3(q, \omega) = -2a^\theta e^{-i\pi\theta} \sin(\pi\theta/2)\Gamma(1-\theta)$$

$$\times \int_0^\infty dX (X+v_s-v_c)^{-1/2-\theta/2}X^{-1/2}(v_c+v_s+X)^{-\theta/2}[\omega-q(X+v_s)]^{\theta-1}$$

$$(19.38)$$

The agreement between this and the spectral function may be confirmed in exactly the same way as before.

References

D. Boies, C. Bourbonnais and A.-M. S. Tremblay, *Phys. Rev. Lett.* **74**, 968 (1995).

V. Meden and K. Schönhammer, *Phys. Rev.* **B46** 15 753 (1992).

Y. Ren and P. W. Anderson, *Phys. Rev.* **B48**, 16 662 (1993).

J. Voit, *Phys. Rev.* **B45**, 4027 (1992).

X. G. Wen, *Phys. Rev.* **B42**, 6623 (1990).

20

Two weakly coupled Tomonaga–Luttinger liquids; spinless case

'Proletariat has nothing to lose except his chains.'
  K. Marx and F. Engels, *The Communist Manifesto*.

In this chapter we shall consider a simple model of two spinless Tomonaga–Luttinger chains with a weak single-particle interchain hopping $t_\perp$. Our main goal is to demonstrate how single-particle hopping generates an effective interchain exchange which interferes with the single-particle processes and eventually creates a spectral gap. The model we consider is also interesting from the formal point of view providing a nontrivial example of the Gaussian model perturbed by an operator with a nonzero conformal spin. As explained in Chapter 8, the standard criterion of relevance is inapplicable to this case. To make reliable conclusions about infrared instabilities of the system, one has to indentify perturbations of zero conformal spin generated in higher orders in $t_\perp$.

The second feature of the model is that the underlying physical picture reflects an interplay between the suppression of coherent single-particle delocalization between the two chains and the development of two-particle interchain correlations, taking place on increasing the in-chain interaction. It is this 'dimensional crossover' between the gapless, Tomonaga–Luttinger-liquid regime and massive phases characterized by pair coherence in the transverse direction, either in the particle–hole or particle–particle channel, that we intend to discuss in detail below. Applying, as in Chapter 19, the two-cut-off scaling method, we shall see how the original model crosses over to a massive sine-Gordon model, and what are the dominant correlations in the system. For application of this method to the realistic case of two coupled chains with spin-1/2 fermions see Khveshchenko and Rice (1993), and Khveshchenko (1993).

The model we consider reads:

$$H = \sum_a \int dx[-iv_F(R_a^\dagger \partial_x R_a - L_a^\dagger \partial_x L_a) + (gJ_{R,a}J_{L,a} + g'J_{R,a}J_{L,-a})]$$

$$+ t_\perp \sum_\mu \int dx(R_a^\dagger R_{-a} + L_a^\dagger L_{-a}) \tag{20.1}$$

Here $a = \pm 1$ is the chain index, g and g' are coupling constants characterizing the in-chain and interchain forward scattering, respectively, and $J_{R,a} = :R_a^\dagger R_a: \quad J_{L,a} = :L_a^\dagger L_a:$ are the density operators for the right and left particles on each chain, or equivalently the Abelian currents satisfying the U(1) Kac–Moody algebra. The interaction included in (20.1) preserves the Tomonaga–Luttinger behaviour of the system at $t_\perp = 0$. Having in mind to single out sole effects of the $t_\perp$ hopping, we did not include the interchain backscattering processes. Their role will be briefly discussed in the end of the chapter.

It is natural to introduce the total and relative fermion densities

$$J_j^{c,s} = \frac{J_{j,+} \pm J_{j,-}}{\sqrt{2}}, \quad (j = R, L) \tag{20.2}$$

When the chain index a is formally treated as a spin-1/2 variable, $J^{c,s}$ in (20.2) are recognized as the Abelian 'charge' and 'spin' currents. Applying Abelian bosonization to the model (20.1) yields decomposition of the Hamiltonian into two independent contributions:

$$H = H_c^0 + H_s \tag{20.3}$$

Here

$$H_c^0 = \pi v_F \int dx[:J_R^c J_R^c : + :J_L^c J_L^c : +(g_c/\pi v_F)J_R^c J_L^c] \tag{20.4}$$

is a Tomonaga–Luttinger Hamiltonian which describes gapless excitations of the total ('charge') density and can be diagonalized by a canonical transformation of the currents. All nontrivial effects caused by the interchain hopping $t_\perp$ are incorporated in the 'spin' part of the model

$$H_s = \pi v_F \int dx[:J_R^s J_R^s : + :J_L^s J_L^s : +(g_s/\pi v_F)J_R^s J_L^s] + H_\perp \tag{20.5}$$

which deals with the relative degrees of freedom. The coupling constants appearing in the charge and spin channels are, respectively, $g_{c,s} = g \pm g'$.

Introducing a scalar field $\Phi(x)$ and its conjugate momentum $\Pi(x)$ in the standard way,

$$J_R^s + J_L^s = \sqrt{\frac{K}{\pi}}\partial_x\Phi, \quad J_R^s - J_L^s = -\sqrt{\frac{\tilde{K}}{\pi}}\Pi$$

with

$$K = \frac{1}{\tilde{K}} = \left(\frac{1 - g_s/2\pi v_F}{1 + g_s/2\pi v_F} \right)^{1/2} \tag{20.6}$$

and bosonizing the fermion bilinears in $H_\perp$, one arrives at the following field-theoretical model (Kusmartsev *et al.* (1992)):

$$H_s = \int dx \left[\frac{u}{2}(\Pi^2 + (\partial_x\Phi)^2) + \frac{2t_\perp}{\pi a} \cos \sqrt{2\pi K}\Phi \, \cos \sqrt{2\pi\tilde{K}}\Theta \right] \tag{20.7}$$

where $\Theta(x)$ is the dual field. As in other cases considered in previous chapters, the relationship (20.6) between the parameter K and coupling constant g_s is universal only for weak interaction ($|g_s| \ll 1$).

The model (20.7) possesses the duality symmetry: H_s remains invariant under transformations $\Phi \leftrightarrow \Theta$, $K \to \tilde{K}$ ($g_s \to -g_s$). The parameters K and $\tilde{K}$ satisfy the Kramers–Wannier-like duality relation $K\tilde{K} = 1$, the self-duality point $K = \tilde{K} = 1$ corresponding to the case of noninteracting particles. According to this relation, if the two-chain system undergoes a phase transition at some value of K, say $K = K_+$, there should be another transition, dual to the former one, at $K = K_- = 1/K_+$.

The perturbation ($t_\perp$) term in (20.7) has the scaling dimension

$$d_\perp = \frac{1}{2}(K + \tilde{K}) \tag{20.8}$$

and, what is most important, a *nonzero* conformal spin

$$S = \frac{\beta\tilde{\beta}}{8\pi} = 1 \tag{20.9}$$

As already discussed in Chapter 8, the latter circumstance makes it necessary to consider higher-order effects in $t_\perp$. For this reason it is instructive to obtain a Coulomb gas representation of the Euclidean version of the model (20.7). The corresponding partition function is given by $Z = Z_0 Z_\perp$ Here Z_0 is the partition function of the unperturbed, Gaussian part of the action, while $Z_\perp$ is given by a perturbation series expansion

$$Z_\perp = \sum_{n=0}^{\infty} \frac{1}{n!} \left(-\frac{2t_\perp}{\pi a} \right)^n \prod_{j=1}^{n} \int d^2x_j Q(1, 2, ..., n) \tag{20.10}$$

where

$$Q(1, 2, ..., n) = \langle C(1)\tilde{C}(1) \cdots C(n)\tilde{C}(n) \rangle_0 \tag{20.11}$$

$$C(j) = \cos \sqrt{2\pi K}\Phi(\mathbf{x}_j), \quad \tilde{C}(j) = \cos \sqrt{2\pi\tilde{K}}\Theta(\mathbf{x}_j), \quad \mathbf{x}_j = (u\tau_j, x_j)$$

To estimate the Gaussian average in (20.11) one needs to use the Baker–Haussdorf formula

$$\langle \exp \hat{A} \rangle_0 = \exp \left(\frac{1}{2}\langle \hat{A}^2 \rangle_0 \right)$$

together with the Green's functions of the fields Φ and Θ:

$$\langle\Phi(\mathbf{x}_1)\Phi(\mathbf{x}_2)\rangle - \langle\Phi^2(0)\rangle = \langle\Theta(\mathbf{x}_1)\Theta(\mathbf{x}_2)\rangle - \langle\Theta^2(0)\rangle = \frac{1}{2\pi}l_{12} \quad (20.12)$$

$$\langle\Phi(\mathbf{x}_1)\Theta(\mathbf{x}_2)\rangle = -\frac{i}{2\pi}\varphi(\mathbf{x}_{12}) \quad (20.13)$$

where

$$l_{12} = \ln(|\mathbf{x}_i - \mathbf{x}_j|/a)$$

and the angle

$$\varphi(\mathbf{x}_{12}) = \tan^{-1}[(x_1 - x_2)/u(\tau_2 - \tau_2)]$$

determines the orientation of the vector $\mathbf{x}_{12} = \mathbf{x}_1 - \mathbf{x}_2$. Notice that (20.13) follows from (20.12) by virtue of the duality relation (3.21).

As a result, $Z_\perp$ takes the form of the grand partition function of a two-dimensional gas of charge–monopole composites (Kusmartsev *et al.* (1992)):

$$Z_\perp = \sum_{n=0}^{\infty} \frac{z_\perp^{2n}}{(2n)!} \int \prod_{j=1}^{2n} \frac{d^2 x_j}{a^2} \sum_{\{\sigma\}} \sum_{\{\mu\}}$$

$$\times \exp\left[K\sum_{i<j} \sigma_i\sigma_j l_{ij} + \tilde{K}\sum_{i<j} \mu_i\mu_j \tilde{l}_{ij} + i\sum_{i\neq j} \sigma_i\mu_j\varphi_{ij} \right] \quad (20.14)$$

with a small dimensionless interchain hopping amplitude

$$z_\perp = \frac{t_\perp a}{2\pi u}$$

being the fugacity. Each composite carries an 'electric' charge $\sigma_j = \pm 1$ and a 'magnetic' charge, or monopole, $\mu_j = \pm 1$. The four-component system of charge–monopole composites is neutral with respect to each kind of charges, $\sum\sigma_i = \sum\mu_i = 0$. The charges and monopoles interact with logarithmic potential l_{ij} among themselves, and there is also a statistical Aharonov–Bohm phase φ_{ij} which couples the charges and monopoles belonging to different composites. The parameters K and $\tilde{K}$ appear as inverse dimensionless temperatures for the charges and monopoles, respectively.

The composites appear as elementary excitations associated with interchain hoppings of the right and left particles, each composite being described by a product of two vertex operators

$$V_{\sigma,\mu}(\mathbf{x}) = V_\sigma(\mathbf{x})\tilde{V}_\mu(\mathbf{x}) \quad (20.15)$$

where $V_\sigma(\mathbf{x}) = \exp[i\sigma\sqrt{2\pi K}\,\Phi(\mathbf{x})]$ and $\tilde{V}_\mu(\mathbf{x}) = \exp[i\mu\sqrt{2\pi\tilde{K}}\,\Theta(\mathbf{x})]$ are the spin-wave (order) and vortex (disorder) operators creating the 'electric' charge σ and 'magnetic' charge μ at the point $\mathbf{x}$. Each of the four vertex operators (20.15) describes interchain hopping of a right or left fermion, i.e.

$$R_+^\dagger R_- \sim V_{-1} \tilde{V}_1, \quad R_-^\dagger R_+ \sim V_1 \tilde{V}_{-1}$$
$$L_+^\dagger L_- \sim V_1 \tilde{V}_1, \quad L_-^\dagger L_+ \sim V_{-1} \tilde{V}_{-1} \tag{20.16}$$

The composites are Bose particles; the vertex operators (20.15) have critical dimension and conformal spin given by formulas (20.8) and (20.9), respectively.

Having the grand partition function (20.14) at our disposal, we can immediately estimate the linear response of two Tomonaga–Luttinger liquids to the $t_\perp$-perturbation. Up to $t_\perp^4$ terms,

$$Z_\perp = 1 - z_\perp^2 \int \frac{d^2 \mathbf{R}}{a^2} \int \frac{d^2 \mathbf{r}}{a^2} \left(\frac{a}{r}\right)^{K+\tilde{K}} \cos 2\varphi_\mathbf{r} + 0(z_\perp^4)$$

where we have used the relation $\varphi_{21} = \varphi_{12} + \pi$. Integrating over the center-of-mass coordinate $\mathbf{R}$ yields the factor Lu/T. The remaining integral over the relative coordinate $\mathbf{r}$ determines the susceptibility (per unit volume)

$$\chi_\perp = \frac{T}{L} \left(\frac{\partial^2}{\partial t_\perp^2} \ln Z_\perp\right)_{t_\perp = 0} = \frac{2}{\pi^2 u} \int_0^\infty dx \int_0^{u\beta} dt \left(\frac{x^2 - t^2}{x^2 + t^2}\right) \left(\frac{a^2}{x^2 + t^2}\right)^{d_\perp} \tag{20.17}$$

The same formula for $\chi_\perp$ could have been obtained by calculating a particle–hole loop diagram, using the one-particle Green's functions for a spinless Tomonaga–Luttinger liquid:

$$G_{R,L}(k, i\varepsilon) \sim \frac{1}{i\varepsilon \mp ku} \left(\frac{\varepsilon^2 + k^2 u^2}{\Lambda^2}\right)^{(d_\perp - 1)/2}$$

where $\Lambda \sim a^{-1}$.

Notice that, due to the antisymmetry of the integrand in (20.17) under interchange of x and t, the integration range $0 < x < u\beta$ gives zero contribution to $\chi_\perp$. Clearly, this is a direct consequence of the existence of the Aharonov–Bohm phase. Making scale transformations $x \to u\beta x$ and $t \to u\beta t$ one singles out a prefactor which determines the temperature dependence of the susceptibility:

$$\chi_\perp = \frac{2}{\pi^2 u} \left(\frac{aT}{u}\right)^{2(d_\perp - 1)} J(K) \tag{20.18}$$

where the convergent integral

$$J(K) = \int_1^\infty dx \int_0^1 dt \left(\frac{x^2 - t^2}{x^2 + t^2}\right)^{d_\perp + 1} > 0$$

By definition (20.8), $d_\perp \geq 1$, the equality sign corresponding to the case of noninteracting particles. We observe that the point $K = 1$ ($d_\perp = 1$), $T = 0$ is *singular* due to noncommutativity of two limits: $K \to 1$ and $T \to 0$. The noninteracting system is characterized by a finite response

$$\lim_{K \to 1} \chi(T,K) \simeq \frac{2}{\pi v_F}$$

(using the spin analogy, this corresponds to the usual Pauli paramagnetism of free one-dimensional fermions with spin $1/2$). On the other hand, the susceptibility of two *Tomonaga–Luttinger* chains ($K \neq 1$, $d_\perp > 1$) vanishes with the temperature following a power law

$$\lim_{T \to 0} \chi_\perp(T,K) \sim T^{2(d_\perp - 1)} \tag{20.19}$$

Thus we conclude that, even in the spinless case, when the Tomonaga–Luttinger liquid has a single (charge) collective mode in the excitation spectrum, two such liquids at zero temperature have a vanishing response to the $t_\perp$-perturbation. This result is a consequence of infrared catastrophe that transforms the Fermi-liquid pole of the one-particle Green's function into a branch cut. The suppression of quasiparticle states is reflected in the energy dependence of the one-particle density of states: $N_{LL}(\omega) \sim |\omega|^{d_\perp - 1}$, the susceptibility (20.19) being simply proportional to $N_{LL}^2(T)$.

One might naively conclude that the problem is 'solved', interpreting the result (20.19) as the manifestation of Anderson's confinement (Anderson (1990, 1991)), i.e. absence of coherent delocalization of the fermions between the two chains at arbitrarily weak in-chain interaction. However, as pointed out in Chapter 8, the linear response theory is not adequate when studying the stability of the Gaussian critical point against various perturbations, since the latter undergo strong renormalizations and acquire anomalous dimensions. The standard criterion of relevance would imply that the region, where the effective single-particle interchain hopping amplitude increases upon renormalization, should be determined from inequality $d_\perp < 2$, i.e.

$$K_- < K < K_+, \quad K_\pm = 2 \pm \sqrt{3} \tag{20.20}$$

This inequality defines the deconfinement (delocalization) phase, characterized by a finite Zeeman-like splitting of two degenerate bands, with the magnitude of the splitting estimated as

$$t_\perp^* \simeq \Lambda |z_\perp|^{\frac{1}{2-d_\perp}} \tag{20.21}$$

$t_\perp^*$ reaches its maximum value ($= t_\perp$) in the noninteracting case ($d_\perp = 1$), and decreases on increasing the in-chain interaction, vanishing at $K \to K_+$ or $K \to K_-$. At these boundary points deconfinement–confinement transitions take place. Thus, contrary to naive expectations following from Eq.(20.19), realization of the confinement regime requires sufficiently strong interaction in the system.

However, this picture is still incomplete since the tendency towards suppression of the single-particle transport in the transverse direction is not the only effect of $t_\perp$. (Remember that the conformal spin of the $t_\perp$-perturbation is nonzero!) To get a qualitative understanding of how new correlations are generated, let us go back to the Coulomb gas representation (20.14).

As follows from (20.14), there is a logarithmic attraction between the opposite 'electric' charges belonging to different composites, dominant at $K > 1$. Similarly, there is attraction between the opposite 'monopoles' dominant at $K < 1$. Thus, at any value of $K \neq 1$, there is a tendency towards creation of bound pairs of composites, either with zero total 'electric' charge or zero 'magnetic' charge. Each bound pair of composites has a total 'free' charge, 'electric' or 'magnetic', equal to $Q = 0, \pm 2$. So, there exist totally neutral pairs with $Q = 0$, and pairs with a doubled charge, $Q = \pm 2$. (Bound states of more than two composites are energetically less favourable, since the potential of a bound pair falls off with distance as $1/r$.)

Using the correspondence between elementary interchain hoppings of right and left fermions and four types of composites, Eqs. (20.16), one can visualize the character of correlations in the system. In the deconfinement regime, a typical trajectory of a right or left fermion represents a sequence of domains connected by 'kinks'; the widths of the upper-chain and lower-chain domains being of the same order of magnitude. Clearly, such a phase suggests the existence of unpaired composites. Pairing of composites with zero total 'electric' and 'magnetic' charge leads to significant difference between the widths of the domains on the upper and lower chains. In such a phase, the fermions spend most of their time on one of the two chains, a situation corresponding to the confinement regime.

On the other hand, binding of composites in pairs with total 'electric' or 'magnetic' charge ± 2 is described by two operators

$$O_{ph} = V_a \tilde{V}_{-a} V_{-a} \tilde{V}_{-a} \sim (R_a^\dagger R_{-a})(L_{-a}^\dagger L_a)$$
$$\rightarrow -\frac{1}{2(\pi a)^2} \cos \sqrt{8\pi K}\, \Phi \qquad (20.22)$$

$$O_{pp} = V_a \tilde{V}_{-a} V_{-a} \tilde{V}_{-a} \sim (R_a^\dagger R_{-a})(L_a^\dagger L_{-a})$$
$$\rightarrow -\frac{1}{2(\pi a)^2} \cos \sqrt{8\pi \tilde{K}}\, \Theta \qquad (20.23)$$

These operators describe interchain *particle–hole* and *particle–particle* hoppings, respectively. The two-particle processes, although absent in the original Hamiltonian (20.7), are generated upon renormalization. Notice that the conformal spin of the operators O_{ph} and O_{pp} is zero. There-fore, now one can safely apply the usual criterion of relevance. Since the critical dimensions of these operators are, respectively, $2K$ and $2\tilde{K}$, at any $K \neq 1$ ($g_s \neq 0$) one of them is always relevant. For this reason the two-particle hopping processes must be included into the renormalization scheme (Yakovenko (1992)).

At the l-th step of the renormalization procedure (l can be identified as a continuous logarithmic variable $l = \ln(a/a_0)$ or $l = \ln(\Lambda/|\omega|)$), the Hamiltonian density takes the form

$$H_l(x) = \frac{u_l}{2}[\Pi^2 + (\partial_x \Phi)^2] + \frac{4u_l z_l}{a^2} \cos\sqrt{2\pi K_l}\Phi \cos\sqrt{2\pi \tilde{K}_l}\Theta$$
$$-\frac{2\pi u_l}{(2\pi a)^2}[G_l \cos\sqrt{8\pi K_l}\Phi + \tilde{G}_l \cos\sqrt{8\pi \tilde{K}_l}\Theta] \quad (20.24)$$

where $z_l \equiv z_\perp(l)$ and $u_l = u_0$. Notice that Eq. (20.24) represents a gener-alization of the Hamiltonian (18.17), extended to include a term with a nonzero conformal spin. The parameters of the Hamiltonian H_l satisfy the following renormalization group equations (Yakovenko (1992); Nersesyan et al. (1993)) which can be derived by the method described in Chapter 10 (below the subscript l is suppressed):

$$dz/dl = \left[2 - \frac{1}{2}(K + K^{-1})\right]z \quad (20.25)$$

$$dG/dl = 2(1 - K)G + (K - K^{-1})z^2 \quad (20.26)$$
$$d\tilde{G}/dl = 2(1 - K^{-1})\tilde{G} + (K^{-1} - K)z^2 \quad (20.27)$$

$$d\ln K/dl = \frac{1}{2}(\tilde{G}^2 K^{-1} - KG^2) \quad (20.28)$$

The initial conditions are $K = K_0$, $G_0 = \tilde{G}_0 = 0, z = t_\perp/\epsilon_F$.

Without loss of generality we may consider the case $K < 1$ and study Eqs. (20.25–20.28) in various regimes of K_0 and z_0. There are three possible regimes.

(i) **Strong confinement regime.** $K < 2 - \sqrt{3} \approx 0.28$. In this case $2 < (K^{-1} + K)/2$ and z_l decreases with $l \to \infty$. This means that one can put $z_l = 1$ in the effective Hamiltonian (20.24) and put $G_0 \sim z_0^2$.

(ii) **Weak confinement regime.** $2 - \sqrt{3} < K < \sqrt{2} - 1$. In this region, as we shall demonstrate in a moment, both z_l and G_l grow, but G_l reaches the strong coupling regime first. As we shall see in a moment, the criterion $K < \sqrt{2} - 1$ coincides with criterion (8.13).

(iii) **Band theory regime.** $K > \sqrt{2} - 1$. The best understanding of this regime may be achieved when one diagonalizes the quadratic fermion

Hamiltonian first to obtain two Fermi surfaces, and then takes into account the interactions.

Let us consider the most interesting case first:

$$K^{-1} - K > 2 \tag{20.29}$$

This case corresponds to criterion (8.13). The results obtained in Chapter 8 suggest that at these values of K only even-particle processes give significant contributions to the system's dynamics. We shall see that this also follows from the RG analysis. Since condition (20.29) corresponds to rather small K, where the operator $\cos(\sqrt{8\pi/K}\Theta)$ is strongly irrelevant, we can neglect it altogether and put $\tilde{G} = 0$ in RG equations (20.25, 20.26, 20.27, 20.28). We shall also neglect renormalization of K from its initial value K_0, which is justified since the bare value of G is small. In this approximation we can solve Eq. (20.28):

$$z_l = z_0 \exp\{[2 - (K + K^{-1})/2]l\} \tag{20.30}$$

Substituting this into Eq. (20.27) and taking into account the initial condition $G_0 = 0$ we get

$$G_l = -\frac{z_0^2(K^{-1} - K)}{K^{-1} - K - 2}\left\{1 - \exp[-l(K^{-1} - K - 2)/2]\right\}\exp[2(1 - K)l] \tag{20.31}$$

Let us suppose that $(K^{-1} - K - 2)$ is not very small. Then in Eq.(20.31) one can neglect the exponent in the curly brackets and obtain that G_l becomes of order of one at

$$l^* \approx -\frac{1}{(1 - K)}\ln z_0 \tag{20.32}$$

Substituting it into Eq. (20.30) we get

$$z(l^*) \approx z_0^{\frac{K^{-1} - K - 2}{2(1 - K)}} \tag{20.33}$$

This means that if condition (20.29) is fulfilled then at the point (20.32) where the effective exchange is already strong and presumably opens a spectral gap, the single particle tunneling is still weak.

Now let us consider the regime of weak interaction (Band Theory regime) when one can set

$$K \simeq 1 - \frac{1}{2}\lambda + 0(\lambda^2), \quad K^{-1} \simeq 1 + \frac{1}{2}\lambda + 0(\lambda^2), \quad d = 1 + 0(\lambda^2) \tag{20.34}$$

and rewrite the RG equations (20.25)–(20.28) as follows:

$$dz/dl = z, \tag{20.35}$$
$$dG/dl = \lambda G - \lambda z^2, \tag{20.36}$$
$$d\tilde{G}/dl = -\lambda\tilde{G} + \lambda z^2 \tag{20.37}$$

$$d\lambda/dl = G^2 - \tilde{G}^2 \tag{20.38}$$

We shall solve these equations, using a two-cut-off scaling procedure (Chapter 18). First, from Eq. (20.35) we determine the value of the variable $l_0 = \ln(1/|z_\perp|) = \ln(\Lambda/|t_\perp^*|)$, at which z becomes ~ 1. This sets a characteristic energy scale in the problem – the effective band splitting $t_\perp^* \simeq \Lambda z_0 \sim t_\perp$. Two regions, $l < l_0$ and $l > l_0$, should be considered separately. In the 'high-energy' region $l \leq l_0$ ($t_\perp^* < |\omega| < \Lambda$) λ_l is not renormalized (up to second-order corrections in g_s), $\lambda(l_0) = g_s/2\pi v_F$, while the charges G_l and $\tilde{G}_l$ acquire small nonzero values:

$$G_l = -\tilde{G}_l = -\frac{g_s}{2\pi v_F} z_0^2 [\exp(2l) - \exp(g_s l/2\pi v_F)] \simeq -\frac{g_s}{2\pi v_F} z_l^2. \tag{20.39}$$

In the region $l < l_0$, the effective amplitude $t_\perp(l)$ increases with l until it reaches the value of the order of the cut-off Λ. At $l \sim l_0$ the renormalization of $t_\perp(l)$ stops. Therefore, one can choose the normalization condition $z_{l_0} = 1$ and, passing to the 'low-energy' region $l > l_0$ ($|\omega| < t_\perp^*$), drop the z^2-terms in the r.h.s. of Eqs. (20.36) and (20.37). The resulting RG equations valid at $l \geq l_0$ read:

$$d\lambda/dl = G^2 - \tilde{G}^2, \quad dG/dl = \lambda G, \quad d\tilde{G}/dl = -\lambda\tilde{G} \tag{20.40}$$

The boundary conditions

$$\lambda_{l_0} = g_s/\pi v_F, \quad G(l_0) = -\tilde{G}(l_0) = -g_s/2\pi v_F \tag{20.41}$$

define new 'bare' couplings in the region $l > l_0$. The fact that these couplings are small ensures the applicability of the two-cut-off scaling method.

We recognize equations (20.40) as the RG equations (18.21) of the XYZ Thirring model analyzed in detail in Chapter 18. Introduce linear combinations $G_\pm = G \pm \tilde{G}$ and rewrite these equations as follows

$$d\lambda/dl = G_+ G_-, \quad dG_\pm/dl = \lambda G_\mp \tag{20.42}$$

From the boundary conditions (20.41) it follows that in the whole region $l > l_0$ we have $\lambda = -G_-$. Eqs. (20.42) then reduce to a pair of Kosterlitz–Thouless equations for the sine-Gordon model:

$$dG_+/dl = -G_-^2, \quad dG_-/dl = -G_+ G_- \tag{20.43}$$

The conditions $G_+(l_0) = 0$, $G_-(l_0) = -g_s/\pi v_F$ define a scaling trajectory which originates from the initial point, lying on the vertical axis of the Kosterlitz–Thouless phase plane $(G_+, |G_-|)$, and approaches asymptotically the line $G_+ = -|G_-|$. The Hamiltonian (20.24) scales towards a sine-Gordon model, either for the field Φ at $K < 1$ ($g_s > 0$), or for the dual field Θ at $K > 1$ ($g_s < 0$). Therefore, for any sign of g_s, a strong-coupling regime develops in the infrared limit, and the Tomonaga–Luttinger liquid fixed point turns out to be infrared *unstable*. The new fixed point is characterized by a nonzero mass gap

$$M \simeq t_\perp \exp(-\pi^2 v_F / 2g_s) \ll t_\perp \qquad (20.44)$$

dynamically generated in the spectrum of relative ('spin') density excitations and signalling the onset of strong interchain two-particle correlations. The fact that the mass does not depend on the sign of the coupling constant is related to the duality symmetry of the model.

Exercise.   Rederive the above results using an alternative description in terms of symmetric (bonding) and antisymmetric (antibonding) one-fermion states,

$$\Psi_\pm = (\pm\psi_+ + \psi_-)/\sqrt{2}$$

Find the resulting bosonized Hamiltonian and show that, at small interaction ($g_s/\pi v_F \ll 1$), strong interchain pair coherence coexists with a finite single-particle band splitting, proportional to 'magnetization':

$$\langle S^z \rangle = -\sqrt{\frac{2}{\pi}} \langle \partial_x \chi \rangle = \frac{2t_\perp}{\pi u} \left[1 - \frac{1}{8} \left(\frac{g_s}{\pi v_F} \right)^2 \ln \left(\frac{\Lambda}{t_\perp} \right) + \cdots \right]$$

The symmetry of dominant correlations can be established in the same way as it has been done in Section 18.IV. The resulting phase diagram is shown in Fig. (20.1). Here CDW stands for two one-dimensional charge-density waves with the relative phase π, S1 and S2 denote the in-chain and interchain Cooper pairings, respectively, and OAF denotes orbital antiferromagnetic state. The phase diagram reflects the duality symmetry of the model: the states OAF and S2 are dual to CDW and S1, respectively, and transform to each other under $g_s \rightarrow -g_s$.

The above renormalization-group analysis was confined to the case of weak interaction when the mass gap M, Eq. (20.44), characterizing strong interchain interactions is exponentially small compared to the interband splitting $t_\perp^*$. In this case, the dimensional crossover from two decoupled Tomonaga–Luttinger liquids ($|t_\perp^*| \ll |\omega| \ll \Lambda$) to a massive phase of strongly coupled chains ($|\omega| < M$) occurs via formation of an intermediate regime of the *two-band* Luttinger liquid ($M \ll |\omega| \ll |t_\perp^*|$). On increasing g_s, the difference between $t_\perp^*$ and M decreases, and at large enough interaction the two-band Tomonaga–Luttinger-liquid regime no longer exists. In this situation, one of the two-particle hopping amplitudes, G_l or $\tilde{G}_l$, goes to strong coupling at such values of l when z_l is still small. Nevertheless, it can be shown that, as long as the model (20.1) is considered, a finite splitting of the two bands survives the presence of the large mass gap indicating strong interchain pair coherence. The

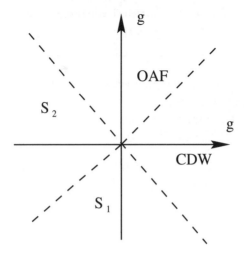

Fig. 20.1. Phase diagram of the two-chain model.

renormalization of the single-fermion interchain hopping is still determined by the Tomonaga–Luttinger liquid effects (infrared catastrophe),
and the deconfinement-confinement transitions are still expected to be of
the Kosterlitz–Thouless nature.

As a counter example violating this picture, consider an extension of
the original model (20.1) to include the interchain backscattering

$$g_{\text{back}} R_a^{\dagger} L_{-a}^{\dagger} R_{-a} L_a \sim g_{\text{back}} \cos \sqrt{8\pi K}\, \Phi$$

which makes the bare value of the coupling G_l nonzero. Then one can
find conditions when the interchain backscattering drives the system to a
new strong-coupling fixed point characterized by two 'ferromagnetically'
correlated charge-density waves. The fact that this type of ordering *does*
tend to suppress the single-particle interchain hopping becomes easily seen
in the symmetric–antisymmetric basis of one-electron states. One finds
that, like in the attractive one-dimensional Fermi system in a magnetic
field (see Chapter 18), at $t_\perp$ less than the mass gap, the $t_\perp$ processes will be
damped. A similar situation occurs in the Hubbard two-chain ladder (or
a double-chain $t - J$ model) where the interchain exchange and Coulomb
interaction make the *relative* charge and spin bosonic modes massive. This
leads to the suppression of the single-particle interchain hopping and, at
the same time, opens an intriguing possibility for strong superconducting
correlations in systems with entirely repulsive interactions (Khveshchenko
and Rice (1993); Khveshchenko (1993)). A nonperturbative treatment of
the problem is presented in Chapter 21.

References

P. W. Anderson, *Phys. Rev. Lett.* **67**, 3844 (1991).

P. W. Anderson, *Phys. Rev. Lett.* **64**, 1839 (1990); **65**, 2306 (1990).

T. Giamarchi and H. J. Schulz, *J. Phys.* (Paris) **49**, 819 (1988).

D. V. Khveshchenko, *Phys. Rev.* B**50**, 380 (1993).

D. V. Khveshchenko and T. M. Rice, *Phys. Rev.* B**50**, 252 (1993).

F. V. Kusmartsev, A. Luther and A. A. Nersesyan, *JETP Lett.* **55**, 692 (1992).

A. A. Nersesyan, A. Luther and F. V. Kusmartsev, *Phys. Lett.* A **176**, 363 (1993).

H. J. Schulz, *Phys. Rev.* B**53**, R2959 (1996).

V. M. Yakovenko, *JETP Lett.* **56**, 5101 (1992).

21

Spin liquids in one dimension: example of spin ladders

In this chapter we continue to study the systems containing coupled chains. Now we shall introduce spin degrees of freedom and as a preparatory step consider the situation of an insulator, i.e. a model of two weakly coupled isotropic spin-1/2 Heisenberg chains with an antiferromagnetic coupling along the chains. We shall also add to the Hamiltonian the four-spin interaction

$$u(\mathbf{S}_n\mathbf{S}_{n+1})_1(\mathbf{S}_n\mathbf{S}_{n+1})_2$$

which has the same scaling dimension as the conventional exchange. Such a term can be generated, for example, by phonons. We suppose that $J_{\parallel} > 0, J_{\parallel} \gg |J_{\perp}|, |u|$ where $J_{\perp}$ is the interchain exchange. We call this system a *spin ladder*.

The problem of a spin ladder is interesting in its own right because such systems exist in nature. However, our discussion is mostly motivated by the fact that weakly coupled spin ladders are systems intermediate between spin $S = 1/2$ and $S = 1$ chains. Observation of the crossover will give us a better understanding of a difference which exists in one dimension between antiferromagnets with half-integer and integer spins. This difference was originally recognized by Haldane (1983) who predicted that antiferromagnetic Heisenberg chains with half-integer spin are gapless, whereas those with integer spin are gapped. Later Dagotto *et al.* (1992) and Rice *et al.* (1993, 1994) expanded these arguments to spin ladders.

A very remarkable fact about the spin $S = 1/2$ Heisenberg chain is that its excitation spectrum consists of spin-1/2 particles (spinons) (see the discussion in Chapter 11). Physically such excitations can be created only in pairs because upon flipping one spin the total spin projection is changed by one: $|\Delta S_z| = 1$. Thus, in the $S = 1/2$ Heisenberg chain, the conventional magnons carrying spin 1 are deconfined into spin-1/2 spinons. Putting two $S = 1/2$ chains together one can observe how spinons are confined back into magnons by measuring the dynamical susceptibility $\chi''(\omega, q)$.

The interchain exchange $J_\perp$ serves here as a control parameter: at $|J_\perp| \ll J_\parallel$ there is a wide energy range where χ'' is dominated by incoherent multiparticle processes, and a narrow region at low energies where χ'' may exhibit a single-magnon peak around $q = \pi$ (as we shall see later the single-magnon peak appears if the exchange is stronger than the charge charge interaction).

In order to obtain a qualitative understanding of the spinon confinement we may put $u = 0$ and consider the strong-coupling limit of the spin-ladder problem. As frequently happens in one-dimensional models, the strong-coupling limit gives a correct qualitative picture of the low-lying excitations. One should, however, be careful to define this limit properly. The proper definition assumes that it is possible to perform a perturbative expansion about the strong-coupling fixed point in negative powers of the coupling constant. In the spin ladder problem there are two candidates for the strong- coupling fixed point: the limits of strong antiferromagnetic $(J_\perp \gg J_\parallel)$ and ferromagnetic $(-J_\perp \gg J_\parallel)$ interchain coupling, respectively. It is clear that only the former case constitutes the correctly defined strong-coupling limit. At $J_\perp / J_\parallel \to +\infty$ the spin ladder is decomposed into an array of decoupled rungs, each rung representing a 'molecule' whose singlet ground state is separated from the triplet excited state by a large gap of the order of $J_\perp$. When one makes $J_\parallel$ finite, the triplet excitations form a band with bandwidth $\sim J_\parallel$. The properties of such a system can be analysed perturbatively, with $J_\parallel / J_\perp$ being the small parameter.

On the other hand, strong ferromagnetic interchain coupling leads to the formation of local spins $S = 1$ associated with each rung of the ladder, thus producing a conventional spin $S = 1$ Heisenberg antiferromagnet with a nonzero Haldane gap in the excitation spectrum. In contrast to the previous case, the bandwidth of the triplet excitations and the spectral gap are of the same order of magnitude, $\sim J_\parallel$. This problem lacks a small parameter and cannot be analysed by perturbation theory. A variety of approximate methods have been suggested to study the $S = 1$ antiferromagnetic spin chain (see Tsvelik (1990) and references therein); however, it is not our purpose to review them here.

Below we present the analysis of a weakly coupled spin ladder $J_\parallel \gg |J_\perp|, |u|$ based on our earlier work (Shelton *et al.* (1996), Nersesyan and Tsvelik (1997)). Despite the fact that our results have only qualitative validity for the presently available experimental realizations of double chain ladders ($Sr_{n-1}Cu_{n+1}O_{2n}$, Azuma *et al.* (1994), Ishida *et al.* (1996)) where both exchange integrals are of the same order, we hope that weakly interacting spin ladders will be synthesized in the future.

An interesting fact about the weak-coupling limit is that at $u = 0$ the emerging physical picture is independent of the sign of $J_\perp$. As follows from the above discussion, this universality is not so obvious at $|J_\perp| \gg J_\parallel$. Therefore comparing our results with the strong coupling analysis one can see that the main universal features of the spectrum are its symmetry and the persistence of the gap.

I   Coupling of identical chains; the Abelian bosonization

In this Section we apply the Abelian bosonization method to the spin-ladder model

$$H = J_\parallel \sum_{j=1,2} \sum_n \mathbf{S}_j(n) \cdot \mathbf{S}_j(n+1) + \sum_n [J_\perp \mathbf{S}_1(n) \cdot \mathbf{S}_2(n) + u\epsilon_1(n)\epsilon_2(n)] \quad (21.1)$$

where

$$\epsilon_j(n) = (-1)^n \mathbf{S}_j(n)\mathbf{S}_j(n+1)$$

is the staggered part of the energy density for a given chain. This model describes two antiferromagnetic ($J_\parallel > 0$) spin-1/2 Heisenberg chains with weak interchain couplings ($|J_\perp|, |u| \ll J_\parallel$) of arbitrary sign. In the continuum limit, the critical properties of isolated $S = 1/2$ Heisenberg chains are described in terms of massless Bose fields $\phi_j(x)$ ($j = 1, 2$):

$$H_0 = \frac{v_s}{2} \sum_{j=1,2} \int dx[\Pi_j^2(x) + (\partial_x \phi_j(x))^2] \quad (21.2)$$

where the velocity $v_s \sim J_\parallel a_0$ and Π_j are the momenta conjugate to ϕ_j. The interchain coupling

$$H_\perp = a_0 \int dx\{J_\perp [\mathbf{J}_1(x) \cdot \mathbf{J}_2(x) + \mathbf{n}_1(x) \cdot \mathbf{n}_2(x)] + u\epsilon_1(x)\epsilon_2(x)\} \quad (21.3)$$

is expressed in terms of the operators $\mathbf{J}_j(x)$, $\mathbf{n}_j(x)$ and $\epsilon_j(x)$ which represent, respectively, the slowly varying and staggered parts of local spin density and energy density operators. These operators have been defined in Chapter 13. According to (13.10), the current–current term in (21.3) is marginal, while interactions of the staggered parts of the spin and energy densities are strongly relevant. So we start our analysis by dropping the former term (its role will be discussed later). Using then bosonization formulas (13.9) for $\mathbf{n}_j(x)$ and (13.10) for $\epsilon_j(x)$, we get

$$H_\perp = \frac{J_\perp \lambda^2}{\pi^2 a_0} \int dx\ [\frac{1}{2}(u/J_\perp - 1) \cos \sqrt{2\pi}(\phi_1 + \phi_2)$$

$$+ \frac{1}{2}(u/J_\perp + 1) \cos \sqrt{2\pi}(\phi_1 - \phi_2) + \cos \sqrt{2\pi}(\theta_1 - \theta_2)] \quad (21.4)$$

where $\theta_j(x)$ is the field dual to $\phi_j(x)$. Denote

$$m = \frac{J_\perp \lambda^2}{2\pi}$$

and introduce linear combinations of the fields ϕ_1 and ϕ_2:

$$\phi_\pm = \frac{\phi_1 \pm \phi_2}{\sqrt{2}} \qquad (21.5)$$

The total (ϕ_+) and relative (ϕ_-) degrees of freedom decouple, and the Hamiltonian of two identical Heisenberg chains transforms to a sum of two independent contributions (Schulz (1986)):

$$H = H_+ + H_- \qquad (21.6)$$

$$H_+(x) = \frac{v_s}{2}\left(\Pi_+^2 + (\partial_x\phi_+)^2\right) + \frac{m}{\pi a_0}(u/J_\perp - 1)\cos\sqrt{4\pi}\phi_+ \qquad (21.7)$$

$$H_-(x) = \frac{v_s}{2}\left(\Pi_-^2 + (\partial_x\phi_-)^2\right) + \frac{m}{\pi a_0}(u/J_\perp + 1)\cos\sqrt{4\pi}\phi_-$$

$$+\frac{2m}{\pi a_0}\cos\sqrt{4\pi}\theta_- \qquad (21.8)$$

Exercise. Consider a generalization of the two-chain spin-ladder model (21.3) in which the interchain coupling is extended to include the exchange between the spins across diagonals of the plaquettes (for simplicity, put $u = 0$):

$$H_\perp = J_\perp \sum_n \mathbf{S}_1(n)\cdot\mathbf{S}_2(n) + J'_\perp\sum_n \mathbf{S}_1(n)\cdot[\mathbf{S}_2(n) + \mathbf{S}_2(n+1)]$$

Show that at the special point, $J_\perp = (1/2)J'_\perp$, the continuum limit of the model is described by two critical, $SU_1(2)$-symmetric WZNW models coupled by a marginal interaction (neglecting the velocity renormalization):

$$g\int dx\,[\mathbf{J}_1(x)\cdot\bar{\mathbf{J}}_2(x) + \mathbf{J}_2(x)\cdot\bar{\mathbf{J}}_1(x)], \quad g\sim J_\perp$$

Derive the RG equation for the coupling constant g and analyze the infrared limit of the model. Show that the model can be reduced to two independent and identical sine-Gordon models, each of them having a strong-coupling, massive phase at $J_\perp > 0$ (strongly coupled chains) and a massless phase at $J_\perp < 0$ (asymptotically decoupled chains.)

In the above derivation, the $\mathbf{J}_1\cdot\mathbf{J}_2$-term has been omitted as being only marginal, as opposed to the retained, relevant $\mathbf{n}_1\cdot\mathbf{n}_2$-term. It is worth mentioning that there are modifications of the original two-chain lattice model for which the $\mathbf{J}_1\cdot\mathbf{J}_2$-term does not appear at all in the continuum limit, and mapping onto the model (21.6) becomes exact. In two such modifications, the interchain coupling is changed to

$$H_\perp^{(A)} = \frac{J_\perp}{2} \sum_n \mathbf{S}_1(n) \cdot [\mathbf{S}_2(n) - \mathbf{S}_2(n+1)] \qquad (21.9)$$

or

$$H_\perp^{(B)} = \frac{J_\perp}{4} \sum_n [\mathbf{S}_1(n) - \mathbf{S}_1(n+1)] \cdot [\mathbf{S}_2(n) - \mathbf{S}_2(n+1)] \qquad (21.10)$$

The structure of these models explains why the low-energy physics of two *weakly* coupled Heisenberg chains must not be sensitive to the sign of the interchain coupling $J_\perp$.

Let us turn back to Eqs. (21.7) and (21.8). One immediately realizes that the critical dimension of all the cosine-terms in Eqs. (21.7), (21.8) is 1; therefore the model (21.6) is a theory of free massive fermions. The Hamiltonian H_+ describes the sine-Gordon model at $\beta^2 = 4\pi$; so it is equivalent to a free massive Thirring model. Let us introduce a spinless Dirac fermion related to the scalar field ϕ_+ via identification

$$\psi_{R,L}(x) \simeq (2\pi a_0)^{-1/2} \exp\left(\pm i\sqrt{4\pi}\ \phi_{+;R,L}(x)\right) \qquad (21.11)$$

Using

$$\frac{1}{\pi a_0} \cos\sqrt{4\pi}\ \phi_+(x) = i\ [\ \psi_R^\dagger(x)\psi_L(x) - \text{H.c.}\ ]$$

we get

$$H_+(x) = -i v_s(\psi_R^\dagger \partial_x \psi_R - \psi_L^\dagger \partial_x \psi_L) + i m(u/J_\perp - 1)(\psi_R^\dagger \psi_L - \psi_L^\dagger \psi_R) \quad (21.12)$$

For future purposes, we introduce two real (Majorana) fermion fields

$$\xi_\nu^1 = \frac{\psi_\nu + \psi_\nu^\dagger}{\sqrt{2}}, \quad \xi_\nu^2 = \frac{\psi_\nu - \psi_\nu^\dagger}{\sqrt{2}i}, \quad (\nu = \text{R, L}) \qquad (21.13)$$

to represent H_+ as a model of two degenerate massive Majorana fermions

$$H_+ = H_{m_t}[\xi^1] + H_{m_t}[\xi^2] \qquad (21.14)$$

where

$$H_m[\xi] = -\frac{i v_s}{2}(\xi_R\ \partial_x \xi_R - \xi_L\ \partial_x \xi_L) - i m\ \xi_R \xi_L \qquad (21.15)$$

and

$$m_t = m(1 - u/J_\perp) \qquad (21.16)$$

Now we shall demonstrate that the Hamiltonian H_- in (21.8) reduces to the Hamiltonian of two *different* Majorana fields. As before, we first introduce a spinless Dirac fermion

$$\chi_{R,L}(x) \simeq (2\pi a_0)^{-1/2} \exp\left(\pm i\sqrt{4\pi}\ \phi_{-;R,L}(x)\right) \qquad (21.17)$$

$$\frac{1}{\pi a_0} \cos \sqrt{4\pi} \, \phi_-(x) = i \, [\, \chi_R^\dagger(x)\chi_L(x) - \text{H.c.}],$$

$$\frac{1}{\pi a_0} \cos \sqrt{4\pi} \, \theta_-(x) = -i \, [\, \chi_R^\dagger(x)\chi_L^\dagger(x) - \text{H.c.}]$$

Apart from the usual mass bilinear term ('CDW' pairing), the Hamiltonian H_- also contains a 'Cooper-pairing' term originating from the cosine of the dual field:

$$H_-(x) = -v_s(\chi_R^\dagger \partial_x \chi_R - \chi_L^\dagger \partial_x \chi_L)$$

$$+im(1 + u/J_\perp) \, (\chi_R^\dagger \chi_L - \chi_L^\dagger \chi_R) + 2im \, (\chi_R^\dagger \chi_L^\dagger - \chi_L \chi_R) \quad (21.18)$$

We introduce two Majorana fields

$$\xi_v^3 = \frac{\chi_v + \chi_v^\dagger}{\sqrt{2}}, \quad \rho_v = \frac{\chi_v - \chi_v^\dagger}{\sqrt{2}i}, \quad (v = \text{R, L}) \quad (21.19)$$

The Hamiltonian H_- then describes two massive Majorana fermions, $\xi_{R,L}^3$ and $\rho_{R,L}$, with masses m_t and m_s, respectively:

$$H_- = H_{m_t}[\xi^3] + H_{m_s}[\rho] \quad (21.20)$$

where

$$m_s = -m(3 + u/J_\perp) \quad (21.21)$$

Now we observe that ξ^a, $a = 1, 2, 3$, form a triplet of Majorana fields with the same mass m_t. There is one more field ρ with a different mass, m_s. So, the total Hamiltonian

$$H = H_{m_t}[\xi] + H_{m_s}[\rho] \quad (21.22)$$

with

$$H_{m_t}[\xi] = \sum_{a=1,2,3} \{ -\frac{iv_s}{2}(\, \xi_R^a \, \partial_x \xi_R^a - \xi_L^a \, \partial_x \xi_L^a) - im_t \, \xi_R^a \xi_L^a \} \quad (21.23)$$

The $O(3)$-invariant model $H_{m_t}[\xi]$ was suggested as a description of the $S = 1$ Heisenberg chain by Tsvelik (1990). This equivalence follows from the fact that, in the continuum limit, the integrable $S = 1$ chain with the Hamiltonian (Takhtajan (1982); Babujan (1983))

$$H = \sum_n [(\mathbf{S}_n \mathbf{S}_{n+1}) - (\mathbf{S}_n \mathbf{S}_{n+1})^2] \quad (21.24)$$

is described by the critical WZNW model on the SU(2) group at the level $k = 2$, and the latter is in turn equivalent to the model of three massless Majorana fermions, as follows from the comparison of conformal charges of the corresponding theories:

$$C_{\text{SU}(2),k=2}^{WZNW} = \frac{3}{2} = 3C_{Major.fermion}$$

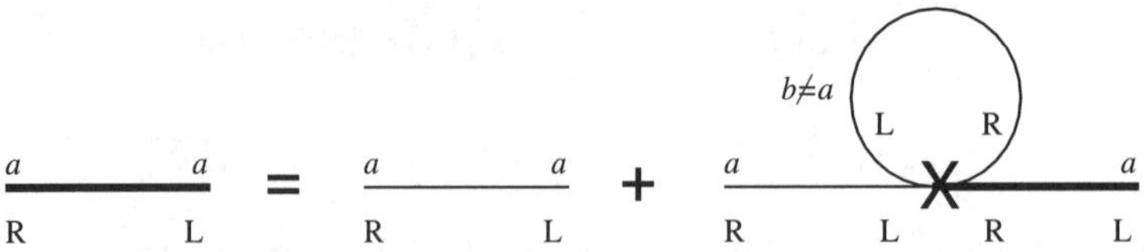

Fig. 21.1. The Dyson equation for the Green's function of the Majorana fermions with Hamiltonian (21.26).

The $k = 2$ level, SU(2) currents expressed in terms of the fields ξ^a are given by

$$I_{R,L}^a = -\frac{i}{2}\epsilon^{abc}\,\xi_{R,L}^b\xi_{R,L}^c \tag{21.25}$$

When small deviations from criticality are considered, no single-ion anisotropy ($\sim D(S^z)^2$, $S = 1$) is allowed to appear due to the original SU(2) symmetry of the problem. So, the mass term in (21.23) turns out to be the only allowed relevant perturbation to the critical SU(2), $k = 2$ WZNW model.

Thus, the fields ξ^a describe triplet excitations related to the effective spin-1 chain. Remarkably, completely decoupled from them are singlet excitations described in terms of the field ρ. Another feature is that this picture is valid regardless of signs of $J_\perp$ and u, in agreement with the effective lattice models (21.9) and (21.10) which we are actually dealing with.

Since the spectrum of the system is massive, the role of the so far neglected (marginal) part of the interchain coupling (21.3) is exhausted by renormalization of the masses and velocity. Neglecting the latter effect, this interaction can be shown to have the following invariant form

$$H_{\text{marg}} = \frac{1}{2}J_\perp a_0 \int dx\,[\,(I_R^a I_L^a) - (\xi_R^a\xi_L^a)\,(\rho_R\rho_L)\,]$$

$$= \frac{1}{2}J_\perp a_0 \int dx[\,(\xi_R^1\xi_L^1)\,(\xi_R^2\xi_L^2) + (\xi_R^2\xi_L^2)\,(\xi_R^3\xi_L^3) + (\xi_R^3\xi_L^3)\,(\xi_R^1\xi_L^1)$$

$$-(\xi_R^1\xi_L^1 + \xi_R^2\xi_L^2 + \xi_R^3\xi_L^3)\,(\rho_R\rho_L)\,] \tag{21.26}$$

In a theory of N massive Majorana fermions, with masses m_a ($a = 1, 2, ..., N$) and a weak four-fermion interaction

$$H_{\text{int}} = \frac{1}{2}\sum_{a\neq b} g_{ab}\int dx(\xi_R^a\xi_L^a)\,(\xi_R^b\xi_L^b), \qquad (g_{ab} = g_{ba})$$

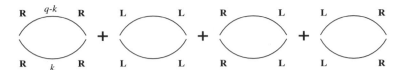

Fig. 21.2. The fermionic loops.

renormalized masses $\tilde{m}_a$ estimated in the first order in g are given by

$$\tilde{m}_a = m_a + \sum_{b(\neq a)} \frac{g_{ab}}{2\pi v} m_b \ln \frac{\Lambda}{|m_b|} \qquad (21.27)$$

Using (21.26) and (21.27), we find renormalized values of the masses of the triplet and singlet excitations:

$$m_{\mathrm{t}} = m_{\mathrm{t}}^{(0)} + \frac{J_\perp a_0}{4\pi v} \ln \frac{\Lambda}{|m|} (2m_{\mathrm{t}}^{(0)} - m_{\mathrm{s}}^{(0)}) \qquad (21.28)$$

$$m_{\mathrm{s}} = m_{\mathrm{s}}^{(0)} + \frac{3J_\perp a_0}{4\pi v} \ln \frac{\Lambda}{|m|} m_{\mathrm{t}}^{(0)} \qquad (21.29)$$

Exercise. Using the fact that uniform magnetic field couples to the fermionic current as

$$i\epsilon^{abc} h^a (\xi_{\mathrm{R}}^b \xi_{\mathrm{R}}^c + (\xi_{\mathrm{L}}^b \xi_{\mathrm{L}}^c)$$

thus leaving the Hamiltonian quadratic and diagonalizable, find a temperature dependent magnetic susceptibility. Show that at $T \ll m_{\mathrm{t}}$

$$\chi(T) \sim T^{-1/2} \exp(-m_{\mathrm{t}}/T)$$

II   Correlation functions for the identical chains

Since the singlet excitation with mass m_{s} does not carry spin, its operators do not contribute to the slow components of the total magnetization. The latter is expressed in terms of the $k = 2$ SU(2) currents (21.25):

$$m^a \sim I_{\mathrm{R}}^a + I_{\mathrm{L}}^a \qquad (21.30)$$

Therefore the two-point correlation function of spin densities at small wave vectors ($|q| \ll \pi/a_0$) is given by the simple fermionic loop (see Fig. 21.2).

A simple calculation gives the following expression for its imaginary part:

$$\Im m\chi^{(R)}(\omega, q) = \frac{2q^2 m^2 v_{\mathrm{s}}^2}{s^3 \sqrt{s^2 - 4m^2}} \qquad (21.31)$$

where $s^2 = \omega^2 - v_s^2 q^2$. Thus the dynamical magnetic susceptibility at small wave vectors has a threshold at $2m_t$.

It turns out that it is possible to calculate exactly the two-point correlation functions of the staggered magnetization. This is due to the fact that the corresponding operators of the Heisenberg chains are related (in the continuum limit) to the order and disorder parameter fields of two-dimensional Ising models; the correlation functions of the latter operators are known exactly even out of criticality (see Chapter 12).

Using formulas (13.9) the components of the total $(\mathbf{n}^{(+)} = \mathbf{n}_1 + \mathbf{n}_2)$ and relative $(\mathbf{n}^{(-)} = \mathbf{n}_1 - \mathbf{n}_2)$ staggered magnetization can be represented as

$$n_x^{(+)} \sim \cos\sqrt{\pi}\theta_+ \cos\sqrt{\pi}\theta_-, \quad n_x^{(-)} \sim \sin\sqrt{\pi}\theta_+ \sin\sqrt{\pi}\theta_-$$
$$n_y^{(+)} \sim \sin\sqrt{\pi}\theta_+ \cos\sqrt{\pi}\theta_-, \quad n_y^{(-)} \sim \cos\sqrt{\pi}\theta_+ \sin\sqrt{\pi}\theta_-$$
$$n_z^{(+)} \sim \sin\sqrt{\pi}\phi_+ \cos\sqrt{\pi}\phi_-, \quad n_z^{(-)} \sim \cos\sqrt{\pi}\phi_+ \sin\sqrt{\pi}\phi_- \qquad (21.32)$$

The fields ϕ_+, θ_+ and ϕ_-, θ_- are governed by the Hamiltonians (21.7) and (21.8), respectively. Let us first consider exponentials $\exp(\pm i\sqrt{\pi}\phi_+)$, $\exp(\pm i\sqrt{\pi}\theta_+)$. Their correlation functions have been extensively studied in the context of the *noncritical* Ising model (see Chapter 12). It has been shown there that these bosonic exponents with scaling dimension $1/8$ are expressed in terms of the order (σ) and disorder (μ) parameters of two Ising models as follows:

$$\cos(\sqrt{\pi}\phi_+) = \mu_1\mu_2, \quad \sin(\sqrt{\pi}\phi_+) = \sigma_1\sigma_2$$
$$\cos(\sqrt{\pi}\theta_+) = \sigma_1\mu_2, \quad \sin(\sqrt{\pi}\theta_+) = \mu_1\sigma_2 \qquad (21.33)$$

Below we shall give an alternative derivation of Eq. (21.33).

As already discussed, the $\beta^2 = 4\pi$ sine-Gordon model H_+, Eq. (21.7), is equivalent to a model of two degenerate massive Majorana fermions, Eqs. (21.14), (21.15). As is explained in Chapter 12, a theory of a massive Majorana fermion field describes long-distance properties of the two-dimensional Ising model, the fermionic mass being proportional to $m \sim t = (T - T_c)/T_c$. So, H_+ is equivalent to two decoupled two-dimensional Ising models. Let σ_j and μ_j $(j = 1, 2)$ be the corresponding order and disorder parameters. At criticality (zero fermionic mass), four products $\sigma_1\sigma_2$, $\mu_1\mu_2$, $\sigma_1\mu_2$ and $\mu_1\sigma_2$ have the same critical dimension $1/8$ as that of the bosonic exponentials $\exp(\pm i\sqrt{\pi}\phi_+)$, $\exp(\pm i\sqrt{\pi}\theta_+)$. Therefore there must be some correspondence between the two groups of four operators which should also hold at small deviations from criticality. To find this correspondence, notice that, as follows from (21.7), at $m > 0$ $\langle\cos\sqrt{\pi}\phi_+\rangle \neq 0$, while $\langle\sin\sqrt{\pi}\phi_+\rangle = 0$. Since the case $m > 0$ corresponds to the disordered phase of the Ising systems $(t > 0)$, $\langle\sigma_1\rangle = \langle\sigma_2\rangle = 0$, while $\langle\mu_1\rangle = \langle\mu_2\rangle \neq 0$. At $m < 0$ (ordered Ising systems, $t < 0$) the situation is inverted: $\langle\cos\sqrt{\pi_+}\phi\rangle = 0$, $\langle\sin\sqrt{\pi}\phi_+\rangle \neq 0$,

$\langle \sigma_1 \rangle = \langle \sigma_2 \rangle \neq 0$, $\langle \mu_1 \rangle = \langle \mu_2 \rangle = 0$. This explains the first two formulas of Eq. (21.33).

Clearly, the exponentials of the dual field θ_+ must be expressed in terms of $\sigma_1 \mu_2$ and $\mu_1 \sigma_2$. To find the correct correspondence, one has to take into account the fact that a local product of the order and disorder operators of a single Ising model results in the Majorana fermion operator, i.e.

$$\xi^1 \sim \cos \sqrt{\pi}(\phi_+ + \theta_+) \sim \sigma_1 \mu_1, \quad \xi^2 \sim \sin \sqrt{\pi}(\phi_+ + \theta_+) \sim \sigma_2 \mu_2$$

This leads to the last two formulas of Eq. (21.33).

To derive similar expressions for the exponents of ϕ_- and θ_-, the following facts should be taken into account: (i) the Hamiltonian (21.18) describing '−'-modes is diagonalized by the same transformation (21.19) as the Hamiltonian (21.12) responsible for the '+'-modes; (ii) the Majorana fermions now have different masses m_s and m_t, and (iii) these masses may have different signs. In order to take a proper account of these facts one should recall the following.

(a) A negative mass means that we are below the transition.

(b) It follows from (ii) that '−' bosonic exponents are also expressed in terms of order and disorder parameters of two Ising models, the latter, however, being characterized by different ts. We denote these operators as σ_3, μ_3 (mass m_t) and σ, μ (mass m_s).

(c) Operators corresponding to a negative mass can be rewritten in terms of the ones with the positive mass using the Kramers–Wannier duality transformation

$$m \rightarrow -m, \; \sigma \rightarrow \mu, \; \mu \rightarrow \sigma \qquad (21.34)$$

Taking these facts into account we get the following expressions for the '−'-bosonic exponents:

$$\cos(\sqrt{\pi}\phi_-) = \mu_3 \sigma$$

$$\sin(\sqrt{\pi}\phi_-) = \sigma_3 \mu, \; \cos(\sqrt{\pi}\theta_-) = \sigma_3 \sigma, \; \sin(\sqrt{\pi}\theta_-) = \mu_3 \mu \qquad (21.35)$$

Combining Eqs. (21.33) and (21.35), from (21.32) we get the following, manifestly SU(2) invariant, expressions:

$$n_x^+ \sim \sigma_1 \mu_2 \sigma_3 \sigma, \; n_y^+ \sim \mu_1 \sigma_2 \sigma_3 \sigma, \; n_z^+ \sim \sigma_1 \sigma_2 \mu_3 \sigma \qquad (21.36)$$

$$n_x^- \sim \mu_1 \sigma_2 \mu_3 \mu, \; n_y^- \sim \sigma_1 \mu_2 \mu_3 \mu, \; n_z^- \sim \mu_1 \mu_2 \sigma_3 \mu \qquad (21.37)$$

It is instructive to compare them with two possible representations for the staggered magnetization operators for the $S = 1$ Heisenberg chain which can be derived from the $SU(2)_2$ WZNW model (see Eq. (13.19) and Tsvelik (1990)):

$$S^x \sim \sigma_1 \mu_2 \sigma_3, \quad S^y \sim \mu_1 \sigma_2 \sigma_3, \quad S^z \sim \sigma_1 \sigma_2 \mu_3 \qquad (21.38)$$

or

$$S^x \sim \mu_1 \sigma_2 \mu_3, \quad S^y \sim \sigma_1 \mu_2 \mu_3, \quad S^z \sim \mu_1 \mu_2 \sigma_3 \qquad (21.39)$$

The agreement is achieved if the singlet excitation band is formally shifted to infinity ($m_s m_t \to -\infty$). This implies substitutions $\sigma \simeq \langle \sigma \rangle \neq 0$, $\mu \simeq \langle \mu \rangle \simeq 0$ for $m_s < 0$ (which would correspond to antiferromagnetic interchain coupling at $u = 0$) or $\sigma \simeq \langle \sigma \rangle \simeq 0$, $\langle \mu \rangle \neq 0$ for $m_s > 0$ (a ferromagnetic interchain coupling). In the latter case, as expected, the staggered $S = 1$ magnetization is determined by the sum of staggered magnetizations of both chains.

II.1 Staggered susceptibility of the conventional (Haldane) spin liquid

As we have mentioned above, the correlation function of staggered magnetization depends on a mutual sign of the masses. As we shall see in a moment, the case $m_t m_s < 0$ corresponds to the conventional (we shall call it Haldane) spin liquid which exhibits a sharp single-magnon peak in $\chi''(\omega, q \sim \pi)$. The case $m_s m_t > 0$, which can be realized only in the presence of four-spin interaction, corresponds to *dimerized* spin liquid where $\chi''(\omega, q \sim \pi)$ has a two particle threshold. We emphasize that both these liquids have a spectral gap and therefore have identical thermodynamic properties.

Let us consider the case $m_t m_s < 0$ and consider the asymptotic behaviour of the two-point correlation functions of staggered magnetizations in the two limits $r \to 0$ and $r \to \infty$. In the limit $r \to \infty$ they are given by Eqs. (12.63):

$$\langle \sigma_a(r) \sigma_a(0) \rangle = G_\sigma(\tilde{r}) = \frac{A_1}{\pi} K_0(\tilde{r}) + O(e^{-3\tilde{r}}) \quad (21.40)$$

$$\langle \mu_a(r) \mu_a(0) \rangle = G_\mu(\tilde{r})$$

$$= A_1 \left\{ 1 + \frac{1}{\pi^2} \left[\tilde{r}^2 \left(K_1^2(\tilde{r}) - K_0^2(\tilde{r}) \right) - \tilde{r} K_0(\tilde{r}) K_1(\tilde{r}) + \frac{1}{2} K_0^2(\tilde{r}) \right] \right\}$$

$$+ O(e^{-4\tilde{r}}) \quad (21.41)$$

where $\tilde{r} = rM$ ($M = m_t$ or $-m_s$), A_1 is a nonuniversal parameter, and it has been assumed that M is positive. If M is negative the correlation functions are obtained simply by interchanging σ and μ, and putting $M \to -M$ (the duality transformation (21.34)). Therefore, as might be expected, at large distances, a difference between the ladder and the $S =$

1 chains appears only in $\exp(-m_s r)$-terms due to the contribution of the excitation branch with $M = m_s$ absent in the $S = 1$ chain.

In the limit $\tilde{r} \to 0$ the correlation functions are of power law form:

$$G_\sigma(\tilde{r}) = G_\mu(\tilde{r}) = \frac{A_2}{\tilde{r}^{\frac{1}{4}}} \qquad (21.42)$$

plus nonsingular terms. The ratio of the constants A_1 and A_2 is a universal quantity given by Eq. (12.65).

We conclude this Section by writing down the exact expression for the staggered magnetization two-point correlation functions. The correlation function for spins on the same chain is given by

$$\langle n_1^a(\tau, x) n_1^a(0,0) \rangle$$
$$= G_\sigma^2(m_t r) G_\mu(m_t r) G_\sigma(m_s r) + G_\mu^2(m_t r) G_\sigma(m_t r) G_\mu(m_s r) \qquad (21.43)$$

The interesting asymptotics are

$$\langle n_1^a(\tau, x) n_1^a(0,0) \rangle = \frac{1}{2\pi r} \tilde{Z} \text{ at } |m_t| r \ll 1 \qquad (21.44)$$

$$= \frac{m}{\pi^2} Z K_0(m_t r) \{ 1 + \frac{2}{\pi^2} [(m_t r)^2 (K_1^2(m_t r) - K_0^2(m_t r)) - m r K_0(m_t r) K_1(m_t r)$$

$$+ \frac{1}{2} K_0^2(m_t r)] \} + O \left(\exp[-(2|m_t| + |m_s|) r] \right) ; \text{ at } m_t r \gg 1 \qquad (21.45)$$

where $r^2 = v_s^2 \tau^2 + x^2$ and

$$\frac{\tilde{Z}}{Z} = \frac{2^{4/3} e}{3^{1/4}} A^{-12} \approx 0.264 \qquad (21.46)$$

The complete expressions for the functions $G_{\sigma,\mu}(\tilde{r})$ are given in Wu *et al.* (1976). For the interchain correlation function we get

$$\langle n_1^a(\tau, x) n_2^a(0,0) \rangle =$$

$$G_\sigma^2(m_t r) G_\mu(m_t r) G_\sigma(m_s r) - G_\mu^2(m_t r) G_\sigma(m_t r) G_\mu(m_s r) \qquad (21.47)$$

At $m_t r \ll 1$ it decays as $(m_t r)^{-2}$; the leading asymptotics at $m r \gg 1$ are the same as (21.45) (up to the -1 factor). The difference appears only in terms of order of $\exp[-(2|m_t| + |m_s|) r]$. The important point is that at $m r \gg 1$ the contribution from the singlet excitation appears only in a high order in $\exp(-m_t r)$. Therefore it is unobservable by neutron scattering at energies below $(2|m_t| + |m_s|)$.

Using the above expressions we can calculate the imaginary part of the dynamical spin susceptibility in two different regimes. For $|\pi - q| \ll 1$ we have (supposing that $J_\perp$ is antiferromagnetic):

$$\Im m\chi^{(R)}(\omega, \pi - q; q_\perp) =$$

$$Z \begin{cases} \sin^2(q_\perp/2)[\frac{m_t}{\pi|\omega|}\delta(\omega - \sqrt{v_s^2 q^2 + m_t^2}) + F(\omega, q)]\omega < (2|m_t| + |m_s|) \\ \dfrac{0.264}{\sqrt{\omega^2 - v_s^2 q^2}} \qquad\qquad\qquad\qquad\qquad \omega \gg (2|m_t| + |m_s|) \end{cases} \quad (21.48)$$

where the transverse 'momentum' $q_\perp$ takes values 0 and π. The factor Z is assumed to be m-independent so that at $m \to 0$ we reproduce the susceptibility of noninteracting chains. (See also Fig. 21.3.) We have calculated the function $F(\omega, q)$ only near the $3m_t$ threshold where it is equal to

$$F(\omega, q) \approx \frac{1}{24\sqrt{3}\pi m_t} \frac{(\omega^2 - v_s^2 q^2 - 9m_t^2)}{m_t^2} \qquad (21.49)$$

The easiest way to calculate $F(\omega, q)$ is the following. Let us expand the functions $K_0(r)$ and $K_1(r)$ in powers of r^{-1}:

$$K_0(r) = \sqrt{\pi/2r}e^{-r}\left(1 - 1/8r + 9/128r^2 - 75/1024r^3 + ...\right)$$

$$K_1(r) = \sqrt{\pi/2r}e^{-r}\left(1 + 3/8r - 15/128r^2 + 105/1024r^3 + ...\right) \quad (21.50)$$

Substituting these expressions into Eq. (21.45) where we set $m = 1$ we get after many cancellations

$$\langle n_1^a(\tau, x)n_1^a(0,0)\rangle = \frac{Z}{4\sqrt{2}\pi^{5/2}}r^{-5/2}e^{-r}[1 + O(r^{-1})] \qquad (21.51)$$

Since the correlation function is Lorentz invariant, we can calculate its ω-dependent part and then restore the q-dependence making the replacement $\omega^2 \to \omega^2 - v^2 q^2$ in the final expression. So for $q = 0$ we can write down the Fourier integral for the function (21.51) in polar coordinates

$$\chi = \frac{Z}{4\sqrt{2}\pi^{5/2}} \int d\phi \int_{\sim 1}^{\infty} dr\, r^{-3/2} \exp(-3r + i\omega_n r \cos\phi) \qquad (21.52)$$

The most singular ω-dependent part of the integral comes from large distances $r \sim (3 - i\omega)^{-1}$. With this in mind we can estimate the r-integral as follows:

$$\int d\phi \int_{\sim 1}^{\infty} dr\, r^{-3/2} \exp(-3r + i\omega_n r \cos\phi) = \text{reg.} + \Gamma(-1/2)[3 - i\omega \cos\phi]^{1/2}$$
$$(21.53)$$

where the abbreviation 'reg.' stands for terms regular in $(\omega \cos\phi - 3)$. The regular terms do not contribute to the imaginary part of $\chi^{(R)}$ after the analytical continuation $i\omega_n \to \omega + i0$. Likewise the next terms in the expansion in $r-1$ will give regular terms together with higher powers of $[3 - i\omega \cos\phi]^{1/2}$. After the analytic continuation we arrive at the following

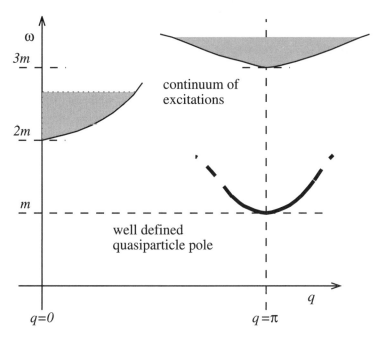

Fig. 21.3. The area of (ω, q) plane where the imaginary part of the dynamical magnetic susceptibility is finite.

expression for the leading singularity:

$$\Im\chi^{(R)}(\omega) = -\Gamma(-1/2)\frac{Z}{4\sqrt{2}\pi^{5/2}}\Re\int_0^{2\pi} d\phi\sqrt{\omega\cos\phi - 3} \quad (21.54)$$

When $\omega \approx 3$ only small ϕ contribute and we can substitute $\cos\phi \approx 1 - \phi^2/2$ and obtain the expression (21.49).

For $|q| \ll 1$ we have

$$\Im\chi^{(R)}(\omega, q; q_\perp) = [1 + \cos^2(q_\perp/2)]f(s, m) \quad (21.55)$$

where $f(s, m)$ is given by Eq. (21.31).

II.2   Dimerized spin liquid

If the masses m_s and m_t have opposite signs, and if the triplet branch of the spectrum remains the lowest, $|m_t| \ll |m_s|$, the two-chain spin ladder is in the Haldane-liquid phase with short-ranged correlations of the staggered magnetization, but with coherent $S = 1$ and $S = 0$ single-magnon excitations. Since we have two interchain couplings, $J_\perp$ and u, the masses may vary independently, and we can ask how the properties of the spin ladder are changed in regions where $m_s m_t > 0$. The thermodynamic properties being dependent on m_a^2 remain unchanged. The symmetry of the ground

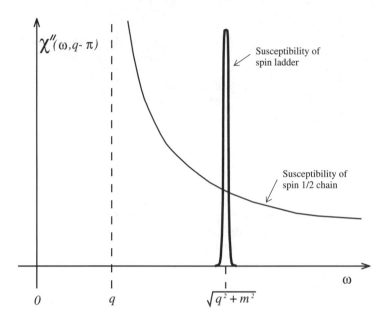

Fig. 21.4.   The optical magnon peak in the Haldane spin liquid.

state and the behaviour of the correlation functions, however, experience a deep change: the ground state turns out to be spontaneously dimerized, and the spectral function of the staggered magnetization displays only incoherent background.

Transitions from the Haldane phase to dimerized phases take place when either the triplet excitations become gapless, with the singlet mode still having a finite gap, or vice versa (see Fig. 21.4). (Notice that in the conventional spin ladder with $u = 0$, both masses vanish simultaneously at $J_\perp \to 0$, and the ladder trivially decouples into a pair of independent $S = 1/2$ Heisenberg chains.) The transition at $m_t = 0$ belongs to the universality class of the critical, exactly integrable, $S = 1$ spin chain (see Eq. (21.24); the corresponding non-Haldane phase with $|m_t| < |m_s|$ represents the dimerized state of the $S = 1$ chain with spontaneously broken translational symmetry and doubly degenerate ground state. The critical point $m_s = 0$ is of the Ising type; it is associated with a transition to another dimerized phase ($|m_t| > |m_s|$), not related to the $S = 1$ chain.

To be specific, let us assume that $u < 0$ and $J_\perp < 0$. We start from the Haldane-liquid phase, $m_t < 0$, $m_s > 0$, increase $|u|$ and, passing through the critical point $m_t = 0$, penetrate into a new phase with $0 < m_t < m_s$. The change of the relative sign of the two masses amounts to the duality transformation in the singlet (ρ) Ising system, implying that in the definitions (21.36) and (21.37) the order and disorder parameters, σ and

μ, must be interchanged. As a result, the spin correlation functions are now given by different expressions:

$$\langle \mathbf{n}^+(\mathbf{r}) \cdot \mathbf{n}^+(\mathbf{0}) \rangle \sim K_0^2(m_t r), \quad \langle \mathbf{n}^-(\mathbf{r}) \cdot \mathbf{n}^-(\mathbf{0}) \rangle \sim K_0(m_t r) K_0(m_s r) \quad (21.56)$$

The total and relative dimerization fields, $\epsilon_{\pm} = \epsilon_1 \pm \epsilon_2$, can be easily found to be

$$\epsilon_+ \sim \mu_1 \mu_2 \mu_3 \mu_0, \quad \epsilon_- \sim \sigma_1 \sigma_2 \sigma_3 \sigma_0, \quad (21.57)$$

and their correlation functions are

$$\langle \epsilon_+(\mathbf{r}) \epsilon_+(\mathbf{0}) \rangle \sim C \left[1 + O(\frac{\exp(-2m_{t,s})}{r^2}) \right], \quad \langle \epsilon_-(\mathbf{r}) \epsilon_-(\mathbf{0}) \rangle \sim K_0^3(m_t r) K_0(m_s r)$$
$$(21.58)$$

From (21.58) it follows that the new phase is characterized by long-range dimerization ordering along each chain, with zero relative phase: $\langle \epsilon_1 \rangle = \langle \epsilon_2 \rangle = (1/2) \langle \epsilon_+ \rangle$. The onset of dimerization is associated with spontaneous breakdown of the Z_2 symmetry related to simultaneous translations by one lattice spacing on both chains.

In the dimerized phase the spin correlations undergo dramatic changes. Using long-distance asymptotics of the MacDonald functions, from (21.56) we obtain:

$$\Im m \, \chi_+(\omega, \pi - q) \sim \frac{\theta(\omega^2 - q^2 - 4m_t^2)}{m_t \sqrt{\omega^2 - q^2 - 4m_t^2}}$$

$$\Im m \, \chi_-(\omega, \pi - q) \sim \frac{\theta[\omega^2 - q^2 - (m_t + m_s)^2]}{\sqrt{m_t m_s} \sqrt{\omega^2 - q^2 - (m_t + m_s)^2}} \quad (21.59)$$

where $\pm$-signs refer to the case where the wave vector in the direction perpendicular to the chains is equal to 0 and π respectively. We observe the disappearance of coherent magnon poles in the dimerized spin fluid; instead we find two-magnon thresholds at $\omega = 2m_t$ and $\omega = m_t + m_s$, similar to the structure of $\chi''(\omega, q)$ at small wave vectors in the Haldane fluid phase. The fact that two massive magnons, each with momentum $q \sim \pi$, combine to form a two-particle threshold, still at $q \sim \pi$ rather than $2\pi \equiv 0$ is related to the fact that, in the dimerized phase with $2a_0$-periodicity, the new Umklapp is just π (see Fig. 21.5).

Exercise. Calculate the spectral function $\Im m \chi(q, \omega)$ at the Ising transition, $m_s = 0$, when $\langle \sigma(r) \sigma(0) \rangle = \langle \mu(r) \mu(0) \rangle \sim r^{-1/4}$. What happens with the pole at $\omega = \sqrt{q^2 + m_t^2}$?

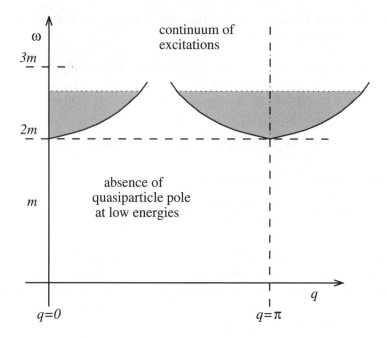

Fig. 21.5.   The area of the ω, q plane where χ'' does not vanish.

III   Inequivalent chains; non-Abelian bosonization

In this Section we consider two interacting spin $S = 1/2$ chains with different intrachain exchange integrals $J_{\|}^1 \neq J_{\|}^2$. It turns out that the most adequate approach in this case is non-Abelian bosonization. The reason for this is that non-Abelian bosonization explicitly preserves the SU(2) symmetry present in the Hamiltonian. The Abelian bosonization approach which does not respect this symmetry encounters difficulties.

As we have already discussed in Chapter 13, the $S = 1/2$ Heisenberg antiferromagnet can be described by the critical $k = 1$, SU(2) WZNW model (Section 13.I.1).

There can in general be marginally irrelevant perturbations to this theory, which generate logarithmic corrections to the correlation function exponents, but do not change their qualitative behaviour (i.e. the power laws).

The bosonized expression for the spin operator of the Heisenberg chain follows from Table 14.1 when the charge field is frozen:

$$\vec{S}_n = \vec{J}_R + \vec{J}_L + \text{const}(-1)^n \text{Tr}(g^+\vec{\sigma} - g\vec{\sigma}) \qquad (21.60)$$

where the currents are given in Table 14.1:

$$J_L^a = +\frac{1}{2\pi}\text{Tr}(\partial_- g)g^+ T^a$$

$$J_R^a = \frac{1}{2\pi}\text{Tr} g^+ \partial_+ g T^a \qquad (21.61)$$

(T^a are the Pauli matrices – generators of the SU(2) group). These currents satisfy the SU(2) Kac–Moody algebra of level $k = 1$.

Consider two Heisenberg chains coupled by an antiferromagnetic nearest neighbour interaction. It can be represented like this:

$$S = W_>(\mathbf{h}) + W_<(\mathbf{g}) + \lambda_1 \left[\vec{H}_R + \vec{H}_L\right]\left[\vec{G}_R + \vec{G}_L\right]$$

$$+\lambda_2 \text{Tr}\left[(\mathbf{g} - \mathbf{g}^+)\,\vec{\sigma}\right]\text{Tr}\left[(\mathbf{h} - \mathbf{h}^+)\,\vec{\sigma}\right] \qquad (21.62)$$

where the dynamics of one chain is represented by the matrix $\mathbf{g}$ and the currents $\vec{G}_{R,L}$ and the other by $\mathbf{h}$ and $\vec{H}_{R,L}$. The subscripts of the WZNW actions distinguish between different spin wave velocities; in order to get explicit expressions for $W_{>,<}$, one has to substitute $x_0 = v_1\tau$ in the general expression for the WZNW action (14.1, 14.2) to get $W_>$ and $x_0 = v_2\tau$ to get $W_<$. Without a loss of generality we can put $v_1 > v_2$. Do not forget that the velocity the measure of integration is modified: $d^2x = v d\tau dx$. Here we encounter a situation where the system as a whole is not Lorentz invariant even in the continuous limit. We need to be careful and remember that velocities cannot be removed from the action by a trivial rescaling of the time coordinate $x_0 = v\tau$ as one can do for a single chain.

The currents have conformal dimensions $(1, 0)$ and $(0, 1)$; in the same time the staggered parts of magnetization have conformal dimensions $(\frac{1}{4}, \frac{1}{4})$. The λ_2 term is therefore a strongly relevant interaction, whereas the current couplings are only marginal. For this reason, the current interaction will be neglected at this stage. Then the interaction can be written as;

$$\text{Tr}[(\mathbf{g} - \mathbf{g}^+)\vec{\sigma}] \cdot \text{Tr}[(\mathbf{h} - \mathbf{h}^+)\vec{\sigma}]$$

$$= \frac{1}{2}\{\text{Tr}[(\mathbf{g} - \mathbf{g}^+)(\mathbf{h} - \mathbf{h}^+)] - \text{Tr}[(\mathbf{g} - \mathbf{g}^+)]\text{Tr}[(\mathbf{h} - \mathbf{h}^+)]\} \qquad (21.63)$$

Our goal is to write an effective action for low-lying excitations. According to the general approach we should integrate over fastest degrees of freedom first. We shall do it by a shift of variables in the action $W_>$. Making the substitution $\alpha = \mathbf{g}\mathbf{h}^+$, which leaves the measure invariant, and using the identity (14.4) we arrive at the following expression for the action:

$$S = [W_>(\mathbf{h}) + W_<(\mathbf{h})] + W_>(\alpha)$$

$$+ \frac{v_>}{2\pi} \int \text{Tr}\alpha^+ \partial_- \alpha \mathbf{h}^+ \partial_+ \mathbf{h} d\tau dx + \lambda_2 \int d\tau dx [\text{Tr}(\alpha + \alpha^+) - \text{Tr}(\alpha^+ \mathbf{h}^{+2} + \text{H.c.})$$
$$+ \text{Tr}(\mathbf{h}^+ - \mathbf{h})\text{Tr}(\mathbf{h}^+ \alpha^+ - \alpha \mathbf{h})] \tag{21.64}$$

where $\partial_\pm = \frac{1}{2}(v_>^{-1} \partial_\tau \mp i \partial_x)$.

We shall consider the most relevant interaction, $\text{Tr}(\alpha + \alpha^+)$ first. The effective action for α is in this approximation;

$$S = W_>(\alpha) + \lambda \text{Tr}(\alpha + \alpha^+) \tag{21.65}$$

From the first order RG equation we get

$$\frac{d\lambda}{d \ln L} \simeq (2 - \frac{1}{2})\lambda \tag{21.66}$$

Integrating up to a scale where the coupling becomes of order 1 and taking this to give some estimate of the dynamically generated mass, one gets $M \sim \lambda^{\frac{2}{3}}$. Much more information can be found by realising that the model (21.65) is equivalent to the $\beta^2 = 2\pi$ sine-Gordon model (see Chapter 23).

Thus on the scale $|x| >> M^{-1}$ the fluctuations of the α-field are frozen and we can approximate

$$\text{Tr}(\alpha \mathbf{h}^+)\text{Tr}(\mathbf{h}) \approx \langle \text{Tr}\alpha \rangle : [\text{Tr}(\mathbf{h})]^2 : \tag{21.67}$$

At this large scale the cross term containing derivatives of h and α gives the irrelevant contribution

$$S_{int} \sim M^{-2} v_> \int d\tau dx \text{Tr} \left[\nabla^2 \mathbf{h}^+ \nabla^2 \mathbf{h} \right] \tag{21.68}$$

where $\nabla^2 = [v_>^{-2} \partial_\tau^2 + \partial_x^2]$. This arises from the fact that in a massive theory such as the one described by the action (21.65) the current–current correlation function at $q^2 << M^2$ has the form

$$\langle\langle J_\mu(-q) J_\nu(q) \rangle\rangle \equiv \Pi_{\mu\nu}(q) \sim \frac{1}{M^2}(q^2 \delta_{\mu\nu} - q_\mu q_\nu) \tag{21.69}$$

This form insures current conservation $q^\mu \Pi_{\mu\nu}(q) = 0$.

The part of the interaction

$$S' = \frac{v_>}{2\pi} \int d\tau dx \text{Tr}\alpha^+ \partial_+ \alpha \mathbf{h} \partial_- \mathbf{h}^+ \tag{21.70}$$

couples the left moving current of the α field $J_L^a = \frac{i}{2\pi}\text{Tr}\left[\alpha^+ \partial_+ \alpha \sigma^a\right]$ to the right-moving current of $\mathbf{h}$ fields. Integrating out the α fields generates an effective interaction for the $\mathbf{h}$ fields. Since $J_L^a = J_0^a - J_1^a$, we have from

(21.69) that $\Pi_{LL} = \langle J_L^a J_L^a \rangle \sim \frac{1}{M^2} q_+^2$. Hence the interaction for the $\mathbf{h}$ fields becomes;

$$S_{int} \sim \int d\tau_1 dx_1 d\tau_2 dx_2 \mathrm{Tr}\,(\mathbf{h}(1)\partial_-\mathbf{h}^+(1)\sigma^a)\,\Pi_{LL}(\mathbf{r}_1 - \mathbf{r}_2)\mathrm{Tr}\,(\partial_-\mathbf{h}(2)\mathbf{h}^+(2)\sigma^a)$$

$$(21.71)$$

which gives the irrelevant contribution described by Eq. (21.68).

Therefore the asymptotic behaviour at large distances is governed by the following action:

$$S = W_>(h) + W_<(h) + c_2 \int d\tau dx : [\mathrm{Tr}(\mathbf{h})]^2 : \qquad (21.72)$$

where $c_2 \sim \lambda^{4/3}$ and which can be further modified by the coordinate rescaling:

$$x_0 = \sqrt{v_1 v_2}\tau, \ x_1 = x \qquad (21.73)$$

such that we finally have

$$S = S_0 + S_1 \quad (21.74)$$

$$S_0 = \frac{1}{2c_1} \int d^2 x \mathrm{Tr}\,(\partial_\mu \mathbf{h}^+ \partial_\mu \mathbf{h}) + 2\Gamma\,(\mathbf{h}) \quad (21.75)$$

$$S_1 = \int d^2 x \tilde{c}_2 \{: [\mathrm{Tr}\mathbf{h}^2] : + : [\mathrm{Tr}(\mathbf{h}^+)^2] : - : [\mathrm{Tr}(\mathbf{h} - \mathbf{h}^+)]^2 : \} \quad (21.76)$$

where

$$1/c_1 = \sqrt{v_1/v_2} + \sqrt{v_2/v_1}$$

$$\tilde{c}_2 = \lambda^{\frac{4}{3}} \sqrt{v_1/v_2} \qquad (21.77)$$

The model with action (21.74) is not critical; coupling constants $c_1, \tilde{c}_2$ undergo further renormalization. Let us show that the coupling $\tilde{c}_2$ grows faster. To show this we shall suppose that this is the case and check that the obtained result is self-consistent. It is easy to check that the effective potential (21.76) vanishes if $\mathbf{h}$ is a traceless matrix and has a fixed determinant:

$$\mathbf{h} \approx i(\sigma \mathbf{n}), \ \mathbf{n}^2 = 1 \qquad (21.78)$$

Writing $\mathbf{h} = n_0 \mathbf{1} + i\sqrt{1 - n_0^2}\sigma \cdot \mathbf{n}$, the effective action for n_0 becomes, in the rescaled coordinates;

$$S_{eff} = \int d^2 x \left\{ \frac{(\partial_\mu n_0)^2}{32\pi c_1} + 2\tilde{c}_2 n_0^2 \right\} \qquad (21.79)$$

Excitations, which correspond to configurations where $\mathrm{Tr}\mathbf{h} \neq 0$, acquire a gap. The estimate for this gap is

$$M_0^2 \sim c_1(v_1 v_2) \, \tilde{c}_2 \sim \frac{\sqrt{\frac{v_1}{v_2}}(v_1 v_2)}{\left(\sqrt{\frac{v_1}{v_2}} + \sqrt{\frac{v_2}{v_1}}\right)} \lambda^{\frac{4}{3}} \sim v_1 v_2 \lambda^{\frac{4}{3}} \sim \frac{v_2}{v_1} M^2 \qquad (21.80)$$

So the assumption that we could ignore fluctuations of the α field in the low energy theory is self-consistently justified; this gap is much smaller than the gap for fluctuations of α for v_1/v_2 sufficiently large. For energies smaller than this gap we can treat the **h**-matrix as traceless. Substituting expression (21.78) into Eqs. (18. 69) and (18. 70) we get the O(3)-nonlinear sigma model as an effective action for small energies:

$$S = \frac{1}{2\tilde{c}} \int d^2x (\partial_\mu \mathbf{n})^2, \quad \mathbf{n}^2 = 1 \qquad (21.81)$$

$$\frac{1}{\tilde{c}} = (\sqrt{v_1/v_2} + \sqrt{v_2/v_1})(1 - \langle n_0^2 \rangle) \qquad (21.82)$$

The reason why the Wess–Zumino term effectively disappears from the action is that after substituting Eq. (21.78) into the expression for $\Gamma(\mathbf{h})$ the latter term reduces to the topological term (the derivation of this nontrivial fact is given in Appendix A):

$$2\Gamma(i\sigma\mathbf{n}) = \frac{i}{4} \int d^2x \epsilon_{\mu\nu} \left(\mathbf{n}[\partial_\mu \mathbf{n} \times \partial_\nu \mathbf{n}]\right) = 2\pi i k \qquad (21.83)$$

where k is an integer number. The factor in front of the topological term is such that its contribution to the action is always a factor of $2\pi i$ and therefore does not affect the partition function. The mass gap of the model (21.81) is given by

$$M = M_0 \tilde{c}^{-1} \exp(-2\pi/\tilde{c}) \approx M[1 - \frac{c}{2\pi} \ln(M/M_0)] \exp(-2\pi/c) \qquad (21.84)$$

As long as this gap is much smaller than M_0, the adopted approach is self-consistent. The latter is achieved for any appreciable difference between the velocities.

As it well known, excitations of the O(3)-nonlinear sigma model are $S = 1$ triplets (Zamolodchikov and Zamolodchikov (1979), Wiegmann (1985)). Thus, the spectrum is qualitatively the same as for identical chains. That is what one might expect because the model of Majorana fermions is a strong coupling limit of the O(3)-nonlinear sigma model (see Tsvelik (1990)). See also Fig. 21.6.

The correlation functions of the O(3)-nonlinear sigma model are known only in the form of the Lehmann expansion (see the book by Smirnov, p. 160, and also Kirillov and Smirnov (1988)).

$$\langle \mathbf{n}(\tau, x)\mathbf{n}(0, 0) \rangle \sim K_0(mr) + O(\exp(-3mr)) \qquad (21.85)$$

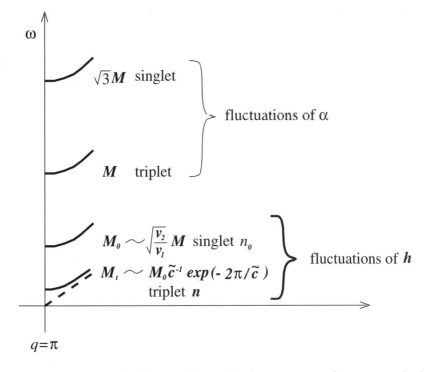

Fig. 21.6. The schematic picture of the excitation spectrum of two nonequivalent coupled Heisenberg chains.

Note that the first term in the expansion coincides with the one for identical chains. Therefore a difference in dynamical magnetic susceptibilities for both cases will become manifest only at energies $\omega > 3m$. The lowest feature in $\Im m\chi^{(R)}(\omega, q)$ is in both cases the sharp peak

$$\Im m\chi^{(R)}(\omega, q) \sim \frac{m}{\sqrt{q^2 + m^2}}\delta(\omega - \sqrt{q^2 + m^2}) \qquad (21.86)$$

corresponding to the triplet excitation.

IV String order parameter in the spin-ladder model

Den Nijs and Rommelse (1989) and Girvin and Arovas (1989) suggested that the gapful Haldane phase of the $S = 1$ spin chain is characterized by a topological order measured by the string order parameter

$$\langle O^\alpha \rangle = \lim_{|n-m|\to\infty} \langle\, S_n^\alpha \exp(i\pi \sum_{j=n+1}^{m-1} S_j^\alpha)S_m^\alpha \,\rangle, \quad (S = 1, \quad \alpha = x, y, z) \qquad (21.87)$$

The nonzero value of $\langle O^\alpha \rangle$ has been related to the breakdown of a hidden $Z_2 \times Z_2$ symmetry (Kohmoto and Tasaki (1992)). In this section we use the Abelian bosonization method to construct the string operator in the continuum limit of the $S = 1/2$ spin-ladder model and identify the corresponding discrete symmetry with that of the related Ising models.

Since spin-rotational invariance remains unbroken, the string order parameter must respect this symmetry. However, Abelian bosonization is not an explicitly SU(2) invariant procedure. For this reason, it turns out that it is the z-component of the string operator that acquires a simple form in the continuum limit. On the other hand, due to the unbroken SU(2) symmetry, the very choice of the quantization (z-) axis is arbitrary; therefore the expectation values for all components of the string operator will coincide.

To construct a string order parameter $O^z(n, m)$ for the spin-ladder model, we shall follow the same route as used for the bond-alternating $S = 1/2$ chain (see Kohmoto and Tasaki (1992), technical details are given in Appendix B). We start from the lattice version of the model, construct a product of two spin-1/2 operators belonging to the j-th rung, $S_1^z(j)S_2^z(j)$, and then take a product over all rungs between $j = n$ and $j = m$:

$$O^z(n, m) = \prod_{j=n}^{m}(-4S_1^z(j)S_2^z(j)) = \exp\left(i\pi \sum_{j=n}^{m}[\, S_1^z(j) + S_2^z(j)\,]\right) \quad (21.88)$$

Assuming that $|m - n| \gg 1$, we pass to the continuum limit in the exponential and retain only the smooth parts of the spin operators expressing them in terms of the spin currents $J_{a;R,L}^z(x)$, $(a = 1, 2)$:

$$O^z(x, y) = \exp\left(\pm i\pi \sum_{a=1,2} \int_x^y dx' S_a^z(x')\right)$$

$$= \exp\left(\pm i\pi \sum_{a=1,2} \int_x^y dx' [J_{a;R}^z(x') + J_{a;R,L}^z(x')]\right) \quad (21.89)$$

Using Eqs. (21.5), (13.1), we find that the exponential is expressed in terms of the field ϕ_+ only. Thus we find a very transparent representation for the string operator:

$$O^z(x, y) = \exp(i\sqrt{\pi}[\phi_+(x) - \phi_+(y)]) \quad (21.90)$$

Using Eq. (21.33),

$$\exp(i\sqrt{\pi}\phi_+(x)) \sim \mu_1\mu_2 + i\sigma_1\sigma_2 \quad (21.91)$$

we find that the string operator is expressed in terms of the Ising order and disorder operators. For either sign of $J_\perp$, we find that, in the limit

$|x - y| \to \infty$, the vacuum expectation value of $O^z(x, y)$ is indeed nonzero:

$$\lim_{|x-y|\to\infty} \langle O^z(x, y) \rangle \sim \langle \sigma_1 \rangle^2 \langle \sigma_2 \rangle^2 = \langle \sigma \rangle^4 \neq 0, \; J_\perp < 0 \quad (21.92)$$

$$\lim_{|x-y|\to\infty} \langle O^z(x, y) \rangle \sim \langle \mu_1 \rangle^2 \langle \mu_2 \rangle^2 = \langle \mu \rangle^4 \neq 0, \; J_\perp > 0 \quad (21.93)$$

As in the case of the bond-alternating spin chain, the nonvanishing expectation value of the string order parameter in the limit of infinite string manifests breakdown of a discrete $Z_2 \times Z_2$ symmetry. This is the symmetry of two decoupled Ising models described by the Hamiltonian H_+ in the Majorana fermion representation (21.14): $H_+ = H_m[\xi^1] + H_m[\xi^2]$ remains invariant with respect to sign inversion of both chiral components of each Majorana spinor, $\xi^a_{R,L} \to -\xi^a_{R,L}$, $(a = 1, 2.)$ Under these transformations, the Ising order and disorder parameters change their signs. On the other hand, since the two Majorana fermions are massive, this symmetry is broken in the *ground state* of H_+: the mass terms break the duality symmetry $\xi^a_L \to -\xi^a_L$, $\xi^a_R \to \xi^a_R$. This amounts to finite expectation values of the Ising variables σ_1 and σ_2 (or μ_1 and μ_2), which in turn results in a nonzero string order parameter, as shown in Eqs. (21.92) and (21.93).

V   Appendix A. The topological term emerging from the Wess–Zumino term

As we know, the Wess–Zumino term has the following form:

$$\Gamma = -\frac{i}{32\pi} \int_0^1 d\xi \int d^2x \epsilon_{\mu\nu} \epsilon_{abc} \Omega^\xi_a \Omega^\mu_b \Omega^\nu_c \quad (21.94)$$

where

$$\Omega^\mu_a = -i\mathrm{Tr}(\sigma^a g^{-1} \partial_\mu g)$$

and

$$g(\xi = 0, x) = g(x), \; g(\xi = 1, x) = I \quad (21.95)$$

Let us take

$$g = n_0(\xi)I + i\sigma \mathbf{n}(x)\sqrt{1 - n_0^2(\xi)}$$

$$g^{-1} = n_0(\xi)I - i\sigma \mathbf{n}(x)\sqrt{1 - n_0^2(\xi)} \quad (21.96)$$

where $n_0(\xi = 0) = 0$ and $n_0(\xi = 1) = 1$ such that the conditions (21.95) are satisfied. Then we get

$$\Omega^\mu_a = 2\left[-n^a \sqrt{1 - n_0^2} \partial_\mu n_0 + n_0 \partial_\mu \left(n_a \sqrt{1 - n_0^2} \right) + \epsilon_{abc} n^b \partial_\mu \left(n_c \sqrt{1 - n_0^2} \right) \right] \quad (21.97)$$

Substituting this expression into (21.94) and treating n_0 as dependent only on ξ, we get

$$\Gamma = \frac{iA}{\pi} \int d^2x \epsilon_{\mu\nu} \left(\mathbf{n}[\partial_\mu \mathbf{n} \times \partial_\nu \mathbf{n}] \right) \qquad (21.98)$$

where the constant A is given by

$$A = \int_0^1 d\xi \left[-\sqrt{1 - n_0^2} \frac{dn_0}{d\xi} + n_0(1 - n_0^2)\frac{d}{d\xi}\sqrt{1 - n_0^2} \right] = -\pi/4 \qquad (21.99)$$

This gives the necessary coefficient at the topological term.

VI　Appendix B. Hidden $Z_2 \times Z_2$ symmetry and string order parameter in the bond-alternating $S = 1/2$ Heisenberg chain

In addition to the $S = 1/2$ spin-ladder model, there is another system which is related to the $S = 1$ spin chain – the spin-1/2 chain with alternating ferromagnetic and antiferromagnetic bonds:

$$H = 4J \sum_{j=1}^{N/2} [(\mathbf{S}_{2j-1} \cdot \mathbf{S}_{2j}) - \beta(\mathbf{S}_{2j} \cdot \mathbf{S}_{2j+1}] \qquad (21.100)$$

This model is instructive in the sense that the string order parameter, whose nonzero expectation value signals breakdown of a hidden discrete symmetry, can be easily constructed. The analogous construction is then directly generalized for the spin-ladder model.

A gap in the excitation spectrum of the model (21.100) persists in the whole range $0 < \beta < \infty$. At $\beta = 0$ the ground state of model represents an array of disconnected singlets. At $\beta \gg 1$, strong ferromagnetic coupling between the spins on the $\langle 2j, 2j + 1 \rangle$-bonds leads to the formation of local triplets, and the model (21.100) reduces to a $S = 1$ Heisenberg chain. Using a nonlocal unitary transformation, Kohmoto and Tasaki (1992) demonstrated equivalence of the model (21.100) to a system of two coupled quantum Ising chains, i.e. two coupled two-dimensional Ising models. This transformation provides an exact representation of the spin operators S_n^α as products of two Ising-like order (σ, τ) and disorder $(\tilde{\sigma}, \tilde{\tau})$ operators, essentially a lattice version of relations (21.33), (21.35) (see, e.g., Nersesyan and Luther (1994)). Nearest neighbour bilinears of the original spin operators take the form

$$4S_{2j}^x S_{2j+1}^x = -\sigma_j^z \sigma_{j+1}^z, \quad 4S_{2j-1}^x S_{2j}^x = -\tau_j^x$$
$$4S_{2j}^y S_{2j+1}^y = -\tau_j^z \tau_{j+1}^z, \quad 4S_{2j-1}^y S_{2j}^y = -\sigma_j^x$$
$$4S_{2j}^z S_{2j+1}^z = -\sigma_j^z \sigma_{j+1}^z \tau_j^z \tau_{j+1}^z, \quad 4S_{2j-1}^z S_{2j}^z = -\sigma_j^x \tau_j^x \qquad (21.101)$$

where

$$\sigma_j^x = \tilde{\sigma}_{j-1/2}^z \, \tilde{\sigma}_{j+1/2}^z, \quad \tau_j^x = \tilde{\tau}_{j-1/2}^z \, \tilde{\tau}_{j+1/2}^z \qquad (21.102)$$

$$\tilde{\sigma}_{j+1/2}^z = \prod_{l=j+1}^{N/2} \sigma_l^x, \quad \tilde{\tau}_{j+1/2}^z = \prod_{l=1}^{j-1} \tau_l^x \qquad (21.103)$$

Relations (21.101) make the Hamiltonian (21.100) equivalent to two coupled quantum Ising chains:

$$H = -J \sum_{j=1}^{N/2} [\, (\beta \sigma_j^z \sigma_{j+1}^z + \sigma_j^x) + (\beta \tau_j^z \tau_{j+1}^z + \tau_j^x) + (\beta \sigma_j^z \sigma_{j+1}^z \tau_j^z \tau_{j+1}^z + \sigma_j^x \tau_j^x) \,] $$

$$(21.104)$$

The model (21.104) is invariant under independent rotations of the σ and τ spins by angle π about the spin x-axis which comprise a $Z_2 \times Z_2$ group. Since this group is discrete, it can be spontaneously broken, in which case the spectrum of the system would be massive. It is easily understood from (21.104) that, in the limit of large positive β when the model reduces to the $S = 1$ chain, the $Z_2 \times Z_2$-symmetry is broken, with

$$\langle \sigma_j^z \rangle = \langle \tau_j^z \rangle = \langle \sigma_j^z \tau_j^z \rangle \neq 0 \qquad (21.105)$$

(It has been used in Eq. (21.105) that, under transformation $\mu_j^z = \sigma_j^z \tau_j^z$ to a new pair of variables, μ_j^z and τ_j^z, the two-chain Hamiltonian (21.104) preserves its form.)

Representation (21.101) hints at the way how an order parameter measuring breakdown of the $Z_2 \times Z_2$-symmetry should be constructed out of the spin operators S_n^α. Following Kohmoto and Tasaki, consider a product

$$\prod_{l=2k}^{2n-1} 2S_l^x = \prod_{j=k}^{n-1} 4S_{2j}^x S_{2j+1}^x = \prod_{j=k}^{n-1} (-\sigma_j^z \sigma_{j+1}^z)$$

$$= (-1)^{n-k} (\sigma_k^z \sigma_{k+1}^z)(\sigma_{k+1}^z \sigma_{k+2}^z) \cdots (\sigma_{n-1}^z \sigma_n^z) = (-1)^{n-k} \sigma_k^z \sigma_n^z \quad (21.106)$$

Using the relation $i\sigma_j^\alpha = \exp(i\pi\sigma_j^\alpha/2)$, we find that

$$O^x(k, n) \equiv \exp \left(i\pi \sum_{l=2k}^{2n-1} S_l^x \right) = \sigma_k^z \sigma_n^z \qquad (21.107)$$

This is the x-component of the string order operator. According to (21.105), in the limit $|k-n| \to \infty$, its vacuum expectation value is nonzero:

$$\langle O^x(k, n) \rangle \to \langle \sigma \rangle^2 \neq 0 \qquad (21.108)$$

It is important that the string always contains an even number of sites, starting at an even site and ending at an odd site. For a string starting at an odd site and ending at an even site, the corresponding string operator is expressed in terms of disorder operators and therefore has zero expectation value:

$$\prod_{l=2k+1}^{2n} 2S_l^x = \prod_{j=k+1}^{n} 4S_{2j-1}^x S_{2j}^x$$
$$= (-1)^{n-k}(\tilde{\tau}_{k+1/2}^z \tilde{\tau}_{k+3/2}^z)(\tilde{\tau}_{k+3/2}^z \tilde{\tau}_{k+5/2}^z)\cdots(\tilde{\tau}_{n-1/2}^z \tilde{\tau}_{n+1/2}^z)$$
$$= (-1)^{n-k}\tilde{\tau}_{k+1/2}^z \tilde{\tau}_{n+1/2}^z$$

The y and z components of the string operator are constructed in a similar manner:

$$O^y(k,n) = \exp\left(i\pi \sum_{l=2k}^{2n-1} S_l^y\right) = \prod_{l=2k}^{2n-1} 2iS_l^y = \prod_{j=k}^{n-1}(-4S_{2j}^y S_{2j+1}^y)$$
$$= \prod_{j=k}^{n-1} \tau_j^z \tau_{j+1}^z = \tau_k^z \tau_n^z \tag{21.109}$$

$$O^z(k,n) = \exp\left(i\pi \sum_{l=2k}^{2n-1} S_l^z\right) = \sigma_k^z \tau_k^z \sigma_n^z \tau_n^z \tag{21.110}$$

The SU(2) invariance of the expectation value of the string order parameter

$$O^\alpha(k,n) = \exp\left(i\pi \sum_{l=2k}^{2n-1} S_l^\alpha\right), \quad (S=1/2, \ \alpha=x,y,z) \tag{21.111}$$

follows from (21.105).

Notice that in the limiting case $\beta \gg 1$, the string order parameter (21.111) for the $S=1/2$ bond-alternating chain automatically transforms to the exponential of the string order parameter (21.87) for the $S=1$ chain.

References

M. Azuma, Z. Hiroi, M. Takano, K. Ishida and Y. Kitaoka, *Phys. Rev. Lett.* **73**, 3463 (1994).

H. M. Babujan, *Nucl. Phys.* **B215**, 317 (1983).

E. Dagotto, J. Riera and D. Scalapino, *Phys. Rev.* **B45**, 5744 (1992).

M. den Nijs and K. Rommelse, *Phys. Rev.* **B40**, 4709 (1989).

S. M. Girvin and D. P. Arovas, *Physica Scripta* **T27**, 156 (1989).

F. D. M. Haldane, *Phys. Lett.* **93A**, 464 (1983).

K. Hida, *J. Phys. Soc. Jpn.* **60**, 1347 (1991).

K. Ishida, Y. Kitaoka, Y. Tokunaga, S. Matsumoto, K. Asayama, M. Azuma, Z, Hiroi and M. Takano, *Phys. Rev.* B**53**, 2827 (1996).

A. N. Kirillov and F. A. Smirnov, *Int. J. Mod. Phys.* A**3**, 731 (1988).

M. Kohmoto and H. Tasaki, *Phys. Rev.* B**46**, 3486 (1992).

A. A. Nersesyan and A. Luther, *Phys. Rev.* B**50**, 309 (1994).

A. A. Nersesyan and A. M. Tsvelik, *Phys. Rev. Lett.* **78**, 3939 (1997).

T. M. Rice, S. Gopalan and M. Sigrist, *Europhys. Lett.* **23**, 445 (1993).

T. M. Rice, S. Gopalan and M. Sigrist, *Physica* B**199-200**, 378 (1994).

H. Schulz, *Phys. Rev.* B**34**, 6372 (1986).

D. G. Shelton, A. A. Nersesyan and A. M. Tsvelik, *Phys. Rev.* B**53**, 8521 (1996).

M. Takano, Z. Hiroi, M. Azuma and Y. Takeda, *Jpn. J. Appl. Phys.* Ser **7**, 3 (1992).

L. A. Takhtajan, *Phys. Lett.* A**87**, 479 (1982).

A. M. Tsvelik, *Phys. Rev.* B**42**, 10 499 (1990).

P. B. Wiegmann, *Phys. Lett.* B**152**, 209 (1985).

T. T. Wu, B. McCoy, C. A. Tracy and E. Barouch, *Phys. Rev.* B**13**, 316 (1976).

A. B. Zamolodchikov and Al. B. Zamolodchikov, *Ann. Phys.* (N.Y.) **120**, 253 (1979).

22
Spin-1/2 Heisenberg chain
with alternating exchange

In this chapter we shall discuss the properties of spin $S = 1/2$ Heisenberg chain with alternating coupling constants $J(1 \pm \delta)$. In reality such dimerization may be present as a permanent feature in the crystalline structure or can emerge via a phase transition as a result of instability of the spin–lattice system (spin–Peierls transition). The spin–Peierls transition is a magnetoelastic transition which occurs in quasi-one-dimensional or strongly anisotropic antiferromagnets because phonons are coupled to the staggered energy density and the latter one has a divergent susceptibility even if the charge field is locked. This instability resolves in dimerization of the lattice at certain temperature T_c and a formation of the spectral gaps in the spectrum of magnetic excitations.

In any realistic system phonons are three-dimensional which determines a three-dimensional nature of the spin–Peierls transition. It can be shown that three-dimensional effects influence magnetic excitations even well below the transition. Since analysis of such interactions would introduce additional complications, we shall not consider the spin–Peierls transition, but restrict our discussion to a simple model where the dimerization is static (that is its origin is independent of spins). Thus we consider the spin-1/2 Heisenberg chain with exchange integrals of alternating magnitude:

$$H = J \sum_n [(\mathbf{S}_n \mathbf{S}_{n+1}) + \delta(-1)^n (\mathbf{S}_n \mathbf{S}_{n+1})] \qquad (22.1)$$

We shall discuss a case of weak dimerization $|\delta| \ll 1$. At $\delta = 0$ the spin-1/2 Heisenberg chain is exactly solvable and, as has been discussed in the previous chapters, its low energy sector is equivalent to the Gaussian model. Therefore the most singular part of the operator $\epsilon(x) = (-1)^n (\mathbf{S}_n \mathbf{S}_{n+1})$ can be expressed in terms of bosonic exponents – relevant fields of the Gaussian model.

$$\epsilon(x) = (-1)^n (\mathbf{S}_n \mathbf{S}_{n+1}) = A \cos(\beta \Phi) + \text{less singular terms} \qquad (22.2)$$

and at finite but small δ the low energy modes are described by a sine-Gordon model. There is a simple way to establish the value of β: we have to use the fact that under translation by one lattice spacing ϵ changes its sign while the scalar field Φ is replaced by $\Phi + \sqrt{\pi/2}$. This fixes the value $\beta = \sqrt{2\pi}$. On the other hand, this can be supported by another observation. Namely, we have to use the fact that the initial Hamiltonian has the SU(2) symmetry and so the sine-Gordon model must also respect it. As we know from Chapter 8, such symmetry is not present for general β which should impose a restriction on its value. The corresponding point in the sine-Gordon spectrum was discovered by Coleman (1976) and Haldane (1982) who pointed out that at $\beta = \sqrt{2\pi}$ there are only two breathers; the first one has the same mass as kink and antikink (let us call it M_t) and the second has mass equal to $\sqrt{3}M_t$. Therefore at $\beta = \sqrt{2\pi}$ kink, antikink and the first breather realize an SU(2) triplet and the second breather becomes an SU(2) singlet. The same value of β was found by the straightforward bosonization by Nakano and Fukuyama (1981). Notice that despite the fact that $\epsilon(x)$ has the same dimensions as the z-component of the staggered magnetization, it is given by the different operator (cosine instead of sine).

As we shall see, the situation is somewhat similar to what we have had for the spin-ladder (see Chapter 21), but there are also certain subtle differences.

Below we first recall the construction of a convenient basis of states for the sine-Gordon theory by means of the Zamolodchikov–Faddeev algebra essentially repeating the arguments of Chapter 10 and then formulate the problem of calculating correlation functions in terms of *formfactors* and finally give explicit results for the first few terms in the formfactor expansion. The analysis given is based on the publication by Essler *et al.* (1997).

The Zamolodchikov–Faddeev algebra for the sine-Gordon model with $\beta^2 = 2\pi$ was derived by Affleck (1986) who suggested a representation which manifestly respects the SU(2) symmetry. The operators $\mathscr{Z}_a^+(\theta), \mathscr{Z}_a(\theta)$ $(a = s, 0, \bar{s})$ are creation and annihilation operators of triplet excitations; $B^+(\theta), B(\theta)$ create and destroy singlet states. As usual, the eigenstates are parametrized by a rapidity θ such that their momentum and energy are equal to

$$p_j = M_j \sinh\theta_j, \quad \epsilon_j = M_j \cosh\theta_j \tag{22.3}$$

where $M_s = \sqrt{3}M$ and $M_t = M$. The FZ operators (and their Hermitian conjugates) satisfy the following algebra

$$\mathscr{Z}_a(\theta_1)\mathscr{Z}_b(\theta_2) = S_{1,1}(\theta_1 - \theta_2)\mathscr{Z}_b(\theta_2)\mathscr{Z}_a(\theta_1)$$
$$\mathscr{Z}_a(\theta_1)B(\theta_2) = S_{1,2}(\theta_1 - \theta_2)B(\theta_2)\mathscr{Z}_a(\theta_1)$$
$$B(\theta_1)B(\theta_2) = S_{2,2}(\theta_1 - \theta_2)B(\theta_2)B(\theta_1) \tag{22.4}$$

where the two-particle scattering matrices $S_{ij}(\theta)$ are given by

$$S_{1,1}(\theta) = -\frac{\sinh\theta + i\sin\frac{\pi}{3}}{\sinh\theta - i\sin\frac{\pi}{3}} = S_0(\theta)$$

$$S_{1,2}(\theta) = S_0(\theta + i\frac{\pi}{6})\, S_0(\theta - i\frac{\pi}{6})$$

$$S_{2,2}(\theta) = \left(\frac{\sinh\theta + i\sin\frac{\pi}{3}}{\sinh\theta - i\sin\frac{\pi}{3}}\right)^3 \tag{22.5}$$

For the creation and annihilation operators we have

$$\mathcal{Z}_a(\theta_1)\mathcal{Z}_b^+(\theta_2) = S_{1,1}(\theta_2 - \theta_1)\mathcal{Z}_b^+(\theta_2)\mathcal{Z}_a(\theta_1) + \delta_{ab}\delta(\theta_1 - \theta_2)$$
$$\mathcal{Z}_a(\theta_1)B^+(\theta_2) = S_{1,2}(\theta_2 - \theta_1)B^+(\theta_2)\mathcal{Z}_a(\theta_1)$$
$$B(\theta_1)B^+(\theta_2) = S_{2,2}(\theta_2 - \theta_1)B^+(\theta_2)B(\theta_1) + \delta(\theta_1 - \theta_2) \tag{22.6}$$

As we see $S_{i,j}(0) = -1$ and $S_{i,j}(\infty) = +1$. Therefore particles with close momenta behave like free fermions and particles far apart in momentum space behave like free bosons. Following Smirnov we introduce the notations

$$Z_{\frac{1}{2}}(\theta) = \mathcal{Z}_s(\theta), \quad Z_{-\frac{1}{2}}(\theta) = \mathcal{Z}_{\bar{s}}(\theta), \quad Z_1(\theta) = \mathcal{Z}_0(\theta), \quad Z_2(\theta) = B(\theta) \tag{22.7}$$

States in the Fock space are constructed by acting with the operators $Z_\varepsilon^\dagger(\theta)$ on the vacuum state $|0\rangle$

$$|\theta_n \dots \theta_1\rangle_{\varepsilon_n \dots \varepsilon_1} = Z_{\varepsilon_n}^\dagger(\theta_n) \dots Z_{\varepsilon_1}^\dagger(\theta_1)|0\rangle, \tag{22.8}$$

where $\varepsilon_j = \pm\frac{1}{2}, 1, 2$. We note that Eq. (22.4) together with Eq. (22.8) implies that states with different ordering of two rapidities *and* indices ε_i are related by multiplication with two-particle S-matrices. The resolution of the identity is given by

$$1 = \sum_{n=0}^{\infty}\sum_{\varepsilon_i}\int \frac{d\theta_1 \dots d\theta_n}{(2\pi)^n n!}|\theta_n \dots \theta_1\rangle_{\varepsilon_n \dots \varepsilon_1}{}^{\varepsilon_1 \dots \varepsilon_n}\langle\theta_1 \dots \theta_n| \tag{22.9}$$

The formfactor approach is based on the idea of inserting (22.9) between the operators in a correlation function

$$\langle\mathcal{O}(x,t)\mathcal{O}(0,0)\rangle = \sum_{n=0}^{\infty}\sum_{\varepsilon_i}\int \frac{d\theta_1 \dots d\theta_n}{(2\pi)^n n!}e^{i\sum_{j=1}^{n} xp_j - t\epsilon_j}|\langle 0|\mathcal{O}(0,0)|\theta_n \dots \theta_1\rangle_{\varepsilon_n \dots \varepsilon_1}|^2 \tag{22.10}$$

and then determining the formfactors

$$F^{\mathcal{O}}(\theta_1 \dots \theta_n)_{\varepsilon_1 \dots \varepsilon_n} := \langle 0|\mathcal{O}(0,0)|\theta_n \dots \theta_1\rangle_{\varepsilon_n \dots \varepsilon_1} \tag{22.11}$$

by taking advantage of their known analytic properties.

From a physical point of view we are interested in the Fourier transforms of the connected retarded two-point correlators of $\cos\sqrt{2\pi}\Phi$ and $\sin\sqrt{2\pi}\Phi$. Their formfactor expansions are of the form

$$D^{cos}(\omega, q) = \int_{-\infty}^{\infty} dx \int_0^{\infty} dt \, e^{i(\omega+i\epsilon)t-iqx} \langle [\cos\sqrt{2\pi}\Phi(t,x), \cos\sqrt{2\pi}\Phi(0,0)] \rangle$$

$$= -2\pi \sum_{n=0}^{\infty} \sum_{\varepsilon_i} \int \frac{d\theta_1 \ldots d\theta_n}{(2\pi)^n n!} |F^{cos}(\theta_1 \ldots \theta_n)_{\varepsilon_1 \ldots \varepsilon_n}|^2$$

$$\times \left\{ \frac{\delta(q - \sum_j M_j \sinh\theta_j)}{\omega - \sum_j M_j \cosh\theta_j + i\epsilon} - \frac{\delta(q + \sum_j M_j \sinh\theta_j)}{\omega + \sum_j M_j \cosh\theta_j + i\epsilon} \right\} \quad (22.12)$$

The Fourier transform $D^{sin}(\omega, q)$ of the connected retarded two-point correlator of $\sin\sqrt{2\pi}\Phi$ is the dynamical staggered susceptibility and will also be denoted by $\chi(\omega, q)$.

In order to implement the formfactor expansion it is very useful to note that like for general values of β, operators from different representations behave differently under the charge conjugation transformation:

$$C\Phi C^{-1} = -\Phi$$

$$C\mathcal{Z}_s(\theta)C^{-1} = \mathcal{Z}_{\bar{s}}(\theta), \quad C\mathcal{Z}_0(\theta)C^{-1} = -\mathcal{Z}_0(\theta)$$

$$CB(\theta)C^{-1} = B(\theta) \quad (22.13)$$

This immediately implies the following expansion

$$\sin[\sqrt{2\pi}\Phi(\tau, x)]|0\rangle = F_1 \int \frac{d\theta}{2\pi} e^{-iM(t\cosh\theta - x\sinh\theta)} \mathcal{Z}_0(\theta)|0\rangle$$

$$+ \int \frac{d\theta_1}{2\pi} \frac{d\theta_2}{2\pi} e^{-iM[t(\cosh\theta_1 + \cosh\theta_2) - x(\sinh\theta_1 + \sinh\theta_2)]} \times$$

$$\times U(\theta_1, \theta_2)[\mathcal{Z}_s^+(\theta_1)\mathcal{Z}_{\bar{s}}^+(\theta_2) - \mathcal{Z}_{\bar{s}}^+(\theta_1)\mathcal{Z}_s^+(\theta_2)] + \ldots \quad (22.14)$$

where $U(\theta_1, \theta_2) = U(\theta_2, \theta_1)$. Significantly, due to the SU(2) symmetry transverse components of the staggered magnetization have the same correlation functions as n_{stag}^z. Therefore we should not worry about calculation of formfactors of the operators $\sin(\sqrt{2\pi}\Theta)$ and $\cos(\sqrt{2\pi}\Theta)$ which appear in the dynamical spin–spin correlation functions (see Chapter 11). The current operator $\partial_x\Phi$ is also odd in Φ and therefore its expansion must begin with $\mathcal{Z}_0$ such that at small q we have

$$\langle\langle S(\omega, q)S(-\omega, -q)\rangle\rangle \sim \frac{q^2}{\omega^2 - v^2 q^2 - M^2} + \ldots \quad (22.15)$$

where dots denote terms which have nonzero imaginary parts at higher energies.

As we will see the threshold of the dynamical spin susceptibility is equal to M for both $q = 0$ and $q = \pi$. This is a distinct feature of the alternating chain. It is related to the fact that kink and antikink create a bound state of the same mass. Recall that for the ladder chain (or $S = 1$ antiferromagnet for that matter) where particles do not have bound states, the value of the energy threshold at $q = 0$ is twice that at $q = \pi$ (see the previous chapter).

At frequencies smaller than $(1 + \sqrt{3})M$ the only contributions to the imaginary part of the magnetic susceptibility come from the first breather and kink–antikink pairs. Kink–antikink formfactors can be calculated in the sine-Gordon model (for any value of the coupling β) along the lines described in Chapter 10. Here we repeat the derivation for completeness. Let us denote by $S_+(\theta)$ ($S_-(\theta)$) the S-matrix eigenvalue corresponding to positive (negative) C-parity obtained by diagonalizing the kink–antikink scattering:

$$S_+ = \frac{\sinh \frac{\pi}{2\gamma}(\theta + i\pi)}{\sinh \frac{\pi}{2\gamma}(\theta - i\pi)} S_0(\theta) \,, \qquad S_- = \frac{\cosh \frac{\pi}{2\gamma}(\theta + i\pi)}{\cosh \frac{\pi}{2\gamma}(\theta - i\pi)x} S_0(\theta) \quad (22.16)$$

where $\gamma = \frac{\pi\beta^2}{8\pi - \beta^2}$ and

$$S_0(\theta) = -\exp(-i \int_0^\infty \frac{dx \sin\theta x \sinh \frac{\pi - \gamma}{2} x}{x \, \cosh \frac{\pi x}{2} \sinh \frac{\gamma x}{2}}) \quad (22.17)$$

Then, general unitarity and crossing arguments imply that the corresponding kink–antikink formfactors $F_\pm(\theta)$ are solutions of the following system of functional equations

$$F_\pm(\theta) = S_\pm(\theta)F_\pm(-\theta) \quad (22.18)$$
$$F_\pm(\theta + 2i\pi) = \pm F_\pm(-\theta) \quad (22.19)$$

The 'minimal' solutions of these equations are

$$F_+(\theta) = \frac{\sinh\theta}{\sinh(\theta - i\pi)\frac{\pi}{2\gamma}} * F_0(\theta)$$

$$F_-(\theta) = \frac{\sinh\theta}{\cosh(\theta - i\pi)\frac{\pi}{2\gamma}} * F_0(\theta) \quad (22.20)$$

F_0 being the exponential part. By minimal solution we mean a solution containing only the expected bound state poles in the physical strip and with the mildest asymptotic behaviour at infinity. This prescription determines the minimal solution *uniquely*. An infinite number of nonminimal solutions corresponding to the all operators in the theory are obtained by multiplying the minimal solution by an analytic function of $\cosh\theta$. However, if we require the formfactor to be power bounded in the momenta and to have only the bound state poles, we conclude that we can actually multiply the minimal solution only by a polynomial in $\cosh \theta$. For a given operator, it is possible to put strong constraints on the asymptotic behaviour of its form factors, and then on the degree of the allowed polynomial (see the reference of Delfino and Mussardo (1996) in Chapter 10). In the sine-Gordon model this procedure is quite complicated due to a very nontrivial behaviour of correlators in the ultraviolet limit. Nevertheless, the result is that for the operators cos and sin the allowed polynomial is of the zero degree, which means that their formfactors coincide with the minimal ones. The same conclusion can be reached in a simpler way going to the free fermion point $\gamma = \pi$, where the formfactors of sin and cos can be easily computed remembering that

$$\cos \beta \Phi \sim \bar{\Psi}\Psi, \quad \varepsilon_{\mu\nu}\partial^{\nu}\Phi \sim J_{\mu} \tag{22.21}$$

and that the sin is related to the elementary field by the equation of motion.

For the operators $\cos \beta \Phi$ and $\sin \beta \Phi$, at the specific value of the coupling we are interested in, we find ($\theta_{12} = \theta_1 - \theta_2$)

$$\langle 0| \sin \sqrt{2\pi}\Phi|\theta_1, \theta_2 \rangle_{-+} = \sqrt{3}(2d)Z^{1/2} \frac{\cosh \theta_{12}/2}{\sinh 3\theta_{12}/2} \zeta(\theta_{12}) = F^{sin}(\theta)_{-+} \tag{22.22}$$

$$\langle 0| \cos \sqrt{2\pi}\Phi|\theta_1, \theta_2 \rangle_{-+} = i\sqrt{3}(2d)Z^{1/2} \frac{\cosh \theta_{12}/2}{\cosh 3\theta_{12}/2} \zeta(\theta_{12}) = F^{cos}(\theta_{12})_{-+} \tag{22.23}$$

$$\zeta(\theta) = c \sinh \theta/2 \exp \left\{ 2 \int_0^\infty dx \frac{\sin^2[x(\theta + i\pi)/2] \cosh \pi x/6}{x \sinh \pi x \cosh \pi x/2} \right\} \tag{22.24}$$

$$c = (12)^{\frac{1}{4}} \exp \left\{ \frac{1}{2} \int_0^\infty dx \frac{\sinh x/2 \cosh x/6}{x \cosh^2 x/2} \right\} \approx 3.494607$$

$$d = \frac{3}{2\pi c} \approx 0.136629 \tag{22.25}$$

where the relative normalization between the two operators can be fixed by exploiting the asymptotic factorization of formfactors discussed in Delfino *et al.* (1996). Z is an overall normalization fixed by Eq. (22.9). We note that $\zeta(\theta)$ is to be analytically continued using the relation

$$\zeta(\theta)S_0(\theta) = \zeta(-\theta) \tag{22.26}$$

The additional factors d and c in (22.22) and (22.23) have been introduced in order to simplify the reduction of multiparticle formfactors using the annihilation-pole condition (for soliton formfactors)

$$\mathrm{Res}F^{0}(\theta_1 \ldots \theta_{2n})_{\varepsilon_1 \ldots \varepsilon_{2n}}\bigg|_{\theta_{2n}-\theta_{2n-1}=i\pi} = F^{0}(\theta_1 \ldots \theta_{2n-2})_{\varepsilon'_1 \ldots \varepsilon'_{2n-2}} \delta_{\varepsilon_{2n}}^{-\varepsilon'_{2n-1}}$$

$$\times \left\{ \delta_{\varepsilon_1}^{\varepsilon'_1} \ldots \delta_{\varepsilon_{2n-1}}^{\varepsilon'_{2n-1}} - S_{\tau_1,\varepsilon_1}^{\varepsilon'_{2n-1},\varepsilon'_1}(\theta_{2n-1}-\theta_1) \ldots S_{\varepsilon_{2n-1},\varepsilon_{2n-2}}^{\tau_{2n-3},\varepsilon'_{2n-2}}(\theta_{2n-1}-\theta_{2n-2}) \right\}$$

(22.27)

where $S_{\varepsilon,\varepsilon}^{\varepsilon,\varepsilon}(\theta) = S_{\varepsilon,\varepsilon'}^{\varepsilon,\varepsilon'}(\theta) = S_{1,1}(\theta)$ and all other components are zero. Multiparticle formfactors are discussed in some detail in the Appendix.

The formfactor (22.22) has a pole at $\theta_{12} = -2i\pi/3$ corresponding to formation of a bound state – the first breather. The breather formfactor F_1 is given by the residue of (22.22) divided by the three-particle coupling:

$$|F_1|^2 = \frac{3^{\frac{3}{2}}}{8\pi^2} \exp\left(-2 \int_0^\infty \frac{dx \sinh \pi x/6 \sinh \pi x/3}{x \sinh \pi x \cosh \pi x/2}\right) Z \approx 0.0533Z \quad (22.28)$$

Similarly (22.23) has a pole at $\theta_{12} = -i\pi/3$ corresponding to the second breather. The absolute square $|F_2|^2$ of the breather formfactor is found to be

$$|F_2|^2 = \frac{3^{\frac{3}{2}}}{8\pi^2} \exp\left(-4 \int_0^\infty \frac{dx \cosh \pi x/6 \sinh^2 \pi x/3}{x \sinh \pi x \cosh \pi x/2}\right) Z \approx 0.0262Z$$

(22.29)

The expression for the imaginary part of the dynamical staggered susceptibility $\chi(\omega, q)$ at $s^2 = \omega^2 - q^2 < (1+\sqrt{3})^2 M^2$ is given by

$$\Im m\chi(\omega, q) = 2\pi |F_1|^2 \delta(s^2 - M^2) + 2\Re e \frac{|F^{\sin}[\theta(s)_{+-}]|^2}{s\sqrt{s^2 - 4M^2}} \quad (22.30)$$

where $\theta(s) = 2\ln(s/2M + \sqrt{s^2/4M^2 - 1})$. Note that all other formfactors do not contribute to this expression in the specified range of s as their thresholds are above $(1 + \sqrt{3})M$. Also the normalization Z enters Eq. (22.30) only as an overall factor. Since the function $\zeta(\theta)$ vanishes at $\theta = 0$, the entire formfactor is also finite. Thus the two-particle contribution to $\chi''(\omega, q)$ exhibits a square-root singularity at the threshold as a function of s.

The breather and $s\bar{s}$ contributions to the real part are found to be

$$\Re e\chi(\omega, q) = -\Re e \frac{2|F_1|^2}{s^2 - M^2 + i\varepsilon}$$

$$-2 \int_0^\infty \frac{d\theta}{\pi} \frac{s^2 - 4M^2 \cosh^2 \frac{\theta}{2}}{(s^2 - 4M^2 \cosh^2 \frac{\theta}{2})^2 + \varepsilon^2} |F^{\sin}(\theta)_{+-}|^2 \quad (22.31)$$

where the factor of 2 stems from the sum over $+$ and $-$. In Figs. 22.1 and 22.2 we plot both the imaginary and real parts of χ.

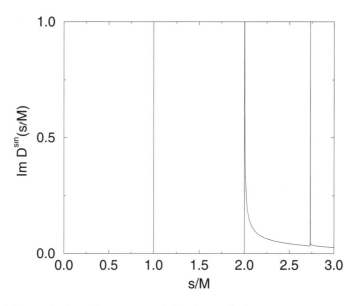

Fig. 22.1.   The imaginary part of the dynamical staggered susceptibility.

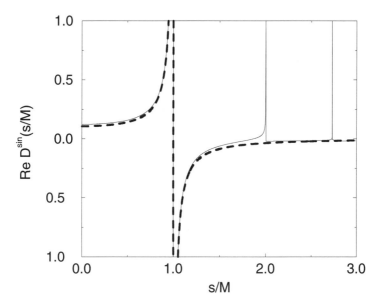

Fig. 22.2.   The real part of the dynamical staggered susceptibility.

It is straightforward to repeat the above analysis for the current operator $\partial_x \Phi$ using the explicit expressions for the formfactors given in Smirnov's book. The contribution of the first breather leads to Eq. (22.15) with some normalization factor. As there is very little spectral weight at $q \approx 0$ we concentrate on $q \approx \pi$ and do not repeat the above analysis for the current operator.

For practical purposes it is convenient to have an expression interpolating between the small q (22.15) and $q \approx \pi$ (22.30) behaviour. Such an expression giving the dynamical spin susceptibility in the entire range of q at frequencies below the continuum may look like

$$\chi(\omega, q) = \frac{g(q) \sin^2(q/2)}{\omega^2 - v^2 \sin^2 q - M^2} + \cdots \tag{22.32}$$

where $g(q)$ is a smooth function interpolating between the normalizations at $q = 0$ and $q = \pi$. The mode $\omega = \sqrt{v^2 \sin^2 q + M^2}$ is separated from the particle continuum by the gap of order of M (see Fig. 22.1).

Let us now turn to the two-point correlator of cosines. The contributions of the second breather and the soliton–antisoliton continuum are given by (see Fig. 22.3)

$$\Im m D^{cos}(\omega, q) = 2\pi |F_2|^2 \delta(s^2 - 3M^2) + 2\Re e \frac{|F^{cos}[\theta(s)_{+-}]|^2}{s\sqrt{s^2 - 4M^2}} \tag{22.33}$$

where the ratio of the single particle residues is universal:

$$\gamma = \frac{|F_2|^2}{|F_1|^2} = \exp \left(- \int_0^\infty dx \frac{\sinh x/3}{x \cosh^2 \frac{x}{2}} \right) \approx 0.49131 \tag{22.34}$$

Note that the threshold of the breather-breather continuum is at $s = 2M$ as well. The corresponding contribution is taken into account in the Appendix. The analogous contributions to the real part of D^{cos} (Fig. 22.4) are given by

$$\Re e D^{cos}(\omega, q) = -\Re e \frac{2|F_2|^2}{s^2 - 3M^2 + i\varepsilon}$$
$$-2 \int_0^\infty \frac{d\theta}{\pi} \frac{s^2 - 4M^2 \cosh^2 \frac{\theta}{2}}{(s^2 - 4M^2 \cosh^2 \frac{\theta}{2})^2 + \varepsilon^2} |F^{cos}(\theta)_{+-}|^2 \tag{22.35}$$

The remaining integrals in Eqs. (22.30), (22.31), (22.33) and (22.35) have to be calculated numerically. We find that at small s the contributions of the two-particle continua to the real parts of both correlators are of the same magnitude as the single-particle contributions from the breather states. As far as a single chain is concerned a single-mode approximation taking into account only the one-particle states is therefore very poor at small s.

In applications one can meet another deformation of the spin-1/2 Heisenberg chain – the Heisenberg chain in a staggered magnetic field

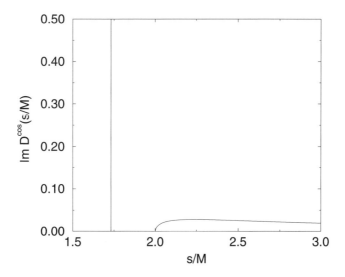

Fig. 22.3. The imaginary part of D^{cos} as a function of $s = \sqrt{\omega^2 - \frac{v^2}{a_0^2}q^2}$.

(see, for example, Schulz (1996)):

$$H = J\sum_n [(\mathbf{S}_n \mathbf{S}_{n+1}) + h(-1)^n S_n^z] \tag{22.36}$$

In the continuous limit this model is also equivalent to the sine-Gordon model with $\beta^2 = 2\pi$, but since the staggered magnetization is proportional to $\sin(\sqrt{2\pi}\Phi)$, one must shift Φ by $\sqrt{\pi/8}$ in the formulas for the alternating chain. This shift interchanges $\sin(\sqrt{2\pi}\Phi)$ and $\cos(\sqrt{2\pi}\Phi)$, but changes neither derivatives of Φ nor the dual field Θ. This means that the correlation function $\langle\langle S^z S^z \rangle\rangle$ at small q is still given by Eq. (22.15), but at $q \approx \pi$ is given by Eq. (22.33). Therefore around $q = \pi$ the pole is now at $s = \sqrt{3}M_t$:

$$\chi^{zz}(\omega, q) = \frac{2Z F_2^2/\pi}{3M_t^2 + (\pi - q)^2 - \omega^2} + \text{incoherent} \tag{22.37}$$

However, since Θ does not change under the shift of Φ, correlation functions of transverse components of the staggered magnetization are still given by D^{sin} and therefore have a pole at $s = M_t$:

$$\chi^{\perp}(\omega, q) = \frac{4Z F_1^2/\pi}{M_t^2 + (\pi - q)^2 - \omega^2} + \text{incoherent} \tag{22.38}$$

Such difference in correlation functions is obviously related to the broken rotational symmetry of the Hamiltonian (22.36). Notice that according to Eq. (22.34) the residues of both functions are almost equal.

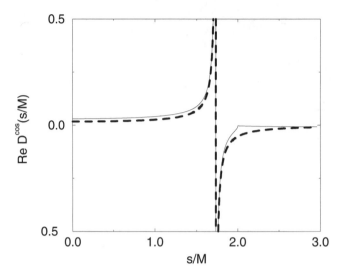

Fig. 22.4.   The real part of D^{cos} as a function of $s = \sqrt{\omega^2 - \frac{v^2}{a_0^2}q^2}$.

I   Appendix. Multiparticle formfactors

In this appendix we consider multiparticle formfactors. We start with 2-soliton 2-antisoliton formfactors and explicitly derive the related three and two-particle formfactors. The extension to n-soliton n-antisoliton formfactors $(n = 3, 4, \ldots)$ is straightforward and will not be discussed here. The formfactor expansion for two-point correlation functions is found to be rapidly converging and for small s it is essentially sufficient to take into account two-particle formfactors only. Note that most of the formulas below are to be understood in terms of analytic continuation of $\zeta(\theta)$. A useful formula is

$$|\zeta(\theta + i\alpha)\zeta(\theta - i\alpha)|^2 = \frac{c^4}{4}(\cosh\theta - \cos\alpha)^2$$

$$\times \exp\left\{4\int_0^\infty dx \frac{\cosh x/6}{x\sinh x\cosh\frac{x}{2}}\left(1 - \cos\frac{x\theta}{\pi}\cosh\frac{\pi - \alpha}{\pi}x\right)\right\}$$

$$\times \frac{\sinh^2\theta\cos^2\alpha + (\cosh\theta\sin\alpha - \sin\frac{\pi}{3})^2}{\sinh^2\theta\cos^2\alpha + (\cosh\theta\sin\alpha + \sin\frac{\pi}{3})^2} \tag{22.39}$$

Multiparticle formfactors of (quasi)local operators were studied in detail by Smirnov. From his work the 2-soliton 2-antisoliton formfactor for

$\cos \sqrt{2\pi}\Phi$ is straightforwardly extracted

$$F^{cos}(\theta_1,\theta_2,\theta_3,\theta_4)_{--++} = \langle 0| \cos \sqrt{2\pi}\Phi|\theta_1,\theta_2,\theta_3,\theta_4\rangle_{--++} =$$

$$2\pi(2d)^2\sqrt{Z} \; 3 \left(\sum_{l=1}^{4} e^{\theta_l}\right)\left(\sum_{m=1}^{4} e^{-\theta_m}\right)\sum_{i<j} e^{\theta_i+\theta_j}\prod_{i<k}\zeta(\theta_i - \theta_k)e^{-\frac{1}{2}\sum_j \theta_j}$$

$$\times \cosh \frac{3}{2}(\theta_3 + \theta_4 - \theta_1 - \theta_2)\prod_{i=1}^{2}\prod_{j=3}^{4}\frac{1}{\sinh 3(\theta_j - \theta_i)} \qquad (22.40)$$

Orderings other than $--++$ are obtained from (22.40) by using the generalization of (22.19) *e.g.*

$$F^{cos}(\theta_1,\theta_2,\theta_3,\theta_4)_{-+-+} = S_0(\theta_3 - \theta_2) \; F^{cos}(\theta_1,\theta_3,\theta_2,\theta_4)_{--++} \qquad (22.41)$$

It is easy to verify that the soliton–antisoliton formfactor of $\cos \sqrt{2\pi}\Phi$ is obtained from (22.40) *via* the annihilation pole condition. The 2-soliton 2-antisoliton formfactor for $\sin \sqrt{2\pi}\Phi$ is very similar to the one for $\cos \sqrt{2\pi}\Phi$

$$F^{sin}(\theta_1,\theta_2,\theta_3,\theta_4)_{--++} = -i\tanh \frac{3}{2}(\theta_3 + \theta_4 - \theta_1 - \theta_2)F^{cos}(\theta_1,\theta_2,\theta_3,\theta_4)_{--++}$$
$$(22.42)$$

The residue at the annihilation pole (times i) now yields the soliton–antisoliton formfactor of $\sin \sqrt{2\pi}\Phi$.

Breather formfactors are obtained from the residues of (22.40) at its poles. In the soliton–antisoliton–even breather sector we find

$$F^{cos}(\theta_1,\theta_2,\theta_3)_{-+2}$$

$$= -2\pi \frac{(2d)^2\sqrt{Z}}{2^{\frac{3}{2}}3^{\frac{1}{4}}}\frac{e^{-\frac{1}{2}(\theta_1+\theta_2)-\theta_3}}{\cosh \frac{3\theta_{21}}{2}\cosh 3\theta_{31}\cosh 3\theta_{32}}\;\zeta(\theta_{12})\zeta\left(-i\frac{\pi}{3}\right)$$

$$\times\zeta\left(\theta_{13}+i\frac{\pi}{6}\right)\zeta\left(\theta_{13}-i\frac{\pi}{6}\right)\zeta\left(\theta_{23}+i\frac{\pi}{6}\right)\zeta\left(\theta_{23}-i\frac{\pi}{6}\right)[e^{\theta_1}+e^{\theta_2}+\sqrt{3}e^{\theta_3}]$$

$$\times[e^{-\theta_1}+e^{-\theta_2}+\sqrt{3}e^{-\theta_3}][e^{\theta_1+\theta_2}+\sqrt{3}e^{\theta_3}(e^{\theta_1}+e^{\theta_2})+e^{2\theta_3}] \qquad (22.43)$$

The corresponding formfactor for $\sin \sqrt{2\pi}\Phi$ is

$$F^{sin}(\theta_1,\theta_2,\theta_3)_{-+2} = -i\coth \frac{3\theta_{21}}{2} \; F^{cos}(\theta_1,\theta_2,\theta_3)_{-+2} \qquad (22.44)$$

and different orderings are obtained by the appropriate generalization of (22.19) *e.g.*

$$F^{cos}(\theta_1,\theta_2,\theta_3)_{2-+} = F^{cos}(\theta_2,\theta_3,\theta_1)_{-+2} \; S_{1,2}(\theta_{21}) \; S_{1,2}(\theta_{31}) \qquad (22.45)$$

The residue at the annihilation pole (times i) in $F^{cos}(\theta_1,\theta_2,\theta_3)_{2-+}$ gives the heavy breather formfactor F_2. The corresponding sin formfactor has no annihilation poles. In the soliton–antisoliton–odd breather sector we obtain

$$F^{cos}(\theta_1, \theta_2, \theta_3)_{-+1}$$

$$= -2\pi \frac{(2d)^2 \sqrt{Z}}{2^{\frac{3}{2}} 3^{\frac{1}{4}}} \frac{e^{-\frac{1}{2}(\theta_1 + \theta_2) - \theta_3}}{\sinh \frac{3\theta_{21}}{2} \sinh 3\theta_{31} \sinh 3\theta_{23}} \zeta(\theta_{12}) \zeta(-i\frac{2\pi}{3})$$

$$\times \zeta(\theta_{13} + i\frac{\pi}{3}) \zeta(\theta_{13} - i\frac{\pi}{3}) \zeta(\theta_{23} + i\frac{\pi}{3}) \zeta(\theta_{23} - i\frac{\pi}{3}) [e^{\theta_1} + e^{\theta_2} + e^{\theta_3}]$$

$$\times [e^{-\theta_1} + e^{-\theta_2} + e^{-\theta_3}] [e^{\theta_1 + \theta_2} + e^{\theta_3}(e^{\theta_1} + e^{\theta_2}) + e^{2\theta_3}]$$

$$= i \coth \frac{3\theta_{21}}{2} F^{sin}(\theta_1, \theta_2, \theta_3)_{-+1} \tag{22.46}$$

From the residues at the poles of (22.46) we can derive the breather-breather formfactors:

$$F^{cos}(\theta_1, \theta_2)_{11} = -2\pi \frac{4(2d)^2 \sqrt{Z}}{3^{\frac{3}{2}}} \frac{(\cosh \frac{\theta_{12}}{2})^2 [\cosh \theta_{12} + \frac{1}{2}]}{(\sinh 3\theta_{12})^2} \zeta^2(\theta_{21}) \zeta^2(-i\frac{2\pi}{3})$$

$$\times \zeta(\theta_{21} + i\frac{2\pi}{3}) \zeta(\theta_{21} - i\frac{2\pi}{3}) S_0(\theta_{21} + i\frac{\pi}{3}) S_0(\theta_{21} - i\frac{\pi}{3}) \tag{22.47}$$

This is identical to the soliton–antisoliton formfactor $F^{cos}(\theta_1, \theta_2)_{-+}$ as can be proved by direct calculation. Some useful identities are $S_0(\theta + i\frac{\pi}{3}) S_0(\theta - i\frac{\pi}{3}) = S_0(\theta)$, $2\zeta(\theta)\zeta(\theta - i\pi) = \frac{\sinh 3\theta}{\sinh \theta + i \sin \frac{\pi}{3}}$ and

$$\exp\left(\int_0^\infty \frac{dx}{x} \frac{\cosh \frac{x}{6} - \cosh \frac{x}{2}}{\sinh x \cosh \frac{x}{2}}\right) = \frac{2\sqrt{2}}{c} \tag{22.48}$$

The other breather–breather formfactors are given by

$$F^{cos}(\theta_1, \theta_2)_{22} = -2\pi \frac{4(2d)^2 \sqrt{Z}}{\sqrt{3}} \frac{(\cosh \frac{\theta_{12}}{2})^2 [\cosh \theta_{12} + \frac{3}{2}]}{(\cosh 3\theta_{12})^2}$$

$$\times \zeta^2(\theta_{12}) \zeta^2(-i\frac{\pi}{3}) \zeta(\theta_{12} + i\frac{\pi}{3}) \zeta(\theta_{12} - i\frac{\pi}{3}),$$

$$F^{sin}(\theta_1, \theta_2)_{21} = -2\pi \frac{4(2d)^2 \sqrt{Z}}{3^{\frac{3}{2}}} \frac{(\cosh \theta_{12} + \frac{\sqrt{3}}{2})[1 + \frac{\sqrt{3}}{2} \cosh \theta_{12}]}{(\cosh 3\theta_{12})^2} \zeta(-i\frac{\pi}{3})$$

$$\times \zeta(-i\frac{2\pi}{3}) \zeta(\theta_{12} + i\frac{\pi}{2}) \zeta(\theta_{12} - i\frac{\pi}{2}) \zeta(\theta_{12} + i\frac{\pi}{6}) \zeta(\theta_{12} - i\frac{\pi}{6}) \tag{22.49}$$

The special values of ζ at the breather poles are given by $\zeta(-i\frac{\pi}{3}) \approx -1.10184i$ and $\zeta(-i\frac{2\pi}{3}) \approx -2.72272i$. We note that $\zeta(-i\frac{\pi}{3})\zeta(-i\frac{2\pi}{3}) = -3$.

It is apparent that n-particle formfactors depend only on $n - 1$ independent rapidity variables. This fact can be used to essentially simplify expressions like (22.12) for correlation functions. For example, 2-particle contributions for $\omega \geq 0$ are given by

$$D^{cos}(\omega, q)\Big|_{2-\text{part.}} =$$

$$-\int_0^\infty \frac{d\theta}{\pi} \left\{ \frac{2|F^{cos}(\theta)_{+-}|^2 + |F^{cos}(\theta)_{11}|^2}{s^2 - 4M^2 \cosh^2 \frac{\theta}{2} + i\varepsilon} + \frac{|F^{cos}(\theta)_{22}|^2}{s^2 - 12M^2 \cosh^2 \frac{\theta}{2} + i\varepsilon} \right\} \quad (22.50)$$

$$D^{sin}(\omega, q)\Big|_{2-\text{part.}} =$$

$$-\int_0^\infty \frac{d\theta}{\pi} \left\{ \frac{2|F^{sin}(\theta)_{+-}|^2}{s^2 - 4M^2 \cosh^2 \frac{\theta}{2} + i\varepsilon} + \frac{2|F^{sin}(\theta)_{21}|^2}{s^2 - 4M^2(1 + \frac{\sqrt{3}}{2} \cosh \theta) + i\varepsilon} \right\} \quad (22.51)$$

where we also have made use of various symmetry properties of the form-factors in order to perform the sum over ϵ_j. Similarly the contribution of formfactors involving one soliton, one antisoliton and one (light) breather of type 1 can be brought to the form

$$-\int_{-\infty}^\infty \frac{d\theta}{2\pi} \int_{-\infty}^\infty$$

$$\times \frac{d\theta_{12}}{2\pi} \frac{2}{s^2 - M^2(1 + 4\cosh\theta\cosh\frac{\theta_{12}}{2} + [2\cosh\frac{\theta_{12}}{2}]^2)} |F^{cos}(\theta, \theta_{12})_{-+1}|^2 ,$$

$$(22.52)$$

where

$$F^{cos}(\theta, \theta_{12})_{-+1} =$$

$$-2\pi \frac{(2d)^2 \sqrt{Z}}{2^{\frac{3}{2}} 3^{\frac{1}{4}}} \frac{(2\cosh\frac{\theta_{12}}{2} + e^\theta)(2\cosh\frac{\theta_{12}}{2} + e^{-\theta})(2\cosh\frac{\theta_{12}}{2} + 2\cosh\theta)}{\sinh 3\frac{\theta_{12}}{2} \sinh 3(\theta + \frac{\theta_{12}}{2}) \sinh 3(\theta - \frac{\theta_{12}}{2})} \zeta(\theta_{12})$$

$$\times \zeta(-i\frac{2\pi}{3}) \zeta(\frac{\theta_{12}}{2} - \theta - i\frac{\pi}{3}) \zeta(\frac{\theta_{12}}{2} - \theta + i\frac{\pi}{3}) \zeta(\frac{\theta_{12}}{2} + \theta - i\frac{\pi}{3}) \zeta(\frac{\theta_{12}}{2} + \theta + i\frac{\pi}{3}).$$

$$(22.53)$$

References

I. Affleck, *Nucl. Phys.* B**265**, 448 (1986).

S. Coleman, *Ann. Phys.* (N.Y.) **101**, 239 (1976).

G. Delfino, P. Simonetti and J. L. Cardy, *Phys. Lett.* B**387**, 327 (1996).

F. H. L. Essler, A. M. Tsvelik and G. Delfino, *Phys. Rev.* B**56**, 11 001, (1997).

F. D. M. Haldane, *Phys. Rev.* B**25**, 4925 (1982).

T. Nakano and H. Fukuyama, *J. Phys. Soc. Jpn.* **50**, 2489 (1981).

H. Schulz, *Phys. Rev. Lett.* **77**, 2790 (1996).

23

Superconductivity in a doped spin liquid

In Chapter 21 we have discussed an example of one-dimensional spin liquid – a system of two coupled spin-1/2 Heisenberg chains. Despite the fact that spins in such a system are disordered, a hidden order exists, the corresponding order parameter being nonlocal with respect to spins. We shall see below that the type of this hidden order determines what will happen under doping (i.e. when one introduces mobile holes into the system). It turns out that only a doped spin ladder of the Haldane type exhibits strongly enhanced superconducting fluctuations. The idea that a doped spin liquid state may favour superconductivity (SC) belongs to Anderson (1987). The first works on doped spin ladders were performed by T. M. Rice *et al.* (1993–4), who suggested that such systems might well be superconducting.[*] These conclusions have been supported by further studies (see Khveshchenko and Rice (1994) and Fujimoto and Kawakami (1995)).

The physics is particularly simple in the strong-coupling limit $J_\perp \gg J_\parallel$ (see Figs. 23.1, 23.2). The holes tend to stay on the same rung of the ladder because in this case they do not break spin singlets gaining the amount of energy of the order of the spin gap $M_s \sim J_\perp - u$, where u is the interchain Coulomb energy. The presence of the latter interaction is very essential: we see that the pairing is possible only if the exchange interaction is greater than u. At energies smaller than M_s one can describe the pairs of holes as bosons with infinite on-site repulsion, that is to represent their creation and annihilation operators by the Pauli matrices $\sigma^\pm$. In this representation the third Pauli matrix σ_z describes the local charge density. At nonzero $J_\parallel$ bosons start to propagate along the ladder with the characteristic hopping matrix element of order of $D \sim J_\parallel^2/J_\perp$.

---

[*] Here and thereafter we define one-dimensional SC as a divergence in the pairing susceptibility at $T = 0$.

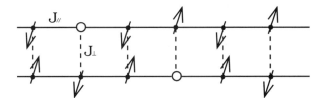

Fig. 23.1.   Holes on the spin ladder.

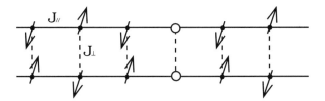

Fig. 23.2.   Pairing of holes on the spin ladder.

Thus the corresponding effective model in the strong coupling limit is the spin-1/2 antiferromagnetic XXZ model (1.1) in the external 'magnetic' field:

$$H = \sum_n [D(\sigma_n^+ \sigma_{n+1}^- + \text{H.c.}) + u\sigma_n^z \sigma_{n+1}^z - \mu\sigma^z] \qquad (23.1)$$

where μ is the chemical potential of the holes and u is the Coulomb interaction between pairs of holes on the nearest rungs. For the case $u < D$, $\mu = 0$ this model has been thoroughly discussed in the previous chapters and we know that its continuous limit describes a Tomonaga–Luttinger liquid with power-law correlations. This result does not change at $\mu \neq 0$, but the case $u > D$ requires a comment. In this case the XXZ model has a gap at $\mu = 0$ and becomes gapless only when μ exceeds some critical value μ_c. However, the region with the spectral gap has $\langle \sigma^z \rangle = 0$ thus corresponding to zero doping. Therefore the charge degrees of freedom of the doped ladder are always described by the Luttinger liquid Hamiltonian. In the Tomonaga–Luttinger liquid state both SC and CDW correlation functions follow power laws, but with different scaling dimensions:

$$\chi_{\text{sc}} = \langle \sigma^+(x, \tau)\sigma^-(0, 0) \rangle \sim (x^2 + v^2\tau^2)^{-1/4K_c}$$
$$\chi_{\text{CDW}} = \langle \sigma^z(x, \tau)\sigma^z(0, 0) \rangle \sim (x^2 + v^2\tau^2)^{-K_c} \qquad (23.2)$$

where K_c is a nonuniversal function of u/D and doping. At $u = 0$ we have the XY model which is equivalent to free fermions (see Chapter 1) and $K_c = 1$. When u/D grows K_c decreases and at $K_c < 1/2$ χ_{CDW} becomes more singular than χ_{sc}.

It is interesting that in the presence of the spin gap the pairing susceptibility has a stronger singularity than in the spin-1/2 Tomonaga–Luttinger
liquid: $d_{sc} = 1/4K_c$ in comparison with $(1/2 + 1/2K_c)$ (see Eq. (15.23)).

All this strong coupling picture essentially survives in the weak coupling
limit. The latter, however, is much more interesting from the mathematical
point of view. In this Chapter we study this limit following the approach
developed by D. G. Shelton and Tsvelik (1996). This approach is a
straightforward generalization of the one discussed in Chapter 21.

We study the model of two Hubbard chains with the exchange and
Coulomb interaction between the chains:

$$H = H_{\text{Hub}} + \sum_r \left(\frac{u}{2} \rho_{1,r} \rho_{2,r} + 2J\mathbf{S}_{1,r}\mathbf{S}_{2,r} \right) \tag{23.3}$$

$$H_{\text{Hub}} = -\frac{D}{2} \sum_{r,a,\alpha} (c^+_{a,\alpha,r+1} c_{a,\alpha,r} + \text{H.c.}) + U \sum_{r,a} c^+_{r,a,\uparrow} c_{r,a,\uparrow} c^+_{r,a,\downarrow} c_{r,a,\downarrow} \tag{23.4}$$

where $c^+_{a,\alpha,r}$, $c_{a,\alpha,r}$ are electron creation and annihilation operators on the
site r, $\alpha = \pm 1/2$ and $a = 1, 2$ are electron spin and chain (orbital) indices
respectively, and

$$\rho_a = \sum_\alpha c^+_{a,\alpha} c_{a,\alpha}; \quad \mathbf{S} = \frac{1}{2} \sum_{\alpha,\beta} c^+_{a,\alpha} \sigma_{\alpha\beta} c_{a,\beta} \tag{23.5}$$

where σ^a are the Pauli matrices.

The model (23.3) does not include a direct hopping between the orbitals.
This is qualitatively justified if the renormalized $t_\perp$ is smaller than the
spectral gaps. As we have mentioned in earlier chapters, in the presence of
strong on-chain interactions the interchain hopping generates effective interchain interactions which may be more relevant than the single-particle
hopping itself. If these interactions form spectral gaps the further renormalization of the single-particle hopping is suppressed and one can adopt
(23.3) as an effective Hamiltonian at the fixed point. In the Appendix we
derive conditions of realization of such scenario.

I   Bosonization and fermionization

As usual, interactions between electrons in one-dimension cannot be
treated perturbatively. Hence we have to resort to nonperturbative methods, the most popular among which is bosonization. Since the Hubbard
model is exactly solvable for any U/D and doping, the bosonization approach requires only smallness of the couplings u and J. We have studied
the behaviour of the model (23.3) at half-filling in Chapter 21, here we
shall assume a fair amount of doping.

We begin by bosonizing the charge and the spin density operators (23.5), taking the corresponding formulas from Chapter 16 (16.3, 16.4). Recall that there are four bosonic fields $\Phi_c^{(a)}$, Φ_s^a ($a = 1,2$) governed by the Hamiltonians (15.17) and the similar Hamiltonian for the spin fields with $K_s = 1$. Recall also that the constant K_c and charge and spin velocities v_c, v_s depend on the interaction and doping such that $K_c = 1$ in the noninteracting case. The fact that the corresponding spin constant remains unrenormalized ($K_s = 1$) reflects the SU(2) symmetry of the individual Hubbard chain (see Chapters 13 and 16).

Substituting expressions (16.3, 16.4) into Eq. (23.3) and keeping only non-oscillatory terms, we get the following expression for the interaction density:

$$V_{int} = V + \frac{u}{2}[J_{R1} + J_{L1}][J_{R2} + J_{L2}] + 2J[\mathbf{J}_{R1} + \mathbf{J}_{L1}][\mathbf{J}_{R2} + \mathbf{J}_{L2}] \quad (23.6)$$

$$V = 2\cos(\sqrt{4\pi}\Phi_c^-)[u|a_c|^2(\cos\sqrt{4\pi}\Phi_s^+ + \cos\sqrt{4\pi}\Phi_s^-) \\ + J|a_s|^2(-\cos\sqrt{4\pi}\Phi_s^+ + \cos\sqrt{4\pi}\Phi_s^- + 2\cos\sqrt{4\pi}\Theta_s^-)] \quad (23.7)$$

Notice that the interaction term is very similar to the one obtained for the spin ladder (21.4); the only difference being the overall factor $\cos(\sqrt{4\pi}\Phi_c^-)$. The $4k_F$-scattering contributes the irrelevant operator $\cos 4\sqrt{\pi}\Phi_c^-$ which we omit. Here we have introduced symmetric and antisymmetric combinations of the bosonic fields: $\Phi_{c,s}^\pm = (\Phi_{c,s}^1 \pm \Phi_{c,s}^2)/\sqrt{2}$. This transformation leaves the bosonic Hamiltonian (15.17) invariant.

The interaction of charge currents can be easily incorporated into the free Hamiltonian (15.17) by the change of K_cs and velocities for the symmetric and antisymmetric charge modes:

$$K_c^\pm = \left[1 \pm (\frac{u}{2\pi v_c})\right]^{-1/2} K_c \quad (23.8)$$

We shall assume that $K_c^- < 1$, i.e. the interaction in the antisymmetric charge channel remains repulsive. Then the scaling dimension of the operator V, $d_V = 1 + K_c^- < 2$ is always smaller than the dimension of the product of spin currents, which is a marginal operator. Therefore we omit the latter term. In the subsequent analysis we repeat the arguments given in Chapter 21 taking advantage of the fact that all bosonic exponents in the square brackets in Eq. (23.7) have the scaling dimension 1 implying that they can be expressed as fermionic bilinears. The corresponding expressions in terms of Majorana (real) fermions are (see also Chapter 21):

$$(\pi a_0)^{-1}\cos\sqrt{4\pi}\Theta_s^- = i(R_1 L_1 - R_0 L_0), \quad (\pi a_0)^{-1}\cos\sqrt{4\pi}\Phi_s^- \\ = i(R_1 L_1 + R_0 L_0), \quad (\pi a_0)^{-1}\cos\sqrt{4\pi}\Phi_s^+ = i(R_2 L_2 + R_3 L_3) \quad (23.9)$$

where the fermion operators satisfy the following anticommutation relations:

$$\{R_a(x), R_b(y)\} = \delta(x-y)\delta_{ab}; \ \{L_a(x), L_b(y)\} = \delta(x-y)\delta_{ab},$$
$$\{R_a(x), L_b(y)\} = 0 \quad (23.10)$$

Being recast in new terms, the Hamiltonian (15.17) becomes

$$H_0 = \frac{1}{2}\sum_{\pm}\int \mathrm{d}x \left[v_c^{\pm}K_c^{\pm}(\partial_x\Theta_c^{\pm})^2 + \frac{v_c^{\pm}}{K_c^{\pm}}(\partial_x\Phi_c^{\pm})^2 \right]$$
$$+ \frac{\mathrm{i}}{2}v_s\sum_{a=0}^{3}\int \mathrm{d}x \,(L_a\partial_x L_a - R_a\partial_x R_a) \quad (23.11)$$

The interaction density acquires the following form:

$$V = \mathrm{i}\cos\sqrt{4\pi}\Phi_c^{-}\left[m_1(1-\delta_{a,0})R_a L_a + m_2 R_0 L_0\right],$$
$$m_1 = 2\pi a_0(u|a_c|^2 + J|a_s|^2), \ m_2 = 2\pi a_0(-3J|a_s|^2 + u|a_c|^2) \quad (23.12)$$

If $K_c^- < 1$ (that is $u/2\pi v_c^- < 1 - K_c^2$, which requires the Coulomb repulsion U to be stronger than u!), the interaction (23.12) is relevant making the Majorana fermions and the field Φ_c^- massive. The modes $a \neq 0$ and $a = 0$ acquire different masses which we call m_t and m_s respectively with 't' and 's' standing for 'triplet' and 'singlet'. The triple degeneracy of the three fermion modes reflects the fact that they realize the spin-1 representation of the SU(2) group. The mass of the bosonic field Φ_c^- is denoted by $\tilde{M}$. The dimensional analysis gives $m_t, m_s, \tilde{M} \sim J^{1/(1-K_c^-)}$. One can safely assume that the following averages do not vanish:

$$\langle\cos\sqrt{4\pi}\Phi_c^-\rangle \neq 0; \ \langle R_a L_a\rangle \neq 0 \quad (23.13)$$

In the limit $K_c^- \ll 1$ this can be rigorously proven by the perturbation expansion in K_c^-. Namely, one should rescale the field Φ_c^-, $\Phi_c^- = \sqrt{K_c^-}\tilde{\Phi}_c^-$, and expand the cosine term in Eq. (23.12): $\cos\sqrt{4\pi K_c^-}\tilde{\Phi}_c^- = 1 - 2\pi K_c^-(\tilde{\Phi}_c^-)^2 + ...$. Then, in the first approximation in K_c^-, one obtains the quadratic effective Hamiltonian (23.14). One can check that corrections coming from higher-order terms are nonsingular and contain powers of the small parameter K_c^-. The appearance of the averages (23.13) breaks the discrete Z_2 symmetry which is not forbidden in $(1 + 1)$-dimensions. The corresponding order parameter is nonlocal in terms of the original fermionic fields (see Chapter 21).

Exercise. Consider a ladder formed by a spin-1/2 antiferromagnetic Heisenberg chain and a repulsive Hubbard chain away from half-filling. In the spin channel the two chains are coupled by exchange interaction $J_\perp$. Show that the strong-coupling massive phase of the system occurs for antiferromagnetic sign of $J_\perp$, while for a ferromagnetic interaction $(J_\perp < 0)$ the chains remain asymptotically decoupled.

II   Superconducting fluctuations

As follows from the previous discussion, the Hamiltonian for the field Φ_c^+ remains unaffected, so that this field is still a free bosonic field. The other fields become massive, and their strong coupling limit can be qualitatively described by the effective Hamiltonian:

$$
H_{\text{eff}} = \frac{i}{2} \int \mathrm{d}x v_s
$$

$$
\times \sum_{a=0}^{3} \{ -R_a \partial_x R_a + L_a \partial_x L_a + 2[m_t(1 - \delta_{a,0}) + m_s \delta_{a,0}] R_a L_a \} \quad (23.14)
$$

$$
+ \frac{1}{2} \int \mathrm{d}x [v_c^-(\partial_x \Theta_c^-)^2 + v_c^-(\partial_x \Phi_c^-)^2 + \tilde{M}^2 (\Phi_c^-)^2]
$$

As we have mentioned above this Hamiltonian becomes exact in the limit $K_c^- \to 0$.

We shall see that the correlation length of the SC order parameter is infinite only provided that in Eq. (23.14) $m_t m_s < 0$. As follows from Eq. (23.12), the masses are

$$
m_t \sim (u|a_c|^2 + J|a_s|^2)\langle \cos \sqrt{2\pi}\Phi_c^- \rangle;
$$
$$
m_s \sim (-3J|a_s|^2 + u|a_c|^2)\langle \cos \sqrt{2\pi}\Phi_c^- \rangle \quad (23.15)
$$

Combining this criterion with $K_c^- < 1$ we obtain the following criterion for the existence of an infinite correlation length for SC fluctuations in the model (23.11, 23.12):

$$
K_c^- < 1; \ (u|a_c|^2/|a_s|^2 + J)(3J - u|a_c|^2/|a_s|^2) > 0 \quad (23.16)
$$

The coefficients a_c, a_s are known only in the weak coupling limit $U/D \ll 1$ where $a_c = a_s$. In this case the physics is very transparent. Namely, the triplet mass vanishes at $u = -J$, i.e. exactly at the point where the inter-chain interaction becomes a projector on the singlet state. Respectively, the singlet mass vanishes at $u = 3J$ when the interaction vanishes on the singlet state. It is unlikely that this picture holds for strong inter-actions. In particular, close to half filling where charge fluctuations are suppressed, we expect that $|a_s| \gg |a_c|$. Thus we can resolve the in-equalities (23.16) explicitly only in the limit of weak interactions, where $K_c \approx 1 - U/4\pi v_c$:

$$
U > u; \ J > u/3 \text{ or } J < -u \quad (23.17)
$$

Instead of this set of inequalities, we suggest a phenomenological criterion for enhancement of SC fluctuations which may be more practical. Namely, according to explanations given in Chapter 21, the inequality $m_t m_s < 0$ means that the dynamical magnetic susceptibility has a sharp single magnon peak. We call a spin liquid supporting such a peak a Haldane spin liquid. Thus the phenomenological criterion is that only Haldane spin liquid supports superconductivity.

Let $\psi_{R,L}$ be right- and left-moving components of the original fermions. Now we shall demonstrate that if the criterion (23.16) is met, the susceptibility of the SC order parameter defined as

$$\Delta = \psi_{R,1,\uparrow}\psi_{L,2,\downarrow} \pm \psi_{R,2,\uparrow}\psi_{L,1,\downarrow} \sim \exp i\sqrt{\pi}(\Phi_s^+ + \Theta_c^+)$$
$$\times\{\exp[i\sqrt{\pi}(\Theta_s^- + \Phi_c^-)] \pm \exp[-i\sqrt{\pi}(\Theta_s^- + \Phi_c^-)]\} \qquad (23.18)$$

is singular. The choice of sign in the above expression depends on the sign of $\langle\cos\sqrt{4\pi}\Phi_c^-\rangle$ (broken Z_2 symmetry). The order parameter includes exponents of the fields Φ_s^+ and Θ_s^- with scaling dimensions 1/4. In the Majorana approach it is most convenient to express these fields in terms of the order (σ) and disorder (μ) parameter fields of the Ising model using the fact that the model of massive Majorana fermions describes a two-dimensional Ising model off the critical point, where the fermionic mass is related to the deviation from T_c: $m \sim (T - T_c)$. The corresponding operators are related as follows (see Chapter 12):

$$\exp(i\sqrt{\pi}\Phi_s^+) \sim \mu_1\mu_2 + i\sigma_1\sigma_2; \quad \exp(-i\sqrt{\pi}\Theta_s^-) \sim \sigma_3\mu_0 - i\sigma_0\mu_3 \quad (23.19)$$

Substituting these expressions into Eq. (23.18) we get

$$\Delta^\pm = \exp(i\sqrt{\pi}\Theta_c^+)\cos\sqrt{\pi}\Phi_c^-(\mu_1\mu_2 + i\sigma_1\sigma_2)\begin{cases}\sigma_3\mu_0\\-i\sigma_0\mu_3\end{cases} \approx \lambda : e^{i\sqrt{\pi}\Theta_c^+} : \quad (23.20)$$

where, depending on the sign of $\langle\cos\sqrt{\pi}\Phi_c^-\rangle$,

$$\lambda = \langle\cos\sqrt{\pi}\Phi_c^-(\sigma_1\sigma_2\sigma_3\mu_0)\rangle \quad \text{or} \quad \langle\cos\sqrt{\pi}\Phi_c^-(\mu_1\mu_2\mu_3\sigma_0)\rangle \qquad (23.21)$$

In the expression (23.20) we have omitted the term proportional to $\sin\sqrt{\pi}\Phi_c^-$, since the sine has a zero average. As is well known, one of the operators σ and μ has a nonzero average off the critical point. Which of the two operators does not vanish depends on the sign of $(T - T_c)$, that is on the sign of mass in the Majorana representation. In order to have $\lambda \neq 0$, the mass term of the zeroth Majorana mode should have a sign opposite to the mass term of the other modes which gives as criterion (23.16).

It is easy to see that if $m_t m_s < 0$ the order parameter describes the charge density wave:

$$O_{CDW}(x) = \sum_\alpha(\psi_{R,1\alpha}^+\psi_{L,1\alpha} \pm \psi_{R,2\alpha}^+\psi_{L,2\alpha}) \qquad (23.22)$$

As follows from Eq. (23.20), at $T = 0$ the correlation function of Cooper pairs has the following asymptotic behaviour at distances $\gg M$ (M is the smallest gap):

$$\langle \Delta(x, \tau)\Delta^+(0,0)\rangle \sim M^2 \frac{1}{(M|\tau + i\frac{x}{v_c}|)^{1/2K_c^+}} \tag{23.23}$$

with K_c^+ defined by Eq. (23.8). As we have mentioned in the beginning of this chapter, the scaling dimension of the order parameter $d = \frac{1}{4K_c^+} > 1/2$ is given by the same expression as in the strong coupling limit. At $K_c^+ > 1/4$ the Fourier transformation of the correlation function (pairing susceptibility) diverges at $\omega, q = 0$. At finite temperatures there will be a finite number of kinks interpolating between vacua with $\langle \cos \sqrt{4\pi}\Phi_c \rangle > 0$ and < 0. Thus the Z_2 symmetry is restored with the finite correlation length $\xi \sim \exp(M_k/T)$, where $M_k \sim M_0/K_c$ is the kink's mass. This exponentially large correlation length will be completely overshadowed by the correlation length of the Φ_c^+-field: $\xi_c \sim 1/T$.

If the masses m_t and m_s have the same sign, the situation becomes unfavourable for superconductivity. The behaviour of spin–spin correlation functions also changes. A possible candidate for such frustrated spin liquid is $Sr_{14}Cu_{24}O$ (Kitaoka *et al.* (1996)).

III   Conclusions

As we see, the described type of SC requires a strong Coulomb repulsion between electrons on the same orbital. It is this repulsion which leads to formation of the spin gap with subsequent freezing out of some electronic degrees of freedom. There is also a charge gap corresponding to antiphase fluctuations of the orbital charge densities.[†] The Coulomb interaction, however, plays a double role since the scaling dimension of the order parameter increases when the Coulomb repulsion grows. The small scaling dimension of the SC order parameter increases the probability of having a relatively large transition temperature in a quasi-three-dimensional system built of weakly interacting chains. Let $t_\perp$ be the interchain Josephson coupling; then the dimensional analysis gives for the temperature of SC transition the following estimate:

$$T_c \sim m_s(t_\perp/m_s)^\eta; \; \eta = \frac{1}{2 - 2d} \tag{23.24}$$

[†] In the situation where the exchange interaction is generated by the Jahn–Teller mechanism, this gap coincides with the spectral gap of vibrons at $q = 2k_F$ (Shelton and Tsvelik (1996)).

where $m_s \sim g^{1+1/(1-K_c^-)}$ is the spin singlet gap. In the conventional case of single chains $d \approx 1$ and the transition temperature is exponentially small in m_s/t, but for the spin ladder it may not be the case.

We conclude this Chapter with a brief description of the physical properties expected for a system described by model (23.3). First, the magnetic excitations are spin triplets and spin singlets, as for the undoped chain. As for a general $S = 1$ magnet, the triplet excitations have a gap (the Haldane gap) which we call m_t. There is a singlet gap m_s as well; the corresponding excitations are spin and charge singlets and can be treated as RVB (resonance valence bond) excitations. Thus doping does not remove spin gaps. Notice, however, that the interchain Coulomb interaction can change the signs of fermionic masses (23.16) and therefore alter the string order parameters changing a topology of the spin liquid state. The remaining gapless mode is the charge free boson which gives a dramatic enchancement to SC fluctuations or, if the criterion $m_t m_s < 0$ is not met, to CDW. In quasi-three-dimensional systems one should expect true SC or CDW orderings.

IV   Appendix. Conditions for suppression of the single-particle tunneling

In this Appendix we derive sufficient conditions for Hamiltonian (23.3) being a fixed point Hamiltonian of the system of two spin-1/2 Luttinger liquid chains coupled by the single-particle interchain tunneling. The calculations presented below are a straightforward generalization of the calculations described in Chapter 20 for the case of fermions with spin.

The corresponding renormalization group (RG) equations for this problem were derived by Khveshchenko and Rice (1994) with a tacit assumption that $u = J$ ($m_s = -m_t$). These equations have to be supplemented by the condition of SU(2) symmetry which, in the original notations corresponds to $g_1 = g_2 = g$, $K_s^- = K_s^+ = 1$. This leads to great simplifications, bringing the number of equations down to three:

$$\frac{dg}{dl} = (1 - K_c^-)g + z(K_c^- - 1/K_c^-) \tag{23.25}$$

$$\frac{d}{dl}(1/K_c^-) = g^2 \tag{23.26}$$

$$\frac{d \ln z}{dl} = 3 - \frac{1}{2}(K_c^- + 1/K_c^-) \tag{23.27}$$

with the initial conditions $z(0) = (t_\perp/D)^2, g(0) = 0, K_c^-(0) = K_c^+$.

Without a loss of generality we may consider the case $K_c^- < 1$. We are interested in the regime where the single-particle tunneling either does not grow at all or grows slower than the dynamically generated interchain backscattering. As we shall see, this happens when $K_c^- < \sqrt{5} - 2 \approx 0.24$. Notice that this requirement is imposed on K_c^- and does not contradict the condition $K_c^+ > 0.25$ necessary for the superconductivity.

Let us consider the case

$$1/K_c^- - K_c^- > 4 \qquad (23.28)$$

It can be shown that at these values of K_c the effective exchange interaction is formed already at small distances. Condition (23.28) corresponds to rather small $K_c^- < 0.24$. Treating these coupling constants as small we shall also neglect renormalization of K_c^- from its initial values $K_c < 1, K_s = 1$.

In this approximation we can solve Eq. (23.27):

$$z_l = z_0 \exp\{[3 - (K_c + 1/K_c)/2]l\} \qquad (23.29)$$

Substituting this into Eq. (23.25) and taking into account the initial condition $g(0) = 0$ we get

$$g_l = -\frac{z_0(1/K_c - K_c)}{1/K_c - K_c - 4}\{1 - \exp[-l(1/K_c - K_c - 4)/2]\}\exp[(1 - K_c)l] \qquad (23.30)$$

Let us suppose that $(K_c^{-1} - K_c - 4)$ is not very small. Then g_l becomes of order of one at

$$l^* \approx -\frac{1}{(1 - K_c)}\ln z_0 \qquad (23.31)$$

Substituting it into Eq. (23.29) we get

$$z(l^*) \approx z_0^{\frac{K_c^{-1} - K_c - 4}{2(1 - K_c)}} \qquad (23.32)$$

Since the exponent is positive and $z_0 \ll 1$, this means that at the point where the effective exchange is already strong and presumably opens a spectral gap, the single-particle tunneling is still weak.

References

P. W. Anderson, *Science* **235**, 1196 (1987).

S. Fujimoto and N. Kawakami, *Phys. Rev.* B**52**, 6189 (1995).

V. Khveshchenko and T. M. Rice, *Phys. Rev.* B**50**, 252 (1994).

Kitaoka *et al.*, preprint (1996).

D. G. Shelton and A. M. Tsvelik, *Phys. Rev.* B**53**, 14 036, (1996).

M. Sigrist, T. M. Rice and F. C. Zhang, preprint ETH-TH/93-35; T. M. Rice *et al.* in *Proc. 16th Tanigushi Symposium,* ed. by A. Okiji and N. Kawakami, Springer-Verlag, (1994).

24

Edge states in the quantum Hall effect

Let us consider an electron gas on a two-dimensional plane in a magnetic field. Everybody knows that magnetic field leads to the Landau quantization and the electronic spectrum becomes discrete:

$$E_{n,\sigma} = \omega_c(n + 1/2) + g_L B\sigma \qquad (24.1)$$

where $\omega_c = eB/mc$ is the cyclotron frequency. However, this result holds only for an infinite system. For finite systems the situation changes and new and interesting physics emerges.

One easy way to model a finite system is to consider a confining parabolic potential $U(x) = m\Omega^2 x^2/2$ which effectively confines the electrons to a strip. Choosing the vector potential as $A_y = Bx$ we get the following Schrödinger equation:

$$\left[-\frac{1}{2m}\partial_x^2 + \frac{1}{2m}(-i\partial_y - eBx/c)^2 + m\Omega^2 x^2/2 \right] \psi(x,y) = E\psi(x,y) \quad (24.2)$$

(we omit the spin part of the Hamiltonian). We seek for the solution in the form

$$\psi(x,y) = e^{iky} f(x)$$

where $f(x)$ satisfies

$$\left[-\frac{1}{2m}\partial_x^2 + \frac{1}{2m}(k - eBx/c)^2 + m\Omega^2 x^2/2 \right] f(x) = Ef(x) \qquad (24.3)$$

Now k cannot be removed by a shift of x as for the infinite system ($\Omega = 0$). The solution has the form

$$E_n(k) = \sqrt{\omega_c^2 + \Omega^2}(n + 1/2) + \frac{k^2\omega^2}{2m(\Omega^2 + \omega_c^2)} \qquad (24.4)$$

$$\psi_{n,k}(x,y) = e^{iky} f_n \left(x - \frac{k\omega_c}{m(\omega_c^2 + \Omega^2)} \right) \qquad (24.5)$$

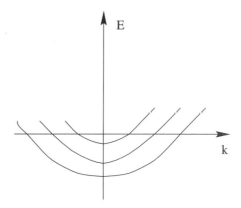

Fig. 24.1. Landau levels for two-dimensional electrons in a confining potential.

where $f_n(x)$ are wave functions for a harmonic oscillator with frequency $\sqrt{\omega_c^2 + \Omega^2}$.

From Eq. (24.4) we see that every Landau level has now became a one-dimensional band (see Fig. 24.1). As follows from Eq. (24.5), states at the opposite Fermi points are centred around

$$x_{\pm} = \pm \frac{k_F \omega_c}{m(\omega_c^2 + \Omega^2)} \qquad (24.6)$$

and therefore are spatially separated.

Spatial separation of states with opposite Fermi velocities is the feature which determines the low energy physics of the problem. Because of this separation the coupling between fermionic modes with different chirality is extremely weak and therefore there is no backscattering. Since all relevant operators include modes with different chirality, there are no relevant perturbations present. Whatever you do with your edges, as long as you keep them far apart, there are always gapless modes in the system.

The pioneer of chiral Tomonaga–Luttinger liquid is Wen (1990–1). A great contribution was made by Kane and Fisher (1992). Recently their predictions were supported experimentally (see Chang *et al.* (1996)).

Bosonization procedure for chiral fermions includes some special features worth discussion.

The action for chiral bosons is (Jackiw and Woo (1975), Moon *et al.* (1993))

$$S = \frac{1}{4\pi v} \int d\tau dx \partial_x \phi(\pm \partial_\tau \phi + \partial_x \phi) \qquad (24.7)$$

where plus and minus signs correspond to different choices of chirality and v is a filling of the Landau level. We assume that the filling is such

Fig. 24.2. Matthew Fisher.

that there is only one edge mode which is not always self-consistent. There are at least two cases, however, when one can be sure there is only mode: $v = 1$ and $v = 1/3$.

In these cases we see that all simple exponents have single-valued correlation functions:

$$O_n(z) = e^{\pm i\sqrt{n}\phi}$$
$$\langle\langle O_n(z_1)O_n^*(z_2)\rangle\rangle = (a/z_{12})^{(n/v)} \tag{24.8}$$

The exponents with $n = \pm 1$ correspond to single-particle creation and annihilation operators. For the case $v = 1$ the edge state is in the universality class of free chiral fermions. For $v = 1/3$ (an example of a fractional Hall effect) it is not the case; the conformal dimension of O_1 is $3/2$, not $1/2$. However, the dual exponents

$$D_n(z) = e^{\pm iv\sqrt{m}\phi}$$

emerge when one considers tunneling processes between edges with different chirality (see, for example, Fendley *et al.* (1995)). These operators create the so-called Laughlin quasi-particles with a fractional electric charge $\pm ve$.

It is very important that for a system consisting of a single edge all operators are chiral. This fact makes it impossible to write down a relevant perturbation. Therefore such state is stable.

References

A. M. Chang, L. N. Pfeiffer and K. W. West, *Phys. Rev. Lett.* **77**, 2538 (1996).

P. Fendley, A. W. W. Ludwig and H. Saleur, *Phys. Rev. Lett.* **74**, 3005 (1995).

R. Jackiw and G. Woo, *Phys. Rev.* **D12**, 1643 (1975).

C. L. Kane and M. P. A. Fisher, *Phys. Rev.* **B46**, 15 233 (1992); *Phys. Rev. Lett.* **68**, 1220 (1992).

K. Moon, H. Yi, C. L. Kane, S. M. Girvin, M. P. A. Fisher, *Phys. Rev. Lett.* **71**, 4381 (1993).

X. G. Wen, *Phys. Rev.* **B43**, 11 025 (1991); *Phys. Rev. Lett.* **64**, 2206 (1990); *Phys. Rev.* **B44**, 5708 (1991).

Part III

Single impurity problems

In the last part of the book we shall discuss applications of the bosonization method to impurity problems. Condensed matter physics is rich in impurity problems of various kinds. Some of them, like the X-ray edge problem, have by now been completely understood. (It should be noted, that it was the X-ray edge problem where the explicit bosonized expression for the electron field operator was used the first time by Schotte and Schotte (1969).) Other problems, like the overscreened multi-channel Kondo problem, are subject to intensive current studies.

It would hardly be possible (and it is not our goal) to give a comprehensive review of this huge and rapidly growing field. We shall limit our discussion to main physical ideas and mathematical methods starting with the classical subject of Fermi edge singularities in metals and then passing to modern problems of impurities in non-Fermi-liquid hosts (Tomonaga–Luttinger liquids) and the multi-channel Kondo effect.

25

Potential scattering

I   Introduction

The simplest impurity problem is the one of electrons scattering from a static center without internal degrees of freedom – the potential scattering problem. It is described by the Hamiltonian:

$$H = H_0 + \hat{V} \tag{25.1}$$

where the first term

$$H_0 = \int \frac{d^3\mathbf{k}}{(2\pi)^3} \xi(\mathbf{k}) c_{\mathbf{k}}^{\dagger} c_{\mathbf{k}} \tag{25.2}$$

is the Hamiltonian of free electrons (we suppress the spin indices) with the spectrum $\xi(\mathbf{k})$ measured from the Fermi energy. The operators $c_{\mathbf{k}}$ and $c_{\mathbf{k}}^{\dagger}$ satisfy anticommutation relations

$$\{c_{\mathbf{k}}, c_{\mathbf{k}'}^{\dagger}\} = (2\pi)^3 \delta(\mathbf{k} - \mathbf{k}'), \quad \{c_{\mathbf{k}}, c_{\mathbf{k}'}\} = 0 \tag{25.3}$$

The second term

$$\hat{V} = \int d\mathbf{r} V(\mathbf{r}) \psi^{\dagger}(\mathbf{r}) \psi(\mathbf{r}) \tag{25.4}$$

corresponds to the electron scattering on the potential $V(\mathbf{r})$, with the electron field operator

$$\psi(\mathbf{r}) = \int \frac{d^3\mathbf{k}}{(2\pi)^3} e^{i\mathbf{k}\mathbf{r}} c_{\mathbf{k}} \tag{25.5}$$

(We have assumed a three-dimensional metal host though our considerations will are also be valid for two-dimensional metals.)

Despite the apparent simplicity of the Hamiltonian (25.1), there are certain subtleties involved in its analysis which are outlined in the following two sections.

292

II Reduction of the local scattering problem to one dimension

The impurity potential $V(\mathbf{r})$ is assumed to be short ranged, $V(\mathbf{r})$ being essentially nonzero in a finite region (of some radius a) around the origin. It is therefore instructive to start the analysis with a point-like scattering potential:

$$V(\mathbf{r}) = V\delta(\mathbf{r}) \tag{25.6}$$

Further simplification results from the assumption that the conduction band spectrum is spherically symmetric:

$$\xi(\mathbf{k}) = \xi(k) \tag{25.7}$$

(the spherical symmetry is, however, not crucial – see below). It is then convenient to use the spherical-waves expansion for the electron field operator instead of the plane-waves representation (25.5):

$$\psi(\mathbf{r}) = \sum_{l,m} \psi_{lm}(\mathbf{r}) = \sum_{l,m} \int_0^\infty \frac{dk}{2\pi} R_{kl}(r) Y_{lm}\left(\frac{\mathbf{r}}{r}\right) c_{klm} \tag{25.8}$$

where Y_{lm} are the conventionally normalized spherical harmonics ($\int d\Omega Y_{lm}^* Y_{l'm'} = \delta_{ll'}\delta_{mm'}$) and the radial functions are given by

$$R_{kl}(r) = 2k j_l(kr) \tag{25.9}$$

$j_l(x) = \sqrt{\pi/2x}J_{l+1/2}(x)$ being the spherical Bessel functions (see e.g. Landau and Lifshits (1982)). In particular, for the s-wave one has

$$R_{k0}(r) = 2\frac{\sin(kr)}{r} \tag{25.10}$$

The partial-waves expansion (25.8) takes the form:

$$\psi(\mathbf{r}) = \int_0^\infty \frac{dk}{2\pi} \frac{\sin(kr)}{\sqrt{\pi}r} c_k + (l \geq 1 \text{ waves}) \tag{25.11}$$

where we have dropped the $l = 0$ index: $c_{k0} \rightarrow c_k$. Since the only radial function nonvanishing at the origin is $R_0(r)$, it is the s-wave only which interacts with the potential (25.6) and the relevant part of the Hamiltonian takes the form:

$$H = \int_0^\infty \frac{dk}{2\pi} \xi(k) c_k^\dagger c_k + V\psi_0^\dagger(0)\psi_0(0) \tag{25.12}$$

This Hamiltonian clearly describes a one-dimensional problem, though formulated in terms of standing waves on a half-line ($r > 0$) instead of conventional right and left movers. One can formally define the right and left movers, restricting consideration by states close to the Fermi momentum ($k \approx k_F$) and isolating slowly varying (in space) parts of the

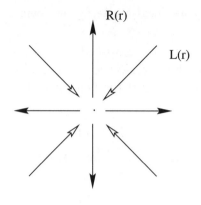

$R(x) = L(-x)$ $R(x) = R(x)$

0

Fig. 25.1. This figure illustrates how the chiral field $R(r = x)$ is constructed out of the incoming $[L(r)]$ and outgoing $[R(r)]$ scattering states.

s-wave field:

$$\psi_0(r) \simeq \frac{1}{2\mathrm{i}\sqrt{\pi r}} \left[\mathrm{e}^{\mathrm{i}k_F r} R(r) - \mathrm{e}^{-\mathrm{i}k_F r} L(r) \right] \tag{25.13}$$

where

$$R(r) = \int_{-\Lambda}^{\Lambda} \frac{\mathrm{d}p}{2\pi} \mathrm{e}^{\mathrm{i}pr} c_{k_F+p} \;; \quad L(r) = \int_{-\Lambda}^{\Lambda} \frac{\mathrm{d}p}{2\pi} \mathrm{e}^{-\mathrm{i}pr} c_{k_F+p} \tag{25.14}$$

Λ being an ultraviolet cut-off. Obviously, $R(r)$ and $L(r)$ should coincide at the origin:

$$R(0) = L(0) \tag{25.15}$$

Thus, we have introduced the right- and the left-moving fields defined on the half-line ($r > 0$) and obeying the boundary condition (25.15). These are equivalent to a single chiral field (right moving, for convention) defined on the whole line ($-\infty < x < \infty$):

$$R(x) = \begin{cases} R(r) \text{ for } & x = r \\ L(r) \text{ for } x = -r \end{cases} \tag{25.16}$$

See Fig. 25.1.

Along with the approximation (25.13, 25.14), it is natural to linearize the spectrum

$$\xi(k) \simeq v_F(k - k_F) \tag{25.17}$$

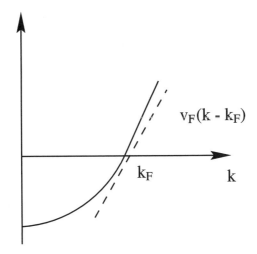

Fig. 25.2. Linearization of the electron spectrum around the Fermi point: $\xi(k) \simeq v_F(k - k_F)$. Note that in this local problem $k > 0$ and there is only one Fermi point.

in the vicinity of the Fermi level, $k_F - \Lambda < k < k_F + \Lambda$, v_F being the Fermi velocity (see Fig. 25.2.).

Then, the Hamiltonian (25.12) can be expressed in terms of the chiral electron field (25.16) as follows:

$$H = v_F \int_{-\infty}^{\infty} dx R^{\dagger}(x)(-i\partial_x)R(x) + V_0 R^{\dagger}(0)R(0) \qquad (25.18)$$

We have defined $V_0 = (k_F^2/\pi)V$, where the coefficient, k_F^2/π, is equal to the ratio of the Fermi-level density of states in the original three-dimensional electron band, $\rho_0 = k_F^2/2\pi^2 v_F$, to the density of states in the effective one-dimensional band of chiral electrons, $\rho_{1D} = 1/2\pi v_F$.

The formula (25.18) completes the reduction of the original problem (25.1) to the one-dimensional one (25.18) for the simplest case of a point-like potential and spherically symmetric conduction band. The equations (25.13) and (25.16) explicitly relate the original electron field operators to the one-dimensional chiral fermions (this allows one to calculate real space correlation functions).

Let us now make a few remarks concerning generalizations of the above results.

- We did not bank our analysis on the fact that the scattering potential is structureless. Imagine that the electron field has an index s ($\psi_s(\mathbf{r})$), which can be changed in the course of the scattering:

$$\hat{V} = \sum_{s,s'} V(s,s') \psi_s^\dagger(0) \psi_{s'}(0)$$

In the simplest case s is the electron spin, $s = \uparrow, \downarrow$. Then $V(s,s') = V\delta_{s,s'} + IS\tau_{s,s'}$ consists of the potential scattering and the exchange term, S being the impurity spin. The reduction to one dimensions proceeds exactly in the same manner as above and it is, therefore, valid for the Kondo problem.

- Once the potential is local, (25.6), it is not important whether the conduction band maintains the spherical symmetry. Indeed, everything happens at the point $\mathbf{r} = \mathbf{0}$, so it is the local dynamics of the conduction electrons which is relevant. Put another way, the $\mathbf{r} \neq \mathbf{0}$ degrees of freedom can be integrated out. The local dynamics of the system is determined by the local conduction electron Green function at $\mathbf{r} = 0$:

$$g_0(t) = G_0(\mathbf{r} = \mathbf{0}, t) = -i\langle T\left\{\psi\ (\mathbf{0}, t)\psi^\dagger(\mathbf{0}, 0)\right\}\rangle \qquad (25.19)$$

This Green's function is the bare one; namely, the average in (25.19) is taken over the ground state of H_0. The important statement that $g_0(t)$ determines the local dynamics of the system implies that if one calculates local electron correlation functions perturbatively in $\hat{V}$, then, due to the Wick theorem and no matter how complex the operator $\hat{V}$ is, one only encounters convolutions of the function $g_0(t)$. In the energy representation one can write

$$g_0(\omega) = \int \frac{d^3k}{(2\pi)^3} G(\mathbf{k}, \omega) = \int \frac{\rho(\epsilon) d\epsilon}{\omega - \epsilon + i\delta \operatorname{sign}\omega} \qquad (25.20)$$

where

$$\rho(\epsilon) = \int \frac{d^3k}{(2\pi)^3} \delta\left[\epsilon - \xi(\mathbf{k})\right] \qquad (25.21)$$

is the energy dependent density of states. It is easy to see from (25.20) that, while the high-energy (short-time) behaviour of the local Green function depends on the details of the band spectrum (i.e., on the shape of the function (25.21)), its low-energy (long-time) form is determined by states close to the Fermi surface. So, for $\omega \ll W$ (W being the bandwidth)

$$g_0(\omega) \simeq A - i\pi\rho_0 \operatorname{sign}\omega \qquad (25.22)$$

ρ_0 being the Fermi level density of states (the constant A reflects the asymmetry of the conduction band with respect to the Fermi level –

the particle–hole asymmetry; it is not important and will be neglected in what follows). Consequently, in the time domain $(t \gg 1/W)$:

$$g_0(t) \simeq -\frac{\rho_0}{t} \tag{25.23}$$

Thus, the problem of the local impurity is essentially zero-dimensional – it develops only in time. The Green function of the form (25.22, 25.23) can as well be reproduced by one-dimensional chiral fermions (25.18). In that sense the reduction to one dimensions still holds for a not spherically symmetric conduction band. Such a reduction may seem artificial at the moment but we shall see in the next chapters that it is very useful indeed, for it allows one to apply powerful techniques from the theory of one-dimensional fermions.

• The Hamiltonian (25.18) describes a single mode of chiral fermions. One can therefore say that the problem involves a single one-dimensional channel. The notion of the number of channels is important – in more complicated cases several channels come into play. Consider instead of the point scatterer a potential $V(\mathbf{r})$ with a d-wave symmetry $(l = 2)$ in a spherically symmetric host. Then, there are five $(m = -2, ..., 2)$ one-dimensional channels,

$$\psi_{2m}(\mathbf{r}) = \int_0^\infty \frac{dk}{2\pi} R_{k2}(r) Y_{2m}\left(\frac{\mathbf{r}}{r}\right) \to R_m(x) = \int_{-\Lambda}^{\Lambda} \frac{dp}{2\pi} e^{ipx} c_{k_F+p,2m} \tag{25.24}$$

which scatter from such a potential. The situation (25.24) is relevant for the case of transition metal impurities, when the conduction electrons are effectively hybridized with the d-shell electrons of the impurity (see Nozières and Blandin (1980) for detail and for the discussion of the hybridization model).

It should be noted that in realistic situations, when the scattering is not exactly local and the host metal is not spherically symmetric (but only has a finite point group symmetry of the crystal lattice), there is always an infinite number of one-dimensional channels involved in the scattering (see Hirst (1978) for a general symmetry analysis). Naturally, one may expect that there is a finite set of channels with predominant couplings and that other channels can, in fact, be neglected. However, the determination of the relevant model very much depends on the concrete impurity structure and requires physical assumptions to be made. This is beyond the scope of this book – we shall only study a few basic models. Here we note that there are several situations in which the electron spin plays the role of the channel index; this happens in the case of two-level systems (Vladár, Zimányi and Zawadowski (1986); Muramatsu and Guinea

(1986)), in the case of the impurity ground state belonging to a nondegenerate nonmagnetic representation of the point group (Cox (1987)), and in some variations of the Coulomb blockade effect. The latter case we shall discuss in Chapter 28. We wish to mention here the review article by Cox and Zawadowski (1997), where the above effects are discussed from a unified perspective.

III The scattering phase

The static potential problem can, of course, be easily solved, and is most elementary in the case of point scatterer (25.6). So, the local electron Green's function in the presence of the potential, $\bar{g}_0$, satisfies the Dyson equation,

$$\bar{g}_0(\omega) = g_0(\omega) + V g_0(\omega) \bar{g}_0(\omega) \qquad (25.25)$$

the solution of which is simply:

$$\bar{g}_0(\omega) = \frac{g_0(\omega)}{1 - V g_0(\omega)} \qquad (25.26)$$

The retarded Green's function $\bar{g}_0^{(r)}(\omega)$ takes the same form as (25.26) with $g_0(\omega)$ substituted by

$$g_0^{(r)}(\omega) = \int \frac{\rho(\epsilon) d\epsilon}{\omega - \epsilon + i\delta} \qquad (25.27)$$

As it is known from the elementary scattering theory (see e.g. the textbook by Landau and Lifshits) the argument of the complex function $[1 - V g_0^{(r)}(\omega)]^{-1}$ determines the energy dependent scattering phase-shift:

$$\delta(\omega) = \mathrm{Arg}[1 - V g_0^{(r)}(\omega)]^{-1} = \tan^{-1} \left[\frac{V \Im m g_0^{(r)}(\omega)}{1 - V \Re e g_0^{(r)}(\omega)} \right] \qquad (25.28)$$

If the density of states $\rho(\epsilon)$ is a nonsingular function of the energy, the scattering phase is also a smooth function, in particular around the Fermi level, where it is equal to

$$\delta = \delta(\omega \to 0) = -\tan^{-1}(\pi \rho_0 V) \qquad (25.29)$$

(For the sake of simplicity, we consider a particle–hole symmetric band, i.e. $\Re e\, g_0^{(r)}(0) = 0$.) The scattering phase (25.28) determines the single-electron wave functions diagonalizing the original Hamiltonian (25.1):

$$\varphi_k(r) = 2 \frac{\sin(kr + \delta)}{r} \qquad (25.30)$$

Notice that the incoming wave, e^{-ikr}, is multiplied by the factor $e^{-i\delta}$, whereas the outcoming one is multiplied by the factor $e^{i\delta}$. Thus, according to the definition (25.16), the single-particle wave functions $\varphi_k(x)$, which diagonalize the equivalent one-dimensional problem (25.18), should have a 2δ phase discontinuity ot the origin:

$$\frac{\varphi_k(x \to 0+)}{\varphi_k(x \to 0-)} = e^{2i\delta} \qquad (25.31)$$

Let us, however, solve the one-dimensional problem separately. Given (25.18), $\varphi_k(x)$ satisfy the Dirac-like equation:

$$v_F k \varphi_k(x) = -iv_F \partial_x \varphi_k(x) + V_0 \delta(x) \varphi_k(x) \qquad (25.32)$$

The solution of (25.32) is obviously given by

$$\varphi_k(x) \sim e^{ikx + 2i\delta_0 \theta(x)} \qquad (25.33)$$

with

$$\delta_0 = -\frac{V_0}{2v_F} = -\pi \rho_0 V \qquad (25.34)$$

Notice that though the scattering phase (25.34) coincides with (25.29) in the first order in V, it differs from (25.29) in higher orders in the potential (which is apparently inconsistent with (25.31)). Essentially, we have come up with two different definitions of the scattering phase-shift: (25.29) and (25.34). Both of them can be found in the literature (to some degree of confusion). In the rest of this paragraph we shall explain the reason for this. The matter is in fact simple. One only has to recall that there are two ultraviolet cut-offs in the problem: the bulk cut-off Λ, which is related to the bandwidth (or to the lattice spacing $\Lambda \sim 1/a_0$), and the local cut-off a – the radius of the scattering potential. Although neither of these cut-offs explicitly appears in the expressions (25.29) and (25.34) for the scattering phase-shifts, the answer turns out to depend on the order of limits.

In order to illustrate this, we consider a one-dimensional scattering problem, keeping both cut-offs finite. It is convenient to represent the Schrödinger equation in the integral form:

$$\varphi_k(x) = e^{ikx} + \int_{-\infty}^{\infty} dy \, G_0^{(r)}(x - y; k) V(y) \varphi_k(y) \qquad (25.35)$$

where $V(x)$ is a potential of the radius a and the retarded Green function is, in the mixed real space–energy ($\omega = v_F k$) representation, defined by (for a finite bandwidth cut-off):

$$G_0^{(r)}(x; k) = \int_{-\Lambda}^{\Lambda} \frac{dp}{2\pi} \frac{e^{ipx}}{v_F(k - p) + i\delta} \qquad (25.36)$$

For an infinite bandwidth cut-off, $\Lambda \to \infty$, the Green function (25.36) is simply

$$G_0^{(r)}(x; k) = \frac{e^{ikx}}{iv_F} \theta(x) \qquad (25.37)$$

proportional to the step function. When Λ is finite, the step is smeared out (on the length-scale of the order of $a_0 \sim 1/\Lambda$)* but the asymptotics (25.37) is maintained:

$$G_0^{(r)}(x;k) \simeq \begin{cases} 0 & \text{for } x \to \infty \\ \dfrac{1}{iv_F}e^{ikx} & \text{for } x \to -\infty \end{cases} \tag{25.38}$$

Given the properties of the functions $G_0^{(r)}(x;k)$ and $V(x)$, one can see that the integral over y in Eq. (25.35) is singular: it is a convolution of two functions, one of which shrinks to a step function and the other one shrinks to a δ-function. The result of the integration thus depends on which limit is taken first. So, in the case of $\Lambda \to \infty$ but a finite, substituting the Green's function (25.37) into Eq. (25.35) and deriving it with respect to x, one finds:

$$v_F k \varphi_k(x) = -iv_F \partial_x \varphi_k(x) + V_0(x)\varphi_k(x)$$

This is equivalent to Eq. (25.32). The scattering phase is therefore given by (25.34). On the other hand, for the case of $a \to 0$ but Λ finite, Eq. (25.35) takes the form:

$$\varphi_k(x) = e^{ikx} + V_0 G_0^{(r)}(x;k)\varphi_k(0)$$

Solving this equation and using (25.38), one finds that the scattering phase is given by (25.29).

To summarize, the scattering phase does in general depend on the cut-offs, $\delta = \delta(\Lambda, a)$, with the following limiting values:

$$\lim_{a \to 0} \lim_{\Lambda \to \infty} \delta(\Lambda, a) = -\pi \rho_0 V \tag{25.39}$$

$$\lim_{\Lambda \to \infty} \lim_{a \to 0} \delta(\Lambda, a) = -\tan^{-1}(\pi \rho_0 V) \tag{25.40}$$

Finally, we notice that the bosonized version of problem (25.18), which will be discussed in Section 26.VI, leads to the scattering phase (25.39). This is natural, for bosonization assumes an infinite bandwidth cut-off. It may therefore be tempting to just substitute the impurity potential by the scattering phase-shift (25.40) in the bosonized version of the model. This

---

* This can be made explicit by replacing the sharp cut-off in (25.36) by a smooth one:

$$\int_{-\Lambda}^{\Lambda} \frac{dp}{2\pi} \to \int_{-\infty}^{\infty} \frac{dp}{2\pi} e^{-a_0|p|} .$$

Then, at the Fermi energy, $k = 0$, the integral over p gives:

$$G_0^{(r)}(x;0) = \frac{1}{iv_F}\left[\frac{1}{2} + \frac{1}{\pi}\tan^{-1}\left(\frac{x}{a_0}\right)\right],$$

which is a particular analytic form of a regularized step-function.

would clearly be correct for the potential scattering problem but may not be so for more complicated cases (see Chapter 28).

References

D. L. Cox, *Phys. Rev. Lett.* **59**, 1240 (1987).

D. L. Cox and A. Zawadowski, Report No. cond-mat/9704103 (1997).

L. L. Hirst, *Adv. Phys.* **27**, 231 (1978).

L. D. Landau and E. M. Lifshits, *Quantum Mechanics*, Pergamon Press, Oxford, (1982).

A. Muramatsu and F. Guinea, *Phys. Rev. Lett.* **57**, 2337 (1986).

P. Nozières and A. Blandin, *J. Phys. (Paris)* **41**, 193 (1980).

K. D. Schotte and U. Schotte, *Phys. Rev.* **182**, 479 (1969).

K. Vladár, G. Zimányi and A. Zawadowski, *Phys. Rev. Lett.* **56**, 286 (1986).

26

X-ray edge problem (Fermi liquids)

I   Introduction

There are important situations in which the impurity potential becomes dynamic, i.e. time dependent. The simplest example of this kind is the so-called X-ray edge problem. This problem is related to the absorption of high-energy electromagnetic radiation, with typical energies ranging from tens to hundreds of electronvolts, by a metallic sample. The oncoming X-ray beam interacts with the sample in different ways, one of the processes being creation of electron–hole pairs: when the electromagnetic wave is absorbed, an electron from a deep level is excited to the conduction band, leaving a hole deep in the valence band. This process is most important for determining the shape of the absorption intensity as a function of energy around the Fermi level.

Since the conduction band is much wider than the hole's band, the hole can be regarded as localized, with an atomic-like wave function; it is therefore often called a core hole. The deep narrow band can be well approximated by a deep level E_0 (see also below).

Ignoring all electron–electron correlations, one expects that the absorption rate is zero below the threshold energy $\omega_{th} = E_F - E_0$ and is proportional to the density of states $\rho(\omega)$ of the conduction electrons above the threshold:

$$I^0_{abs}(\omega) \sim \rho(\omega)\theta(\omega - \omega_{th}) \tag{26.1}$$

where $\theta(x)$ is the step-function (see Fig. 26.1).

Once the hole is created, it can recombine with a conduction electron below the Fermi level, emitting a photon. The intensity of the emission process is therefore expected to be

$$I^0_{em}(\omega) \sim \rho(\omega)\theta(\omega_{th} - \omega) \tag{26.2}$$

302

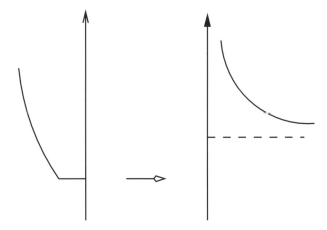

Fig. 26.1. This figure qualitatively shows how the X-ray absorption rate is renormalized by the electron–electron interactions $I_0(\omega) \rightarrow I(\omega)$: it becomes a power-law divergent function at the threshold (which is shifted but remains sharp).

Due to the obvious symmetry between the emission and absorption processes, in what follows only absorption will be considered.

However, the above naive picture fails to be correct because of the highly nontrivial role of the electron correlations drastically changing the behavior of $I(\omega)$ near the threshold. The most important correlations are those between the conduction electrons and the core hole. The electron–electron interactions in the conduction band do not lead to any qualitative changes in the X-ray response, provided that the Fermi-liquid properties of the host metal are maintained (see Section III); the situation, however, is more complicated in Tomonaga–Luttinger liquids (see Chapter 27).

The main point missed in the naive approach is the fact that the local scattering potential experienced by the conduction electrons is not static but *transient*. It suddenly switches on at the moment when the core hole is created. As predicted by Mahan (1967), the transient nature of the scattering potential makes the intensity $I(\omega)$ a singular (power-law) function of the energy near the threshold (see Fig. 26.1):

$$I(\omega) \sim \left(\frac{\omega_c}{\omega - \omega_{th}}\right)^{\alpha} \theta(\omega - \omega_{th}) \tag{26.3}$$

where ω_c is a cut-off frequency (typically of the order of the conduction bandwidth) and α is an exponent which may have either sign. Notice that the threshold remains sharp.

There are two effects influencing the absorption rate near the Fermi edge. Sudden switching on of the core hole potential excites the electron system by creating an infinite number of local low-energy electron–hole pairs in the conduction band. As a result, in the thermodynamic limit, the initial state of the filled Fermi sea with N electrons and no local potential (filled deep level) turns out to be orthogonal to the final state of $N + 1$ electrons in the conduction band experiencing the local potential – the famous Anderson's *orthogonality catastrophe* (Anderson, 1967). The orthogonality catastrophe tends to suppress the absorption rate. There is also multiple scattering of the electrons on the core hole in the final state which accounts for spatial correlations between the electrons and the scattering center. The induced effective attraction between the electrons near the core hole tends to enhance the absorption rate. We shall show in this Chapter that both mechanisms contribute to the critical exponent α which can be expressed in terms of the scattering phase.

The singularity (26.3), also called the Fermi-edge singularity, has been experimentally observed (see Mahan (1981) and Ohtaka and Tanabe (1990) for a recent review, see also references in Chapter 27). A detailed discussion of the experimental results on the Fermi-edge singularity is beyond the scope of this book. We only note here that there are obstacles to its direct observation, mainly due to the fact that the core-hole level E_0 always has a 'natural width' Γ, which effectively smears out the singularity:

$$\frac{\omega_c}{\omega - \omega_{th}} \rightarrow \frac{\omega_c}{\sqrt{(\omega - \omega_{th})^2 + \Gamma^2}} \qquad (26.4)$$

A finite Γ is primarily provided by intra-atomic effects, like the Auger effect; the bandwidth of the deep band also contributes to Γ, though it is usually negligible (but not if one measures $I(\omega)$ in semiconductors in the optical frequency range). The possibility to observe the Fermi-edge singularity is thus very much the question of whether Γ is smaller than ω_c for a given material. We shall now abandon all these questions (and neglect Γ), referring the reader to references at the end of this chapter and Chapter 27.

The bulk of this chapter is devoted to the discussion of physics behind the formula (26.3), and explicit calculation of the exponents α and that of the core-hole propagator. As simple as it looks, the X-ray edge problem turns out to be very rich. First, it is closely related to the Anderson orthogonality catastrophe, which is in fact a local counterpart of infrared catastrophe in one-dimensional systems of interacting fermions where it is responsible for the disappearance of single-fermion excitations from the spectrum. On the other hand, being a local scattering problem and thus allowing to be reformulated as a one-dimensional one (see Chapter 25), the X-ray edge problem naturally suggests straightforward application of bosonization methods, revealing connections with such modern concepts in the theory of one-dimensional fermions as boundary condition changing operators. Additionally, the appearance of transient impurity potentials is not a privilege of the X-ray edge problem, which is just the first in a row. The understanding of the X-ray response has been a milestone in the condensed matter theory. This understanding led to a progress in the theory of the Kondo effect (Yuval and Anderson (1970)), two-level systems and diffusion in solids (for a review on these topics see the collection of articles *Quantum Tunneling in Condensed Matter* (1992)).

II   Statement of the problem

Let d be the operator annihilating the deep electron (alternatively d creates the deep hole). Then, the coupling of the metal to the X-ray field, which describes a transfer of an electron from the deep level (d) to the conduction band ($c_{\mathbf{k}}$) while an X-ray quantum is absorbed, can be written in the form:

$$H_X = \int \frac{d^3k}{(2\pi)^3} W_{\mathbf{k}} c_{\mathbf{k}}^\dagger d e^{-i\omega t} + \text{H.c.} \tag{26.5}$$

where ω is the X-ray frequency, and the matrix element $W_{\mathbf{k}}$ will be assumed **k**-independent in what follows. The second term in (26.5) actually describes the X-ray emission process.

The part of the Hamiltonian which is responsible for the dynamics of the metal is

$$H = \int \frac{d^3k}{(2\pi)^3} \xi(\mathbf{k}) c_{\mathbf{k}}^\dagger c_{\mathbf{k}} + E_0 d^\dagger d + \int d\mathbf{r} V(\mathbf{r}) \psi^\dagger(\mathbf{r}) \psi(\mathbf{r}) dd^\dagger \tag{26.6}$$

Although this Hamiltonian neglects the Auger type processes and the mutual electron interaction in the conduction band, one has good reasons to expect that, at least for Fermi liquids, it captures the main physics of the X-ray edge problem. In what follows, we shall discuss approximations leading to (26.6).

Notice that the number of the core electrons (holes) is conserved:

$$\left[dd^\dagger, H \right] = 0 \tag{26.7}$$

This is another way of saying that the natural life time of the core hole is infinite ($\Gamma = 0$).

From the mathematical point of view, this means that the ground state of H is not unique. If $dd^\dagger = 0$ (i.e. there is no core hole), then H reduces to:

$$H = H_0 = H_i \tag{26.8}$$

while for $dd^\dagger = 1$ (i.e. there is one core hole)

$$H = H_0 + \hat{V} = H_f \tag{26.9}$$

Having in mind the absorption process, we have assigned the subscript 'i' for 'initial' to the Hamiltonian (26.8) and the subscript 'f' for 'final' to the Hamiltonian (26.9). We shall denote the ground states of (26.8) and (26.9) as $|0\rangle$ and $|V\rangle$, respectively:

$$H_i|0\rangle = E_i^0|0\rangle \quad \text{and} \quad H_f|V\rangle = E_f^0|V\rangle \tag{26.10}$$

with E_i^0 and E_f^0 being the ground state energies. Their difference $\Delta E = E_f^0 - E_i^0$ is the energy shift of the core hole level caused by the interaction with the conduction electrons; we shall calculate it in Section V. There are two sectors in the Hilbert space of the problem: $\{|i\rangle\}$ which includes $|0\rangle$, and $\{|f\rangle\}$ which includes $|V\rangle$. It is the coupling (26.5) of the system to the X-ray field which mixes these sectors. As we shall extensively discuss in the following paragraphs, the difficulty and the beauty of the X-ray edge problem is rooted in the fact that these sectors are orthogonal: $\langle i|f\rangle = 0$ for any $|i\rangle$ and $|f\rangle$.

The quantity of interest for the X-ray edge experiments – the X-ray absorption rate $I(\omega)$ – can be standardly written in the form of the Fermi golden rule

$$I(\omega) = 2\pi|W|^2 \sum_f |\langle f|\psi^\dagger(0)|i\rangle|^2 \delta(\omega + E_i^0 - E_0 - E_f) \tag{26.11}$$

(the Born approximation in H_X). In what follows, we shall measure the X-ray frequency from the threshold and set $E_0 = 0$.

At this point we notice that there are two equivalent approaches to the X-ray edge problem. In the first approach one treats the core hole operator d as a dynamical field and develops a standard many-body perturbation theory for the Hamiltonian (26.6). While in the second approach one reduces the problem to a single-particle one by fully making use of the conservation of the number of the core holes (26.7). In the rest of this Section we develop along these two lines.

II.1 Many-body formulation

Consider the casual two-particle Green's function:

$$S(t) = \langle\langle \tilde{d}^\dagger(t)\tilde{\psi}(0, t)\tilde{\psi}^\dagger(0, 0)\tilde{d}(0)\rangle\rangle \tag{26.12}$$

where $\tilde{d}(t)$ and $\tilde{\psi}(0, t)$ are the Heisenberg operators (i.e. $\tilde{d}(t) = e^{iHt}de^{-iHt}$, etc.) and the average is taken over the ground state $|0\rangle$ of the full Hamiltonian H (we chose $|0\rangle$, not $|V\rangle$, for we consider the absorption process). The Lehmann spectral representation for this function is

$$S(\omega) = \int_{-\infty}^{\infty} dt e^{i\omega t} S(t) = -\sum_f \frac{|\langle f|\psi^\dagger(0)|i\rangle|^2}{\omega + E_i - E_0 - E_f + i\delta} \tag{26.13}$$

Notice that, because $d^\dagger|0\rangle = 0$, the function (26.12) is in fact a retarded one: $S(t) \sim \theta(t)$. Its Fourier transform (26.13) is therefore a function of ω, analytic in the upper half-plane.

By comparing (26.13) with (26.11) one finds that the X-ray absorption rate is proportional to the imaginary part of the Green's function (26.13):

$$I(\omega) = 2|W|^2 \, \Im mS(\omega) . \tag{26.14}$$

Let us start the analysis of the absorption rate with the simplest case $V = 0$. The zero-order Green's function (26.12) factorizes,

$$S_0(t) = -ig_0(t)D_0(t) \tag{26.15}$$

where $g_0(t)$ is the (bare) local electron propagator (25.19) we are familiar with from the previous Chapter, while

$$D(t) = -\langle\langle \tilde{d}^\dagger(t)\tilde{d}(0)\rangle\rangle \tag{26.16}$$

is the core hole propagator. The bare hole propagator is simply

$$D_0(t) = -i\theta(t) \tag{26.17}$$

so that its Fourier transform

$$D_0(\omega) = \frac{1}{\omega + i\delta} \tag{26.18}$$

Due to the asymptotic behaviour (25.23) of $g_0(t)$, the time integral determining $S_0(\omega)$ is logarithmically divergent:

$$S_0(\omega) \simeq \rho_0 \int_0^\infty \frac{dt}{t} e^{i\omega t} \simeq \rho_0 \ln\left(\frac{\omega_c}{|\omega|}\right) \tag{26.19}$$

where ω_c is the ultraviolet energy cut-off. The above estimation is valid within the logarithmic accuracy and therefore misses the finite imaginary part of $S_0(\omega)$ which we are interested in. The latter, however, can be easily recovered, using analytic propeties of $S(\omega)$. Thus

$$\Im mS_0(\omega) = \pi\rho(\omega)\theta(\omega) \simeq \pi\rho_0\theta(\omega) \tag{26.20}$$

The zero-order absorption rate is therefore a step function,

$$I_0(\omega) = 2\pi\rho_0|W|^2\theta(\omega) \tag{26.21}$$

as anticipated in the previous Section (formula (26.1)).

The logarithmic divergence (26.19) may seem irrelevant for it does not enter the zero-order absorption rate. Yet it is important because, as we shall see shortly, it dominates the perturbative expansion in the core-hole potential V. Indeed, $S_0(\omega)$ is given by the zero-order diagram in Fig. 26.2.

Fig. 26.2.   The lowest-order graph for the X-ray response function $S(\omega)$ (the 'Peierls bubble'): the conduction electron propagator is shown by the solid line, while the dashed line corresponds to the deep-electron propagator.

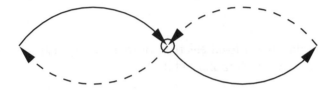

Fig. 26.3.   The first-order graph to $S(\omega)$. The circle corresponds to the bare interaction vertex V [the last term in the many-body Hamiltonian (26.6)].

The only relevant first-order diagram is shown in Fig. 26.3.[*]
This diagram is simply a product of two zero-order diagrams of Fig. 26.2, so its contribution to $S(\omega)$ is:

$$\rho_0 g \ln^2 \left(\frac{\omega_c}{|\omega|} \right)$$

(We have assumed that the core-hole potential in (26.6) is local: $V(\mathbf{r}) \to V\delta(\mathbf{r})$.) The quantity

$$g = \rho_0 V \qquad\qquad (26.22)$$

is a small dimensionless coupling constant in the many-body formulation of the X-ray edge problem. The (first-order) perturbative expansion of $S(\omega)$ reveals a typical logarithmic situation:

$$S(\omega) = \rho_0 \left[\ln \left(\frac{\omega_c}{|\omega|} \right) + g \ln^2 \left(\frac{\omega_c}{|\omega|} \right) + ... \right] \qquad (26.23)$$

Mahan (1967) has conjectured that this expansion extrapolates to

$$S(\omega) = \frac{\rho_0}{2g} \left[\left(\frac{\omega_c}{|\omega|} \right)^{2g} - 1 \right] \qquad (26.24)$$

---

[*] There are two more tadpole diagrams in the first order, both being irrelevant. The self-energy correction to the local electron propagator vanishes because it involves a core-hole loop not contributing due to the conservation of $dd^\dagger$, while the self-energy correction to the core-hole propagator leads only to the shift of the core-hole level – see next Section.

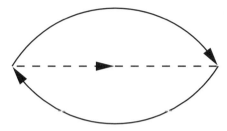

Fig. 26.4. The self-energy diagram for the core-hole propagator (26.16).

Amusingly, this guess turned out to be correct, though the actual proof of (26.24) in the framework of the many-body formulation requires a complex summation of the so-called parquet graphs carried out by Roulet, Gavoret, and Nozières (1969). We are leaving the discussion of the parquet method to the Appendix A, while the expression (26.24) shall again appear in Section IV, where we utilize the one-particle formulation of the X-ray edge problem.

Notice that (26.24) is in fact the real part of the function $S(\omega)$. Although the imaginary part of $S(\omega)$ is out of the scope of the logarithmic approximation used in the many-body perturbation theory, it can easily be restored by applying a Kramers–Kronig type relation for the function $S(\omega)$. We leave as an exercise for the reader to check that

$$\Im m\,(-\omega - i\delta)^{2g} = \sin(2\pi g)\omega^{2g}\theta(\omega) \qquad (26.25)$$

Hence the absorption rate ($g \ll 1$):

$$I(\omega) = 2\pi\rho_0|W|^2 \left(\frac{\omega_c}{\omega}\right)^{2g} \theta(\omega) \qquad (26.26)$$

The expression implies that the X-ray edge exponent α (see (26.3)) is given by

$$\alpha = 2g \qquad (26.27)$$

This is of course a perturbative result for the exponent α, valid at the first order in g. In what follows, we shall show that the power-law behaviour near the threshold is a general feature of $I(\omega)$; the critical exponent, however, is expressed in terms of the phase shift, including contributions of the vertex part and the self-energy.

Logarithmic singularities in the core-hole self-energy are precursors of orthogonality catastrophe. Consider the lowest- (second-)order self-energy diagram (Fig. 26.4).

This diagram can be most easily estimated in the real-time representation:

$$\Sigma^{(2)}(t) = V^2 g_0(t) g_0(-t) D_0(t) \simeq -\mathrm{i}g^2 \frac{\theta(t)}{t^2}$$

(the second equation is asymptotically valid for large t). Fourier transforming this expression we immediately find that

$$\frac{\partial \Sigma^{(2)}(\omega)}{\partial \omega} \simeq g^2 \int_0^\infty \frac{\mathrm{d}t}{t} \mathrm{e}^{\mathrm{i}\omega t} \simeq g^2 \ln\left(\frac{\omega_{\mathrm{c}}}{\omega}\right)$$

and therefore

$$\Sigma^{(2)}(\omega) \simeq g^2 \omega \ln\left(\frac{\omega_{\mathrm{c}}}{\omega}\right) \tag{26.28}$$

It is worth noticing that the self-energy corrections to the core-hole propagator become important at much smaller energies ($\omega \ll \omega_{\mathrm{c}} \exp(-1/g^2)$) than those which matter for the vertex corrections to the absorbtion rate ($\omega \ll \omega_{\mathrm{c}} \exp(-1/g)$, see (26.23)). Though at small energies, the logarithmic singularity in $\Sigma(\omega)$ manifests an apparent disappearance of the Fermi pole in the core-hole Green's function. For the quasi-particle residue,

$$Z^{-1}(\omega) = 1 + \frac{\partial \Sigma(\omega)}{\partial \omega}$$

the second-order result (26.28) corresponds to the expansion

$$Z^{-1}(\omega) = 1 + g^2 \ln\left(\frac{\omega_{\mathrm{c}}}{\omega}\right) + \dots$$

It was shown by Nozières, Gavoret and Roulet (1969) that this expansion converges to

$$Z^{-1}(\omega) = \left(\frac{\omega_{\mathrm{c}}}{\omega}\right)^{g^2} \tag{26.29}$$

This result was originally derived by means of an elegant (though involved) self-consistent parquet calculation, supplemented by Ward's identity. It can also be obtained in a simpler way applying the linked-cluster theorem, as discussed in the next Section. In real time, the result (26.29) means that the core-hole propagator vanishes at large time:

$$D(t) \sim \left(\frac{1}{\omega_{\mathrm{c}} t}\right)^{\alpha_{\mathrm{orth}}} \tag{26.30}$$

with $\alpha_{\mathrm{orth}} = g^2$. Such a behaviour is a manifestation of the orthogonality catastrophe, as we discuss in more detail in Section III.

II.2   One-particle formulation

Because of the conservation of the core-hole number, the field d is not truly dynamic. A one-particle formulation of the problem should therefore be possible. In such a formulation the conduction electrons ought to be subjected to a time-dependent core-hole potential.

Indeed, consider the core-hole Green's function (26.16):

$$D(t) = -\langle\langle\tilde{d}^\dagger(t')\tilde{d}(t)\rangle\rangle = -i\theta(t'-t)e^{iE_i^0(t'-t)}\langle 0|e^{-iH_f(t'-t)}|0\rangle \qquad (26.31)$$

where the second equality is obtained by substituting the explicit expression for $\tilde{d}(t)$ and using the projector properties of the operator $dd^\dagger$ (conservation of the number of the holes). This can be further rewritten as

$$D(t) = -i\theta(t'-t)\langle 0|\hat{S}(t',t)|0\rangle \qquad (26.32)$$

where the evolution operator $\hat{S}(t',t)$ is defined by

$$\hat{S}(t',t) = T\exp\left\{-i\int_t^{t'} d\tau V\psi^\dagger(\mathbf{0},\tau)\psi(\mathbf{0},\tau)\right\} \qquad (26.33)$$

with $\psi(\mathbf{0},\tau) = e^{iH_i t}\psi(\mathbf{0})e^{-iH_i t}$ being the usual Dirac operator. In the above we have utilized the standard operator identity

$$e^{-iH_f(t'-t)} = e^{-iH_i(t'-t)}\hat{S}(t',t)$$

from the elementary $\hat{S}$-matrix theory (see e.g. Abrikosov, Gor'kov and Dzyaloshinskii (1963)). The vacuum average (26.32) of the evolution operator can be understood as an average of the $\hat{S}$-matrix,

$$D(t) = -i\theta(t'-t)\langle 0|\hat{S}|0\rangle \qquad (26.34)$$

of a one-particle problem for which the core-hole potential depends on time in the following way:

$$V \to V(\tau) = V\theta(\tau-t)\theta(t'-\tau) \qquad (26.35)$$

The many-body time-independent Hamiltonian (26.6) should therefore be reduced to the time-dependent one-particle Hamiltonian as follows:

$$H \to H(\tau) = H_0 + V\theta(\tau-t)\theta(t'-\tau)\psi^\dagger(\mathbf{0})\psi(\mathbf{0}) \qquad (26.36)$$

The $\hat{S}$-matrix for this Hamiltonian is simply equal to (26.33). The time t corresponds to the moment when the core hole has been created while t' is the time of the observation when – formally – the core hole is annihilated (the problem is, of course, homogeneous in time, so that we could have set $t = 0$; it is however convenient to keep t arbitrary in order to make the following calculations more symmetric).

Along with the core-hole propagator, the function (26.12), relevant for the X-ray absorption rate, can easily be expressed in terms of the one-particle quantities:

$$S(t'-t) = -i\theta(t'-t)F(t'-t)$$

where the function F, conventionally introduced in the literature, is given by

$$F(t' - t) = -i\langle |T\psi(\mathbf{0}, t')\psi^\dagger(\mathbf{0}, t)\hat{S}|0\rangle \qquad (26.37)$$

At this point, it is instructive to discuss the important notion, introduced by Schotte and Schotte (1969) – a unitary transformation relating the sectors $\{|i\rangle\}$ and $\{|f\rangle\}$ in the Hilbert space of the problem.

Conjecture. There exists a unitary operator U, $U^\dagger U = 1$, such that

$$U^\dagger H_f U = H_i \qquad (26.38)$$

In other words, the conjecture is that there is a unitary transformation which provides exact mapping of all the states of H_f onto those of H_i, thus completely eliminating the scattering potential V from the problem.

Generally speaking, it is not at all obvious that such a transformation does exist. We shall come back to this very important point later in connection with the effective one-dimensional formulation of the problem, quantum anomaly, and the so-called boundary condition changing operator. Here we shall only assume that the required operator U is found, and look at the transformed structure of the X-ray response functions.

Using the operator U in the definitions of F and D, one immediately finds that

$$F(t) = -i\langle 0|T\left\{A(t)A^\dagger(0)\right\}|0\rangle \quad \text{with} \quad A = \psi(\mathbf{0})U \qquad (26.39)$$

Similarly

$$D(t) = -i\langle 0|U(t)U^\dagger(0)|0\rangle \qquad (26.40)$$

for $t > 0$.

III   Linked clusters expansion

In this Section we shall continue the analysis of the X-ray response functions and establish connections of the many-body approach with the one-particle formulation. We shall also introduce the notion of the transient electron Green function, which plays the central role in the exact solutions (Sections IV and V).

Let us focus on the core-hole Green function. The expression (26.32), being expanded in powers of V, becomes a sum of both connected and disconnected diagrams. The well known linked cluster theorem (see e.g. paragraph 15 of Abrikosov, Gor'kov and Dzyaloshinskii (1963)) states that the resulting expansion can be re-summed to an exponential form,

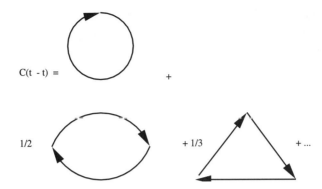

Fig. 26.5. Linked cluster expansion for the core-hole propagator. The vertices correspond to the time-dependent potential scattering operators: $\hat{V}(\tau) = \hat{V}\theta(\tau - t)\theta(t' - \tau)$.

with the exponential containing connected graphs only $(t' > t)$:

$$D(t' - t) = -\mathrm{i}e^{C(t'-t)} \tag{26.41}$$

where

$$C(t' - t) = \langle 0|T \exp\left\{-\mathrm{i}\int_t^{t'} d\tau \hat{V}(\tau)\right\}|0\rangle_{\text{connected}} \tag{26.42}$$

is the sum of connected diagrams, which are closed loops: Fig. 26.5.

The analytic expression, corresponding to Fig. 26.5, is:

$$C(t'-t) = -\sum_{n=1}^{\infty} \frac{V^n}{n} \int_t^{t'} d\tau_1 \int_t^{t'} d\tau_2 \dots \int_t^{t'} d\tau_n g_0(\tau_1-\tau_2)g_0(\tau_2-\tau_3) \dots g_0(\tau_n-\tau_1) \tag{26.43}$$

where $g_0(t)$ is the local electron propagator (25.19). Notice that there is a nonuniversal coefficient, $1/n$, in front of the n-th order graph (this is a general property of linked cluster expansions, which reflects different numbers of topologically equivalent diagrams for closed loops and open-line graphs).

Let us start the discussion by considering a few low-order terms of the expansion (26.43).

- The first-order term in the scattering potential simply is

$$C^{(1)}(t' - t) = -V \int_t^{t'} d\tau g_0(0) = -\mathrm{i}V n_0(t' - t) \tag{26.44}$$

where $n_0 = \langle 0|\psi^\dagger(0)\psi(0)|0\rangle$ is the density of the conduction electrons (in fact, $g_0(0)$ should be understood as $g_0(0-)$; see also below). Note that, in this approximation, C is purely imaginary. The physical meaning of the first-order term (26.44) is transparent: it corresponds to the first-order contribution of the scattering potential to the energy shift of the core-hole level:

$$\Delta E^{(1)} = V n_0 \qquad (26.45)$$

Put another way, $\Delta E = E_f - E_0$ is the difference of total energies of the conduction sea in the presence– and in the absence– of the scattering potential V.

- The second-order contribution to $C(t' - t)$ is given by:

$$C^{(2)}(t' - t) = -\frac{V^2}{2} \int_t^{t'} d\tau_1 \int_t^{t'} d\tau_2 g_0(\tau_1 - \tau_2) g_0(\tau_2 - \tau_1) \qquad (26.46)$$

In the energy representation, (26.46) takes the form:

$$C^{(2)}(t' - t) = -iV^2 \int_{-\infty}^{\infty} \frac{d\Omega}{2\pi} \frac{1 - \cos\left[\Omega(t' - t)\right]}{\Omega^2} \chi(\Omega) \qquad (26.47)$$

where we have introduced the function:

$$\chi(\Omega) = -i \int_{-\infty}^{\infty} \frac{d\omega}{2\pi} g_0(\omega) g_0(\omega + \Omega) \qquad (26.48)$$

The function χ is central to the X-ray edge problem (and, in some sense, to become clear later, it is central to all the impurity problems). Thus, we shall discuss properties of the function χ in some detail. Physically, χ is a local charge susceptibility (polarization). Indeed, (26.48) is nothing but the connected part of the local density–density correlation function. In real time:

$$\chi(t) = -\langle\langle n(0, t)n(0, 0)\rangle\rangle = -i g_0(t) g_0(-t) \qquad (26.49)$$

In order to calculate χ, we substitute the spectral representation (25.20) for the local electron propagator $g_0(\omega)$ into (26.48) and take the ω-integral. That gives:

$$\chi(\Omega) = \int_{-\infty}^{\infty} \rho(\epsilon) d\epsilon \int_{-\infty}^{\infty} \rho(\epsilon') d\epsilon' \left[\frac{\theta(-\epsilon)\theta(\epsilon')}{\Omega + \epsilon - \epsilon' + i\delta} - \frac{\theta(\epsilon)\theta(-\epsilon')}{\Omega + \epsilon - \epsilon' - i\delta} \right] \qquad (26.50)$$

It is instructive to rewrite this expression for χ as a spectral representation:

$$\chi(\Omega) = \int_0^{\infty} \rho_{eh}(E) dE \left[\frac{1}{\Omega - E + i\delta} - \frac{1}{\Omega + E - i\delta} \right] \qquad (26.51)$$

(For simplicity, we have assumed a particle–hole symmetric band, $\rho(\epsilon) = \rho(-\epsilon)$, but this is not really restrictive.) The spectral function in (26.51) is defined by:

$$\rho_{eh}(E) = \int_0^{\infty} \rho(\epsilon) d\epsilon \int_{-\infty}^0 \rho(\epsilon') d\epsilon' \delta(E - \epsilon + \epsilon') \qquad (26.52)$$

One can easily recognize the local density of states of electron–hole excitations in the function $\rho_{\text{eh}}(E)$, especially if one writes it in the form:

$$\rho_{\text{eh}}(E) = \int\limits_{\xi(\mathbf{k})>0} \frac{d^3\mathbf{k}}{(2\pi)^3} \int\limits_{\xi(\mathbf{k}')<0} \frac{d^3\mathbf{k}'}{(2\pi)^3} \delta\left[E - \xi(\mathbf{k}) + \xi(\mathbf{k}')\right] \qquad (26.53)$$

At low energies, $E \to 0$, one immediately finds that the electron–hole pairs' density of states is a linear function of E:

$$\rho_{\text{eh}}(E) = \int_0^E \rho(\epsilon)\rho(E - \epsilon)d\epsilon \simeq \rho_0^2 E \qquad (26.54)$$

On the other hand, at high energies, $\rho_{\text{eh}}(E)$ eventually vanishes, because of the finite bandwidth effects. A widely used model form for the density of states is (smooth cut-off):

$$\rho_{\text{eh}}(E) = \rho_0^2 E e^{-E/\omega_c} \qquad (26.55)$$

As one can see from the formula (26.51), the analytic properties of the function $\chi(\Omega)$ are identical to those of a local Bose-field propagator. Observe that the expression (26.51), together with (26.54), coincides with the local, i.e. integrated over q, charge susceptibility (2.33) for noninteracting fermions in one space dimension. The bosons identified as the electron–hole excitations are in fact the Tomonaga–Luttinger bosons. Thus we have arrived at the local version of the chiral anomaly. This is the reason which makes the bosonization method natural for the impurity problems. In the bosonized version of the X-ray edge problem (Section VI), the correlation function χ explicitly becomes a bosonic one:

$$\chi(t) = \frac{1}{\pi}\langle\langle\partial_x\phi(0, t)\partial_x\phi(0, 0)\rangle\rangle \qquad (26.56)$$

where $\phi(x, t)$ is a chiral Bose field in one dimension.

The real part of $\chi(\Omega)$ is responsible for the imaginary part of $C^{(2)}$ thus contributing the second-order correction $\Delta E^{(2)}$ to the core-hole level energy shift (which is cut-off dependent, just like $\Delta E^{(1)}$ is, (26.45)). It is the imaginary part of $\chi(\Omega)$ which is of interest, for it leads to a nonvanishing real part of $C^{(2)}$, absent at the first-order treatment. As follows from (26.51),

$$\mathfrak{Im}\chi(\Omega) = -\pi\rho_{\text{eh}}\left(|\Omega|\right) \qquad (26.57)$$

(The low-energy behaviour of $\chi(\Omega \ll \omega_c) \simeq -i\pi\rho_0^2|\Omega|$ corresponds to the real time asymptotics of the form: $\chi(t) = -i\rho_0^2/t^2$, which is, of course, consistent with (26.49) and (25.20).) Substituting (26.57) into (26.47), one finds:

$$\mathfrak{Re}C^{(2)}(t' - t) = -g^2 \ln\left[\omega_c(t' - t)\right] \qquad (26.58)$$

The core-hole propagator thus becomes a power-law decaying function of time,

$$D(t' - t) \sim \left[\frac{1}{\omega_c(t' - t)} \right]^{g^2} \tag{26.59}$$

with the exponent equal, in the V^2-approximation, to:

$$\alpha_{\text{orth}}^{(2)} = g^2 = \left(\frac{\delta_0}{\pi} \right)^2 \tag{26.60}$$

where δ_0 is defined by (25.39). This result is, of course, in agreement with the result (26.29) of Nozièrs, Gavoret, and Roulet (1969) who have obtained it in the framework of the many-body formulation of the X-ray edge problem. The physics of the decay (26.59) of the core-hole propagator is the creation of low-energy electron–hole pairs by the time-dependent potential; the rate of the decay being controlled by the density of states of the electron–hole excitations and by the potential strength.

The remarkable feature of the X-ray edge problem in Fermi liquids is that the core-hole propagator is a power-law function of time up to all orders of the perturbation theory,

$$\sum_n \Re C^{(n)}(t' - t) \simeq \alpha_{\text{orth}}(g) \ln \left[\omega_c(t' - t) \right] \tag{26.61}$$

the only thing happening at higher orders is that the exponent gets expressed in terms of the exact scattering phase (25.40):

$$\alpha_{\text{orth}}^{(2)} \rightarrow \alpha_{\text{orth}} = \left(\frac{\delta}{\pi} \right)^2, \quad \delta = -\tan^{-1}(\pi g) \tag{26.62}$$

One can, indeed, derive a closed equation for the sum of all loop diagrams. Let us isolate one vertex of each graph shown in Fig. 26.5. The remaining parts of these graphs are then diagrams for an electron propagator in the time-dependent potential $V(\tau)$. The sum of those diagrams would be the fully renormalized electron propagator if not the nonuniversal $1/n$ coefficient. The latter can easily be taken care of by noticing that $V^n/n = \int_0^1 d\lambda (\lambda V)^n$. Multiplying the result by the vertex $V(\tau)$, which we have isolated at the beginning of the argument, and integrating over τ one finds:

$$C(t' - t) = -V \int_t^{t'} d\tau \int_0^1 d\lambda g_\lambda(\tau, \tau + 0 | t, t') \tag{26.63}$$

where the so-called transient electron Green function g_λ is defined by:[†]

$$g_\lambda(\tau, \tau' | t, t') = -i \frac{\langle 0| T \left\{ \psi(\mathbf{0}, \tau) \psi^\dagger(\mathbf{0}, \tau') \exp\left[-i\lambda \int d\tau'' \hat{V}(\tau'')\right] \right\} |0\rangle}{\langle 0| T \exp\left[-i\lambda \int d\tau'' \hat{V}(\tau'')\right] |0\rangle} \qquad (26.64)$$

The response function $F(t' - t)$, defined by (26.37) which determines the X-ray absorption intensity $I(\omega)$ can also be expressed in terms of g_λ. Indeed, expanding (26.37) in powers of the potential V, one finds connected open-line diagrams, identical to those for the transient electron propagator, and graphs disconnected from the open line, which coincide with those for e^C. The result is the product of both:

$$F(t' - t) = g_{\lambda=1}(t', t | t, t') e^{C(t'-t)} \qquad (26.65)$$

The transient propagator (26.64) satisfies the Dyson equation:

$$g_\lambda(\tau, \tau' | t, t') = g_0(\tau - \tau') + \lambda V \int_t^{t'} d\tau'' g_0(\tau - \tau'') g_\lambda(\tau'', \tau' | t, t'), \qquad (26.66)$$

which in the context of the X-ray edge problem, is also called the Nozières–De Dominicis equation.

The formulas (26.63), (26.65) and (26.66) reduce the X-ray edge problem to solving an integral equation: (26.66). This solution, which in particular proves (26.61) and (26.62), is discussed in the following two Sections.

Before completing this section we remark that Eq. (26.66) does in fact imply a definite ultraviolet regularization of the problem. Namely, the point scatterer limit $(a \to 0)$ is supposed to be taken first, like in (25.40). For the opposite limit, when one starts with an infinite bandwidth $(\Lambda \to \infty)$ all closed loops with more than two external lines vanish (Dzyaloshinskii and Larkin (1974)), so that the second-order results (26.58–26.60) are exact. This situation corresponds to bosonizing the problem and it will be discussed in Section VI.

IV Nozières–De Dominicis solution

The equation (26.66) has been derived and, in the case when the kernel, $g_0(t)$, takes the form (25.23), exactly solved in the pioneering paper by Nozières and De Dominicis (1969). In this Section we shall discuss their solution in detail.

[†] The limit $\tau' \to \tau + 0$ in (26.63) follows from comparison with the perturbation theory. An algebraic derivation of (26.63, 26.64) can be accomplished by applying a standard trick of defining $C_\lambda = \langle T \exp\left[-i\lambda \int d\tau \hat{V}(\tau)\right]\rangle$, taking the λ-derivative, and solving the resulting equation for C_λ.

Let us first re-write the equation (26.66) as:

$$g(\tau) = g_0(\tau - \tau') - \frac{\tan \delta_\lambda}{\pi} \int_t^{t'} \frac{g(\tau'')d\tau''}{\tau'' - \tau}$$

(26.67)

where we have defined the λ-dependent phase-shift:

$$\delta_\lambda = -\tan^{-1}(\pi \lambda g)$$

(26.68)

In order to save writing we only indicate the active argument τ of the transient Green's function $g_\lambda(\tau, \tau'|t, t') \to g(\tau)$; other arguments will be restored at the end of the calculation. Employing the Hilbert transformation

$$f(\tau) = \frac{1}{\pi} \int_{-\infty}^\infty \frac{g(\tau')d\tau'}{\tau' - \tau}$$

(26.69)

where the principal value of the integral is understood, we can represent the equation (26.67) in another, slightly more convenient, form:

$$f(\tau) = f_0(\tau - \tau') + \tan \delta_\lambda S(\tau)g(\tau)$$

(26.70)

Here the function $f_0(\tau)$ is the Hilbert transform of the function $g_0(\tau)$, the function

$$S(\tau) = \theta(\tau - t) - \theta(t' - \tau)$$

(26.71)

is the combination of two step functions, and we have used the inversion formula for the Hilbert transformation:

$$g(\tau) = -\frac{1}{\pi} \int_{-\infty}^\infty \frac{f(\tau')d\tau'}{\tau' - \tau}$$

(26.72)

Nozières and De Dominicis observed that the equation (26.67) belongs to the class of Wiener–Hopf equations, or singular integral equations, and that it is, therefore, exactly solvable. In what follows we shall illustrate the main ideas of the Wiener–Hopf method on the concrete example of solving (26.67). A brief summary of the Wiener–Hopf method is given in the Appendix B, for more detailed and rigorous discussion of the method, we refer the reader to the mathematical literature (see e.g. Muskhelishvili (1953)).

Defining the functions

$$F_\pm(z) = \pm \frac{1}{2\pi i} \int_{-\infty}^\infty \frac{f(\tau)d\tau}{\tau - z \mp i\delta}$$

analytic in the upper (lower) half-plane of the complex variable z, one can see that on the real axis

$$f(\tau) = F_+(\tau) + F_-(\tau)$$

(26.73)

and

$$g(\tau) = -i [F_+(\tau) - F_-(\tau)]$$

(26.74)

The equation (26.70), which is equivalent to (26.67), can therefore be written in the form:

$$e^{2i\delta_\lambda S(\tau)}F_+(\tau) + F_-(\tau) = e^{i\delta_\lambda S(\tau)}\cos\left[\delta_\lambda S(\tau)\right]f_0(\tau - \tau') \tag{26.75}$$

The task is to reduce (26.75) to the Hilbert problem (see Appendix B). In order to accomplish this, we observe that the following factorization

$$e^{-2i\delta_\lambda S(\iota)} = X_+(\tau)X_-(\tau) \tag{26.76}$$

where $X_+(\tau)$ [$X_-(\tau)$] is the boundary value of the function $X_+(z)$ [$X_-(z)$] analytic in the upper (lower) half-plane, is possible, for (26.76) is nothing but the Hilbert problem for the functions $\ln X_\pm(z)$.[‡] The solution to (26.76) reads:

$$X_\pm(\tau) = \exp\left[-i\delta_\lambda S(\tau) \mp \delta_\lambda A(\tau)\right] \tag{26.77}$$

where the function $A(\tau)$ is the Hilbert transform of the function $S(\tau)$. Using (26.77) and (26.76), one can reduce the equation (26.75) to the desired form:

$$\frac{F_+(\tau)}{X_+(\tau)} + X_-(\tau)F_-(\tau) = R(\tau) \tag{26.78}$$

where the right-hand side:

$$R(\tau) = X_-(\tau)e^{i\delta_\lambda S(\tau)}\cos\left[\delta_\lambda S(\tau)\right]f_0(\tau - \tau') = e^{\delta_\lambda A(\tau)}\cos\left[\delta_\lambda S(\tau)\right]f_0(\tau - \tau').$$
$$\tag{26.79}$$

It is the uniqueness of the solution to the Hilbert problem, which with necessity requires:

$$F_+(\tau) = X_+(\tau)R_+(\tau) \quad\text{and}\quad F_-(\tau) = \frac{F_+(\tau)}{X_+(\tau)} \tag{26.80}$$

Given (26.74), this essentially solves the problem. It remains to represent the answer for the Green's function $g(\tau)$ in a transparent form. Noticing that

$$f_0(\tau) = \frac{1}{\pi}\int_{-\infty}^\infty \frac{g_0(\tau')d\tau'}{\tau' - \tau} = \frac{1}{\pi}\int_{-\infty}^\infty \frac{d\tau'}{\tau' - \tau}\left[-\frac{\rho_0}{\tau}\right] = -\pi\rho_0\delta(\tau)$$

one easily finds:

$$R_\pm(\tau) = \pm\frac{\rho_0}{2\pi}e^{\delta_\lambda A(\tau')}\frac{\cos\left[\delta_\lambda S(\tau')\right]}{\tau - \tau' \pm i\delta} \tag{26.81}$$

Plugging (26.81) and (26.77) into (26.80), one obtains:

$$F_\pm(\tau) = \pm\frac{\rho_0}{2\pi}e^{\mp i\delta_\lambda S(\tau) + \delta_\lambda[A(\tau') - A(\tau)]}\frac{\cos\left[\delta_\lambda S(\tau')\right]}{\tau - \tau' \pm i\delta} \tag{26.82}$$

[‡] If the argument of the left-hand side of (26.76) would change more than by 2π while τ moves along the real axis, additional complications arise. This is, however, not the case for the present problem since $\delta_\lambda < \pi/2$.

This formula is valid for any τ. Since we are really interested in the case $t \leq \tau \leq t'$, let us restrict τ (and τ') to this interval. Then, using (26.74, 26.82) and the explicit expression for

$$A(\tau) = \frac{1}{\pi} \int_{-\infty}^{\infty} \frac{S(\tau')d\tau'}{\tau' - \tau} = \frac{1}{\pi} \ln\left(\frac{t' - \tau}{\tau - t}\right)$$

one arrives at the Nozières–De Dominicis result for the electron Green's function in the time-dependent potential $[V(\tau) = \lambda V S(\tau)]$:

$$g_\lambda(\tau, \tau'|t, t') = \bar{g}_0(\tau - \tau') \left[\frac{(\tau - t)(t' - \tau')}{(t' - \tau)(\tau' - t)}\right]^{\delta_\lambda/\pi} \tag{26.83}$$

where

$$\bar{g}_0(\tau) = \frac{\pi\rho_0}{2} \sin(2\delta_\lambda)\delta(\tau) - \frac{\rho_0 \cos^2 \delta_\lambda}{\tau}$$

is the Green function in the static potential λV (c.f. (25.26)).

The result (26.83) contains the main physics of the X-ray edge singularity. The Green's function does tend to the static one for large time intervals $t' - t$, but it has a transient factor (square brackets in (26.83)). The response functions follow.

So, in order to calculate the core-hole Green's function,

$$\ln D(t' - t) = -V \int_0^1 d\lambda \int_t^{t'} d\tau g_\lambda(\tau, \tau + |t, t') \tag{26.84}$$

we need to carefully take the limit $\tau' \to \tau$ in (26.83):

$$g_\lambda(\tau, \tau|t, t') = \bar{g}_0(0) + g_{\text{tr}}(\tau) \tag{26.85}$$

where

$$g_{\text{tr}}(\tau) = -\frac{\rho_0 \delta_\lambda}{\pi} \cos^2 \delta_\lambda \left(\frac{1}{\tau - t} + \frac{1}{t' - \tau}\right) \tag{26.86}$$

The first term in (26.85) is τ-independent, it is responsible for the energy shift, ΔE, of the core-hole level[§] and it is the second term we are interested in. Substituting (26.86) into (26.63) and using

$$\int_0^1 d\lambda\delta_\lambda \cos^2 \delta_\lambda = \frac{1}{\pi g} \int_0^1 d\lambda\delta_\lambda \frac{\partial\delta_\lambda}{\partial\lambda} = \frac{\delta_\lambda^2}{2\pi g}$$

[§] This energy shift is nonuniversal (cut-off dependent) and we can not calculate it in our infinite bandwidth approximation. An exact expression for ΔE in the case of local potential and finite bandwidth is given in the next Section: (26.106). It is clear from this expression that ΔE acquires contributions from all the states below the Fermi energy and not just from those in the vicinity of the Fermi level.

we finally obtain:

$$\ln D(t' - t) = -i\Delta E(t' - t) - \left(\frac{\delta}{\pi}\right)^2 \ln(t' - t) \qquad (26.87)$$

Notice that the orthogonality catastrophe term in (26.87) is universally expressed in terms of the scattering phase shift on the Fermi surface (this phase shift, though, is a function of the impurity potential).

Given the result (26.87) and plugging (26.83) into the expression (26.65) for the X-ray response function $F(t' - t)$, one obtains:

$$\ln F(t' - t) = -i\Delta E(t' - t) + \left[\frac{2\delta}{\pi} - \left(\frac{\delta}{\pi}\right)^2\right] \ln(t' - t) \qquad (26.88)$$

Here the first term is responsible for a shift of the absorption threshold, whereas the second term describes the X-ray edge singularity (26.3). The exponent for the absorption probability is thus found to be:

$$\alpha = \frac{2\delta}{\pi} - \left(\frac{\delta}{\pi}\right)^2 \qquad (26.89)$$

As expected, the exact exponent (26.89), being expanded up to the first order in g, recovers the parquet result (26.27). By Fourier transforming the exponent functions (26.88) and (26.87) one finds the results (26.24) and (26.29) with the exact exponents. The absorption rate turns out to have the same form (26.26), but this time with the exact exponent. Notice that so far we have neglected the electron spin. The effect of the spin is very simple: the closed loops contribution is doubled (for it should be summed over the spin projections), while the open line contribution is unchanged (since its spin projection is fixed). The X-ray edge exponents for the electrons with spin therefore are:

$$\alpha_{\text{orth}} \rightarrow \frac{2\delta^2}{\pi^2}, \quad \alpha \rightarrow \frac{2\delta}{\pi} - \frac{2\delta^2}{\pi^2}$$

Bearing this in mind, we shall continue to neglect the electron spin in the following calculations.

V Exact solution for the overlap integral

The explicit solution to the X-ray edge problem discussed in the previous paragraph, is only valid when the the local electron propagator $g_0(\tau)$ is replaced by its asymptotic form (25.23):

$$g_0(\tau) = -\frac{\rho_0}{\tau}$$

Hence a natural question: should deviations of $g_0(\tau)$ from (25.23), which are sure to occur at short times (where $g_0(\tau)$ is in fact a non-singular function), affect the results? The following argument makes the answer 'No' plausible. The argument goes in reverse. The result $(\delta/\pi)^2$ for the orthogonality exponent, obtained in the framework of the approximation (25.23), only depends on the properties of the Fermi surface. The exponent is therefore determined by large time intervals. This justifies the use of the long-time asymptotics of the electron propagator.

In this Section we shall provide a more rigorous basis to this qualitative argument by analysing the Nozières–De Dominicis equation,

$$g(\tau) = g_0(\tau - \tau') + \lambda V \int_t^{t'} d\tau'' g_0(\tau - \tau'')g(\tau'')$$

with an arbitrary kernel $g_0(\tau)$ for large time intervals $t' - t$. These considerations will be of particular use for the cases when the kernel $g_0(\tau)$ does not assume the form (25.23) at large times (see Chapter 27).

One can get more insight into the problem by examining the result (26.86) for the transient part, $g_{tr}(\tau)$, of the electron Green function $g(\tau)|_{\tau'=\tau}$. The formula (26.86) consists of two terms. Each of them corresponds to a transient decay originating from one of the two steps of the core-hole potential, at times t and t'. Each of these terms, being integrated over τ, gives half a contribution to the orthogonality exponent

$$\frac{1}{2}\alpha_{\text{orth}} \ln(t' - t)$$

For the particular form (25.23) of the kernel $g_0(\tau)$, when the result (26.86) is exact, the transients are independent: the one originated from the potential step at the time moment t does not depend on the distance to another step at t', and vice versa. This is not true for $g_0(\tau)$ different from (25.23). We can, however, make these contributions asymptotically independent by considering steps separated by arbitrary large time intervals $t' - t \to \infty$. So, by setting $t = 0$ and $t' \to \infty$, we expect to lose the contribution of the second step (which takes place in the remote future) and only retain the contribution of the first one:

$$\frac{1}{2}\alpha_{\text{orth}} \int_0^\infty \frac{d\tau}{\tau} \tag{26.90}$$

There is another way to look at the problem. For $t = 0$ and $t' \to \infty$, the expression (26.84) defines the overlap integral $\langle 0|V \rangle$ (we stress that t' is meant to be an actual ∞ here). The latter vanishes for an infinite system (the integral (26.90) is divergent), as it should according to the Anderson orthogonality catastrophe. In a finite system, however, there is a minimal frequency ω_{min} inversely proportional to the linear size L of the system: $\omega_{\text{min}} = \pi v_F/L$ (see Appendix C). The integral (26.90) should therefore be cut at large times $\tau \sim 1/\omega_{\text{min}}$. The short-time divergence of (26.90) is fictitious: it is automatically cut by a short-time scale of the exact $g_0(\tau)$.

One concludes that the overlap integral exponent,

$$|\langle 0|V\rangle| \sim (k_F L)^{\alpha_{OI}} \tag{26.91}$$

is half the orthogonality exponent:

$$\alpha_{OI} = \frac{1}{2}\alpha_{orth}$$

The nice thing about the overlap integral is that it can be computed exactly, without making assumptions about the form of the propagator $g_0(\tau)$.¶ Indeed, the equation (26.66) for $t = 0$ and $t' \to \infty$,

$$g(\tau) = g_0(\tau - \tau') + \lambda V \int_0^\infty d\tau'' g_0(\tau - \tau'')g(\tau'') \tag{26.92}$$

being Fourier transformed, becomes a Wiener–Hopf type equation

$$g(\omega) = g_0(\omega)e^{i\omega\tau'} + \lambda V g_0(\omega)G_+(\omega) \tag{26.93}$$

The functions $G_+(\omega)$ and $G_-(\omega)$, analytic in the upper and lower plane, respectively, are defined as follows:

$$G_\pm(\omega) = \pm\frac{1}{2\pi i}\int_{-\infty}^\infty \frac{g(\omega')d\omega'}{\omega' - \omega \mp i\delta}$$

Note that $g_0(\omega)$ is now simply a multiplier and not a kernel. In order to solve (26.93), we need to reduce it to the Hilbert problem. The procedure of the previous Section fully applies. Namely, using the decomposition

$$g(\omega) = G_+(\omega) + G_-(\omega) \tag{26.94}$$

we rewrite (26.93) as

$$[1 - \lambda V g_0(\omega)] G_+(\omega) + G_-(\omega) = g_0(\omega)e^{i\omega\tau'} \tag{26.95}$$

Then, employing the factorization

$$\frac{1}{[1 - \lambda V g_0(\omega)]} = X_+(\omega)X_-(\omega) \tag{26.96}$$

with

$$X_\pm(\omega) = \exp\left\{\mp\int_{-\infty}^\infty \frac{d\omega'}{2\pi i}\frac{\ln\left[1 - \lambda V g_0(\omega')\right]}{\omega' - \omega \mp i\delta}\right\} \tag{26.97}$$

we arrive at the Hilbert problem:

$$\frac{G_+(\omega)}{X_+(\omega)} + X_-(\omega)G_-(\omega) = X_-(\omega)g_0(\omega)e^{i\omega\tau'} \tag{26.98}$$

¶ For alternative calculations of the overlap integral see Section VI and Appendix C.

Since we only need $g(\tau)$ for $\tau > 0$, where the Fourier transform of the function $G_-(\omega)$ vanishes, we conclude that the solution to (26.92) is given by:

$$g(\tau, \tau') = \int_{-\infty}^{\infty} \frac{d\omega}{2\pi} e^{-i\omega\tau} G_+(\omega)$$

$$= \int_{-\infty}^{\infty} \frac{d\omega}{2\pi} e^{-i\omega\tau} X_+(\omega) \int_{-\infty}^{\infty} \frac{d\omega'}{2\pi i} \frac{X_-(\omega')g_0(\omega')e^{i\omega'\tau'}}{\omega' - \omega - i\delta} \qquad (26.99)$$

The function $g(\tau, \tau')$ [as well as the limit $g(\tau, \tau)$] given by (26.99) is well defined. However, in order to compute the overlap integral, we have got to evaluate

$$\int_0^{\infty} d\tau g(\tau, \tau)$$

which leads to an ill-defined expression:

$$\frac{1}{(\omega' - \omega + i\delta)(\omega' - \omega - i\delta)} .$$

This technical difficulty has caused considerable confusion at the time (see Hamann (1971) for historical references). The difficulty has been overcome by Hamann (1971), who considered a step potential,

$$V(\tau) = V\theta(\tau)e^{-\eta\tau}$$

adiabatically decaying at large times (on the time scale $\sim 1/\eta$). He then kept η finite throughout the calculation, taking the limit $\eta \to 0$ at the very end. Although his result for the overlap integral is perfectly correct, the procedure of distinguishing infinitesimals (η has to be viewed as an infinitesimal, otherwise (26.93) is not a Wiener–Hopf type equation and can not be solved) is somewhat discouraging. The problem can be traced back to the fact that there are two divergent terms in the overlap integral:

$$\ln\langle 0|V\rangle = -i\Delta E \int_0^{\infty} d\tau + \alpha_{OI} \int_0^{\infty} \frac{d\tau}{\tau} \qquad (26.100)$$

We are looking for the next-to-leading divergency. Thus, a separation of the first term should yield a well defined expression for the (logarithmic) second term. Such a separation can be easily accomplished by noticing that the first term in (26.100) is related to the energy shift of the core-hole level, which is a static property determined by the Green function $\bar{g}_0$ in the static potential, whereas the second term in (26.100) is determined by the transient part of the Green function. Representing (26.98) in the form

$$G_+(\omega) = g_0(\omega)e^{i\omega\tau'} - X_+(\omega)X_-(\omega)G_-(\omega) \qquad (26.101)$$

(we have used (26.96): $\bar{g}_0 = X_+X_-g_0$), we find that $g(\tau, \tau)$ is given by

(26.85),

$$g(\tau, \tau) = \bar{g}_0(0) + g_{tr}(\tau)$$

with the transient part,

$$g_{tr}(\tau) = -\int_{-\infty}^{\infty} \frac{d\omega}{2\pi} e^{-i\omega\tau} X_+(\omega) X_-(\omega) G_-(\omega) \qquad (26.102)$$

determined by the Fourier transform of the second term on the right-hand side of (26.101).

The first term in (26.85), which corresponds to the energy shift, can easily be evaluated:

$$\Delta E = -iV \int_0^1 d\lambda \bar{g}_0(0) = -iV \int_0^1 d\lambda \int_{-\infty}^{\infty} \frac{d\omega}{2\pi} \bar{g}_0(\omega) \qquad (26.103)$$

Using the spectral representation,

$$\bar{g}_0(\omega) = -\int_{-\infty}^{\infty} \frac{d\omega'}{\pi} \frac{\Im m \bar{g}_0(\omega') \operatorname{sign}\omega'}{\omega - \omega' + i\delta \operatorname{sign}\omega} \qquad (26.104)$$

analogous to (25.20), and observing that, since $\tau' = \tau + 0$, the ω-integral in (26.103) can be closed in the upper half-plane, one finds:

$$\Delta E = V \int_0^1 d\lambda \int_{-\infty}^0 d\omega \Im m \bar{g}_0(\omega)$$

This can further be simplified by noticing that

$$\Im m \bar{g}_0(\omega) = -\Im m \bar{g}_0^{(r)}(\omega)$$

for $\omega < 0$, and that, according to the definition (25.28) of the scattering phase-shift,

$$\delta_\lambda(\omega) = -\Im m \left\{ \ln \left[1 - \lambda V \bar{g}_0^{(r)}(\omega) \right] \right\}$$

the following useful relation holds:

$$\frac{\partial \delta_\lambda(\omega)}{\partial \lambda} = V \Im m \bar{g}_0^{(r)}(\omega) \qquad (26.105)$$

Evaluating the λ-integral, we arrive at the final result for the energy shift of the core-hole level:

$$\Delta E = -\int_{-\infty}^0 \frac{d\omega}{\pi} \delta(\omega) \qquad (26.106)$$

Having isolated the energy shift contribution to the overlap integral, we can return to the main subject of this Section – calculation of the transient contribution. Substituting the $G_-(\omega)$ solution of the Hilbert problem (26.98) into (26.102) and integrating over time, we obtain:

$$\ln |\langle 0|V \rangle| = -V \int_0^1 d\lambda \int_{-\infty}^{\infty} \frac{d\omega}{2\pi i} \int_{-\infty}^{\infty} \frac{d\omega'}{2\pi i} \frac{X_+(\omega) X_-(\omega') g_0(\omega')}{(\omega' - \omega + i\delta)^2} \qquad (26.107)$$

The frequency integrations in (26.107) are now well defined (that means apart from the logarithmic divergency at low ω, which we are looking for). There is, however, still some work to be done, in order to represent the overlap integral in a transparent form. This involves certain manipulations with the formula (26.107), which are carried out in the rest of this Section.

At the present stage, it is convenient to work exclusively with the functions $X_\pm$ which have simple analytic properties. To this end, we express g_0 in terms of $X_\pm$ using (26.96):

$$g_0(\omega) = \frac{1}{\lambda V}\left[1 - \frac{1}{X_+(\omega)X_-(\omega)}\right]$$

Substituting this into (26.107), one finds two terms:

$$-\int_0^1 \frac{d\lambda}{\lambda} \int_{-\infty}^\infty \frac{d\omega}{2\pi i} \int_{-\infty}^\infty \frac{d\omega'}{2\pi i} \frac{X_+(\omega)X_-(\omega')}{(\omega' - \omega + i\delta)^2}$$
$$+\int_0^1 \frac{d\lambda}{\lambda} \int_{-\infty}^\infty \frac{d\omega}{2\pi i} \int_{-\infty}^\infty \frac{d\omega'}{2\pi i} \frac{X_+(\omega)}{X_+(\omega'))(\omega' - \omega + i\delta)^2}$$

The second term is zero (since the contour in the ω'-integral can be enclosed in the upper half-plane), whereas the first term gives:

$$\ln |\langle 0|V\rangle| = -\int_0^1 \frac{d\lambda}{\lambda} \int_{-\infty}^\infty \frac{d\omega}{2\pi i} X_+(\omega)\frac{\partial X_-(\omega)}{\partial \omega} \qquad (26.108)$$

This expression is convenient for some purposes and we shall make use of it in Chapter 27. However, in order to compare the results for the overlap integral and those of Nozières and De Dominicis for the electron Green's function, we need to express $|\langle 0|V\rangle|$ in terms of the scattering phase-shift. Using the explicit formulas (26.97) for $X_\pm$, we can rewrite the right-hand side of (26.108) as:

$$\int_0^1 \frac{d\lambda}{\lambda} \int_{-\infty}^\infty \frac{d\omega}{2\pi i} \frac{1}{1 - \lambda V g_0(\omega)} \int_{-\infty}^\infty \frac{d\omega'}{2\pi i} \frac{\ln\left[1 - \lambda V g_0(\omega')\right]}{(\omega' - \omega + i\delta)^2} \qquad (26.109)$$

Since

$$\frac{1}{1 - \lambda V g_0(\omega)} = 1 + \lambda V \bar{g}_0(\omega)$$

and the ω-integral vanishes for the constant term, (26.109) takes the form:

$$V \int_0^1 d\lambda \int_{-\infty}^\infty \frac{d\omega}{2\pi i} \bar{g}_0(\omega) \int_{-\infty}^\infty \frac{d\omega'}{2\pi i} \frac{\ln\left[1 - \lambda V g_0(\omega')\right]}{(\omega' - \omega + i\delta)^2} \qquad (26.110)$$

Next we observe that

$$\ln\left[1 - \lambda V g_0(\omega)\right] = 2i\theta(-\omega)\delta_\lambda(\omega) + \ln\left[1 - \lambda V g_0^{(r)}(\omega)\right]$$

and that the ω'-integral over the second term is zero. This simplifies the ω'-integration. Additionally, the ω-integral in (26.110) can be completed in the lower half-plane thus picking up the $\omega > 0$ part of the $\bar{g}_0(\omega)$ spectrum only: $\Im m g_0(\omega) = \Im m g_0^{(r)}(\omega)$ for $\omega > 0$. Using (26.105), we arrive at the final result for the overlap integral:

$$\ln|\langle 0|V\rangle| = -\frac{1}{\pi^2} \int_{-\infty}^{0} d\omega \int_{0}^{\infty} d\omega' \frac{\delta_\lambda(\omega)}{(\omega - \omega')^2} \frac{\partial \delta_\lambda(\omega')}{\partial \lambda} \qquad (26.111)$$

This result has been found by Hamann (1971). It is exact, involving only the phase shift $\delta_\lambda(\omega)$ at a given energy, and therefore valid for arbitrary local electron propagator $g_0(\tau)$.

In Fermi liquids, $\delta_\lambda(\omega)$ is a regular function around $\omega = 0$. Thus, in order to find the divergent part of (26.111), it is sufficient to substitute $\delta_\lambda(\omega) \to \delta_\lambda(0) = \delta_\lambda$ into (26.111). One can easily see that the corrections, $\delta_\lambda(\omega) = \delta_\lambda(0) + \omega \delta'_\lambda(0) + ...$, yield a convergent contribution which determines the pre-exponential factor of the overlap integral. Thus, carrying out the λ-integral, one obtains:

$$\ln|\langle 0|V\rangle| = -\frac{\delta^2}{2\pi^2} \int_{\omega_{min}}^{\Lambda} \int_{\omega_{min}}^{\Lambda} \frac{d\omega d\omega'}{(\omega + \omega')^2} + \text{const}$$

$$= -\frac{1}{2}\left(\frac{\delta}{\pi}\right)^2 \ln\left(\frac{\Lambda}{\omega_{min}}\right) + \text{const} \qquad (26.112)$$

in agreement with the result by Nozières and De Dominicis.

One may wonder what kind of next-to-leading corrections the asymptotic result (26.111) acquires when $t' - t$ is finite. To our knowledge, such corrections have not been evaluated for the X-ray problem. Though the equation (26.66) can not be exactly solved, the required asymptotic expansion has been studied, for a different problem, by Japaridze, Nersesyan and Wiegmann (1984).

VI   Bosonization approach to the X-ray edge problem

In Section II.2, we put forward a conjecture that there exists a unitary operator U that effectively eliminates the local deep-hole potential V from the problem. Under this assumption, the absorption rate $F(t)$ and deep-hole Green's function $D(t)$ acquire very simple and suggestive forms, Eqs. (26.39) and (26.40). In this Section we shall explicitly construct this unitary operator and discuss some rather general and important related issues.

VI.1   Boundary condition changing operator (chiral anomaly)

The very possibility of bosonizing the X-ray edge problem is due to the equivalence, extensively discussed in Section II, of the original three-dimensional Hamiltonian to a one-dimensional one. In this Section we shall therefore study a one-dimensional version of the X-ray edge problem, in which the final state Hamiltonian takes the form:

$$H = \int_{-\infty}^{\infty} dx \left\{ R^{\dagger}(x)(-iv_F\partial_x)R(x) + V(x)R^{\dagger}(x)R(x) \right\} \qquad (26.113)$$

We remember that $R(x)$ is a chiral (right-moving) Fermi field. The local version (25.18) of (26.113) is obtained by setting $V(x) = V_0\delta(x)$ (see Section II for details). Physically, it is clear that, since the potential $V(x)$ does not scatter the fermions backwards, the only effect of the forward scattering is exhausted by a phase shift which a right-moving fermion picks up upon crossing the impurity point without changing the velocity.

Let us now consider this problem both classically and quantum-mechanically. We start with classical description. First, we observe that any local potential $V(x)$ in (26.113) appears as a gauge field:

$$H = -iv_F \int dx R^{\dagger}(x)[\partial_x + iA(x)]R(x) \qquad (26.114)$$

with $A(x) = V(x)/v_F$. The Hamiltonian (26.114) is invariant under simultaneous local transformations

$$A(x) \rightarrow A(x) + \partial_x\alpha(x)$$
$$R(x) \rightarrow e^{-i\alpha(x)}R(x) \qquad (26.115)$$

It follows from (26.115) that the potential $V(x)$ can be completely eliminated ('gauged away') from H by choosing $\alpha(x) = (1/v_F)\int^x V(y)dy$; namely

$$R(x) = \tilde{R}(x)e^{-\frac{i}{v_F}\int^x V(y)dy} \qquad (26.116)$$

Thus, we observe that classically the potential $V(x)$ has no physical effect whatsoever, and the chiral symmetry would remain unbroken.

Now let us turn to the quantum case. Of course, we can still eliminate the potential, making transformation (26.116) to a new field $\tilde{R}$. As a result, the Hamiltonian $H[\tilde{R}]$ is free. However, we cannot conclude that the Fermi system does not respond to the potential as we did in the classical case. The reason is that the potential $V(x)$ affects the boundary condition for the Fermi field.

First of all, we realize that, in the present formulation, the operator U introduced in Section II.2 is the one that generates the gauge transformation (26.116):

$$UH[R]U^{-1} = H_0[\tilde{R}]$$

This relation implies that the field $\tilde{R}$ satisfies the usual (periodic) boundary condition: $\tilde{R}(-L/2) = \tilde{R}(L/2)$, where L is the length of our one-dimensional system. In terms of the in and out asymptotic states, this means that

$$\tilde{R}_{\text{out}} = \tilde{R}_{\text{in}} \tag{26.117}$$

However, according to (26.116), for a δ-potential

$$R(x) = \tilde{R}(x) e^{-\frac{iV_0}{v_F}\theta(x)} \tag{26.118}$$

and therefore the boundary condition for the original field $R(x)$ turns out to be nontrivial

$$R_{\text{out}} = e^{2i\delta_0} R_{\text{in}} \tag{26.119}$$

and contains information about the phase shift

$$\delta_0 = -V_0/2v_F \tag{26.120}$$

(c.f. (25.29) and (25.34)). So, quantum-mechanically, the system does respond to the potential, and the operator U should be understood as a boundary condition changing operator. From the mathematical point of view, this means that the unitary operator U can only be defined in an enlarged Hilbert space which includes all the states with different boundary conditions (in particular, it includes $|i\rangle$ and $|f\rangle$) (see Affleck and Ludwig (1994)).

Let us now look at the problem of potential scattering of chiral fermions, Eq. (26.113), using bosonization. We introduce a right-moving (chiral) Bose field $\phi(x) \equiv \phi_R(x)$ and use the correspondence

$$R(x) \simeq \frac{1}{\sqrt{2\pi a_0}} \exp\left[i\sqrt{4\pi}\phi(x)\right]$$

and, for the electron density,

$$\rho(x) =: R^\dagger(x)R(x) := \frac{1}{\sqrt{\pi}}\partial_x\phi(x) \tag{26.121}$$

Then, for $V(x) = V_0\delta(x)$,

$$H_f \Rightarrow H_{\text{Bose}} = \int dx[\, v_F(\partial_x\phi)^2 + \frac{V_0}{\sqrt{\pi}}\delta(x)\partial_x\phi(x)\,] \tag{26.122}$$

The potential term is eliminated by a shift

$$\partial_x\phi(x) = \partial_x\phi(x) - \frac{V_0}{2\sqrt{\pi v_F}}\delta(x) \tag{26.123}$$

or equivalently

$$\phi(x) = \phi(x) - \frac{V_0}{4\sqrt{\pi v_F}}\text{sign}x \tag{26.124}$$

As a result of this transformation,

$$H_f \to H_i = v_F \int dx (\partial_x \phi)^2 \tag{26.125}$$

$$R(x) \to \frac{1}{\sqrt{2\pi a_0}} e^{i\sqrt{4\pi}\phi(x)+2i\delta_0\theta(x)} \tag{26.126}$$

in agreement with (26.119).

So, we have arrived at a clear definition of U in bosonic language:

$$U^\dagger \phi(x) U = \phi(x) + \frac{\delta_0}{\sqrt{\pi}} \theta(x) \tag{26.127}$$

or equivalently (changing the value of the field at infinity)

$$U^\dagger \phi(x) U = \phi(x) + \frac{\delta_0}{\sqrt{4\pi}} \mathrm{sign}x \tag{26.128}$$

Using the algebra of chiral Bose fields

$$[\phi(x), \phi(y)] = \frac{i}{4}\mathrm{sign}(x-y)$$

one easily finds U:

$$U = \exp\left[-\frac{2i\delta_0}{\sqrt{\pi}}\phi(0)\right] \tag{26.129}$$

The important observation one makes when looking at the transformation (26.127) of the scalar field ϕ is that, in bosonic language, the change of the boundary condition for fermions appears as formation of a 'topological kink' of the field ϕ in the ground state of H_f. Using the definition (26.121) of the fermion density (measured from the average density), we define the total fermionic particle number (measured from its vacuum value), i.e. the total 'charge' of the system, as the topological charge of the field ϕ:

$$Q = N - \langle N \rangle = \frac{1}{\sqrt{\pi}} \int_{-\infty}^{\infty} \partial_x \phi(x) = \frac{1}{\sqrt{\pi}} [\phi(\infty) - \phi(-\infty)]$$

For the free Hamiltonian $H_i = H_0$, $\phi(x)$ satisfies the standard periodic boundary condition; therefore its vacuum topological charge is zero. Then from (26.127) it immediately follows that, for H_f, there is a kink of ϕ at the point $x = 0$:

$$\phi(\infty) = \delta_0/\sqrt{\pi}, \qquad \phi(-\infty) = 0$$

This kink contributes to a local redistribution of the electron density in the presence of the deep-hole potential (screening effect)

$$\langle \rho(x) \rangle = \frac{\delta_0}{\pi} \delta(x)$$

which induces the corresponding change of the boundary condition.

The appearance of a nonzero 'vacuum charge' is the content of quantum (chiral) anomaly. The anomaly shows up as nonconservation of the chiral fermionic current which would be conserved classically. This is easily seen from the equation of motion for the right-moving scalar field ϕ_R at $V(x) \neq 0$. Starting from

$$i\partial_t \phi_R(x, t) = [\phi_R(x, t), H]$$

one finds:

$$(\partial_t + v_F \partial_x)\phi_R(x, t) = -\frac{V_0}{\sqrt{\pi}} \delta(x) \qquad (26.130)$$

The nonzero right-hand-side in this equation implies (local) breakdown of the continuous chiral symmetry (chiral anomaly).

VI.2   X-ray response functions via bosonization

Now we are in a position to estimate the long-time asymptotics of the deep-hole Green's function. Indeed, we have the representation (26.40) for $D(t)$ in terms of the boundary condition changing operator U, and representation (26.129) for U as a phase exponential of the bosonic field $\phi(x)$. So,

$$D(t) \sim \langle 0| \exp\left\{-\frac{2i\delta_0}{\sqrt{\pi}} [\phi(0, t) - \phi(0, 0)]\right\} |0\rangle \qquad (26.131)$$

Recall that the field $\phi(x)$ is chiral (right moving). For such a field, the critical dimension of a phase exponential $\exp[i\beta\phi(x, t)]$ coincides with its conformal dimension and equals $\beta^2/8\pi$ (rather than $\beta^2/4\pi$ as it would be for a true scalar field, that is, a sum of the right- and left-moving fields). So, the critical dimension of the exponential in the right-hand-side of Eq. (26.131) is

$$h = \frac{1}{2}\left(\frac{\delta_0}{\pi}\right)^2 \qquad (26.132)$$

Therefore

$$D(t) \sim t^{-2h} \sim t^{-\left(\frac{\delta_0}{\pi}\right)^2} \qquad (26.133)$$

(The response function $F(t)$ can be calculated in the same manner, with the help of the formula (26.39). We shall skip this elementary calculation.) Notice that, because $\phi(x, t) = \phi(x - v_F t)$, the formula (26.131) can also be written in the form (26.47) with the density–density correlation function defined as in (26.56). The bosonic field entering equation (26.56) is thus identical to that we have introduced in this Section. This establishes a link between the bosonization approach and the perturbation theory: by the very essence of the bosonization, the electron density is treated as a Gaussian variable so that a one-loop approximation suffices.

Given (26.40), the overlap integral can easily be found

$$|\langle 0|V\rangle| = |\langle 0| \exp\left[-\frac{2i\delta_0}{\sqrt{\pi}}\phi(0)\right]|0\rangle| = \exp\left[-\frac{2\delta_0^2}{\pi}\langle\phi^2(0)\rangle\right]$$

In a finite system $\langle\phi^2(0)\rangle \simeq (1/4\pi)\ln L$, so that

$$|\langle 0|V\rangle| = \left(\frac{\omega_{\min}}{\omega_c}\right)^{\frac{\delta_0^2}{2\pi^2}}$$

The bosonization thus gives $\alpha_{OI} = \frac{1}{2}\alpha_{orth} = \delta_0^2/2\pi^2$ for the overlap integral exponent (in accordance with the results of Section V).

VII   Appendix A. Parquet approximation

The so-called parquet approximation corresponds to summing a given class of diagrams. This approximation has been invented in the late 1950s by Diatlov, Sudakov and Ter Martirosian (1957) in the context of meson theory. The parquet approximation has also been applied to various condensed matter problems, such as the Kondo effect (Abrikosov (1965)) and one-dimensional interacting electrons (Buchkov, Gor'kov and Dzyaloshinskii (1966)). Later on, it has been understood that the results of the parquet approximation were equivalent to those of the one-loop renormalization group. Thus, after the renormalization group methods invaded the condensed matter theory (Wilson (1975)), the significance of the parquet approach has faded. That is why, investigating interacting electron systems in the bulk, we have employed a renormalization group – rather than a parquet – derivation of the coupling constants flow equations. Yet the parquet method is an important tool of the theory, for it not only establishes a direct relation between the perturbation theory and the renormalization group, but also, being merely a summation of diagrams, allows one to treat nonrenormalizable theories.[‖] Therefore, in this Appendix, we are giving an account of the parquet approximation for the X-ray edge problem, which serves as an excellent illustration of the general method. We believe that the reader who digests this material will be in a position to apply the parquet method to new problems which shall undoubtedly arise in the future.

[‖] So, one of the recent attempts to understand high-T_c materials (Dzyaloshinskii and Yakovenko (1988)) led to a $\log^2$ theory causing a return of the parquet methods to the pages of Physics journals.

Without loss of generality we can combine the terms of the perturbative expansion for the X-ray response function (26.13) as follows:

$$S(\omega) = \frac{\rho_0}{g} \left[s_0(gl_\omega) + g s_1(gl_\omega) + g^2 s_2(gl_\omega) + ... \right] \qquad (26.134)$$

where we have utilized the notation

$$l_\omega = \ln\left(\frac{\omega_c}{|\omega|}\right)$$

and made explicit the coupling constant dependence of $S(\omega)$; s_0, s_1, and s_2 are dimensionless functions.

In this Appendix we shall calculate the first term of the expansion (26.134) and show that

$$s_0(x) = \frac{1}{2}\left(e^{2x} - 1\right)$$

the result anticipated in Section II.1 (see Eq. (26.24)). This will be done by summing up main logarithms, which is the essence of the parquet approximation. The procedure is based on the assumption that $gl_\omega \sim 1$. Therefore one has to retain all diagrams containing the maximal power of the logarithm in each given order of perturbation theory, i.e. retain all terms proportional to $g^n l_\omega^n$ and ignore less singular terms (like $g^n l_\omega^{n-1}$ and smaller).

In Section II.1, we have already calculated a few perturbative contributions to $S(\omega)$. Let us now return to this, recalculating diagrams in the energy representation and defining what simplifications are implied by the leading logarithmic approximation.

The zero-order diagram is the singular bubble in Fig. 26.2:

$$S_0(\omega) = \int \frac{d\epsilon}{2\pi i} G_0(\epsilon) g_0(\epsilon + \omega) \qquad (26.135)$$

In the diagrammatic technique, it is conventional (and convenient) to use the deep-electron propagator $G(t)$ rather than the deep-hole propagator $D(t)$; they are simply related by $G(t) = -D(-t)$. As we know $S_0(\omega) = \rho_0 l_\omega + O(1)$. We have to retain the logarithmic term and neglect the $O(1)$ contribution. One can therefore substitute:

$$g_0(\omega) \rightarrow -i\pi \,\text{sign}\,\omega$$
$$G_0(\omega) \rightarrow -1/\omega \qquad (26.136)$$

for the bare propagators (c.f. (25.22) and (26.18)). That gives

$$S_0(\omega) = \frac{\rho_0}{2} \int\limits_{-\omega_c}^{\omega_c} \frac{d\epsilon}{\epsilon} \,\text{sign}(\epsilon + \omega) \qquad (26.137)$$

At this point we should make a technical but important remark. Because of the properties of the log-function, any energy variable entering the

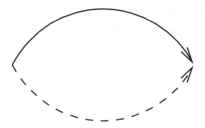

Fig. 26.6. The 'Cooper bubble'.

calculations can be multiplied by arbitrary constant factor, $\omega \to C\omega$, the logarithm will only change by $\ln|C|$, which is $\sim O(1)$ and can be neglected. The energy integral in (26.137) can therefore be approximated as

$$\frac{1}{2} \int_{-\omega_c}^{\omega_c} \frac{\mathrm{d}\epsilon}{\epsilon} \mathrm{sign}(\epsilon + \omega) \to \int_0^{l_\omega} \mathrm{d}l_\epsilon \qquad (26.138)$$

and all the other energy integrals, which will appear in the due course, can be dealt with in the same manner. It is also worth noticing that whenever we encounter a log-function of a sum of two arguments it can be written as

$$\ln|\omega + \omega'| \to \ln\left[\max(|\omega|, |\omega'|)\right] \qquad (26.139)$$

Analogously, one can approximate

$$\ln\left(\frac{\omega_c}{|\omega + \omega'|}\right) \to \min(l_\omega, l_{\omega'}) \qquad (26.140)$$

The bubble in Fig. 26.2 is referred to in the literature as the Peierls bubble (or the electron–hole bubble). A bubble of another type is shown in Fig. 26.6.

It is referred to as the Cooper bubble (or the electron–electron bubble) and it is also log-divergent:

$$\int_{-\omega_c}^{\omega_c} \frac{\mathrm{d}\epsilon}{2\pi i} G_0(\epsilon)g_0(\epsilon - \omega) \simeq \frac{\rho_0}{2} \int_{-\omega_c}^{\omega_c} \frac{\mathrm{d}\epsilon}{\epsilon} \mathrm{sign}(\epsilon + \omega) \simeq -\rho_0 \int_0^{l_\omega} \mathrm{d}l_\epsilon \quad (26.141)$$

with the sign opposite to that of the Peierls bubble. The fact that the Cooper bubble is log-divergent turns out to be very important. The Cooper bubble first shows up in the second order for $S(\omega)$. Indeed, in this order we have a ladder type diagram, Fig. 26.3, which contributes

$$\rho_0 g^2 l_\omega^3$$

to $S(\omega)$. The anticipated result (26.24) would however suggest that the second-order term of $S(\omega)$ is equal to $\frac{2}{3}\rho_0 g^2 l_\omega^3$. Thus, if (26.24) is to be

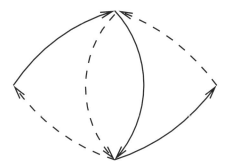

Fig. 26.7. Second-order graph for $S(\omega)$. This is the lowest order contribution to this function which reveals an interference of the divergencies in the Cooper and Peierls channels.

correct, a contribution of $-\frac{1}{3}\rho_0 g^2 l_\omega^3$ is missing. In order to account for this contribution one has to consider another second-order graph, shown in Fig. 26.7.

This graph contains a Cooper bubble placed in between two Peierls bubbles:

$$\int \frac{d\epsilon}{2\pi i} G_0(\epsilon) g_0(\epsilon + \omega) \int \frac{d\epsilon'}{2\pi i} G_0(\epsilon') g_0(\epsilon' + \omega) \left[\int \frac{d\Omega}{2\pi i} G_0(\Omega) g_0(\Omega - \omega) \right]$$

Substituting (26.141) for the expression in the square brackets (the Cooper bubble) and using (26.136) along with the logarithmic approximation (26.138–26.140), one easily finds

$$-\frac{\rho_0 g^2}{4} \int\limits_{-\omega_c}^{\omega_c} \frac{d\epsilon}{\epsilon} \text{sign}(\epsilon + \omega) \int\limits_{-\omega_c}^{\omega_c} \frac{d\epsilon'}{\epsilon'} \text{sign}(\epsilon' + \omega) \ln \left(\frac{\omega_c}{\epsilon + \epsilon' + \omega} \right)$$

$$\simeq -\rho_0 g^2 \int_0^{l_\omega} dl \int_0^{l_\omega} dl' \min(l, l') = -\frac{1}{3}\rho_0 g^2 l_\omega^3$$

This calculation shows that the Cooper bubbles are important and should be treated on the same footing with the Peierls bubbles.

First we need to calculate the fully renormalized interaction vertex Γ shown in Fig. 26.8.

Considering the series of diagrams shown in Fig. 26.9, or using the equations of motion for the field operators, one finds that the response function $S(\omega)$ is related to Γ by

$$S(\omega) = \int \frac{d\epsilon}{2\pi i} G(\epsilon) g(\epsilon + \omega) + \tag{26.142}$$

$$\int \frac{d\epsilon_1}{2\pi i} G(\epsilon_1) g(\epsilon_1 + \omega) \int \frac{d\epsilon_2}{2\pi i} \Gamma(\epsilon_1, \epsilon_2, \epsilon_2 + \omega, \epsilon_1 + \omega) G(\epsilon_2) g(\epsilon_2 + \omega)$$

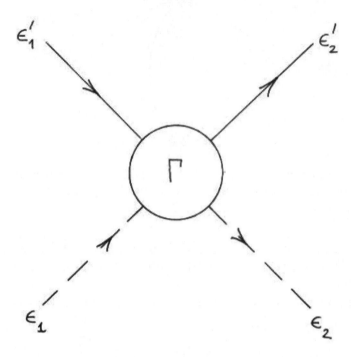

Fig. 26.8. The fully renormalized interaction vertex $\Gamma(\epsilon_1, \epsilon_2, \epsilon_1', \epsilon_2')$.

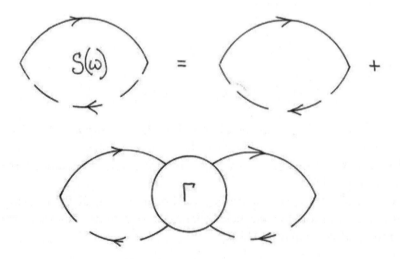

Fig. 26.9. The relation of the response function S to the interaction vertex Γ (all the propagators are renormalized).

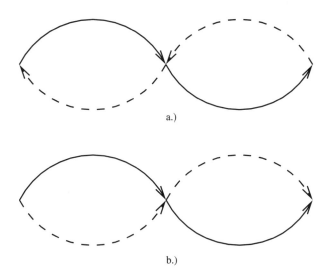

Fig. 26.10. The ladder type graphs.

We already know the interaction vertex in the lowest nontrivial approximation:

$$\Gamma(\epsilon_1, \epsilon_2, \epsilon_1', \epsilon_2') = V \left[1 + g l_{\epsilon_1 - \epsilon_2'} - g l_{\epsilon_1 + \epsilon_1'} + \ldots \right] \tag{26.143}$$

Notice that the Peierls bubble is singular with respect to the relative energy

$$\omega = \epsilon_1 - \epsilon_2' = \epsilon_2 - \epsilon_1'$$

in the channel with antiparallel full and dashed lines (the Peierls channel), while the Cooper bubble is singular with respect to the total energy

$$\Omega = \epsilon_1 + \epsilon_1' = \epsilon_2 + \epsilon_2'$$

in the channel with parallel full and dashed lines (the Cooper channel). We shall therefore use one of the two equivalent forms

$$\Gamma(\epsilon_1, \epsilon_2, \epsilon_1', \epsilon_2') = \Gamma(\epsilon_1, \epsilon_2, \omega) = \bar{\Gamma}(\epsilon_1, \epsilon_2, \Omega)$$

of the interaction vertex at our convenience.

We remember that our goal is to account for all $g^n l^n$ terms in the perturbative expansion of Γ. At this point we make a crucial observation: at the second order in g, not only the ladder type graphs (Fig. 26.10) contribute $g^2 l^2$ to Γ but the graphs in Fig. 26.11 are of the same order of magnitude.

Consider, for example, the graph in Fig. 26.11a:

$$V^3 \int \frac{d\epsilon}{2\pi i} G_0(\epsilon) g_0(\epsilon + \omega) \int \frac{d\epsilon'}{2\pi i} G_0(\epsilon') g_0(\epsilon + \epsilon_1 + \epsilon' + \omega)$$

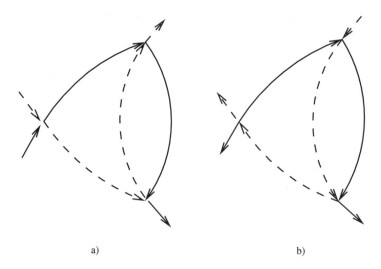

a)   b)

Fig. 26.11. The simplest (second-order) parquet graphs.

In our logarithmic approximation this gives

$$\frac{Vg^2}{4} \int_{-\omega_c}^{\omega_c} \frac{d\epsilon}{\epsilon} \operatorname{sgn}(\epsilon + \omega) \int_{-\omega_c}^{\omega_c} \frac{d\epsilon'}{\epsilon'} \operatorname{sgn}(\epsilon' + \epsilon_1 + \epsilon + \omega) \simeq$$

$$g^2 V \int_0^{l_\omega} dl \int_0^{l_{\epsilon_1} + \epsilon + \omega} dl' \simeq V \int_0^{l_\omega} dl \, \min(l, l_{\epsilon_1 + \omega}) = \frac{1}{2} V g^2 l_\omega^2$$

for $\epsilon_1 < \omega$. (The careful reader has, of course, noticed that the graph in Fig. 26.11a forms a part of the graph in Fig. 26.7. Hence both of them contribute a maximal divergence $g^n l^n$.) We therefore conclude that summation of any kind of ladder diagrams will not be sufficient. The graphs in Fig. 26.11 are in fact the simplest parquet graphs. Note that the graph Fig. 26.11a is constructed by inserting the Cooper bubble in place of a basic interaction vertex of the Peierls bubble. This process can be continued to give higher-order parquet graphs: a given g^n-graph is obtained from a g^{n-1}-graph by inserting a basic bubble in the place of a basic interaction vertex.

In order to achieve the programme of summing up the parquet graphs, we have to classify the diagrams according to their topological structure. The key concept is 'reducibility'. We shall call a given graph reducible (irreducible) in the Cooper channel if it can be (can not be) divided into two separate parts by cutting two parallel lines – one full and one dashed; analogously for the Peierls channel. So, the graph in Fig. 26.11a is reducible in the Peierls channel but irreducible in the Cooper channel, while the graph in Fig. 26.12 is irreducible in both channels (or totally irreducible).

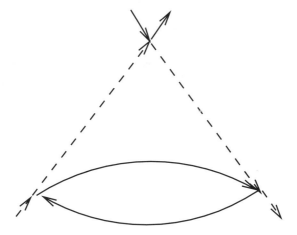

Fig. 26.12. The lowest-order nontrivial graph for the totally irreducible interaction vertex R.

Further, we define

$I =$ (the sum of all graphs, reducible in the Cooper channel)
$J =$ (the sum of all graphs, reducible in the Peierls channel)
$C =$ (the sum of all graphs, irreducible in the Cooper channel)   (26.144)
$P =$ (the sum of all graphs, irreducible in the Peierls channel)
$R =$ (the sum of all graphs, irreducible in both channels)

Two simple topological facts,

• each graph is either reducible or irreducible in a given channel,

• there are no graphs simultaneously reducible in both channels,

enable one to relate the interaction vertices defined above to Γ and to each other:

$$\Gamma = I + C = J + P = R + C + P \qquad (26.145)$$

and

$$C = R + J, \quad P = R + I \qquad (26.146)$$

Additionally, the interaction vertices satisfy the standard Bethe–Salpeter equations (see Fig. 26.13):

$$C(\epsilon_1, \epsilon_2, \Omega) = \int \frac{d\epsilon}{2\pi i} I(\epsilon_1, \epsilon, \Omega) G(\epsilon) g(\Omega - \epsilon) \Gamma(\epsilon, \epsilon_2, \Omega)$$

$$(26.147)$$

$$\bar{P}(\epsilon_1, \epsilon_2, \omega) = \int \frac{d\epsilon}{2\pi i} \bar{J}(\epsilon_1, \epsilon, \omega) G(\epsilon) g(\omega + \epsilon) \bar{\Gamma}(\epsilon, \epsilon_2, \omega)$$

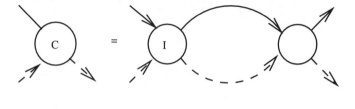

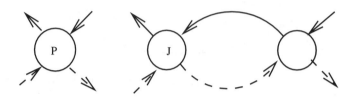

Fig. 26.13.   Bethe–Salpeter equations.

Eqs. (26.145), (26.146) and (26.147) are exact though they do not form a closed set. Yet they do, provided that the totally irreducible interaction (R) and the renormalized propagators $(G$ and $g)$ are known. The parquet summation can be carried out because the deviations of the latter quantities from their bare values stay beyond the leading logarithmic approximation, as follows.

- The conduction electron propagator does not have any self-energy corrections at all, because such corrections would involve dashed loops which are prohibited by the conservation of the number of deep electrons.[**]

- The lowest-order correction to the deep electron self-energy (26.28) was found in Section II.1 (Fig. 26.4). It is of the order of $g^2 l$.

- The lowest order nontrivial diagram for the totally irreducible interaction R is shown in Fig. 26.12. A straightforward calculation shows that this diagram is of the order of $Vg^2 l$. Indeed,

$$R^{(3)} = iV^3 \int \frac{d\Omega}{2\pi} G_0(\epsilon_1 + \Omega) G_0(\epsilon_2 + \Omega) \chi(\Omega) \qquad (26.148)$$

[**] To avoid any confusion we notice that the many-body conduction electron propagator g is different from the transient propagator we defined in Section III. The transient propagator is rather related to the two-particle Green function S of the many-body theory.

where $\chi(\Omega)$ is local charge polarization which we have calculated in Section III. The logarithmic singularity is clearly related to the imaginary part of this function: (26.57). So, substituting $\chi(\Omega) = -i\pi\rho_0^2|\Omega|$ into (26.148), one finds:

$$R^{(3)} \simeq \frac{1}{2}Vg^2 \int_{-\omega_c}^{\omega_c} \frac{|\Omega|d\Omega}{(\epsilon_1 + \Omega)(\epsilon_2 + \Omega)}$$

$$= Vg^2 \frac{\epsilon_1 \ln\left(\frac{\omega_c}{|\epsilon_1|}\right) - \epsilon_2 \ln\left(\frac{\omega_c}{|\epsilon_2|}\right)}{\epsilon_1 - \epsilon_2} \simeq Vg^2 \min(l_{\epsilon_1}, l_{\epsilon_2}) \quad (26.149)$$

One can therefore replace G and g by their bare values (26.136) and substitute V for R. Strictly speaking, one has to prove that Σ and R do not contain contributions like $g^n l^n$ for any n and not just for $n = 2$. A qualitative argument in favour of this is simply that the diagrams for Σ and R do not involve, by construction, enough independent singular bubbles. A more rigorous analysis may be found in Nozières, Gavoret and Roulet (1969). Upon doing so and making use of the logarithmic approximation (26.138–26.140), one finds the so-called parquet equations (which now form a closed set):

$$C(l_1, l_2, \xi) = -\rho_0 \int_0^\xi dl I \, [\min(l_1, l)] \, \Gamma(l, l_2, \xi) \quad (26.150)$$

$$\bar{P}(l_1, l_2, \eta) = -\rho_0 \int_0^\eta dl \bar{J} \, [\min(l_1, l)] \, \bar{\Gamma}(l, l_2, \eta) \quad (26.151)$$

with

$$I(l) = V + \bar{P}(l), \quad \bar{J}(l) = V + C(l) \quad (26.152)$$

and

$$\Gamma(l_1, l_2, \xi) = I \, [\min(l_1, l_2)] + C(l_1, l_2, \xi)$$
$$\quad (26.153)$$
$$\bar{\Gamma}(l_1, l_2, \xi) = \bar{J} \, [\min(l_1, l_2)] + \bar{P}(l_1, l_2, \xi)$$

We have used the shorthand notation $I(l, l, l) = I(l)$, etc. for the interaction vertices and $l_{1,2} = l_{\epsilon_{1,2}}$, $\xi = l_\Omega$, $\eta = l_\omega$ for the logarithmic variables. Notice that, because of the energy conservation,

$$\epsilon_1 + \epsilon_2 = \Omega + \omega$$

the latter variables are not independent but satisfy

$$\min(l_1, l_2) = \min(\xi, \eta) \quad (26.154)$$

With the derivation of the parquet equations (26.150–26.153) we have essentially completed the programme of this Appendix. In order to obtain the X-ray absorption rate $S(\omega)$ one simply has to solve the equations (26.150–26.153) with respect to $\bar{\Gamma}(l_1, l_2, \eta)$ and substitute the result into the equation (26.142), which, in our leading logarithmic approximation, reads:

$$S(\eta) = \rho_0\eta + \rho_0^2 \int_0^\eta dl_1 \int_0^\eta dl_2 \bar{\Gamma}(l_1, l_2, \eta)$$

$$= \rho_0\eta + 2\rho_0^2 \int_0^\eta dl_1 \int_0^{l_1} dl_2 \bar{\Gamma}(l_1, l_2, \eta) \qquad (26.155)$$

the last equality following from the symmetry of $\bar{\Gamma}$ with respect to the interchange of the variables l_1 and l_2.

Solving the parquet equations (26.150–26.153) involves yet another technical trick (generally used in the parquet calculations), introduced by Sudakov (1956). Namely, Sudakov noticed that the equation (26.150) can be written in the equivalent form

$$C(l_1, l_2, \xi) = -\rho_0 \int_0^\xi dl\, \Gamma(l_1, l, l)\Gamma(l, l_2, l) \qquad (26.156)$$

and analogously for (26.151). His method is also referred to as 'the maximal cross section method', since it is based on picking up the bubble with maximal logarithm in the iteration series for C and $\bar{P}$. A possibility of diagrams' doubling caused some confusion at the time (see Roulet, Gavoret and Nozières (1969) for references), so we shall give here a purely analytic proof of this equivalence.

Indeed, consider (26.151) as an integral equation for the unknown function C. The iteration then yields

$$C(l_1, l_2, \xi) = \sum_{n=1}^\infty A_n(l_1, l_2, \xi) \qquad (26.157)$$

where we have defined

$$A_n(l_1, l_2, \xi) = (-\rho_0)^n \int_0^\xi d\kappa_1 \int_0^\xi d\kappa_2 ... \int_0^\xi d\kappa_3 I\,[\min(l_1, \kappa_1)]\, I\,[\min(\kappa_1, \kappa_2)]$$
$$... I\,[\min(\kappa_n, l_2)] \qquad (26.158)$$

Notice that, due to (26.153), we can also write

$$\Gamma(l_1, l_2, \xi) = \sum_{n=0}^\infty A_n(l_1, l_2, \xi)$$

provided that $A_0(l_1, l_2, \xi) = I\,[\min(l_1, l_2)]$ is defined. Because $A_n(l_1, l_2, 0) = 0$

for $n \geq 1$, we can identically rewrite (26.157) as

$$C(l_1, l_2, \xi) = \int_0^\xi dl \sum_{n=1}^\infty \frac{\partial}{\partial l} A_n(l_1, l_2, l) \qquad (26.159)$$

Since ξ only appears as a limit of integration in (26.158), the derivative can be easily found:

$$\frac{\partial}{\partial l} A_n(l_1, l_2, \xi) = (-\rho_0)^n \int_0^\xi d\kappa_1 ... \int_0^\xi d\kappa_{i-1} \int_0^\xi d\kappa_{i+1} ... \int_0^\xi d\kappa_n I \, [\min(l_1, \kappa_1)] \, ...$$

$$I \, [\min(\kappa_{i-1}, l)] \, I \, [\min(l, \kappa_{i+1})] \, ...I \, [\min(\kappa_n, l_2)] = \sum_{i=1}^n A_{i-1}(l_1, l, l) A_{n-i}(l, l_2, l)$$

Substituting this into (26.159) and re-arranging the double summation $[(n, i) \rightarrow (k = n - 1, q = i - 1)]$ one immediately finds (26.156).

Thus, in the framework of the Sudakov method the parquet equations are brought to the form

$$C(l_1, l_2, \xi) = -\rho_0 \int_0^\xi dl \, \{I \, [\min(l_1, l)] + C(l_1, l, l)\}$$

$$\{I \, [\min(l, l_2)] + C(l, l_2, l)\} \qquad (26.160)$$

$$\bar{P}(l_1, l_2, \eta) = \rho_0 \int_0^\eta dl \, \{\bar{J} \, [\min(l_1, l)] + \bar{P}(l_1, l, l)\}$$

$$\{\bar{J} \, [\min(l, l_2)] + \bar{P}(l, l_2, l)\} \qquad (26.161)$$

As we shall see shortly, these equations are much easier to deal with than the original ones (26.151, 26.150). So, it immediately follows from (26.160, 26.161) that, in the simplest case when $l_1, l_2 \geq \xi, \eta$, the vertices C and $\bar{P}$ do not depend on l_1 and l_2. Eqs. (26.160, 26.161) then reduce to

$$C(\xi) = -\rho_0 \int_0^\xi dl \, [V + C(l) + \bar{P}(l)]^2$$

$$\qquad (26.162)$$

$$\bar{P}(\eta) = -\rho_0 \int_0^\eta dl \, [V + C(l) + \bar{P}(l)]^2$$

where we have used (26.152). The solution of (26.162) is

$$C(l) = -Vgl, \quad \bar{P}(l) = Vgl \qquad (26.163)$$

Note that, for comparable energies in the two channels, the logarithmic singularities in the renormalized interaction vertex Γ completely cancel[tt]

$$\Gamma(l) = V + C(l) + \bar{P}(l) = V \qquad (26.164)$$

This is a consequence of the potential scattering operator being structure-less (i.e. not having internal degrees of freedom, like, for example, impurity spin).

Let us now consider a more complicated situation when $l_1 > \eta \geq l_2$. Then, the vertex $\bar{P}(l_1, l_2, \eta) = \bar{P}(l_2, \eta)$ depends on two variables and one has to distinguish two regions of integration, $0 < l < l_2$ and $l_2 < l < \eta$, in the equation (26.161). (We focus on calculating $\bar{P}$, for this is the vertex we need to obtain the absorption rate.) The equation (26.161) takes the form

$$\bar{P}(l_2, \eta) = Vg\eta - Vg^2 l_2(\eta - l_2) + g \int_{l_2}^{\eta} dl \bar{P}(l_2, l)$$

This can be reduced to the differential equation

$$\frac{\partial \bar{P}(l_2, \eta)}{\partial \eta} = Vg(1 - gl_2) + g\bar{P}(l_2, \eta)$$

(Generally, such a reduction of the integral parquet equations to the differential ones is the mathematical reason for the success of the Sudakov method.) Integrating this equation with the boundary condition (26.163): $\bar{P}(l_2, l_2) = Vgl_2$, one finds

$$\bar{P}(l_2, \eta) = Vgl_2 + V \left[e^{g(\eta - l_2)} - 1 \right] \qquad (26.165)$$

Finally, we consider a general case of $\eta > l_1 > l_2$. Now we must distinguish three integration ranges $0 < l < l_1$, $l_1 < l < l_2$, and $l_2 < l < \eta$ in (26.161). Because, in the right-hand-side of (26.161) two arguments of $\bar{P}$ always coincide, one can simply substitute (26.165) and (26.163) into (26.161) in order to obtain $\bar{P}(l_2, l_2, \eta)$. Then, using (26.153) we find

$$\bar{\Gamma}(l_2, l_2, \eta) = \frac{V}{2} \left[e^{g(l_1 - l_2)} + e^{g(2\eta - l_1 - l_2)} \right]$$

Substituting this into (26.155) one arrives at the final result for the X-ray response function

$$S(\eta) = \frac{\rho_0}{2g} \left(e^{2g\eta} - 1 \right)$$

This is the result (26.24).

To conclude, we notice that the simple version of the parquet method discussed in this Appendix requires only minor modifications in order

---

[tt] In quantum field theory such a situation is known as the zero-charge infrared regime.

to became applicable to more complicated problems. So, for the Kondo problem (Abrikosov (1965)), one has to take into account the spin structure of the interaction vertices. This will essentially produce numerical factors in the parquet equations (26.160, 26.161) so that, for instance, the cancellation (26.164) no longer occurs. On the other hand, applying the method to translationally invariant systems (Buchkov *et al.* (1966)), one has to treat the momentum variables on the same footing as the energy variables.

VIII   Appendix B. The Wiener–Hopf method

The Wiener–Hopf method is based on simple propositions from the theory of analytic functions. Let us start with highlighting the notion of sectionally holomorphic functions.

- The function $\Phi(z)$ of the complex variable z is sectionally holomorphic if it is analytic everywhere except a finite set of contours and arcs (the boundary). In what follows we shall only consider the case when the boundary consists of the real axis.

  The heart of the Wiener–Hopf method is the solution to the following problem (the Hilbert problem).

- It is required to find a sectionally holomorphic function $\Phi(z)$ analytic in the upper (S^+) and in the lower (S^-) half-plane and vanishing at infinity, given the discontinuity of $\Phi(z)$ on the real axis ($z = \tau$):

$$\Phi(\tau + i0) - \Phi(\tau - i0) = \varphi(\tau) \tag{26.166}$$

The solution to this problem is given by the following Cauchy integral:

$$\Phi(z) = \frac{1}{2\pi i} \int_{-\infty}^{\infty} \frac{\varphi(\tau)d\tau}{\tau - z} \tag{26.167}$$

The fact that (26.167) is a solution to (26.166) can easily be verified. First, due to the basic properties of the Cauchy integral, (26.167) defines a function analytic everywhere except the real axis. (Of course, the integrand function $\varphi(\tau)$ should be 'well behaved'; rigorous conditions are formulated in Muskhelishvili (1953). The functions we consider below do satisfy those conditions.) Second, when z tends to the real axis, the integral (26.167) reduces to its principal value plus it picks up half a residual at the pole $z = \tau$ (with the sign determined by the sign of the imaginary part of z), so that:

$$\Phi(\tau \pm i0) = \pm \frac{1}{2}\varphi(\tau) + \frac{1}{2\pi i} \int_{-\infty}^{\infty} \frac{\varphi(\tau')d\tau'}{\tau' - \tau} \tag{26.168}$$

The most important point is that (26.167) is the unique solution. Indeed, assume that there are two solutions. Then their difference $\Psi(z)$ is analytic in S^+ and S^-, and has no discontinuity on the the the real axis:

$$\Psi(\tau + i0) - \Psi(\tau - i0) = 0$$

The function $\Psi(z)$ is thus analytic in the whole plane and vanishes at infinity. The only such function is $\Psi(z) = 0$. (All the above statements can be generalized to the case of boundaries more complicated than just the real axis; the conditions to be satisfied by $\Phi(z)$ at infinity can also be relaxed (Muskhelishvili (1953)). We shall not use these generalizations.)

In practical calculations, it is sometimes convenient to use a pair of functions,

$$\Phi_+(z) = \Phi(z) \quad \text{for } z \in \bar{S}^+$$
$$\Phi_-(z) = -\Phi(z) \, \text{for } z \in \bar{S}^-$$

instead of the function $\Phi(z)$ ($\bar{S}^{\pm}$ means $S^{\pm}$ with the real axis included). Obviously, Φ_+ (Φ_-) is analytic in S^+ (S^-) and, at the boundary,

$$\Phi_{\pm}(\tau) = \frac{1}{2}\varphi(\tau) \pm \frac{1}{2\pi i} \int_{-\infty}^{\infty} \frac{\varphi(\tau')d\tau'}{\tau' - \tau} \qquad (26.169)$$

(of course, one can analytically continue Φ_+ (Φ_-) to S^- (S^+), but it will, in general, have singularities there). Using infinitisimals ($\delta = 0+$), one can re-state the formula (26.169) in the form:

$$\Phi_{\pm}(z) = \pm \frac{1}{2\pi i} \int_{-\infty}^{\infty} \frac{\varphi(\tau)d\tau}{\tau - z \mp i\delta} \qquad (26.170)$$

In terms of the functions $\Phi_{\pm}(z)$, the Hilbert problem can be reformulated as follows.

- A function $\varphi(\tau)$, given on the real axis, can be decomposed into a sum of two functions,

$$\varphi(\tau) = \Phi_+(\tau) + \Phi_-(\tau) \qquad (26.171)$$

which are the boundary values of functions $\Phi_{\pm}(z)$ analytic in $S^{\pm}$ (and given by (26.170)). Moreover, this decomposition is unique.

This decomposition is a key to solving singular integral equations. Any singular integral equation, which is solvable, can in one or another way be reduced to the problem (26.171).

- Another way to formulate the Hilbert problem is to represent a function $G(\tau)$, defined on the real axis, as a product of two functions,

$$G(\tau) = X_+(\tau)X_-(\tau) \qquad (26.172)$$

which are the boundary values of functions $X_{\pm}(z)$ analytic in $S^{\pm}$. Clearly one simply has to take the logarithm in order to reduce the problem (26.172) to the problem (26.171) (one should be careful, though, if the argument of the left-hand-side of (26.172) exceeds 2π; see Muskhelishvili (1953)). (In the literature the problem (26.166) is often referred to as the Dirixlet problem, while its version (26.172) is called the Hilbert problem; in this book, we use the term 'Hilbert problem' for both equivalent versions.)

IX   Appendix C. Orthogonality of Slater determinants

In this Chapter, we have described two calculations of the overlap $\langle 0|V\rangle$ of the Fermi system ground states, with and without the impurity potential. The result is striking: $\langle 0|V\rangle$ vanishes in the limit of an infinite system. One calculation was based on the Hamann exact solution (Section V), while the other resorted to the bosonization of the electron system (Section VI). Both calculations involved tedious mathematical methods, so there is a need to establish the orthogonality by simple qualitative arguments. In this Appendix we shall return to the original Anderson paper (Anderson (1967)) and estimate the overlap of the Slater determinants $|0\rangle$ and $|V\rangle$.

First of all we confine our system to a spherical box of radius L (the shape of the confining potential is clearly unimportant). Then the momentum (and the energy) will be quantized. For free electrons

$$k = k_n = \pi n/L$$

n being an integer, while the normalized single-electron eigenfunctions are spherical waves (Section II)

$$\varphi_n(r;0) = \sqrt{\frac{2}{L}} \frac{\sin(k_n r)}{r} \tag{26.173}$$

As the system is finite, there is a minimal frequency ω_{min} (i.e., the dimensional quantization energy), which is equal to $\pi v_F/L$. Note that the number of electrons confined to the box is related to its radius by $N \sim (k_F L)^d$, d being the dimension of the system. In the presence of the local potential V at the centre of the sphere, the eigenfunctions are modified as

$$\varphi_n(r;V) = \sqrt{\frac{2}{L}} \frac{\sin(k_n' r + \delta)}{r} \tag{26.174}$$

with

$$k_n' = (\pi n - \delta)/L$$

(we neglect the energy dependence of the scattering phase). The matrix elements between the single-electron states are

$$a_{nn'} = \int_0^L r^2 dr \varphi_n(r;0)\varphi_{n'}(r;V) \simeq \frac{\sin\delta}{\pi(n-n') + \delta/L} \tag{26.175}$$

where we have only retained the part singular at $n \to n'$.

Now, let n_f label the closest to the Fermi energy (from above) single-electron level. Then all the states with $n < n_f$ are occupied and the many-body state $|0\rangle$ is noting but the Slater determinant constructed using the single-electron functions (26.173) with $n < n_f$. Analogously, $|V\rangle$ is the Slater determinant made of the functions (26.174) with $n < n_f$. The overlap integral we are interested in clearly is

$$\langle 0|V\rangle = \det[a_{nn'}] \tag{26.176}$$

(with $n, n' < n_f$).

We want to find an upper bound for the overlap integral, so a suitable inequality is the Adamar inequality:[‡‡]

$$|\det[a_{nn'}]| \le \prod_{n<n_f} \sqrt{\sum_{n'<n_f} |a_{nn'}|^2} \tag{26.177}$$

We therefore find

$$|\langle 0|V\rangle| \le \exp\{\frac{1}{2}\sum_{n<n_f} \ln \sum_{n'<n_f} |a_{nn'}|^2\}$$

Using the completeness of the electron eigenstates and an obvious property $\ln(1-x) \le -x$ for $0 \le x \le 1$, one estimates the expression in the exponent as

$$\frac{1}{2}\sum_{n<n_f} \ln \left[1 - \sum_{n'>n_f} |a_{nn'}|^2\right] \le \sum_{n<n_f,\, n'>n_f} |a_{nn'}|^2 \simeq \frac{\sin^2 \delta}{2\pi^2} \ln \left(\frac{\omega_c}{\omega_{min}}\right)$$

This gives the desired estimate for the overlap integral

$$|\langle 0|V\rangle| \le \left(\frac{\omega_{min}}{\omega_c}\right)^{\frac{\sin^2 \delta}{2\pi^2}}$$

first obtained by Anderson. (Notice that this method gives an approximate exponent coinciding with the exact one for a weak potential only.)

References

A. A. Abrikosov, *Physics* **2**, 5 (1965).

A. A. Abrikosov, L. P. Gor'kov and I. E. Dzyaloshinskii, *Methods of Quantum Field Theory in Statistical Physics*, ed. A. R. Silvermann, Prentice–Hall, Englewood, NJ (1963).

[‡‡] The ratio J of the left-hand-side of (26.177) to its right-hand-side can be written as $J = \det[u_{nn'}]$, where $u_{nn'} = a_{nn'}/\sqrt{\sum_{n',n_f} |a_{nn'}|^2}$ is a set of unit vectors. J is a Jacobian equal to the volume of the parallelepiped formed by these unit vectors and it clearly is ≤ 1.

P. W. Anderson, *Phys. Rev. Lett.* **18**, 1049 (1967).

I. Affleck and A. W. W. Ludwig, *J. Phys.* A**27**, 5375 (1994).

Yu. A. Buchkov, L. P. Gor'kov and I. E. Dzyaloshinskii, *Soviet Phys. JETP* **23**, 489 (1966).

I. T. Diatlov, V. V. Sudakov and K. A. Ter Martirosian, *Soviet Phys. JETP* **5**, 631 (1957).

I. E. Dzyaloshinskii and A. I. Larkin, *Soviet Phys. JETP,* **38**, 202 (1974).

I. E. Dzyaloshinskii and V. M. Yakovenko, *Soviet Phys. JETP* **67**(4), 844 (1988).

D. R. Hamann, *Phys. Rev. Lett.* **26**, 1030 (1971).

G. I. Japaridze, A. A. Nersesyan, and P. B. Wiegmann, *Nucl. Phys.* B**230**, 511 (1984).

Quantum Tunneling in Condensed Matter, ed. by Yu. Kagan and A. J. Leggett, Elsevier (1992).

D. C. Langreth, *Phys. Rev.* B**1**, 471 (1970).

G. D. Mahan, *Phys. Rev.* **163**, 612 (1967).

G. D. Mahan, *Many-Particle Physics*, Plenum, New York (1981).

N. I. Muskhelishvili, *Singular Integral Equations*, P. Nordhoff Ltd., Groningen, The Netherlands (1953).

P. Nozières and C. T. De Dominicis, *Phys. Rev.* **178**, 1097 (1969).

P. Nozières, J. Gavoret and B. Roulet, *Phys. Rev.* **178**, 1084 (1969).

K. Ohtaka and Y. Tanabe, *Rev. Mod. Phys.* **62**, 929 (1990).

B. Roulet, J. Gavoret and P. Nozières, *Phys. Rev.* **178**, 1072 (1969).

K. D. Schotte and U. Schotte, *Phys. Rev.* **182**, 479 (1969).

V. V. Sudakov, *Soviet Phys. Doklady* **1**, 662 (1956).

K. G. Wilson, *Rev. Mod. Phys.* **47**, 773 (1975).

K. Yamada and K. Yoshida, *Prog. Theor. Phys.* **60**, 353 (1978); **62**, 363 (1979); **68**, 1504 (1982).

G. Yuval and P. W. Anderson, *Phys. Rev.* B**1**, 1522 (1970).

27

Impurities in a Tomonaga–Luttinger liquid

I Introduction

The importance of a bulk disorder in one-dimensional interacting Fermi systems was realized long ago (Mattis (1974)). In this Chapter we turn to some specific problems of the potential scattering in Tomonaga–Luttinger liquids, which arise when the scattering potential is localized in some region in space so that it represents a single scatterer (or a few scatterers). This situation should be opposed to the case of bulk disorder, when the scattering potential is randomly distributed over the whole system. Apart from earlier perturbative approaches, the local scattering problem has only recently attracted the attention of theorists (Kane and Fisher (1992); Furusaki and Nagaosa (1993)); originally it was in response to the progress in the nano-fabrication technology (see review articles by Kastner (1992), Gogolin (1994), and Voit (1995)).

The Hamiltonian of the problem is

$$H = H_{\mathrm{TL}} + \hat{V} \tag{27.1}$$

where H_{TL} is the Tomonaga–Luttinger liquid Hamiltonian including the kinetic energy and the interaction terms, while the potential scattering

$$\hat{V} = \int \mathrm{d}x V(x) \psi_s^\dagger(x)\psi_s(x) \tag{27.2}$$

is assumed to be local, i.e. $V(x)$ is essentially nonzero in a region of a finite radius a (the index s denotes the electron spin and summation over repeating indices is implied). In what follows we shall often simply take $V(x) = V\delta(x)$ for a single scatterer (and an appropriate sum of δ-functions for the case of several scatterers).

The basic physics of the single scatterer problem is captured by the following two statements.

(i) The local potential operator $\hat{V}$ is a relevant operator; namely, it grows under the renormalization group transformations (provided that the electron–electron interactions are predominantly repulsive – see below).

(ii) At low energies (temperatures), the operator $\hat{V}$ effectively cuts the systems into two separate parts, the tunneling between which is, on the contrary, represented by an irrelevant operator.

In this chapter we shall prove (i) (weak-coupling analysis: Section II) and (ii) (strong-coupling analysis: Section III). We shall also discuss the case of a dynamic local potential and relate the problem of the impurity scattering in the Tomonaga–Luttinger liquid to some other condensed matter problems.

We stress that we only consider isolated quantum impurities in this book leaving behind the complex issues related to the interplay of a bulk disorder and interactions. This is a separate and fast growing field of research which is, unfortunately, outside the scope of this book. To the reader, interested in these issues, we recommend to start the adventure into this fascinating field with the pioneering paper by Giamarchi and Schulz (1988).

II   Weak-coupling analysis of a single impurity

Conventionally, we isolate the slow-varying in space chiral parts of the conduction electron field:

$$\psi_s(x) \simeq e^{ik_Fx}R_s(x) + e^{-ik_Fx}L_s(x)$$

In terms of the chiral fields $R_s(x)$ and $L_s(x)$, the potential scattering operator (27.2) is explicitly divided into the forward scattering and the backward scattering parts

$$\hat{V} = \hat{V}_{\mathrm{fs}} + \hat{V}_{\mathrm{bs}}$$

where

$$\hat{V}_{\mathrm{fs}} = \int \mathrm{d}x \left[V_{\mathrm{fs}}(x)R_s^\dagger(x)R_s(x) + (R \to L) \right]$$

and

$$\hat{V}_{\mathrm{bs}} = \int \mathrm{d}x \left[V_{\mathrm{bs}}(x)R_s^\dagger(x)L_s(x) + \mathrm{H.c.} \right]$$

The forward- and backscattering amplitudes are related to the original scattering potential via $V_{\mathrm{fs}}(x) = V(x)$ and $V_{\mathrm{bs}}(x) = V(x)\exp(2ik_Fx)$, though it is sometimes convenient to consider them as independent potentials.

II.1   Bosonization of the impurity Hamiltonian

We now employ the standard bosonization procedure:

$$R_s(x) \simeq \frac{1}{\sqrt{2\pi a_0}} e^{i\sqrt{4\pi}\phi_{sR}(x)} , \quad L_s(x) \simeq \frac{1}{\sqrt{2\pi a_0}} e^{-i\sqrt{4\pi}\phi_{sL}(x)} \tag{27.3}$$

ϕ_{sR} and ϕ_{sL} being the right- and left-moving Bose fields.

Let us consider, for simplicity, the case of spinless electrons (when the index s is suppressed). We shall leave the generalizations to the spinful case for the exercises at the end of the Chapter. The potential scattering then only involves the combination

$$\Phi(x) = \phi_R(x) + \phi_L(x) \tag{27.4}$$

we are familiar with from the first part of this book. As we have already learned, interactions between the electrons in the conduction band are accounted for by rescaling the Bose field

$$\Phi(x) \to \sqrt{K}\Phi(x)$$

where K is the Tomonaga–Luttinger liquid parameter ($K < 1$ for repulsive interactions and $K > 1$ for attractive interactions). Thus, the total Hamiltonian for the case of spinless electrons takes the form:

$$H = H_0[\Phi] + \sqrt{\frac{K}{\pi}} V_{\text{fs}} \partial_x \Phi(0) + \frac{V_{\text{bs}}}{\pi a_0} \cos\left[\sqrt{4\pi K}\Phi(0)\right] \tag{27.5}$$

where

$$H_0[\Phi] = \frac{v}{2} \int dx \left[\Pi^2(x) + (\partial_x \Phi(x))^2\right]$$

$\Pi(x)$ is the momentum, canonically conjugate to $\Phi(x)$, and v is the (renormalized) Fermi velocity. Notice that we have assumed the scattering potentials to be strictly local: $V_{\text{fs}}(x) \to V_{\text{fs}}\delta(x)$ and $V_{\text{bs}}(x) \to V_{\text{bs}}\delta(x)$. This assumption simplifies the calculations but it is not really restrictive: as it will become clear later, the deviations from locality are irrelevant.

Before we proceed with the analysis of the Hamiltonian (27.5), we should make the following important remark: the forward- and the backscattering processes are independent (commuting), they are related to different chiral fields. In order to support this statement, we consider combinations of the Bose field $\Phi(x)$ which have a given parity with respect to the origin $x = 0$:

$$\Phi_\pm(x) = \frac{1}{\sqrt{2}} [\Phi(x) \pm \Phi(-x)] \tag{27.6}$$

The fields (27.6), which are actually defined on the semi-axis $0 < x < \infty$, can be decomposed into the left- and right-moving parts analogously to

(27.4):

$$\Phi_\pm(x) = \phi_{R\pm}(x) + \phi_{L\pm}(x)$$

These fields are defined for $x > 0$ and satisfy $\phi_{R\pm}(0) = \phi_{L\pm}(0)$, so we can replace them with two chiral (right-moving) fields defined on the whole axis. One of them is related to the even parity combination (27.6)

$$\phi(x) = \begin{cases} \phi_{R+}(x), & \text{for} \quad x > 0 \\ \phi_{L+}(-x), & \text{for} \quad x < 0 \end{cases}$$

while the other is related to the odd parity combination (27.6)

$$\phi_a(x) = \begin{cases} \phi_{R-}(x), & \text{for} \quad x > 0 \\ \phi_{L-}(-x), & \text{for} \quad x < 0 \end{cases}$$

In terms of these new chiral fields

$$\partial_x \Phi(0) = \sqrt{2} \partial_x \phi_a(0) , \quad \text{and} \quad \Phi(0) = \sqrt{2} \phi(0)$$

The total Hamiltonian (27.5) can therefore be decomposed into two independent parts, describing the forward- and the backscattering processes,

$$H = H_{\text{fs}} + H_{\text{bs}}$$

with

$$H_{\text{fs}} = H_0[\phi_a] + \sqrt{\frac{2K}{\pi}} V_{\text{fs}} \partial_x \phi_a(0) \tag{27.7}$$

$$H_{\text{bs}} = H_0[\phi] + \frac{V_{\text{bs}}}{\pi a_0} \cos\left[\sqrt{8\pi K} \, \phi(0)\right] \tag{27.8}$$

and

$$H_0[\phi] = v \int dx \, [\partial_x \phi(x)]^2$$

The backscattering model (27.8) we arrived at is referred to in the literature as the boundary sine-Gordon model. The boundary sine-Gordon model differs from the bulk one by the cos-interaction term being present at a single point in space (i.e. at the boundary; in some variations of the model a bulk interaction term is also included). This model is exactly solvable; the solution has been obtained by Ghoshal and Zamolodchikov (1994).

The above derivation of (27.7, 27.8) is similar to the one we have outlined in Chapter 25 while considering the reduction of a three-dimensional impurity model to a model involving a chiral one-dimensional fermion. Notice however that, in this Section, we have worked with the bosonized model. Alternatively we could have introduced the symmetric- and antisymmetric fermionic states before bosonizing the model (this is equivalent to the partial-wave expansion of Chapter 25, for the parity is the only symmetry operation of the 'group of rotations' in one dimension). However, upon a subsequent bosonization, the interaction terms in the Hamiltonian (27.1) would become complicated (nonlocal). So, such a procedure would have led to a representation of the impurity problem which is interesting in itself, but less convenient for our purposes.

Exercise. Using the basis of symmetric and antisymmetric one-fermion states, show that the problem of a single impurity in a spinless Tomonaga–Luttinger liquid can alternatively be formulated in terms of the following Hamiltonian defined on a semiaxis:

$$H = H_+ + H_-$$

$$H_+ = H_+^0 + \frac{g}{(2\pi a_0)^2} \int_0^\infty dx[\cos\sqrt{8\pi}\phi_+(x) - \cos\sqrt{8\pi}\theta_+(x)] + \frac{V_{bs}}{\sqrt{2\pi}}\partial_x\theta_+(0)$$

$$H_- = H_-^0 + \frac{g}{4\pi} \int_0^\infty dx[(\partial_x\phi_-(x))^2 - (\partial_x\theta_-(x))^2] + \frac{V_{fs}}{\sqrt{2\pi}}\partial_x\phi_-(0)$$

Here $\phi_\pm$ are massless Gaussian fields, $\theta_\pm$ are their dual counterparts, g is the coupling constant of the Tomonaga–Luttinger model.

We also note that, in the literature, the boundary sine-Gordon problem is often formulated in the following (non-chiral) form:

$$H_{BSG} = \frac{v}{2} \int dx \left[\Pi^2(x) + (\partial_x\Phi(x))^2\right] + \frac{V_{bs}}{\pi a_0}\cos\left[\sqrt{4\pi K}\Phi(0)\right] \qquad (27.9)$$

As we have just learned, this is equivalent to the formulation (27.8) involving a chiral Bose field (the only difference being that (27.9) also includes a free chiral Bose field corresponding to the antisymmetric mode).

Thus, a single impurity in the spinless Tomonaga–Luttinger liquid is described by the boundary sine-Gordon model (27.8) (or (27.9)). Before we proceed with studying this model, let us discuss the decoupled forward scattering processes (27.7). We immediately recognize in (27.7) the Hamiltonian (26.122) which we encountered studying the X-ray edge problem.[*] By adopting the same course of action as in Section 26.VI, we diagonalize the forward scattering Hamiltonian (27.7) by a canonical transformation

$$U_{fs}^\dagger H_{fs} U_{fs} = H_0$$

with the unitary operator U_{fs} defined by

$$U_{fs} = \exp\left[-i\sqrt{\frac{2K}{\pi}\frac{V_{fs}}{v}}\phi_a(0)\right] \qquad (27.10)$$

---

[*] One could thus say that the three-dimensional potential scattering problem is equivalent to the forward scattering problem in one dimension.

(As we have discussed in Chapter 26, U_{fs} is in fact a boundary condition changing operator.) The canonical transformation (27.10) solves the forward scattering problem (which is therefore quite a simple one). Thus, discussing the backscattering problem in what follows, we shall imply that the transformation (27.10) has been performed so that the forward scattering term is eliminated (if applied to the Hamiltonian (27.5), the transformation (27.10) clearly yields (27.9)). Note that (27.10) only affects the antisymmetric bosonic mode so it will drop out of most of the correlation functions of interest but the X-ray response functions (see Section VI).

II.2   Lagrangian formulation: local action

The local nature of the impurity problem can be best exploited using the Lagrangian formulation. The partition function for the Hamiltonian (27.9) is given by the functional integral

$$ Z = \int D\Phi(x, \tau) e^{-S_0[\Phi(x,\tau)] - S_1[\Phi(0,\tau)]} \tag{27.11} $$

over the field $\Phi(x, \tau)$, where

$$ S_0[\Phi(x, \tau)] = \frac{1}{2} \int_0^\beta d\tau \int_{-\infty}^\infty dx \left[v^{-1}(\partial_\tau \Phi)^2 + v(\partial_x \Phi)^2 \right] \tag{27.12} $$

is the Gaussian action (3.1) for the infinite (in the real space) system, while

$$ S_1[\Phi(0, \tau)] = -\frac{V_{bs}}{\pi a_0} \int_0^\beta d\tau \cos \left[\sqrt{4\pi K} \Phi(0, \tau) \right] $$

is the impurity potential term.[†]
    It is the dynamics of $\Phi(\tau) = \Phi(0, \tau)$ which is important, so it is instructive to integrate over all $x \neq 0$ degrees of freedom to obtain the local action S_1 governing the dynamics of $\Phi(\tau)$:

$$ Z = \int D\Phi(\tau) e^{-S_1[\Phi(\tau)]} \tag{27.13} $$

In order to formally accomplish this integration and determine $S_1[\Phi]$, we use the rather standard trick of introducing the Lagrange multiplier with the auxiliary field $\lambda(\tau)$:

$$ \delta[\Phi(0, \tau) - \Phi(\tau)] = \int D\lambda(\tau) \exp \left\{ i \int_0^\beta \lambda(\tau) d\tau [\Phi(0, \tau) - \Phi(\tau)] \right\} $$

[†] We work with the Hamiltonian (27.9) rather (27.8) in order to avoid the use of a chiral action. This does not change the results.

The partition function can thus be represented as

$$Z = \int D\Phi(\tau)e^{-S_1[\Phi(\tau)]} \int D\lambda(\tau) \exp\left\{-i\int_0^\beta d\tau\lambda(\tau)\Phi(\tau)\right\} Z[\lambda(\tau)] \quad (27.14)$$

with

$$Z[\lambda(\tau)] = \int D\Phi(x,\tau) \exp\left\{-S_0[\Phi(x,\tau)] + i\int_0^\beta d\tau\lambda(\tau)\Phi(0,\tau)\right\}$$

As $Z[\lambda(\tau)]$ is nothing but the generating functional (3.8) with $\eta(x,\tau) = -i\delta(x)\lambda(\tau)$, and we immediately find

$$Z[\lambda(\tau)] = Z[0] \exp\left[-\frac{1}{2\beta}\sum_\omega \lambda(\omega)D_0(\omega)\lambda(-\omega)\right]$$

Where $D_0(\omega)$ is the local Green's function of the Bose field in an infinite system:

$$D_0(\omega) = \int_{-\infty}^\infty \frac{dq}{2\pi}\frac{1}{v^{-1}\omega^2 + vq^2} = \frac{1}{2|\omega|} \quad (27.15)$$

Performing the remaining elementary (shift of variables) integration over $\lambda(\tau)$ in (27.14) one arrives at (27.13) with the local action given by

$$S_l[\Phi] = S_{0l}[\Phi] + S_1[\Phi]$$

$$\quad (27.16)$$

$$S_{0l}[\Phi] = \frac{1}{2\beta}\sum_\omega \Phi(\omega)D_0^{-1}(\omega)\Phi(-\omega)$$

The local action will prove useful in Section II.3, where we discuss the renormalization group analysis of the impurity problem. We also note that the action (27.16) is similar to the one describing a quantum dissipative particle (in Section V we shall discuss this analogy in some detail).

II.3   *Renormalization group analysis of local operators*

In Chapter 10 we performed the renormalization group analysis of the conventional (bulk) sine-Gordon model. Here we shall study the boundary model along the same lines and introduce the notion of the scaling dimension of local operators.

As we have repeatedly stressed in Chapters 25 and 26, any local problem develops only in time.[‡] Therefore there is only one variable to be scaled in the framework of the renormalization group approach – the energy (while

---

[‡] For the impurity in the Tomonaga–Luttinger liquid, the local action, depending on the Bose field at the origin $\Phi(\tau) = \Phi(x = 0, \tau)$, has been derived in the previous section: (27.16).

studying the bulk problem in Chapter 10 we have scaled both the energy
and the momentum; or the two-dimensional momentum of the Euclidean
formulation). Hence we define

$$\Phi(t) \to \Phi_\Lambda(\tau) = \Phi_{\Lambda'}(\tau) + h(\tau) = \int\limits_{|\omega|<\Lambda'} \frac{d\omega}{2\pi} e^{i\omega\tau}\Phi(\omega) + \int\limits_{\Lambda'<|\omega|<\Lambda} \frac{d\omega}{2\pi} e^{i\omega\tau}\Phi(\omega)$$

Notice that we have introduced the cut-off Λ and the field $\Phi_\Lambda(t)$; $\Phi_{\Lambda'}(t)$
and $h(t)$ being its slow and fast components respectively, and $\Lambda' = \Lambda - d\Lambda$
(also, we work at zero temperature in what follows).

In the spirit of Chapter 10, our goal is to integrate out the slow
components of the field $\Phi_\Lambda(t)$ in order to define the effective (renormalized)
parameters of the problem. At this stage we notice another important
feature of the boundary problem: the local operator can not affect the
bulk Hamiltonian (action). Therefore the bulk Hamiltonian does not scale
and it is just the local operator itself which is subject to the renormalization
group flow (the renormalization of the scaling dimension, so important
for the bulk model, does not occur in the boundary case). So, for the local
problem, the definition (10.7) of the effective action becomes a definition
of the effective potential and reads:

$$S_{1\text{eff}}[\Phi_{\Lambda'}] = -\int d\tau\, V_{\text{eff}}[\Phi_{\Lambda'}(\tau)] = -\ln\langle e^{-S_{1\text{eff}}[\Phi_{\Lambda'}+h]}\rangle_h$$

(27.17)

$$\langle S_{1\text{eff}}[\Phi_{\Lambda'}+h]\rangle_h + \frac{1}{2}\left\{\langle S_{1\text{eff}}^2[\Phi_{\Lambda'}+h]\rangle_h - \langle S_{1\text{eff}}[\Phi_{\Lambda'}+h]\rangle_h^2\right\} + \dots$$

Formally this definition is valid for an arbitrary boundary operator, but
let us now study it for the particular case of the boundary sine-Gordon
problem (27.8, 27.16). Then

$$V(\Phi_\Lambda) = \frac{V_{\text{bs}}}{\pi a_0}\cos(\beta\Phi_\Lambda)$$

(27.18)

and it is easy to calculate the first-order term in (27.17):

$$\langle V[\Phi_{\Lambda'}(\tau) + h(\tau)]\rangle_h = \frac{V_{\text{bs}}}{2\pi a_0}\left\{\exp[\beta\Phi_{\Lambda'}(\tau)]\,\langle\exp[\beta h(\tau)]\rangle_h + \text{c.c.}\right\}$$

Using the unperturbed action in (27.16), one immediately finds

$$\langle\exp[\beta h(t)]\rangle_h = \exp\left[-\frac{\beta^2}{2}\langle h^2(t)\rangle_h\right] =$$

$$\exp\left[-\frac{\beta^2}{2}\int_{\Lambda'<|\omega|<\Lambda}\frac{d\omega}{2\pi}D_0(\omega)\right] = 1 - \frac{\beta^2}{4\pi}dl + O(dl^2)$$

where $dl = d\Lambda/\Lambda$. Since we want to restore the original cut-off Λ, we
rescale the energy $\omega' = (\Lambda/\Lambda')\omega \simeq (1+dl)\omega$ and the time $t' = (1-dl)t$,

so that the effective potential operator is of the same form as the bare one in (27.18) but with a renormalized potential strength:

$$V'_{bs} = \left[1 + \left(1 - \frac{\beta^2}{4\pi} \right) dl \right] V_{bs} \qquad (27.19)$$

In the differential form this reads

$$\frac{dV_{bs}}{dl} = \left(1 - \frac{\beta^2}{4\pi} \right) V_{bs} \qquad (27.20)$$

The scaling dimension (which we shall denote by Δ) of the local operator (27.18) is therefore equal to

$$\Delta = \frac{\beta^2}{4\pi} \qquad (27.21)$$

Notice the difference with the bulk sine-Gordon model: as we have determined in Chapter 10, the bulk $\int dx \cos[\beta\Phi(x)]$ operator is marginal when $\beta^2 = 8\pi$ while the local $\cos[\beta\Phi(x = 0)]$ operator is marginal if $\beta^2 = 4\pi$.

For the impurity in a Tomonaga–Luttinger liquid $\beta = \sqrt{4\pi K}$, the scaling dimension Δ simply coincides with the Luttinger liquid parameter K in this case and the first-order renormalization group equation (27.20) reads:

$$\frac{dV_{bs}}{dl} = (1 - K)V_{bs} \qquad (27.22)$$

Thus, for attractive electron–electron interactions ($K > 1$), the impurity scattering operator is irrelevant: the weak-coupling fixed point is stable. On the other hand, for repulsive interactions ($K < 1$), the impurity scattering operator is relevant: the weak-coupling fixed point is unstable.[§] The system should therefore cross over to the intermediate – and eventually to the strong coupling. This is the most interesting regime, which we shall study in the next Section. (We again stress that, for a critical boundary problem, the dimensionality of space-time is reduced to 1, i.e. only the time axis remains. This reduction affects the criterion of relevancy, as is clearly seen from Eq. (27.20).)

III   Strong-coupling analysis

In the previous section we have have discovered that, for repulsive interactions in the conduction band ($K < 1$), the effective impurity potential

---

[§] For noninteracting electrons ($K = 1$), the impurity operator is marginal and, moreover, one can show that the right-hand-side of (27.22) vanishes in all orders in V: the impurity potential does not scale.

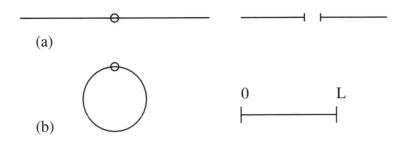

Fig. 27.1.   The electron system is effectively cut at the impurity position; (a) infinite system, (b) periodic system.

increases upon lowering the cut-off, i.e. the weak-coupling fixed point $(V = 0)$ is unstable. This situation is common for all the impurity prob-lems of interest (an impurity operator can be relevant, like in the case of the impurity in a Tomonaga–Luttinger liquid, or 'marginally relevant', like in the case of the Kondo problem, see Chapter 28).

In this Section we approach the central question: what happens to the system when the impurity operator is relevant?

As we have learned throughout the book, a relevant bulk operator completely changes physical properties of the system, leading to the open-ing of a gap in the excitations spectrum. We can not expect, of course, that an impurity operator, present at a single point in the space, opens a gap in the bulk spectrum. We can, however, follow a natural course of action assuming that the impurity potential continues to increase grad-ually with decreasing the cut-off, so that it eventually becomes infinite: $V(\Lambda \to 0) \to \infty$. Such an assumption allows a simple physical interpreta-tion. Indeed, an infinite impurity potential simply means that the system is disconnected at the impurity site, so that the original electron system is cut into two independent semi-infinite pieces: Fig. 27.1(a). Alternatively, if we consider a system with periodic boundary conditions (a circle of the length L), then an infinite impurity potential at the origin will make a segment out of it, with open ends: Fig. 27.1(b). The two pictures are clearly identical for $L \to \infty$.

Thus, the strong-coupling fixed point $(V = \infty)$ is identified as two independent semi-infinite Tomonaga–Luttinger liquids. It is not yet clear, though, whether the renormalization group trajectories really end up at the point $V = \infty$. So, the next step is to study the stability of the strong-coupling fixed point. Physically, deviations from $V = \infty$ mean that a residual electron tunneling between two semi-infinite pieces in Fig. 27.1(a) is allowed. Thus, the vicinity of the strong coupling fixed point is described by a weak link between two Tomonaga–Luttinger liquid leads.

The Hamiltonian of the weak link model can be written as:

$$H_{WL} = H_{LL1} + H_{LL2} + t \left[\psi_1^\dagger(0)\psi_2(0) + \text{H.c.} \right] \qquad (27.23)$$

where $\psi_{1(2)}$ is the electron field operator corresponding to the lead 1(2), H_{LL1} and H_{LL2} stand for the interacting Tomonaga–Luttinger liquid Hamiltonians of the leads, and t is the electron tunneling amplitude. In order to study the stability of the strong-coupling fixed point, we need to determine the scaling dimension of the electron tunneling operator in (27.23). This is not a trivial task, for the scaling dimension of the electron field operator is influenced by the boundary. The role of the boundary conditions and their effects on the scaling dimensions are discussed in the next section.

III.1   Open boundary bosonization

Generally speaking, the strong-coupling fixed points of the impurity problems are subject to what is called boundary conformal field theory (Cardy (1990)). A comprehensive discussion of this theory is outside the scope of this book. Yet in this Section we shall discuss some particular results of the boundary conformal field theory. These results, which we derive from first principles following Eggert and Affleck (1992), Wang and Affleck (1994), and Fabrizio and Gogolin (1995), are sufficient to describe the strong-coupling fixed point of the impurity problem in the Tomonaga–Luttinger liquid and, in particular, determine the scaling dimension of the tunneling operator (27.23).

We shall consider the system shown in the Fig. 27.1(b): interacting electrons on the segment $0 \leq x \leq L$. The electron field operator satisfies vanishing (open) boundary conditions on the edges of the segment:

$$\psi(0) = \psi(L) = 0 \qquad (27.24)$$

These boundary conditions are essentially different from the periodic (and anti-periodic) boundary conditions and can not be dealt with in the framework of the bosonization method we have used so far. So we start from scratch by considering the single-electron eigenstates of the noninteracting Fermi system with the boundary conditions (27.24); and these are standing waves:

$$\psi(x) = \sqrt{\frac{2}{L}} \sum_k \sin(kx) c_k \qquad (27.25)$$

with $k = \pi n / L$, n being a positive integer. Just like for the impurity problem of Chapter 25, the Fermi surface consists of the single point $k = k_F$. In the spirit of the bosonization method, we concentrate on the vicinity of this point and define slow varying right and left moving fields:

$$R(x) = -\frac{i}{\sqrt{2L}} \sum_p e^{ipx} c_{k_F+p}$$

$$L(x) = \frac{i}{\sqrt{2L}} \sum_p e^{-ipx} c_{k_F+p}$$

(27.26)

such that $(p = \pi n/L)$

$$\psi(x) = e^{ik_F x} R(x) + e^{-ik_F x} L(x)$$

(27.27)

These fields, however, are not independent, as in the case of periodic boundary conditions, but satisfy the condition

$$R(x) = -L(-x)$$

(27.28)

Therefore, one can actually work with the right-moving field only, the left-moving one is then defined by the above relation. The technical trick is the same one which we have used in Chapter 25; namely, we define a new chiral (right-moving) Fermi field by

$$\Psi(x) = \begin{cases} R(x), & \text{for} \quad x > 0 \\ L(-x), & \text{for} \quad x < 0 \end{cases}$$

We now have to investigate how the boundary conditions (27.24) translate in terms of the field Ψ: the boundary condition

$$\psi(0) = 0$$

is automatically satisfied (simply stating the continuity of Ψ at $x = 0$), whereas the condition

$$\psi(L) = 0$$

implies that the operator $\Psi(x)$ should obey

$$\Psi(-L) = \Psi(L)$$

Thus, we can regard the field $\Psi(x)$ as defined for all x but obeying the periodicity condition with the period $2L$:

$$\Psi(x + 2L) = \Psi(x)$$

(27.29)

In terms of the right-moving field, the kinetic energy takes the form:

$$H_0 = v_F \int_0^L dx \left[R^\dagger(x)(-i\partial_x)R(x) + L^\dagger(x)(i\partial_x)L(x) \right]$$

(27.30)

$$= v_F \int_{-L}^L dx \Psi^\dagger(x)(-i\partial_x)\Psi(x)$$

Since $\Psi(x)$ is a chiral Fermi field obeying the usual periodic boundary conditions, it can be standardly bosonized:

$$\Psi(x) = \frac{1}{\sqrt{2\pi a_0}} e^{i\sqrt{4\pi}\phi(x)}$$

with $\phi(x)$ being a chiral (right-moving) Bose field. In particular

$$J(x) = \Psi^\dagger(x)\Psi(x) = \frac{1}{\sqrt{\pi}}\partial_x\phi(x)$$

is a $U(1)$ (level 1) Kac–Moody current (identical, in this simplest case, with the fermion density),

$$[J(x), J(y)] = \frac{i}{4\pi}\partial_x\delta(x-y) \tag{27.31}$$

and the free Hamiltonian (27.30) takes the simple form:

$$H_0 = \pi v_F \int\limits_{-L}^{L} dx : J^2(x) :$$

Notice that the current $J(x)$ is related to the chiral currents $J_R(x) = : R^\dagger(x)R(x) :$ and $J_L(x) =: L^\dagger(x)L(x) :$ by

$$J(x) = \begin{cases} J_R(x), & \text{for} \quad x > 0 \\ J_L(-x), & \text{for} \quad x < 0 \end{cases}$$

We have thus discovered that the task of bosonizing the free fermions with open boundary conditions is quite a trivial one. It may seem that taking into account the electron–electron interactions will greatly complicate things, for, in the chosen representation, these interactions suffer from nonlocality. Namely, for the interacting Tomonaga–Luttinger liquid Hamiltonian on a segment, we obtain

$$H_{LL} = H_0 + \int\limits_{0}^{L} dx : \left\{ g_4\left[J^2(x) + \bar{J}^2(x)\right] + gJ(x)\bar{J}(x) \right\} :$$

$$\tag{27.32}$$

$$= \int\limits_{-L}^{L} dx \left[(\pi v_F + g_4) : J^2(x) : + \frac{1}{2}g : J(x)J(-x) : \right]$$

The problem is, however, not as difficult as it looks, because the interaction terms are still quadratic in the Kac–Moody current (even though nonlocal in space). Therefore the Hamiltonian (27.32) can straightforwardly be diagonalized by a Bogol'yubov rotation:

$$J(x) \rightarrow cJ(x) + sJ(-x) \tag{27.33}$$

It is easy to check that the algebra (27.31) is preserved, provided that the parameters c and s satisfy $c^2 - s^2 = 1$, so we shall write them as $c = \cosh(\varphi)$ and $s = \sinh(\varphi)$ (φ being the Bogol'yubov rotation angle). Substituting (27.33) into (27.32), one finds that if

$$\tanh(2\varphi) = -\frac{g}{2(\pi v_F + g_4)} \tag{27.34}$$

then

$$H_{LL} = \pi v \int_{-L}^{L} dx \, : J^2(x) : \tag{27.35}$$

with the renormalized Fermi velocity given by: $v = (v_F + g_4/\pi)/\cosh(2\varphi)$. It is important to note that, provided φ is defined by (27.34), the quantity $K = \exp(2\varphi)$ is identical to the bulk Tomonaga–Luttinger liquid exponent (i.e., the exponent we would have obtained considering periodic system with the same interaction constants g_4 and g). We now have everything needed to find the representation for the electron field operators in terms of the free bosons. Observing that the transformation (27.33) for the current translates into the transformation

$$\phi(x) \rightarrow c\phi(x) - s\phi(-x)$$

for the phase field, we obtain

$$\Psi(x) \rightarrow \frac{1}{\sqrt{2\pi a_0}} \exp\left\{ i\sqrt{\frac{\pi}{K}} \left[\phi(x) + \phi(-x)\right] + i\sqrt{\pi K} \left[\phi(x) - \phi(-x)\right] \right\} \tag{27.36}$$

The representation (27.36) is the central result of the open boundary bosonization method. It allows one to calculate all the correlation functions of the Fermi field (see problems IV and V at the end of the Chapter) as well as describe various tunneling processes. In what follows we shall use this representation to study the strong-coupling fixed point of the impurity problem.

III.2   Strong-coupling fixed point

At the beginning of this Section we have made the conjecture that the impurity potential flows to infinity, eventually resulting in a weak link between two semi-infinite Tomonaga–Luttinger liquids (Fig. 27.1(a)). Now we are in a position to bosonize each of these semi-infinite leads, i.e. to bosonize the weak-link Hamiltonian (27.23). Close to the edge the electron field operators are given by[¶]

$$\psi_{1(2)}(0) \sim e^{i\sqrt{\frac{4\pi}{K}}\phi_{1(2)}(0)}$$

Defining the symmetric and the antisymmetric combinations of the Bose fields $\phi_{1(2)}$, we find that the symmetric combination decouples, while the relevant part of the weak-link Hamiltonian involves the antisymmetric field (which we denote as $\tilde{\phi}$) and takes the form:

$$H_{\mathrm{WL}} \to H_0[\tilde{\phi}] + \frac{\bar{t}}{\pi a_0} \cos\left[\sqrt{\frac{8\pi}{K}}\tilde{\phi}(0)\right] \qquad (27.37)$$

where $\bar{t}$ is proportional to the tunneling amplitude t in (27.23), with a nonuniversal coefficient. We recall that the field $\tilde{\phi}$ is chiral, so the Hamiltonian (27.37) is exactly of same form as the backscattering Hamiltonian (27.8). We conclude that, amazingly, the strong-coupling fixed point is described by the same model as the weak-coupling fixed point (but with an inverse exponent and, of course, different amplitude of the cos-interaction term).[‖] Recalling the discussion of Section (II.3), we conclude that the tunneling operator in (27.37) obeys the renormalization group equation

$$\frac{d\bar{t}}{dl} = \left(1 - \frac{1}{K}\right)\bar{t} \qquad (27.38)$$

so that its scaling dimension is given by the quantity

$$\bar{\Delta} = \frac{1}{K} \qquad (27.39)$$

inverse to the scaling dimension K of the potential scattering operator. Consequently, if the potential scattering operator is relevant, then the tunneling operator is irrelevant, and vice versa.

The conclusion is that, for repulsive interaction ($K < 1$), the strong-coupling fixed point is stable. So, there are good reasons to believe that the renormalization group flow continues to an infinite coupling, at low energies, breaking the system up into two independent leads (as we shall see in Chapter 28 there are also more interesting possibilities when both the strong-coupling and the weak-coupling fixed points are unstable). Of course, the above argument are qualitative; the exact solution of Section IV shall provide further support for them.

[¶] Strictly speaking, the electron field operator vanishes at the boundary. Thus, more accurately, we should have allowed the electron tunneling in (27.23) to take place in a small but finite vicinity of the edges. The corrections to the bosonized Hamiltonian, generated in this way, are highly irrelevant.

[‖] This is closely related to the so-called duality properties of the partition function, known from the quantum dissipation theory (Schmidt (1983)).

IV Exact solution at $K = 1/2$ and the conductance

In this Section we shall discuss the particular case ($K = 1/2$) when the boundary sine-Gordon problem has a rather simple nonperturbative solution. This solution was first devised for the equivalent problem of quantum dissipation (see Section V) by Guinea, Hakim and Muramatsu (1985).

For $K = 1/2$, the scaling dimension Δ of the backscattering operator (27.8) is equal to $1/2$ (see Section II.3) – the same as for a Fermi field operator. It is therefore tempting to substitute the boundary cos-term in the Hamiltonian (27.8) by

$$\cos\left[\sqrt{4\pi}\phi(0)\right] \rightarrow \sqrt{\frac{\pi a_0}{2}}\left[\Psi(0) + \Psi^\dagger(0)\right] \tag{27.40}$$

where Ψ is a new Fermi field. The free part of (27.8) will then take a quadratic form in terms of Ψ. The impurity term is, however, linear in the Fermi operators.

This leads to considerable technical difficulties, so instead of following the original solution, we employ the following trick. Namely, we replace the original Hamiltonian (27.8) by:

$$H_{bs} \rightarrow \bar{H}_{bs} = H_0[\phi] + \frac{2V_{bs}}{\pi a_0}\hat{s}_x \cos\left[\sqrt{4\pi}\phi(0)\right] \tag{27.41}$$

where $\hat{s}_x$ is the x-component of a spin-1/2 operator. Clearly the operator $\hat{s}_x$ is conserved, $\left[\hat{s}_x, \bar{H}_{bs}\right] = 0$, so that one can set $\hat{s}_x = 1/2$ (or $\hat{s}_x = -1/2$). The new Hamiltonian $\bar{H}_{bs}$ is therefore equivalent to the old one, H_{bs}. (From the mathematical point of view we have doubled the Hilbert space of the problem.)

Now, instead of (27.40), we choose the following way to re-fermionize the problem:

$$\hat{s}_+ = d^\dagger$$

$$\tag{27.42}$$

$$\Psi(x) = \frac{1}{\sqrt{2\pi a_0}}e^{i\pi d^\dagger d}e^{i\sqrt{4\pi}\phi(x)}$$

where d is a Fermi operator and $\Psi(x)$ is a (chiral) Fermi field. (The Jordan–Wigner type phase stands to assure the correct commutation rules; see also Chapter 28.)

In terms of the new Fermi fields (27.42), the Hamiltonian (27.41) takes the form

$$\bar{H}_{bs} = H_0[\Psi] + \frac{V_{bs}}{\sqrt{2\pi a_0}}(d^\dagger - d)\left[\Psi^\dagger(0) + \Psi(0)\right] \tag{27.43}$$

where $H_0[\Psi]$ is a free Hamiltonian for the chiral Fermi field. The advantage of this representation is that the Hamiltonian (27.43) is quadratic in the Fermi operators. Therefore, it can be straightforwardly diagonalized.

It is instructive to introduce the Majorana components of the Fermi fields,

$$d = \frac{1}{\sqrt{2}}(a + ib)$$

$$\Psi(x) = \frac{1}{\sqrt{2}}[\xi(x) + i\zeta(x)]$$

(27.44)

Then the relevant (coupled to the impurity) part of the Hamiltonian (27.43) reads

$$\bar{H}_{bs} = H_0[\xi] - i\sqrt{\frac{2}{\pi a_0}}V_{bs}b\xi(0)$$

(27.45)

(where we have omitted the decoupled $H_0[\zeta]$ term).

Thus, the $K = 1/2$ boundary sine-Gordon problem reduces to what can be identified as a Majorana resonant-level model, (27.45). The Majorana resonant-level model also plays an important role in the theory of the two-channel Kondo effect: Chapter 28. The Green's functions can easily be found. So, the local Green's function for the Majorana fermion ξ,

$$G(\tau) = \langle\langle\xi(0,\tau)\xi(0,0)\rangle\rangle$$

satisfies

$$G(i\omega_n) = G^{(0)}(i\omega_n) + i\sqrt{\frac{2}{\pi a_0}}V_{bs}G^{(0)}(i\omega_n)\bar{F}(i\omega_n)$$

$$\bar{F}(i\omega_n) = -i\sqrt{\frac{2}{\pi a_0}}V_{bs}D^{(0)}(i\omega_n)G(i\omega_n)$$

where the 'anomalous' Green function $\bar{F}$ is defined by

$$\bar{F}(\tau) = \langle\langle b(0)\xi(0,\tau)\rangle\rangle$$

ω_n are fermionic Matsubara frequencies, and $D^{(0)}(\tau) = \langle\langle b(\tau)b(0)\rangle\rangle|_{V_{bs}=0}$ so that $D^{(0)}(i\omega_n) = 1/i\omega_n$. As immediately follows from the considerations of Chapter 25 and the definition (27.44), the bare local Green's function for a chiral Majorana fermion is

$$G^{(0)}(i\omega_n) = -\frac{i}{2v}\text{sign}\omega_n$$

(27.46)

and therefore

$$G(i\omega_n) = -\frac{i}{2v}\frac{\omega_n}{|\omega_n| + \gamma}$$

(27.47)

where $\gamma = V_{bs}^2/\pi v_F a_0$ is the energy scale characterizing the strength of the backscattering potential.

Conductance

In the experimental application of the boundary sine-Gordon problem to quantum wires, the quantity of interest is conductance. If the wire is connected to reservoirs with different chemical potentials (the difference being the external voltage V), then there is a static current $\langle j \rangle$ flowing through the system. The ratio of the current to the voltage, $G = \langle j \rangle/V$, is called conductance. In the rest of this Section we shall (exactly) calculate the conductance for the $K = 1/2$ case and (qualitatively) discuss it for the general case.

We shall make the physical assumption that the whole voltage drop occurs at the impurity site (see also the discussion at the end of this Section). Then the voltage term in the Hamiltonian can be written as

$$\frac{V}{2}(Q_+ - Q_-) \tag{27.48}$$

where Q_+ is the total electron charge to the right of the impurity,

$$Q_+ = e \int_0^\infty dx \left[R^\dagger(x)R(x) + (R \rightarrow L) \right] \tag{27.49}$$

and analogously Q_- is the total electron charge to the left of the impurity. These expressions along with bosonization formulas allow one to identify the current operator satisfying the continuity equation as

$$j = \frac{\partial Q_+}{\partial t} = -i\,[Q_+, H_{bs}] = ev\Psi^\dagger(0)\Psi(0) \tag{27.50}$$

Given the current operator (27.50), the linear conductance can be calculated by the Kubo formula

$$G = \lim_{\omega \to 0} \frac{1}{i\hbar\omega} K^{(R)}(\omega) \tag{27.51}$$

where $K^{(R)}$ is the retarded current–current correlation function,

$$K^{(R)}(t) = -\theta(t)\langle [j(t), j(0)]\rangle \tag{27.52}$$

calculated at a finite temperature (in real time). We shall find $K^{(R)}(\omega)$ as an analytic continuation of the Matsubara current–current correlation function $K(i\omega_n)$. Since $\Psi^\dagger(0)\Psi(0) = i\xi(0)\zeta(0)$, the latter is given by

$$K(i\omega_n) = e^2v^2T\sum_{\Omega_m} G(i\Omega_m)G^{(0)}(i\omega_n - i\Omega_m)$$

(note that Ω_m are fermionic Matsubara frequencies, while ω_n are the bosonic ones). Substituting the expression (27.47) for the Green's function and achieving the ultraviolet regularization by subtracting an infinite quantity $K(0)$ (which, being real, does not contribute to the conductance anyway) one finds (for $\omega_n > 0$):

$$K(i\omega_n) = -\frac{e^2}{4\pi}\left\{\omega_n - \gamma\left[\psi\left(\frac{1}{2} + \frac{\gamma + \omega_n}{2\pi T}\right) - \psi\left(\frac{1}{2} + \frac{\gamma}{2\pi T}\right)\right]\right\} \quad (27.53)$$

where $\psi(z)$ is the digamma function. According to Chapter 17 of Abrikosov, Gor'kov and Dzyaloshinskii (1963), the retarded correlator can be obtained by finding such a function $K^{(R)}(\omega)$, which is analytic in the upper half-plane and coincides with $K(i\omega_n)$ for the set of frequencies: $\omega = i\omega_n$ ($\omega_n > 0$). Clearly, the required function is

$$K^{(R)}(\omega) = \frac{e^2}{4\pi}\left\{i\omega + \gamma\left[\psi\left(\frac{1}{2} + \frac{\gamma - i\omega}{2\pi T}\right) - \psi\left(\frac{1}{2} + \frac{\gamma}{2\pi T}\right)\right]\right\} \quad (27.54)$$

Taking the limit $\omega \to 0$ in the formula (27.51), one finally obtains the conductance for the $K = 1/2$ case (Kane and Fisher (1992); Weiss, Egger and Sassetti (1995))

$$\frac{G}{G_0} = 1 - \frac{\gamma}{2\pi T}\psi'\left(\frac{1}{2} + \frac{\gamma}{2\pi T}\right) \quad (27.55)$$

with $G_0 = e^2/2h$ being the so-called perfect conductance (namely, the conductance of the pure system, $\gamma = 0$).

Analysing Eq. (27.55), one finds that $G \propto T^2$ for $T \ll \gamma$, i.e. the conductance vanishes with temperature. This result is consistent with the qualitative strong coupling analysis of the previous Section. Indeed, according to the latter, at low temperatures the impurity effectively cuts the system into two leads connected by an (irrelevant) electron tunneling operator t. The conductance of such a system is expected to scale as t^2 (which immediately follows from the second-order perturbation theory in the tunneling) and, since the scaling dimension of t is $1/K$, the low-temperature behaviour of the conductance is

$$G \propto T^{2/K-2} . \quad (27.56)$$

which indeed agrees with (27.55) at $K = 1/2$. This power-law asymptotics is also consistent with the exact result for $T = 0$ but finite V conductance found by Tsvelik (1995) by means of the Bethe ansatz calculation.

In conclusion we would like to stress that the conductance, as we have defined it above, is not trivially related to the conductance measured in the real experiments (Tarucha, Honda and Saku (1995)). The latter quantity is currently the subject to an intensive discussion in the literature. Since so far there is no definite conclusion to this discussion, we refrain from going into detail. We only note that while the important role is played by the 'contact regions' (Maslov and Stone (1995); Safi and Schultz (1995); the finite frequency effects also being of interest: Guinea *et al.* (1995)), the general strong-coupling prediction of a conductance vanishing with temperature according to a power-law (27.56) survives (provided the experimentally used wires are long enough for a given strength of the bare backscattering potential).

V Relation of the impurity backscattering model to the Caldeira–Leggett model

The dynamics of a particle coupled to a dissipative environment is a long standing problem in condensed matter physics. In its most general form, this problem has been formulated by Caldeira and Leggett (1983). They studied the Hamiltonian

$$H_{\text{CL}} = H_0[X] + H_{\text{XB}}[x, \{x_q\}] \qquad (27.57)$$

where the Hamiltonian $H_0[X]$ describes the dynamics of the particle X, while $H_{\text{XB}}[x, \{x_q\}]$ describes the environment ($\{x_q\}$ being the the variables of the environment to be specified shortly) and the coupling of the particle to the environment. Caldeira and Leggett considered a particular case when the particle is linearly coupled to a bath of harmonic oscillators:

$$H_{\text{XB}}[x, \{x_q\}] = \sum_q \left\{ \frac{1}{2} \left(P_q^2 + \omega_q^2 x_q^2 \right) + C_q x_q X + \frac{C_q^2}{2\omega_q^2} X^2 \right\} \qquad (27.58)$$

where P_q and x_q are the momenta and the coordinates of the oscillators, ω_q are their frequencies, and C_q characterizes the coupling of the q-th oscillator to the particle X.**

Somewhat surprisingly, the simple formulation of the problem, banked on a linear coupling and a harmonic bath, covers most of the problems of interest, ranging from physics of SQUIDs and two-level systems to the X-ray edge problem (which is a particular case of (27.57), with the electron–hole excitations playing the role of bath's oscillators). So, the effects due to a nonlinearity of the coupling and/or an anharmonicity of the bath are not of a primary interest (see, however, next Section). We have not yet specified the particle Hamiltonian $H_0[X]$. The case of a free particle $H_0[X] = P_X^2/2M$ can, of course, be easily solved, with the coupling (27.58) merely being a quantum generalization of the linear friction force. The situation is more involved when the particle is subjected to an external potential $V(X)$. Various forms of the external potential correspond to different physical problems. So, the potential $V(X)$ can be

---

** The last term in (27.58) is needed to compensate the adiabatic energy shift, i.e. to preserve the invariance of (27.58) with respect to translations $X \to X + X_0$, $x_q \to x_q - C_q X_0/\omega_q^2$.

of the form $V(X) \propto X^2 - X^3$ to describe a dissipative tunneling out of a quasi-stationary state (e.g., slipping of a Josephson phase), it can have two minima to describe a two-level system, or it can be periodic (see below). A discussion of the various versions of the problem (27.57) is clearly outside the scope of this book. We refer the interested reader to the extensive literature on these subjects (apart from Caldeira and Leggett (1983), see Leggett *et al.* (1987), Fisher and Zwerger (1985), and also Kagan and Prokof'ev (1992) for a discussion of the quantum diffusion in real metals, which though not exactly captured by the Caldeira–Leggett model has many similar features). In the rest of this Section, we shall focus on a particular but important issue of the 'localization' of the particle due to its coupling to the bath and establish the relation of the Caldeira–Leggett problem to the boundary sine-Gordon problem.

It is worth noting that, as far as the dynamics of the particle is concerned, only a certain combination of the individual parameters of the oscillators enters the calculations. The simplest way to realize this is to integrate out the degrees of freedom belonging to the bath. As a result the action of the particle will acquire the following correction (see Problem X):

$$S_{\text{diss}} = \int\limits_{-\infty}^{\infty} d\tau' \int\limits_{0}^{\beta} d\tau K(\tau - \tau') X(\tau) X(\tau') \tag{27.59}$$

with the kernel

$$K(\tau - \tau') = \frac{1}{2\pi} \int\limits_{0}^{\infty} d\omega e^{-\omega|\tau - \tau'|} I(\omega)$$

where the function

$$I(\omega) = \frac{\pi}{2} \sum_{q} \frac{C_q^2}{\omega_q} \delta(\omega - \omega_q) \tag{27.60}$$

is referred to as the spectral function. It is the asymptotic form of the spectral function at low frequencies which is important. It turns out that in most cases the spectral function is linear at small ω:

$$I(\omega) \simeq \eta\omega \tag{27.61}$$

where the quantity η is called 'friction coefficient'. The case when the spectral function is of the form (27.61) is referred to as Ohmic dissipation.

Consider now the particle moving on a lattice with the period X_0, so that $X = nX_0$ (n being an integer) and

$$H_0[X] = \bar{\omega} \sum_{n} [|nX_0\rangle\langle nX_0 + X_0| + \text{H.c.}] \tag{27.62}$$

$\bar{\omega}$ being the bare tunneling amplitude. In general, the high-energy degrees of freedom of the environment ($\omega > \omega_0$) adiabatically follow the moving particle while those with low energies ($\omega < \omega_0$) are essentially free, not being able to adjust to the moving particle. The frequency ω_0, which separates the low- and the high-energy degrees of freedom, can be identified as the actual, renormalized tunneling amplitude of the particle. This amplitude can be found by considering the overlap integral of the oscillators' wave functions, corresponding to the particle being fixed at the nearby sites of the lattice:

$$\omega_0 = \bar{\omega} \langle nX_0; \{x_q\} | nX_0 + X_0; \{x_q\} \rangle = \bar{\omega} \exp \left\{ -\frac{X_0^2}{2\pi} \int\limits_{\omega_0}^{\infty} \frac{d\omega}{\omega^2} I(\omega) \right\} \quad (27.63)$$

According to what was said above, the low-frequency modes ($\omega < \omega_0$) do not enter the expression (27.63) for the overlap integral. Yet, in the Ohmic dissipation regime, (27.61), the overlap integral is a vanishing function of ω_0, as $\omega_0 \to 0$:

$$\omega_0 = \bar{\omega} \left(\frac{\omega_0}{\omega_c} \right)^{\alpha} \quad (27.64)$$

We have defined a dimensionless constant

$$\alpha = \frac{\eta X_0^2}{2\pi}$$

characterizing the dissipation strength, and introduced the ultraviolet cut-off frequency ω_c (i.e., the maximal frequency of the bath oscillators).

The renormalized tunneling amplitude can be found by solving the self-consistency equation (27.64):

$$\omega_0 = \begin{cases} \omega_c \left(\dfrac{\bar{\omega}}{\omega_c} \right)^{\frac{1}{1-\alpha}} & \text{for} \quad \alpha < 1, \\ 0, & \text{for} \quad \alpha > 1 \end{cases} \quad (27.65)$$

Thus, the effective tunneling amplitude decreases with increasing dissipation strength α, eventually approaching the threshold $\alpha = 1$, above which the particle is localized. Though this simple argument in favour of the existence of a localized phase looks very qualitative, it has been confirmed by means of more elaborate methods such as duality (Schmidt (1983)) and renormalization group (Fisher and Zwerger (1985), see also Bulgadaev (1984)).

A keen reader may have already noticed a similarity between the low-frequency behaviour of the dissipative action (27.59) (with the spectral function (27.61)) and the action (27.16) we derived in Section II.2. One can therefore expect that the Caldeira–Leggett model and the boundary sine-Gordon model are related. In the rest of this Section we establish an exact equivalence between them. Bearing in mind the backscattering Hamiltonian (27.8), we start by considering the Fourier mode expansion of the Bose field:

$$\phi(0) = -\frac{\theta}{\sqrt{8\pi K}} + \sum_{q>0} \frac{1}{\sqrt{2qL}} \left(b_q + b_q^\dagger \right) \tag{27.66}$$

This is, of course, a particular case of the expansion (4.22) of Chapter 4, with the definition $b_q = (\hat{a}_q + \hat{a}_{-q})/\sqrt{2}$. In this paragraph we use a rescaled phase variable corresponding to the zero mode, $\theta = -\sqrt{4\pi K}\Phi_0$, so that θ is defined modulo 2π. The operator J, satisfying

$$[\theta, J] = 2i \tag{27.67}$$

describes the total electron current through the system (the difference between the number of electrons in the right- and left-moving branches of the spectrum). The backscattering Hamiltonian can therefore be written in the form

$$H_{bs} = \sum_{q>0} vq b_q^\dagger b_q + \frac{\pi K v}{2L} J^2 + \frac{V_b}{\pi a_0} \cos\left\{ -i\theta + i\sum_{q>0} \sqrt{\frac{4\pi K}{qL}} \left(b_q + b_q^\dagger \right) \right\} \tag{27.68}$$

Notice that the total current can only be changed by two units by transferring an electron from the left to the right branch (or vice versa). Thus one can, in agreement with the commutation relation (27.67), interpret the quantity J as a coordinate of a 'particle' on the lattice with the period 2: $J = 2n$. It is the backscattering operator which changes the total current and causes the 'particle' to hop from one site of the lattice to another. Each event of hopping is, however, accompanied by a complicated pattern of excited electron density modes – bosonic oscillators (the last term in (27.68)). This can be helped by employing the canonical transformation[††]

$$U_{CL} = \exp\left\{ i\sum_{q>0} \sqrt{\frac{\pi K}{qL}} \left(b_q + b_q^\dagger \right) J \right\} \tag{27.69}$$

The transformed Hamiltonian takes the desired form, i.e. describes a 'particle', J, linearly interacting with a bath of harmonic oscillators:

$$\tilde{H}_{bs} = U_{CL} H_{bs} U_{CL}^\dagger = \frac{V_b}{\pi a_0} \left[|J\rangle\langle J+2| + \text{H.c.} \right] + \frac{\pi K v}{2L} J^2$$

$$+ \sum_q \left\{ \frac{1}{2} \left(P_q^2 + \omega_q^2 x_q^2 \right) + C_q x_q X + \frac{C_q^2}{2\omega_q^2} X^2 \right\} \tag{27.70}$$

[††] A similar transformation has been used in the theory of polarons long ago. In the context of dissipation theory it was first applied by Guinea *et al.* (1985).

The variables corresponding to the harmonic oscillators are identified by $x_q = i\left(b_q^\dagger - b_q\right)/\sqrt{2\omega_q}$, $P_q = -\sqrt{\omega_q/2}\left(b_q + b_q^\dagger\right)$ with $\omega_q = vq$ and $C_q = \sqrt{2\pi g v^3 q^2/L}$. For the spectral function one easily finds that it corresponds to the Ohmic case (27.61) with the friction coefficient $\eta = \pi K/2$ and ($X_0 = 2$) the dimensionless dissipation strength being simply $\alpha = K$.

We therefore conclude that the boundary sine-Gordon problem is exactly equivalent to the Caldeira–Leggett problem on a one-dimensional lattice with Ohmic bath. The role of the dissipative particle is played by the total electron current, the kinetic energy of which is given by the backscattering potential. (Note that J is also subjected to a weak harmonic potential: the second term in (27.70), which vanishes in the limit $L \to \infty$.) The different parameter regions of the two models map onto each other as follows:

- $K < 1$ (backscattering is relevant) – J is delocalized,

- $K > 1$ (backscattering is irrelevant) – J is localized.

Indeed, without the backscattering the total current is simply conserved. As for when the backscattering is switched on, the total current becomes a dynamic variable, still localized when the backscattering is irrelevant. On the other hand, if the backscattering operator is relevant it qualitatively changes the properties of the system: the total current becomes delocalized. The delocalization means that the discreteness of the variable J is not so important and it may be made continuous. The continuity approximation essentially means substituting the hopping term in (27.70) by the mass term:

$$\frac{V_0}{\pi a_0}\left[|J\rangle\langle J + 2| + \text{H.c.}\right] \to \frac{1}{2}\omega_0\theta^2 \tag{27.71}$$

with the parameter ω determined, in the order of magnitude, from Eq. (27.65) with $\bar{\omega} = V_0/\pi a_0$. It is worth noting that, in terms of the original backscattering Hamiltonian (i.e. before the transformation (27.69)), the continuity approximation reads:

$$H_{\text{BSG}}[\Phi] \to H_{\text{eff}}[\Phi] = H_0[\Phi] + \frac{1}{2}\omega_0\Phi^2(0) \tag{27.72}$$

from which one can see that the field Φ acquires a finite expectation value. The approximation leading to the the the effective Hamiltonian (27.72) is also referred to in the literature as a 'self-consistent harmonic approximation'. Though one should be careful while calculating the correlation functions from (27.72) (for which the open boundary approach of Section III is better suited), it correctly describes the fixed point itself and will prove useful in the next Section.

VI   X-ray edge problem in Tomonaga–Luttinger liquids

In Chapter 26 we have extensively discussed the X-ray edge response functions of noninteracting Fermi systems. We have also brought forward qualitative arguments supporting a notion that, provided the dimension of the system is higher than one, the electron–electron interactions in the conduction band (which make the system a Fermi liquid) do not qualitatively change the X-ray response. As we shall see in this Section, the interactions play a much more important (and interesting) role in one-dimensional systems (Tomonaga–Luttinger liquids). The X-ray edge problem in Tomonaga–Luttinger liquids has been widely debated in the literature, partly because it is of certain theoretical interest and partly in response to the measurements of Fermi edge singularities in quantum wires (Calleja *et al.* (1991); Fritze *et al.* (1994)). We shall not attempt to discuss these experiments in any detail but rather proceed with the basic theory assuming, like in Chapter 26, a vanishing natural width of the core-hole level.

Then, in full analogy to the discussion of Section II of the previous Chapter, the Hamiltonian of the problem is:

$$H = H_{\mathrm{LL}} + \hat{V} dd^{\dagger}$$

(*d* being the core-hole creation operator), so that the initial state Hamiltonian is simply the Tomonaga–Luttinger liquid Hamiltonian,

$$H_{\mathrm{i}} = H_{\mathrm{LL}}$$

while the final state Hamiltonian corresponds to the Tomonaga–Luttinger liquid with impurity (27.1):

$$H_{\mathrm{f}} = H_{\mathrm{LL}} + \hat{V}$$

As we have shown in Section II.1, this Hamiltonian, being bosonized, splits into two independent parts describing the forward and the backward scattering from the impurity potential: $H_{\mathrm{f}} = H_{\mathrm{fs}} + H_{\mathrm{bs}}$, with H_{fs} and H_{bs} given by (27.7) and (27.8) respectively. It follows that the X-ray response functions simply factorize with respect to the forward and the backscattering processes and these factors can be computed separately.

The forward scattering Hamiltonian which is, of course, of the same form as the effective one-dimensional Hamiltonian (26.122) for the impurity scattering in Fermi liquids, can be diagonalized by the canonical transformation (27.10). Therefore, the forward scattering contribution to the X-ray edge response functions can be easily calculated. So, for the core-hole Green function one obtains (26.40):

$$D_{fs}(t) = -i\langle 0|U(t)U^{\dagger}(0)|0\rangle \sim \left(\frac{1}{\omega_c t}\right)^{\frac{\delta_{fs}}{2\pi^2}} \tag{27.73}$$

so that the forward scattering contribution to the orthogonality exponent is given by

$$\alpha_{orth}^{fs} = \frac{\delta_{fs}}{2\pi^2} \quad \text{with} \quad \delta_{fs} = \frac{\sqrt{K}V_{fs}}{v} = 2\sqrt{K}\frac{v_F}{v}\delta_0 \tag{27.74}$$

(note the factor of two difference in the definition of the forward scattering phase shift as compared to the chiral phase shift (26.120)). The forward scattering phase is thus renormalized by the interaction (Ogawa, Furusaki and Nagaosa (1992); Lee and Chen (1992)), though the latter does not influence the X-ray response in a qualitative way.

It is the backscattering contribution which is most interesting. Thus, in the rest of this Section we study the problem of a transient backscattering potential:

$$H_{bs}(t) = H_0[\Phi] + \frac{V_{bs}(t)}{\pi a_0}\cos\left[\sqrt{4\pi K}\Phi(0)\right] \tag{27.75}$$

It is instructive to start with perturbative considerations like the linked cluster expansion of Section III (previous Chapter). The analogue of the one-loop contribution to $C(t)$ for the problem (27.75) (with $V_{bs}(t) = V_{bs}\theta(t)$) is determined by the formula (26.47),

$$C_{bs}^{(2)}(t' - t) = -iV^2 \int_{-\infty}^{\infty} \frac{d\Omega}{2\pi} \frac{1 - \cos\left[\Omega(t' - t)\right]}{\Omega^2}\chi_{bs}(\Omega)$$

Here χ_{bs} is the local charge polarization related to backscattering processes:

$$\chi_{bs}(t) = -i\langle T\left\{\cos\left[\sqrt{4\pi K}\Phi(t)\right]\cos\left[\sqrt{4\pi K}\Phi(0)\right]\right\}\rangle$$

As we know this function decays at large time as $\chi_{bs}(t) \sim -i/|t|^{2K}$. Therefore, the local electron–hole density of states $[\Im m\chi_{bs}(\Omega) = -\pi\rho_{eh}(|\Omega|)]$ is enhanced by repulsive interactions

$$\rho_{eh} \sim |\Omega|^{2K-1} \tag{27.76}$$

This is related (due to a Kramers–Krönig-type relation) to the enhancement of the Peierls susceptibility (given by the real part of the same function, χ_{bs}) in a system with repulsive interactions, discovered in the early applications of the bosonization method (Luther and Peshel (1974)). Notice that it is the backscattering contribution to the total charge susceptibility which is enhanced, the forward scattering contribution remains linear in Ω. As follows from the discussion of Section III (previous Chapter), the linearity of the electron–hole density of states is required for the X-ray response functions to behave according to power-law and enables one to define exponents. On the other hand, combining (27.76) with (26.47), one finds that in the repulsive TL liquid (Meden (1992))[‡‡]

$$\Re C_{\text{bs}}^{(2)}(t' - t) = -\frac{1}{2}\left(\frac{V_{\text{bs}}}{\pi v}\right)^2 \gamma \frac{[\omega_{\text{c}}(t' - t)]^{2(1-K)} - 1}{2(1 - K)} \qquad (27.77)$$

where γ is a numerical coefficient, an explicit expression for which is not of interest here (but $\gamma \to 1$ for $K \to 1$). The time dependence of the contribution (27.77) is unusual – unlike the Fermi liquid, in the Tomonaga–Luttinger liquid the core-hole propagator does not decay as a power-law (in the second order in V_{bs}). The X-ray response has therefore an anomalous character. This is, of course, a consequence of the fact that the impurity backscattering operator is relevant in the TL liquid but only marginal in the Fermi liquid. It is important to realize, however, that the above results are not applicable at large time intervals (again in contrast with the Fermi liquid case). So, the fourth-order contribution $C^{(4)}$ grows as $(t' - t)^{4(1-K)}$. There is, therefore, a crossover time in the problem,

$$t_0 \sim \frac{1}{\omega_{\text{c}}}\left(\frac{\omega_{\text{c}}}{V_{\text{bs}}}\right)^{1/(1-K)} \qquad (27.78)$$

(notice that $t_0 \sim 1/\omega_0$, as expected) so that for $t \ll t_0$ the result (27.77) is correct but in order to determine the behaviour of the system at large times $t \gg t_0$, one needs to treat the backscattering potential nonperturbatively.

It is not *a priori* clear that the core-hole propagator resumes a power-law form in the nonperturbative regime ($t \gg t_0$). Yet it does, as we shall derive in what follows. Moreover, the exponent characterizing this resumed power-law turns out to be a universal number. It is important to realize that the long-time behaviour of the X-ray response is the property of the strong-coupling fixed point itself: it is determined by a projection of the eigenstates of the homogeneous system onto those of the system cut at the impurity site. It is therefore of no importance how the fixed point is being approached and one can choose any Hamiltonian interpolating between the weak-coupling and the strong-coupling fixed points. An evidently convenient choice would be Eq. (27.72) – the self-consistent harmonic approximation, the transient version of which reads

$$H_{\text{eff}}(t) = H_0[\Phi] + \frac{1}{2}\omega_0(t)\Phi^2(0) \qquad (27.79)$$

The advantage of the Hamiltonian (27.79) is that it is quadratic in the

---

[‡‡] For an attractive TL liquid the integral (26.47) would not diverge, indicating the absence of the orthogonality catastrophe with respect to the backscattering potential. This is not surprising as we know that the latter is irrelevant for $K < 1$. Yet the orthogonality catastrophe related to the forward scattering potential will survive.

Bose field Φ ((27.79) is applicable for large time intervals only, $t \gg t_0$; it leads to a totally wrong short-time behaviour). We can therefore express the core-hole propagator in terms of the transient Green's function (see Section III of Chapter 26) of the Bose field

$$C_{bs}(t' - t) = -\frac{\omega_0}{2} \int_t^{t'} d\tau \int_0^1 d\lambda D_\lambda(\tau, \tau + 0|t, t') \qquad (27.80)$$

where D_λ is determined by the analogue of the Nozières–De Dominicis equation,

$$D_\lambda(\tau, \tau'|t, t') = D_0(\tau - \tau') + \lambda \omega_0 \int_t^{t'} d\tau'' D_0(\tau - \tau'') D_\lambda(\tau'', \tau'|t, t') \qquad (27.81)$$

with D_0 being the bare propagator of the field Φ. Unfortunately, the equation (27.81) does not admit an exact solution. Yet, according to the logic of Section V (Chapter 26), it can be solved for an infinite time interval $(t = 0, t' = \infty)$, when Eq. (27.79) defines the overlap integral

$$\ln |\langle 0|V_{bs}\rangle| = -\frac{\omega_0}{2} \int_0^\infty d\tau \int_0^1 d\lambda \Re e D_\lambda(\tau, \tau + 0|0, \infty) \qquad (27.82)$$

Moreover, writing the bare Bose propagator in the form

$$D_0(\omega) = \int_0^\infty P(E) dE \left[\frac{1}{\omega - E + i\delta} - \frac{1}{\omega + E - i\delta} \right]$$

with $P(E) = 1/2\pi E$, we observe that the local Bose propagator $D_0(\omega)$ has exactly the same analytic properties as the local Fermi propagator $g_0(\omega)$ has (with $P(|\omega|)\mathrm{sgn}(\omega)$ playing the role of $\Im m g_0(\omega)$). As the bare propagator completely determines the structure of the the transient integral equation, we can spare our energy refraining from repeating the calculations of Section V (Chapter 26) – they are literally applicable to the present case, provided that the substitutions $\Im m g_0(\omega) \to P(|\omega|)\mathrm{sgn}(\omega)$ and $V \to \omega_0$ are made in all the formulas of that Section. (Then, the only difference between the two calculations is a trivial one: the $1/2$ factor in Eq. (27.82), which is due to the hermicity of the Bose field operator.) We can thus directly pass to the answer – the Hamman-type expression (26.111) for the overlap integral, which for the case of bosons takes the form

$$\ln |\langle 0|V_b\rangle| = -\frac{1}{2\pi^2} \int_0^1 d\lambda \int_0^\infty dE \int_0^\infty dE' \frac{\Delta_\lambda(E) \partial_\lambda \Delta_\lambda(E')}{(E + E')^2} \qquad (27.83)$$

where $\Delta_\lambda(E) = \tan^{-1}[\pi \lambda \omega_0 P(E)]$ is the bosonic scattering phase. Regularizing the integral at low energy by introducing $\omega_{min} = v/L$ (L being the length of the system), one finally obtains (Gogolin (1993); Prokof'ev

(1994))

$$\ln |\langle 0 | V_b \rangle| = -\frac{1}{16} \ln \left(\frac{\omega_0}{\omega_{min}} \right) \tag{27.84}$$

This expression tells us that the backscattering contribution to the overlap integral exponent is $\alpha_{OI}^{bs} = 1/16$. According to the analysis of Section V (Chapter 26), the orthogonality exponent is twice larger. The core-hole Green function therefore takes the form

$$D_{bs} \sim \left(\frac{1}{\omega_c t} \right)^{\frac{1}{8}} \tag{27.85}$$

We stress that this result is only valid for asymptotically large times $t \gg t_0$ (so, for vanishing interactions $K \to 1$, the region of applicability of (27.85) disappears: $t_0 \to \infty$).

Combining the results (27.73) and (27.85), one finds the total orthogonality exponent:

$$\alpha_{orth} = \alpha_{orth}^{fs} + \alpha_{orth}^{bs} == \frac{\delta_{fs}}{2\pi^2} + \frac{1}{8} \tag{27.86}$$

Exercises.

Problem I. Consider the impurity potential (27.2) for the electrons with spin and use the bosonized expression (27.3) for the electron field operators to show that the impurity potential takes the form

$$\frac{2V_{bs}}{\pi a_0} \cos \left[\sqrt{8\pi K_c} \Phi_c(0) \right] \cos \left[\sqrt{8\pi K_s} \Phi_s(0) \right] \tag{27.87}$$

Derive the local action, analogous to (27.16), for the spinful case.

Problem II. Show that the first-order renormalization group equation for the potential (27.87) is

$$\frac{dV_{bs}}{dl} = \left[1 - \frac{1}{2}(K_c + K_s) \right] V_{bs}$$

What operators will be generated in the higher orders of the renormalization group? Derive the second-order renormalization group equations for the impurity operator (with higher harmonics included) both for the spinless and spinful cases.

Problem III. Consider the impurity scattering problem in the vicinity of the point $K_c = 1/2$, $K_s = 3/2$. Show that the most relevant operators in the vicinity of this point are

$$\hat{V} = V \left[R_\uparrow^\dagger(0) L_\uparrow(0) + R_\downarrow^\dagger(0) L_\downarrow(0) + \text{H.c.} \right]$$

and

$$\hat{W} = W \left[R_\uparrow^\dagger(0) L_\uparrow(0) R_\downarrow^\dagger(0) L_\downarrow(0) + \text{H.c.} \right]$$

and that these operators obey the renormalization group equations:

$$\frac{dV}{dl} = \left[1 - \frac{1}{2}(K_c + K_s)\right] V - VW$$
$$\frac{dW}{dl} = (1 - 2K_c)W - V^2$$

(27.88)

Study the fixed points of these equations.

Problem IV. Calculate the electron Green's function for a semi-infinite Tomonaga–Luttinger liquid. Show that

$$G(x, y; t) = -i\langle \psi(x, t)\psi^\dagger(y, 0)\rangle = \sum_{a,b=\pm 1} ab e^{ik_F(ax - by)} \bar{G}(ax, by; t)$$

with

$$\bar{G}(x, y; t) = -i\langle \Psi(x, t)\Psi^\dagger(y, 0)\rangle$$

and, using formula (27.36), derive:

$$\bar{G}(x, y, t) = -\frac{i}{2\pi\alpha}\left[\frac{(x + y - vt)(x + y + vt)}{4xy}\right]^{sc}\left[\frac{a_0}{x - y - vt}\right]^{c^2}\left[\frac{a_0}{x - y + vt}\right]^{s^2}$$

(27.89)

Consider also the spatial dependence of the average electron density – the Friedel oscillation. Show that, at a large distance x from the boundary, the Friedel oscillation decays according to:

$$\delta n(x) = -\frac{1}{\pi a_0}\left[\frac{a_0}{\sqrt{a_0^2 + 4x^2}}\right]^K \sin(2k_F x)$$

(27.90)

How are the expressions (27.89) and (27.90) modified in the case of a TL liquid on a finite segment $(0 < x < L)$?

Problem V. Generalize the approach of Section III.1 to the case of electrons with spin. Show that, in particular, formula (27.36), should be generalized as:

$$\Psi_{\uparrow(\downarrow)}(x) \to \frac{1}{\sqrt{2\pi a_0}}$$

$$\exp\left\{i\sqrt{\frac{\pi}{2K_c}}[\phi_c(x) + \phi_c(-x)] + i\sqrt{\frac{\pi K_c}{2}}[\phi_c(x) - \phi_c(-x)]\right\}$$ (27.91)

$$\exp\left\{\pm i\sqrt{\frac{\pi}{2K_s}}[\phi_s(x) + \phi_s(-x)] \pm i\sqrt{\frac{\pi K_s}{2}}[\phi_s(x) - \phi_s(-x)]\right\}$$

Using this expression, recalculate the electron Green's function and the Friedel oscillation (see Problem IV).

Problem VI. Discuss the electron tunneling between two semi-infinite spin-1/2 TL liquid leads. Show that the tunneling operator is renormalized as

$$\frac{dt}{dl} = \left[1 - \frac{1}{2}\left(\frac{1}{K_c} + \frac{1}{K_s}\right)\right] t$$

Consider other tunneling processes and work out the following table.

Boundary operator $\hat{O}(t)$	bosonized form	exponent Δ_0
$\Psi_s(t)$	$\exp\{i\frac{\phi_c(t)}{\sqrt{2K_c}} \pm i\frac{\phi_s(t)}{\sqrt{2K_s}}\}$	$\frac{1}{4}\left(\frac{1}{K_c} + \frac{1}{K_s}\right)$
$\Psi_s^\dagger(t)\Psi_s(t)$	$\partial\phi_c(t) + \partial\phi_s(t)$	1
$\Psi_s^\dagger(t)\Psi_{-s}(t)$	$\exp\{\mp i\sqrt{\frac{2}{K_s}}\phi_s(t)\}$	K_s^{-1}
$\Psi_s(t)\Psi_s(t)$	$\exp\{i\sqrt{\frac{2}{K_c}}\phi_c(t) \pm i\sqrt{\frac{2}{K_s}}\phi_s(t)\}$	$K_c^{-1} + K_s^{-1}$
$\Psi_{-s}(t)\Psi_s(t)$	$\exp\{i\sqrt{\frac{2}{K_c}}\phi_c(t)\}$	K_c^{-1}

Problem VII. As in the case of a translationally invariant system, the spin backscattering interaction is not quadratic in terms of the Bose fields for a finite system. Show that, in the bosonized version of the model, the spin backscattering interaction,

$$\frac{g_{bs}}{2}\sum_s \int_0^L dx \left[R_s^\dagger(x)L_s(x)L_{\bar{s}}^\dagger(x)R_{\bar{s}}(x) + (R \to L)\right]$$

$$= \frac{g_{bs}}{2}\sum_s \int_{-L}^L dx\,\Psi_s^\dagger(x)\Psi_s(-x)\Psi_{\bar{s}}^\dagger(-x)\Psi_{\bar{s}}(x)$$

where $\bar{s}$ denotes $-s$, assumes the following form in terms of the spin phase field:

$$\frac{g_{bs}}{(2\pi a_0)^2} \int_{-L}^L dx\, e^{-i\sqrt{8\pi K_s}\phi_s(x)} e^{i\sqrt{8\pi K_s}\phi_s(-x)} e^{-i\pi K_s \text{sgn}(x)} \tag{27.92}$$

Problem VIII. Consider whether the Luther–Emery solution at $K_s = 1/2$ can be applied in the case of a finite system. Show that the Hamiltonian (27.92) indeed takes a quadratic form in terms of a new Fermi field at $K_s = 1/2$. Diagonalize the Hamiltonian at this point and calculate the excitation spectrum. What is the difference of this spectrum from the one of a translationally invariant problem?

Problem IX. Derive the expression (27.59) for the dissipative action. For this purpose consider the path integral

$$Q[q] = \int_{-\infty}^{\infty} dx_i \int_{x(0)=x_i}^{x(\beta)=x_i} Dx(\tau) \exp\left\{ -\int_0^\beta d\tau \left[\frac{1}{2} \left((\partial_\tau x)^2 + \omega^2 x^2 \right) + CX(\tau)x(\tau) \right] \right\}$$

and show that

$$\frac{Q[q]}{Q[0]} = \exp\left\{ \frac{C^2}{4\omega} \int_{-\infty}^{\infty} d\tau' \int_0^\beta e^{-\omega|\tau-\tau'|} X(\tau)X(\tau') \right\}$$

Problem X. Consider an electron system with the long-range Coulomb interaction: $U(r) = e^2/\kappa r$ in the real space. Rederive the bosonization procedure for such a system. Show that, in particular, the Caldeira–Leggett problem corresponding to the impurity in a system with Coulomb interactions falls into a 'sub-Ohmic' dissipation regime characterized by the spectral function

$$I(\omega) \sim \omega \ln^{1/2}(1/\omega)$$

Problem XI. Discuss the response of a Tomonaga–Luttinger liquid ring of length L with an impurity at the origin to a magnetic flux Φ passing through it. The presence of the flux leads to a finite current in the ground state of the system ('persistent current'):

$$j = -\frac{\partial \langle H[\varphi] \rangle}{\partial \varphi}$$

where $\varphi = 2\pi\Phi/\Phi_0$ (Φ_0 being the elementary flux quantum) and $H[\varphi]$ is obtained from the Tomonaga–Luttinger liquid Hamiltonian without the flux by substituting

$$J \to J - \varphi/\pi$$

Show that, at zero temperature, the persistent current is given by

$$j \sim \omega_0 \left(\frac{v}{\omega_0 L} \right)^{1/K} \sin \varphi$$

ant that, at finite temperatures, it scales as $j \propto L^{-1} T^{1/K-1}$.

Problem XII. Rederive the result (27.84) for the overlap integral exponent using the representation (26.108) of Section V (Chapter 26).

Problem XIII. Generalize the consideration of Section VI to the case of electrons with spin. Also find the exponent α for the X-ray absorption rate.

Problem XIV. Discuss the situation when the backscattering potential changes sign at $t = 0$, i.e. calculate the response functions for the transient potential of the form: $V_{bs}(t) = V_{bs}\text{sign}(t)$. Notice that, in this case, the system remains at the same (strong coupling) fixed point but the average value of the Bose field is shifted: $\langle\Phi(0)\rangle \to \langle\Phi(0)\rangle + \sqrt{\frac{\pi}{4K}}$. Show that

$$\langle V_{bs};t| - V_{bs};t\rangle \sim \left(\frac{1}{\omega_c t}\right)^{\frac{1}{2K}}$$

i.e. the dynamic exponent is nonuniversal (being dependent on the interaction yet not on the potential) and unbounded from above (which makes it impossible to interpret in terms of a scattering phase).

Problem XV. Discuss the X-ray edge response functions at the exactly solvable point $K = 1/2$: derive the exact equation for the transient response function and analyse them in the limit of large time intervals. Show that, in this limit, the results are consistent with those based on the effective Hamiltonian (27.79).

References

A. A. Abrikosov, L. P. Gor'kov and I. E. Dzyaloshinskii, *Methods of Quantum Field Theory in Statistical Physics*, ed. A. R. Silvermann, Prentice–Hall, Englewood, NJ (1963).

S. A. Bulgadaev, *Pis'ma Zh. Eksp. Teor. Fiz.* **39**, 264 (1984) [*JETP Lett.* **39**, 315 (1984)].

A. O. Caldeira and A. J. Leggett, *Ann. Phys.* **149**, 374 (1983).

J. M. Calleja et al., *Solid State Commun.* **79**, 911 (1991).

J. Cardy in *Fields, Strings, and Critical Phenomena*, ed. E. Brézin and J. Zinn-Justin, North-Holland, Amsterdam (1990).

S. Eggert and J. Affleck, *Phys. Rev.* **B46**, 10 866 (1992).

M. Fabrizio and A. O. Gogolin, *Phys. Rev.* **B51**, 17 827 (1995).

M. P. A. Fisher and W. Zwerger, *Phys. Rev.* **B32**, 6190 (1985).

M. Fritze et al., *Surface Science* **305**, 580 (1994).

A. Furusaki and N. Nagaosa, *Phys. Rev.* **B47**, 4631 (1993).

S. Ghoshal and A.B. Zamolodchikov, *Int. J. Mod. Phys.* A **9**, 3841 (1994).

T. Giamarchi and H. J. Schulz, *Phys. Rev.* **B37**, 325 (1988).

A. O. Gogolin, *Phys. Rev. Lett.* **71**, 2995 (1993).

A. O. Gogolin, *Ann. Phys. (Paris)* **19**, 411 (1994).

F. Guinea, J. Gómez–Santos, M. Sassetti, and M. Ueda, *Europhys. Lett.* **30**, 561 (1995).

F. Guinea, V. Hakim and A. Muramatsu, *Phys. Rev. Lett.* **54**, 263 (1985).

Yu. Kagan and N. V. Prokof'ev in *Quantum Tunneling in Condensed Matter*, ed. Yu. Kagan and A. J. Leggett, Elsevier (1992).

C. L. Kane and M. P. A. Fisher, *Phys. Rev. Lett.* **68**, 1220 (1992); *Phys. Rev.* **B46**, 1220 (1992).

M. A. Kastner, *Rev. Mod. Phys.* **64**, 849 (1992).

D. K. K. Lee and Y. Chen, *Phys. Rev. Lett.* **69**, 1399 (1992).

A. J. Leggett, S. Chakravarty, A.T. Dorsey, M. P. A. Fisher, A. Garg, and W. Zwerger, *Rev. Mod. Phys.* **59**, 1 (1987).

A. Luther and I. Peshel, *Phys. Rev.* **B9**, 2911 (1974).

D. Maslov and M. Stone, *Phys. Rev.* **B52**, R5539 (1995).

D.C. Mattis, *Phys. Rev. Lett.* **32**, 714 (1974).

V. Meden, Diploma-Thesis, Universität Göttingen (1992).

T. Ogawa, A. Furusaki and N. Nagaosa, *Phys. Rev. Lett.* **68**, 3638 (1992).

N. Prokof'ev, *Phys. Rev.* **B49** 2148 (1994).

I. Safi and H.J. Schultz, *Phys. Rev.* **B52**, R17040 (1995).

A. Schmidt, *Phys. Rev. Lett.* **51**, 1506 (1983).

S. Tarucha, T. Honda and T. Saku, *Sol. State Commun.* **94**, 413 (1995).

A. M. Tsvelik, *J. Phys. A.* **28**, 625-L (1995).

J. Voit, *Rep. Prog. Phys.* **58**, 977 (1995).

E. Wang and J. Affleck, *Nucl. Phys.* **B417**, 403 (1994).

U. Weiss, R. Egger, and M. Sassetti, *Phys. Rev.* **B52**, 16 707 (1995).

28

Multi-channel Kondo problem

I   Introduction

The Kondo problem has been discussed by condensed matter physicists for over three decades. Historically, it is the so-called single-channel Kondo model (or the $s - d$ exchange model), which was first proposed and investigated. The Hamiltonian of the single channel model is:

$$H = H_0[\psi] + I s^a \sum_{ss'} \psi_s^\dagger(0) \tau_{ss'}^a \psi_{s'}(0) \tag{28.1}$$

where s^a is the impurity spin-1/2 operator ($a = x, y, z$), τ^a are the spin-1/2 matrices, I is the exchange coupling, and ψ_s are the conduction electron field operators. The basic physics of the single-channel Kondo model has been well understood. At low temperatures, the impurity spin is screened by conduction electrons: the ground state is a nondegenerate singlet.[*] Consequently, the impurity magnetic susceptibility crosses over from the Curie–Weiss law $\chi \sim 1/T$ at high-T to a constant at $T = 0$ and the impurity contributes a T-linear term to the specific heat of the system. This is reminiscent of the bulk magnetic susceptibility and specific heat behaviour in Fermi liquids. It is therefore often said that the single-channel Kondo model has Fermi liquid properties (Fermi liquid ground state).

The model (28.1) is an effective low-energy model, the derivation of which involves several important assumptions. If (28.1) is devoted to describe an impurity ion embedded into a metallic host, the main assumptions are: (i) the electron scattering from the impurity ion is supposed to be local (i.e., any lattice effects, leading to a momentum dependence of the exchange coupling, are ignored); (ii) the localized electron wave functions, spanning the representation of the impurity spin operator $\mathbf{s}$, are considered

---

[*] From the renormalization group point of view this means that the system flows to a trivial, $I = \infty$, strong coupling fixed point.

as orbitally nondegenerate (i.e., it is only the Kramers degeneracy which is taken into account). Then, the operators ψ_s can be regarded as one-dimensional right-moving electron field operators. For real systems, both assumptions (i) and (ii) may turn out to be oversimplifying. The exchange coupling can never be exactly local, which involves, strictly speaking, an infinite number of one-dimensional electron channels. The channels with smaller couplings, though, have a tendency to die out in the course of the renormalization group transformations. A violation of the assumption (ii) can be more serious, for it requires several electron channels to be considered on equal footing.

In order to account for qualitatively new types of behaviour of the impurity ions that may arise from the above assumptions being violated, the following generalization of (28.1)

$$ H = \sum_{i=1}^{k} \left\{ H_0[\psi_i] + IS^a \sum_{ss'} \psi_{is}^\dagger(0)\tau_{ss'}^a \psi_{is'}(0) \right\} \qquad (28.2) $$

has been proposed, which is known as the multi-channel Kondo model. Here **S** is the spin-S ($\mathbf{S}^2 = S(S+1)$) operator and the operators ψ_{is}, with $i = 1, ..., k$, correspond to k conduction electron channels.[†] For $2S \geq k$, the low-temperature properties of the model (28.2) are very similar to those of the model (28.1) (in the so-called underscreened case, $2S > k$, a residual spin $S - k/2$ is left over, which is essentially free). A novel low-temperature behaviour arises in the (overscreened) case $2S < k$ where the renormalization group trajectory flows to a nontrivial finite coupling fixed point. This fixed point is characterized by a nonanalytic (in the temperature and magnetic field) free energy and power-law behaviour of the correlation functions. So, both the impurity magnetic susceptibility and the impurity specific heat coefficient,

$$ \chi_{\text{imp}} \propto \frac{\partial C_{\text{imp}}}{\partial T} \propto \left(\frac{1}{T}\right)^{\frac{k-2}{k+2}} $$

diverge upon lowering the temperature ($\chi_{\text{imp}} \propto \partial C_{\text{imp}}/\partial T \propto \ln(1/T)$ for $k = 2$). This behaviour is often referred to as non-Fermi-liquid behaviour.

There is a vast literature on the Kondo problem, including several good reviews: Wilson (1975); Nozières (1974); Tsvelik and Wiegmann (1983); Andrei, Furuya and Lowenstein (1983); Hewson (1993). So, why do we again address it in this book? The reason is that, in the past few years, there has been a breakthrough in the understanding of the non-Fermi-liquid properties of the multi-channel Kondo problem, primarily achieved

---

[†] The exchange coupling may be anisotropic in both spin and channel space, with obvious modifications of the Hamiltonian (28.2).

by the Abelian and non-Abelian bosonization methods. It is this recent development, which we intend to reflect in this chapter.

II   Qualitative analysis

Why should the multi-channel Kondo model (28.2) display a behaviour qualitatively different from the one of the single-channel model (28.1)? As we shall see shortly, the difference shows up in the vicinity of the strong-coupling fixed point and is due to the so-called overscreening phenomenon discovered by Nozières and Blandin (1980).

Let us, however, follow a conventional route and first explore what can be inferred from the perturbative analysis. The latter is almost synonymous to deriving the weak-coupling renormalization group. The first terms of the perturbative expansion for the β-function read (see Problem I):

$$\frac{\mathrm{d}\lambda_{\mathrm{K}}}{\mathrm{d}l} = \beta(\lambda_{\mathrm{K}}) = \lambda_{\mathrm{K}}^2 - \frac{k}{2}\lambda_{\mathrm{K}}^3 + \ldots \qquad (28.3)$$

where $\lambda_{\mathrm{K}} = I/2\pi v_{\mathrm{F}}$ is the dimensionless coupling constant. (The normalization to $2\pi v_{\mathrm{F}}$ corresponds to one-dimensional conduction electrons. If the effective one-dimensional model is derived for a higher-dimensional problem then the normalization should be changed as explained in Chapter 25.) For the sake of simplicity we focus on the case of the impurity spin $s = 1/2$ in what follows.

The first term of this expansion, originally derived by Abrikosov (1965) (see discussion and references in Appendix A to Chapter 26) in the framework of the parquet method highlights the difference with the problem of the impurity in the Tomonaga–Luttinger liquid (Chapter 27). Unlike (27.22), the expansion (28.3) starts with the second power of the coupling constant indicating that the exchange is a marginal operator: marginally relevant for $\lambda_{\mathrm{K}} > 0$ (antiferromagnetic case) but marginally irrelevant for $\lambda_{\mathrm{K}} < 0$ (ferromagnetic case that is almost trivial and will not be considered in what follows).

As noticed by Abrikosov and Migdal (1970), who derived the third-order term for $k = 1$, the β-function has, in the approximation (28.3), an intermediate-coupling fixed point at $\lambda_{\mathrm{K}} = \lambda_{\mathrm{K}}^* = 2/k$. For a large number of channels ($k \gg 1$), $\lambda_{\mathrm{K}}^* \to 0$ suggesting that the approximation (28.3) may be sufficient, which means that λ_{K}^* is a true fixed point. (A proof of this statement would require an analysis of the high-order terms of the β-function, see Gan, Andrei and Coleman (1993) and references therein.) Notice that a fixed point of this type is different from the infinite-coupling fixed point $\lambda_{\mathrm{K}} \to \infty$ (which is the only type of fixed points we have encountered so far studying the impurity problems). So, it is characterized by a finite slope of the β-function, $\beta'(\lambda_{\mathrm{K}}^*) = -2/k$, leading to a nontrivial

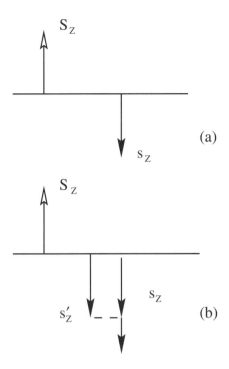

Fig. 28.1. The arrangement of spins at the impurity site in the strong-coupling limit for (a) $k = 1$ and (b) $k > 1$.

(power-law) scaling of the coupling constant and the free energy with the temperature and hence to anomalous exponents for observable quantities (these exponents coincide with the large-k limit of the exact ones, see e.g. Gan *et al.* (1993)).

If the number of channels is not large ($k = 1, 2$, etc.), there is no reason to trust the approximation (28.3) for the β-function at $\lambda_K = \lambda_K^*$ since the high-order terms of the perturbative expansion have a role to play. The very existence of the intermediate-coupling fixed point, $\lambda_K = \lambda_K^*$ is therefore questionable (so, for $k = 1$, such a fixed point is definitely absent). The criterion as to whether the β-function has a zero at a finite coupling was given by Nozières and Blandin (1980) following the line of reasoning earlier developed for the single-channel Kondo problem (Nozières (1974)). The main idea is very simple: one has to model the infinite-coupling fixed point ($\lambda_K = \infty$) and analyse its stability as it has been done for the weak-coupling fixed point. If it turns out that the $\lambda_K = \infty$ fixed point is stable then there are good reasons to expect the system just to flow to it; otherwise, there must exist an intermediate-coupling fixed point $\lambda_K = \lambda_K^*$ (λ_K^* being finite). An attentive reader has noticed that the same concept was made use of in our analysis of the impurity in the Tomonaga–Luttiger liquid (Chapter 27) – the infinite-coupling fixed point turned out to be stable in that case (for repulsive interactions).

In order to model the $\lambda_K = \infty$ fixed point, we consider our multi-channel Kondo problem on a lattice (the way the ultraviolet regularization is performed should not matter at low energies). At $\lambda_K = \infty$ the situation is very simple – the impurity traps as many electrons (let us denote the number of those electrons by n) as allowed by the Pauli principle: $n = k$ (see Fig. 28.1). In order to analyse the stability of the fixed point we have to perturb around it. Namely, we consider a very large but finite λ_K. Then a possibility for the conduction electron to hop on (and from) the impurity site opens up.

For $k = 1$, just one electron is trapped (Fig. 28.1(a)). It forms a singlet with the impurity spin. The $\lambda_K = \infty$ ground state is therefore nondegenerate. The perturbation expansion around it is regular – elaborating on it one arrives at the Nozières Fermi-liquid theory for the single-channel Kondo effect (Nozières (1974)). We conclude that the $\lambda_K = \infty$ fixed point is stable.

On the contrary, for $k > 1$, there is a residual spin $s' = k/2 - 1$ left over (Fig. 28.1(b)) in the $\lambda_K = \infty$ ground state which is thus degenerate and a subject to the renormalization group flow. By symmetry, the coupling of the residual spin s' to the conduction sea can only be of the exchange form:

$$I's' \sum_i \psi_i^\dagger \tau \psi_i$$

The stability of the $\lambda_K = \infty$ fixed point is determined by the sign of the coupling I'. The argument is straightforward: consider a conduction electron on a site of the lattice neighbouring to the impurity site. If the spin projection of that electron is parallel to the one of the residual spin s' then, this electron is not allowed to hop, by the Pauli principle. However, if the spin projections are anti-parallel then the hopping takes place and the energy of the system is lowered (as always in the second-order perturbation theory). One concludes that the residual spin and the conduction electron spin are preferably anti-aligned – the coupling must be antiferromagnetic ($I' > 0$).

To summarize, the $\lambda_K = \infty$ fixed point turns out to be unstable for any number of channels but one. (The argument can easily be extended to the case of the impurity spin $S > 1/2$ when the instability criterion becomes $k > 2S$, see Problem II.) A nontrivial fixed point at a finite coupling does therefore exist for the multi-channel Kondo problem – the rest of this Chapter is devoted to the analysis of this fixed point.

III   The Toulouse limit

It is natural to start studying how the Abelian bosonization methods work
in the Kondo problem with the discussion of the Toulouse limit for the
simplest, one-channel model. The Toulouse limit corresponds to the longitudinal exchange coupling taking a given value: $I_z = I_z^*$. For this value of
the coupling, the long-time dynamics of the Kondo model is equivalent to
that of some exactly solvable model (namely, the resonant-level model).
The Kondo model can therefore be easily solved at $I_z = I_z^*$ and depar-
tures from this limit are subject to a perturbative analysis. Historically,
Toulouse (1969) has established the equivalence of the models, analysing
the partition function (for more detail see Wiegmann and Finkelstein
(1978)) by means of the Yuval–Anderson (1970) expansion, based, in turn,
on the Nozières–De Dominicis (1969) solution to the X-ray edge problem
(the latter we have discussed in Chapter 26). The bosonization approach
not only provides a short-cut to this equivalence but also reveals certain
useful relations between the operators involved (Schlottmann (1978)).

The Hamiltonian of the spin-anisotropic Kondo model is given by:

$$H = H_0 + \frac{I_\perp}{2}\left[s_+ J^-(0) + \text{H.c}\right] + I_z s_z J^z(0) \tag{28.4}$$

where H_0 describes free electrons and the spin currents (densities) $J(x)$,
are, in the one-channel case, defined as:

$$J^-(x) =: \psi_\downarrow^\dagger(x)\psi_\uparrow(x): \quad \text{and} \quad J^z(x) = \frac{1}{2}:\left[\psi_\uparrow^\dagger(x)\psi_\uparrow(x) - \psi_\downarrow^\dagger(x)\psi_\downarrow(x)\right]:$$

Throughout this chapter, ψ_s ($s =\uparrow,\downarrow$) is understood to be a one-dimensional
right-moving Fermi field. The reader will undoubtedly recognize the $SU(2)$
level-1 Kac–Moody currents in $J(x)$. In what follows, we shall only consider impurity spin-1/2, denoted as s.

Bosonizing electron fields in a standard way,

$$\psi_s(x) = \frac{1}{\sqrt{2\pi a_0}}e^{i\sqrt{4\pi}\phi_s(x)}$$

where $\phi_s(x)$ are the right-moving Bose fields, one finds:

$$J^-(x) = \frac{1}{2\pi a_0}e^{i\sqrt{8\pi}\phi(x)} \quad \text{and} \quad J^z(x) = \frac{1}{\sqrt{2\pi}}\partial_x\phi(x) \tag{28.5}$$

Notice that it is the combination describing the spin degrees of freedom,

$$\phi(x) = \frac{1}{\sqrt{2}}\left[\phi_\uparrow(x) - \phi_\downarrow(x)\right]$$

which enters the bosonized expression (28.5) for the spin currents. The charge field,

$$\phi_c(x) = \frac{1}{\sqrt{2}} \left[\phi_\uparrow(x) + \phi_\downarrow(x)\right]$$

is completely decoupled from the impurity spin – the first achievement of the bosonization approach.

Thus, the bosonized Hamiltonian reads:

$$H = H_0[\phi] + \frac{I_\perp}{4\pi a_0} \left[s_+ e^{i\sqrt{8\pi}\phi(0)} + \text{H.c.}\right] + \frac{I_z}{\sqrt{2\pi}} s_z \partial_x \phi(0) \qquad (28.6)$$

where

$$H_0[\phi] = v_F \int dx \, [\partial_x \phi(x)]^2$$

describes a free right-moving Bose field ϕ (we have omitted the Hamiltonian $H_0[\phi_c]$ of the decoupled charge field).

Now we shall introduce a technical but important trick, which is commonly used in the theory of Kondo-type problems. Consider a canonical transformation accomplished by the unitary operator:

$$U = e^{i\sqrt{4\pi}\alpha s_z \phi(0)} \qquad (28.7)$$

Obviously the operator U commutes with s_z. However, it does not commute with other spin components and, using standard commutation relations for the spin operators, one easily finds that

$$U^\dagger s_+ U = s_+ e^{-i\sqrt{4\pi}\alpha\phi(0)}$$

Thus, the operator (28.7) generates a rotation around the s_z-axes in the spin space:

$$U^\dagger s_x U = s_x \cos\left[\sqrt{4\pi}\alpha\phi(0)\right] + s_y \sin\left[\sqrt{4\pi}\alpha\phi(0)\right]$$
$$U^\dagger s_y U = -s_x \sin\left[\sqrt{4\pi}\alpha\phi(0)\right] + s_y \cos\left[\sqrt{4\pi}\alpha\phi(0)\right]$$

Upon transformation (28.7), the transverse exchange term takes the form:

$$\frac{I_\perp}{4\pi a_0} s_+ e^{i\sqrt{4\pi}(\sqrt{2}-\alpha)\phi(0)} + \text{H.c.}$$

Therefore its scaling dimension has been changed and now is equal to:

$$\Delta = \frac{1}{2} \left(\sqrt{2} - \alpha\right)^2$$

This change of the scaling dimension is the purpose of the transformation (28.7). We can now choose a particular α such that the scaling dimension of the transverse exchange operator Δ is equal to $1/2$, i.e. to the scaling dimension of a single Fermi operator. Such a choice would require:

$$\alpha = \sqrt{2} - 1 \tag{28.8}$$

For this α, the transformed Hamiltonian takes the form:

$$H \to U^\dagger H U = H_0[\phi] + \frac{I_\perp}{4\pi a_0}\left[s_+ e^{i\sqrt{4\pi}\phi(0)} + \text{H.c.}\right] + \frac{\lambda}{\sqrt{\pi}}s_z\partial_x\phi(0) \tag{28.9}$$

where

$$\lambda = \frac{I_z}{\sqrt{2}} - 2(\sqrt{2} - 1)\pi v_F$$

The longitudinal exchange term has been modified since the canonical commutation relation

$$[\partial_x\phi(x), \phi(y)] = \frac{i}{2}\delta(x - y)$$

results in the shift

$$U^\dagger \partial_x\phi(x)U = \partial_x\phi(x) - \sqrt{\pi}\alpha s_z$$

As we have said, the operator $e^{i\sqrt{4\pi}\phi(0)}$ in the Hamiltonian (28.9) has the properties of a Fermi operator. Thus, defining a new Fermi field

$$\psi\,(x) = \frac{1}{\sqrt{2\pi a_0}}e^{i\pi d^\dagger d}\, e^{i\sqrt{4\pi}\phi(x)} \tag{28.10}$$

and using the fermionic representation for the impurity spin operators

$$s_+ = d^\dagger \quad \text{and} \quad s_z = d^\dagger d - \frac{1}{2} \tag{28.11}$$

one can refermionize the problem. Notice, that the definition (28.10) of the $\psi(x)$ field includes the phase factor, $e^{i\pi d^\dagger d}$, necessary to assure the correct anticommutation relations:[‡]

$$\{\psi(x), d\} = 0$$

In terms of the new fermions, (28.10) and (28.11), the Hamiltonian (28.9) reads:

$$H = H_0[\psi] + \frac{I_\perp}{2\sqrt{2\pi a_0}}\left[d^\dagger\psi(0) + \text{H.c}\right] + \lambda\left(d^\dagger d - \frac{1}{2}\right) : \psi^\dagger(0)\psi(0) : \tag{28.12}$$

Thus, we arrive at the model describing a resonant level d which is situated at the Fermi energy; being hybridized with the conduction

---

[‡] This phase factor is essentially similar to the one entering the standard Jordan–Wigner transformation.

electrons ($I_\perp$-term) it also interacts with the conduction sea (λ-term). The model radically simplifies at the particular value of the longitudinal exchange coupling

$$I_z = I_z^* = 2\sqrt{2}(\sqrt{2}-1)\pi v_F$$

where the interaction term in (28.12) vanishes (the Toulouse limit). The solution at the Toulouse limit produces a correct ('Fermi-liquid') low-energy behaviour. We do not dwell on this solution here, since we shall again meet the Hamiltonian (28.12) in the next section, where it appears as a limiting case of a more complicated problem (and where the free energy and the magnetic field response will be calculated).

Before proceeding any further, the following remark should be made. The position of the Toulouse limit (i.e. the quantity I_z^*) is not universal – as we have discussed at length in Chapter 25. A more accurate discussion of the mapping (28.12) in terms of electron scattering phase shifts can be found in Toulouse (1969). It should be stressed that these nonuniversal features of the Toulouse limit solution do not affect the results for low-temperature properties of the model.

IV   The Emery–Kivelson solution

It took just 23 years to generalize the Toulouse approach to the two-channel Kondo model (Emery and Kivelson (1992)). Of course, many interesting results have been found in the meanwhile, the most important of which was the prediction, made by Nozières and Blandin (1980), of a non-Fermi-liquid behaviour for multi-channel Kondo models. The Emery–Kivelson solution to the two-channel model has in a simple way revealed the reasons for such nonanalytic properties – the ground state degeneracy, the decoupling of certain degrees of freedom. In what follows we shall give a detailed account of the Emery–Kivelson solution in the form generalized to the channel anisotropic case (Fabrizio, Gogolin and Nozières (1995)).

For the case of the most general exchange coupling, anisotropic in the both channel and spin space, the Hamiltonian of the k-channel Kondo model takes the form:

$$H_K = \sum_{i=1}^{k} \left\{ H_0[\psi_i] + \frac{I_{\perp i}}{2}\left[s_+ J_i^-(0) + \text{H.c.}\right] + I_{zi}s_z J_i^z(0) \right\}$$

where

$$J_i^-(x) =: \psi_{i\downarrow}^\dagger(x)\psi_{i\uparrow}(x) : \quad \text{and} \quad J_i^z(x) = \frac{1}{2} : \left[\psi_{i\uparrow}^\dagger(x)\psi_{i\uparrow}(x) - \psi_{i\downarrow}^\dagger(x)\psi_{i\downarrow}(x)\right] :$$

are the spin currents in each individual channel (i).

Bosonizing the Fermi-fields as before,

$$\psi_{is}(x) = \frac{1}{\sqrt{2\pi a_0}} e^{i\sqrt{4\pi}\phi_{is}(x)}$$

we find

$$J_i^-(x) = \frac{1}{2\pi a_0} e^{i\sqrt{8\pi}\phi_i(x)} \quad \text{and} \quad J_i^z(x) = \frac{1}{\sqrt{2\pi}}\partial_x \phi_i(x)$$

for the spin currents. Here $\phi_i(x)$ are the spin fields:

$$\phi_i(x) = \frac{1}{\sqrt{2}} [\phi_{i\uparrow}(x) - \phi_{i\downarrow}(x)]$$

the charge fields decouple from the impurity in the same way as they do in the one-channel case. Then the bosonized Hamiltonian is:

$$H_K = \sum_{i=1}^{k} \left\{ H_0[\phi_i] + \frac{I_{\perp i}}{4\pi a_0} \left[s_+ e^{i\sqrt{8\pi}\phi_i(0)} + \text{H.c.} \right] + \frac{I_{zi}}{\sqrt{2\pi}} s_z \partial_x \phi_i(0) \right\} \quad (28.13)$$

(as usual, we have dropped the part of the Hamiltonian describing the decoupled charge fields).

From now on we focus on the two-channel case ($k = 2$). It will prove convenient to introduce the channel symmetric and the channel antisymmetric combinations of the spin fields:

$$\phi_s(x) = \frac{1}{\sqrt{2}} [\phi_1(x) + \phi_2(x)]$$

$$\phi_{sf}(x) = \frac{1}{\sqrt{2}} [\phi_1(x) - \phi_2(x)] \quad (28.14)$$

(we follow the notations of Emery and Kivelson; the Roman index s refers here to the total spin, not to be confused with the italic $s = \uparrow, \downarrow$ spin index, while sf abbreviates 'spin-flavour'). In terms of the fields (28.14), the Hamiltonian (28.13) reads:

$$H_K = H_0[\phi_s] + H_0[\phi_{sf}] + \frac{1}{2\pi a_0} \left(s_+ e^{i\sqrt{4\pi}\phi_s(0)} \{ I_{1+} \cos[\sqrt{4\pi}\phi_{sf}(0)] \right.$$

$$\left. + i I_{\perp-} \sin[\sqrt{4\pi}\phi_{sf}(0)] \} + \text{H.c.} \right) + \frac{1}{\sqrt{\pi}} s_z [I_{z+}\partial_x \phi_s(0) + I_{z-}\partial_x \phi_{sf}(0)]$$

where

$$I_{\perp\pm} = \frac{1}{2}(I_{\perp 1} \pm I_{\perp 2}) \quad \text{and} \quad I_{z\pm} = \frac{1}{2}(I_{z1} \pm I_{z2})$$

We shall now make use of the canonical transformation

$$U = e^{i\sqrt{4\pi}\alpha_z \phi_s(0)} \quad (28.15)$$

similar to the one (28.7) we applied in the previous Section studying the Toulouse limit for the one-channel model. It is easy to see that, after the transformation (28.15), the scaling dimension of the transverse exchange coupling becomes

$$\Delta = \frac{1}{2} + \frac{1}{2}(1 - \alpha)^2$$

Thus, in order to achieve the same goal as in Section II, i.e. to make $\Delta = 1/2$, we need to choose $\alpha = 1$. The transformed Hamiltonian is:

$$H_K \rightarrow U^\dagger H_K U = H_0[\phi_s] + H_0[\phi_{sf}] + \frac{1}{\pi a_0} \{ s_x I_{\perp +} \cos[\sqrt{4\pi} \phi_{sf}(0)]$$
$$- s_y I_{\perp -} \sin[\sqrt{4\pi} \phi_{sf}(0)] \}$$
$$+ \frac{1}{\sqrt{\pi}} s_z [\lambda_+ \partial_x \phi_s(0) + \lambda_- \partial_x \phi_{sf}(0)] \qquad (28.16)$$

with

$$\lambda_+ = I_{z+} - 2\pi v_F \quad \text{and} \quad \lambda_- = I_{z-}$$

The next step, as for the one-channel case, is to refermionize the problem. We use the fermionic representation for the impurity spin

$$s_+ = d^\dagger \quad \text{and} \quad s_z = d^\dagger d - 1/2$$

and define two new Fermi fields by:[§]

$$\psi_s(x) = \frac{1}{\sqrt{2\pi a_0}} e^{i\sqrt{4\pi}\phi_s(x)} \quad \text{and} \quad \psi_{sf}(x) = e^{i\pi d^\dagger d} \frac{1}{\sqrt{2\pi a_0}} e^{i\sqrt{4\pi}\phi_{sf}(x)}$$

In terms of these new Fermi fields, the Hamiltonian (28.16) takes the form:

$$H_K = H_0[\psi_s] + H_0[\psi_{sf}] + \frac{I_{\perp+}}{\sqrt{8\pi a_0}}(d^\dagger - d) \left[\psi_{sf}^\dagger(0) + \psi_{sf}(0) \right]$$
$$+ \frac{I_{\perp-}}{\sqrt{8\pi a_0}}(d^\dagger + d) \left[\psi_{sf}^\dagger(0) - \psi_{sf}(0) \right]$$
$$+ (d^\dagger d - 1/2) \left[\lambda_+ : \psi_s^\dagger(0)\psi_s(0) : + \lambda_- : \psi_{sf}^\dagger(0)\psi_{sf}(0) : \right]$$

Thus, we have again arrived at a resonant-level type model, with the resonant level of d-particles hybridized and interacting with the conduction electrons. It is instructive to reformulate the problem in terms of the

---

[§] Being pedantic we should have attached a Jordan–Wigner type phase, $\exp(i\pi d^\dagger d + i N_{sf})$, also to the operator $\psi_s(x)$, N_{sf} being the total number of the ψ_{sf}-fermions. This is, however, not necessary, since the operator ψ_s appears in bilinear combinations only.

Majorana components of the Fermi fields:

$$d = \frac{1}{\sqrt{2}}(a + ib)$$

$$\psi_{sf}(x) = \frac{1}{\sqrt{2}}[\xi_{sf}(x) + i\zeta_{sf}(x)] \tag{28.17}$$

$$\psi_s(x) = \frac{1}{\sqrt{2}}[\xi_s(x) + i\zeta_s(x)]$$

The result is:

$$\begin{aligned} H_K = {} & H_0[\xi_{sf}] + H_0[\zeta_{sf}] + H_0[\xi_s] + H_0[\zeta_s] \\ & - i\frac{I_{\perp+}}{\sqrt{2\pi a_0}}b\xi_{sf}(0) - i\frac{I_{\perp-}}{\sqrt{2\pi a_0}}a\zeta_{sf}(0) \\ & - ab\,[\lambda_+\xi_s(0)\zeta_s(0) + \lambda_-\xi_{sf}(0)\zeta_{sf}(0)] \end{aligned} \tag{28.18}$$

As we can see from the above equation, the conditions

$$\lambda_+ = 0, \quad \lambda_- = 0 \tag{28.19}$$

correspond to a generalization of the Toulouse limit to the case of the (channel-anisotropic) two-channel Kondo model. The Toulouse limit Hamiltonian (i.e. the Hamiltonian (28.18) with $\lambda_{\pm} = 0$) interpolates between the usual resonant level model,

$$H_{RL} = H_0[\psi_{sf}] + \frac{I_{\perp 1}}{\sqrt{8\pi a_0}}\left[d^\dagger\psi_{sf}(0) + \text{H.c.}\right] \tag{28.20}$$

for the case of $I_{\perp+} = I_{\perp-} = I_{\perp 1}$ (i.e. for the one-channel case: $I_{\perp 2} = 0$) and the Emery–Kivelson Hamiltonian (a Majorana resonant level model),

$$H_{MRL} = H_0[\xi_{sf}] - i\frac{I_\perp}{\sqrt{2\pi a_0}}b\xi_{sf}(0) \tag{28.21}$$

for the channel symmetric case: $I_{\perp+} = I_\perp$, $I_{\perp-} = 0$.

IV.1   Green's functions and zero-field free energy

Let us first discuss the channel symmetric model. The Toulouse limit Hamiltonian (28.21) is quadratic and it is therefore straightforward to diagonalize it. However, it will be most convenient to calculate the Green's functions. We shall mainly need the following ones:

$$D_b(\tau) = \langle\langle b(\tau)b(0)\rangle\rangle$$
$$F_b(\tau) = -\langle\langle \xi_{sf}(0,\tau)b(0)\rangle\rangle$$

Either by means of the equations of motion method or using the diagram technique, one easily finds that these Green's functions satisfy the equations:

$$D_b(i\omega_n) = D^{(0)}(i\omega_n) - \frac{iI_\perp}{\sqrt{2\pi a_0}} D^{(0)}(i\omega_n) F_b(i\omega_n)$$

$$F_b(i\omega_n) = \frac{iI_\perp}{\sqrt{2\pi a_0}} G^{(0)}(i\omega_n) D_b(i\omega_n)$$

ω_n being the Matsubara frequencies. Here

$$D^{(0)}(i\omega_n) = \frac{1}{i\omega_n}$$

is the Green function of a free Majorana fermion and

$$G^{(0)}(i\omega_n) = -\frac{i}{2v_F} \text{sign}\omega_n \qquad (28.22)$$

is the local conduction electrons Green function (see Section IV of the previous Chapter, where we have first encountered the Majorana resonant level model). Thus, we find

$$D_b(i\omega_n) = \frac{1}{i\omega_n + i\Gamma \text{sign}\omega_n} \qquad (28.23)$$

The quantity

$$\Gamma = \frac{I_\perp^2}{4\pi v_F a_0} \qquad (28.24)$$

is the characteristic energy scale (Γ should not be confused with the Kondo temperature of the spin-isotropic model, it rather describes how fast the system moves along the Emery–Kivelson line towards the $I_\perp = \infty$ fixed point provided the initial parameters have been chosen to satisfy (28.19)). $D_b(\tau)$ is a complicated function in the time domain. At zero temperature it behaves as $-1/\pi\Gamma\tau$ at large τ (since it has a $1/\pi\Gamma$ discontinuity at the origin in the ω-representation). We are mainly interested in the low-temperature properties ($T \ll \Gamma$) when $i\omega_n$ in (28.23) can be neglected as compared to Γ. In the time domain this leads to

$$D_b(\tau) \simeq -\frac{T}{\Gamma \sin(\pi T \tau)} \qquad (28.25)$$

This asymptotic form of the D_b-function is correct for such τ that $\tau \gg 1/\Gamma$ and that $\tau \ll \beta - 1/\Gamma$, and it might, or might not, require a small-τ regularization for calculating observable quantities.

Since

$$s_z = iab$$

we can find at once the spin–spin correlation function:

$$\langle\langle s_z(\tau)s_z(0)\rangle\rangle = -\frac{1}{2}\text{sign}\tau D_b(\tau) \qquad (28.26)$$

Therefore, the local impurity susceptibility

$$\chi_1(T) = \int_0^\beta d\tau \langle\langle s_z(\tau) s_z(0)\rangle\rangle \qquad (28.27)$$

i.e. the susceptibility of the system with respect to a local, acting only on the impurity spin, magnetic field, is logarithmically divergent at low temperatures $T \ll 1$:

$$\chi_1(T) = \frac{T}{2} \int_{\tau_0}^{1/T-\tau_0} \frac{d\tau}{\Gamma \sin(\pi T \tau)} \simeq \frac{1}{\pi\Gamma} \ln\left(\frac{\Gamma}{T}\right) \qquad (28.28)$$

Notice that we have had to regularize the τ-integration by introducing an ultraviolet cut-off $\tau_0 \sim 1/\Gamma$ in order to account for the regions of integration where the asymptotics (28.25) fails. Alternatively we could have performed the calculation in the frequency domain using the exact Green's function (28.23). Clearly, the cut-off procedure does not affect the leading, log-divergent, part of the local susceptibility.

A more complicated (and more interesting) situation arises when the magnetic field acts equally on the impurity spin and the spin of the conduction electrons (as it is normally supposed to do). We shall consider this shortly after we have discussed the specific heat and the free energy of the system in zero external field.

The impurity correction to the free energy can be calculated in a standard way by integration over the coupling constant. Performing this simple exercise with our quadratic Hamiltonian (28.20), one finds for the total free energy:

$$F(T) = F_0(T) + \int \frac{d\omega}{2\pi} f(\omega) \tan^{-1}\left(\frac{\Gamma}{\omega}\right) \qquad (28.29)$$

where $F_0(T)$ is the free energy in absence of coupling between the impurity and conduction electrons, $f(\omega)$ is the Fermi distribution function. In order to obtain a finite result for the free energy, the ω-integral in (28.29) should be limited to the conduction bandwidth. Therefore, it is more convenient to make an explicit calculation for the impurity entropy (rather than the free energy). The result is:

$$S(T) = \ln(2) + \bar{S}\left(\frac{T}{\Gamma}\right) \qquad (28.30)$$

where we have defined the function

$$\bar{S}(z) = \frac{1}{2\pi z} \left[\psi\left(\frac{1}{2} + \frac{1}{2\pi z}\right) - 1\right] - \ln\Gamma\left(\frac{1}{2} + \frac{1}{2\pi z}\right) + \frac{1}{2}\ln\pi \qquad (28.31)$$

with $\psi(z)$ being the psi-function and $\Gamma(z)$ being the gamma-function. The second term in (28.30) follows by deriving the second term of (28.29) at finite temperatures, whereas the first one simply describes the entropy of a free spin (decoupled from the conduction).

Taking the $T \to 0$ limit in (28.30), we obtain:

$$S(0) = \frac{1}{2} \ln 2 \qquad (28.32)$$

Thus, there is a nonvanishing residual entropy. The presence of the residual entropy is the most peculiar feature of non-Fermi-liquid Kondo models. It indicates the degeneracy of the ground state. This degeneracy is, however, a non-integer one, so it is of a nontrivial nature. The latter can be understood from the Hamiltonian (28.21). Indeed, while the impurity spin s is represented in terms of the two Majorana fermions (a and b), only one of them (b) enters the Emery–Kivelson Hamiltonian (28.21), the other one (a) being decoupled from the conduction electrons. Put another way, it is only 'half' of the impurity spin that is interacting with the conduction sea. Hence the ground state degeneracy and the fractional residual entropy.¶

The next to leading term of the low-temperature expansion of the entropy (28.30) is proportional to T. This might lead to an incorrect conclusion that the impurity specific heat is linear in temperature ($C_{\text{imp}}(T) = \pi T / 6\Gamma$). However, as found by Sengupta and Georges (1994), it is the λ_+-operator in (28.18) which is responsible for the generic behaviour of the impurity specific heat. We shall denote this operator as:

$$O^{(+)}(\tau) = -a(\tau)b(\tau)\xi_s(0,\tau)\zeta_s(0,\tau)$$

It is worth investigating especially since it couples the 'a-half' of the impurity spin back to the conduction electrons. The point is, though, that this operator is irrelevant. Indeed, we have already calculated the $s_z - s_z$ correlation function: (28.26). At the Emery–Kivelson line the impurity spin s_z is decoupled from the ψ_s electrons and, therefore, the $O^{(+)} - O^{(+)}$ correlation factorizes into the spin–spin correlation function and the density–density correlation function of free electrons:

$$\langle\langle O^{(+)}(\tau)O^{(+)}(0)\rangle\rangle = -\frac{1}{2}\text{sign}\tau D_b(\tau)\left[G^{(0)}(\tau)\right]^2 \qquad (28.33)$$

Since the local free electron Green's function (28.22) is of the form

$$G^{(0)}(\tau) \simeq -\frac{T}{2v_F \sin(\pi T \tau)}$$

¶ This scenario may sound similar to what is sometimes referred to in the literature as the 'slaves liberation'. In this relation we note that, in the theory of the two-channel Kondo problem, the Majorana fermions appear naturally in the course of the exact transformations.

in the time domain, we conclude that the $O^{(+)}$-operator has the scaling dimension $3/2$ and is irrelevant. Since we do not have other perturbations in the Hamiltonian (28.18) around the Emery–Kivelson line (for the channel symmetric model), the latter is stable in the renormalization group sense. The contribution of the λ_+-perturbation to the free energy can therefore be calculated in the second-order perturbation theory.

We now pause to study how, in general, a local operator contributes to the free energy of the system. Consider the operator

$$\lambda O$$

of scaling dimension Δ. Its correlation function is of the form:

$$\langle\langle O(\tau)O(0)\rangle\rangle \simeq \left[\frac{\pi T}{\sin(\pi T\tau)}\right]^{2\Delta} \tag{28.34}$$

where the normalization of the correlation function (28.34) is understood to be included in the coefficient λ. We shall first consider the case $\Delta > 1$ when the operator λO is irrelevant in the renormalization group sense. Then, the standard perturbation theory applies and the second order correction to the free energy reads:

$$\delta F^{(\lambda)}(T) = -\frac{\lambda^2}{2}\int_0^\beta d\tau \langle\langle O(\tau)O(0)\rangle\rangle = -\lambda^2 \int_{\tau_0}^{\beta/2} d\tau \left[\frac{\pi T}{\sin(\pi T\tau)}\right]^{2\Delta} \tag{28.35}$$

where the cut-off, τ_0, is necessary since the integral diverges at small τ. Physically, the cut-off time τ_0 is inversely proportional to the characteristic energy scale of the system (an example is the $D_b(\tau)$-function, which is of the form (28.34) with $\tau_0 \sim 1/\Gamma$). It is convenient to introduce a new integration variable $x = \tan(\pi T\tau)$ in (28.35) and integrate by parts:

$$\delta F^{(\lambda)}(T) = -\frac{\lambda^2}{\pi}T^{2\Delta-1}\int_{x_0}^\infty \frac{dx}{x^{2\Delta}}(1+x^2)^{\Delta-1} =$$

$$-\frac{\lambda^2}{(2\Delta-1)\pi^{2\Delta}\tau_0^{2\Delta-1}}\left[1+O(T^2)\right] \tag{28.36}$$

$$-\frac{2(\Delta-1)}{\pi(2\Delta-1)}\lambda^2 T^{2\Delta-1}\int_{x_0}^\infty \frac{dx}{x^{2(\Delta-1)}}(1+x^2)^{\Delta-2}$$

where $x_0 = \tan(\pi\tau_0 T)$. The remaining integral in (28.36) is convergent for $\Delta < 3/2$ and divergent for $\Delta \geq 3/2$. We therefore find that there always is a nonuniversal (i.e. cut-off dependent) constant term in the free energy (which contributes a constant to the ground state energy of the system), and that there is a nonuniversal T^2 term (which contributes a constant to the specific heat coefficient). Additionally, there are universal (cut-off independent) nonanalytic in T corrections, which actually are, for

$\Delta \leq 3/2$, the leading ones:

$$\delta F^{(\lambda)}(T) - \delta F^{(\lambda)}(0) = \tag{28.37}$$

$$\begin{cases} -\dfrac{3(\Delta-1)\Gamma\left(\frac{1}{2}-\Delta\right)}{4\sqrt{\pi}(2\Delta-1)\Gamma(1-\Delta)}\lambda^2 T^{2\Delta-1} + O(T^2) & \text{for } \Delta < \frac{3}{2} \\[2ex] -\dfrac{\lambda^2}{2\pi}T^2\ln T + O(T^2) & \text{for } \Delta = \frac{3}{2} \\[2ex] O(T^2) & \text{for } \Delta > \frac{3}{2} \end{cases}$$

Now let us discuss the case when λO is relevant ($\Delta < 1$). Of course, the perturbation expansion in λ will diverge, and the system develops a new λ-related low-energy scale (so that the temperature dependence of the free energy below this scale is only accessible by a strong-coupling theory). If, however, we are interested in the temperature dependence of the linear susceptibility of the system with respect to the operator λO, we have still got to calculate (28.35). The result (28.37) clearly holds for $\Delta > 1/2$ (the integral (28.35) diverges at small τ).[‖] For $\Delta = 1/2$, the τ-integral in (28.35) is log-divergent; we have already calculated it considering the local susceptibility: (28.28). Finally, for $\Delta < 1/2$, the τ-integral in (28.35) is convergent and equal to:

$$\delta F^{(\lambda)}(T) = -\frac{\Gamma\left(\frac{1}{2}-\Delta\right)}{\sqrt{\pi}\Gamma(1-\Delta)}\lambda^2 T^{2\Delta-1} \tag{28.38}$$

Returning to the free energy contribution of the λ_+-operator, we just need to compare the normalizations of the correlation function (28.33) and (28.34) to find:

$$\delta F^{\lambda_+}(T) - \delta F^{\lambda_+}(0) \simeq -\frac{\lambda_+^2 T^2}{16\pi\Gamma v_F^2}\ln\left(\frac{\Gamma}{T}\right) \tag{28.39}$$

Thus, due to the $O^{(+)}$-operator, the specific heat behaves as $T\ln T$ at low temperatures, i.e. the specific heat coefficient is log-divergent:

$$\frac{\partial C_{\text{imp}}(T)}{\partial T} \simeq \frac{\lambda_+^2}{8\pi\Gamma v_F^2}\ln\left(\frac{\Gamma}{T}\right) \tag{28.40}$$

Let us now return to the Hamiltonian (28.18) in order to investigate the effects of the channel anisotropy. As we can see from (28.18), the

---

[‖] The interval $1/2 < \Delta$, where (28.37) applies, covers the particular case $\Delta = 1$ of the operator λO being marginal. Note that the coefficient in front of the T-linear term in (28.37) vanishes for $\Delta = 1$. So do the coefficients in front of all the terms with odd powers of temperature. The integral (28.35) is simply:

$$\delta F^{(\lambda)}(T) = -\frac{\lambda^2}{\pi}\frac{T}{\tan(\pi\tau_0 T)},$$

so that the free energy is a function of T^2, as one does expect for a marginal operator.

nonvanishing transverse exchange anisotropy (i.e., a finite $I_{\perp-}$) couples the local a Majorana fermion back to the conduction sea already on the level of the quadratic Hamiltonian in the Toulouse limit. This is going to have a drastic effect on our previous conclusions – however small the anisotropy is, it will lift the degeneracy of the ground state, remove the residual entropy, and make the thermodynamic functions analytic in the temperature (and in the magnetic field).

In order to study how this happens, we first notice that though the anisotropy does couple the a fermion back to the conduction electrons, it does this in a very simple way. Namely, the Toulouse limit Hamiltonian is the sum of the two independent Majorana resonant level models:

$$H_{\mathrm{TL}} = H_0[\xi_{\mathrm{sf}}] - \mathrm{i}\frac{I_{\perp+}}{\sqrt{2\pi a_0}} b \xi_{\mathrm{sf}}(0)$$

$$+ H_0[\zeta_{\mathrm{sf}}] - \mathrm{i}\frac{I_{\perp-}}{\sqrt{2\pi a_0}} a \zeta_{\mathrm{sf}}(0) \qquad (28.41)$$

associated with the pairs of the operators $b - \xi_{\mathrm{sf}}(x)$ and $a - \zeta_{\mathrm{sf}}(x)$.

The D_b Green function is not at all affected (under the understanding that $I_\perp$ is substituted by $I_{\perp+}$ in the definition (28.24) of the resonance width Γ), whereas the D_a-function takes the same form as (28.23)

$$D_a(\mathrm{i}\omega_n) = \frac{1}{\mathrm{i}\omega_n + \mathrm{i}\gamma \mathrm{sign}\omega_n} \qquad (28.42)$$

but with a different (smaller) resonant width, related to the anisotropy:

$$\gamma = \frac{I_{\perp-}^2}{4\pi v_F a_0} \qquad (28.43)$$

It is instructive to return now to the (complex) resonant-level fermion field d, (28.17), and to have a look at the corresponding Green function. Since the number of the d-particles is not conserved, we have to work in the Nambu representation:

$$\mathbf{d} = \begin{bmatrix} d \\ d^\dagger \end{bmatrix}$$

The resonant-level fermion Green function,

$$\hat{D}(\tau) = \langle\langle \mathbf{d}^\dagger(\tau) \mathbf{d}(0) \rangle\rangle$$

is then a 2×2 matrix. Using (28.17) and noticing that a and b Majorana components of the operator d are uncorrelated in the Toulouse limit, one finds:

$$\hat{D}(\tau) = \frac{1}{2}(\hat{\tau}_0 + \hat{\tau}_x) D_a(\tau) + \frac{1}{2}(\hat{\tau}_0 - \hat{\tau}_x) D_b(\tau) \qquad (28.44)$$

where $\hat{\tau}_x$ is the Pauli matrix and $\hat{\tau}_0$ is the unit matrix. Assuming a standard spectral representation,

$$\hat{D}(i\omega_n) = \int_{-\infty}^{\infty} dx \frac{\hat{A}(x)}{i\omega_n - x} \tag{28.45}$$

one obtains

$$\hat{A}(x) = \frac{1}{2\pi} (\hat{\tau}_0 + \hat{\tau}_x) \frac{\gamma}{x^2 + \gamma^2} + \frac{1}{2\pi} (\hat{\tau}_0 - \hat{\tau}_x) \frac{\Gamma}{x^2 + \Gamma^2} \tag{28.46}$$

for the spectral function. The impurity spectral weight is therefore equally shared by two Lorentzians with different widths γ and Γ. In the channel symmetric case $\gamma \to 0$ and one of the two Lorentzians shrinks to a δ-function, representing the impurity degree of freedom decoupled from the conduction electrons. Any finite channel anisotropy ($\gamma \neq 0$) lifts this degeneracy.

It immediately follows that the local magnetic susceptibility $\chi_l(T)$ is not divergent any more and tends to a constant as $T \to 0$ (this is because we now have the extra $1/\tau$ factor in the spin–spin correlation function which derives from the D_a-function and makes the integral (28.27) infrared convergent). We can make use of the exact Green's functions in the Toulouse limit in order to see how the susceptibility crossovers from the $\ln T$ behaviour (28.28) to a constant:

$$\chi_l(T) = T \sum_n D_a(-i\omega_n) D_b(i\omega_n)$$

$$= \frac{1}{\pi(\Gamma - \gamma)} \left[\psi \left(\frac{1}{2} + \frac{\Gamma}{2\pi T} \right) - \psi \left(\frac{1}{2} + \frac{\gamma}{2\pi T} \right) \right]$$

$$\simeq \begin{cases} \dfrac{1}{\pi(\Gamma - \gamma)} \ln \left(\dfrac{\Gamma}{\gamma} \right), & T \ll \gamma \\[2ex] \dfrac{1}{\pi(\Gamma - \gamma)} \ln \left(\dfrac{\Gamma}{T} \right), & \gamma \ll T \ll \Gamma \\[2ex] \dfrac{1}{4T}, & T \gg \Gamma \end{cases} \tag{28.47}$$

The free energy is additive with respect of the contributions of the two Majorana resonant level models (28.19), and so is the entropy:

$$S(T) = \ln(2) + \bar{S} \left(\frac{T}{\Gamma} \right) + \bar{S} \left(\frac{T}{\gamma} \right) \simeq \begin{cases} \dfrac{\pi T}{6} \left(\dfrac{1}{\Gamma} + \dfrac{1}{\gamma} \right), & T \ll \gamma \\[2ex] \ln \sqrt{2}, & \gamma \ll T \ll \Gamma \\[2ex] \ln 2 - \dfrac{\Gamma + \gamma}{2\pi T}, & T \gg \Gamma \end{cases} \tag{28.48}$$

The role of the second, anisotropy term in (28.48) is to quench the residual entropy:

$$S(0) = 0$$

The scaling dimension of the operator $O^{(+)}$ is now equal to 2 since we again have the extra $1/\tau$-factor in (28.33). Its contribution to the free energy is therefore proportional to T^2 at low T. Consequently, the specific heat coefficient is a nonuniversal (cut-off dependent) constant. Despite this nonuniversality we can still estimate the specific heat coefficient for the case of a small exchange anisotropy: $\gamma \ll \Gamma$. Indeed, the specific heat coefficient is summed up of two contributions – the contribution of the $O^{(+)}$-operator and the contribution coming from the temperature dependence of the Toulouse limit entropy (28.48). The first one obviously is of the order of $\lambda_+^2 \ln(1/\gamma)$ (see Eq. (28.40)). However, the second contribution dominates, because (28.48) drastically depends on temperature quenching the residual entropy (28.32) on a very small energy scale ($\sim \gamma$). Thus, for $\gamma \ll \Gamma$, we have:

$$\frac{\partial C_{\text{imp}}(T)}{\partial T} = \frac{\pi}{6\gamma} + O\left[\ln(1/\gamma)\right] \tag{28.49}$$

Notice, that for a model with a longitudinal exchange anisotropy ($I_{z1} \neq I_{z2}$ and $\lambda_- \neq 0$), an additional operator $\lambda_- O^{(-)} = -\lambda_- ab\xi_{\text{sf}}(0)\zeta_{\text{sf}}(0)$ appears in the Hamiltonian (28.18). We shall not discuss this operator in detail. We only mention that (i) it is a relevant operator with scaling dimension $1/2$ at the channel symmetric point $\gamma = 0$ (meaning that the longitudinal exchange anisotropy is as relevant as the transverse one is) but (ii) $O^{(-)}$ becomes irrelevant for $\gamma \neq 0$.

Andrei and Jerez (1995) have recently solved the channel anisotropic two-channel model generalizing to this case the Bethe-ansatz method. The Bethe-ansatz approach is beyond the scope of this book; we only note that whereas the thermodynamic properties of this model are by now well understood, the dynamic correlation functions require further studies.

IV.2   Magnetic field effects

Before we close the discussion of the Emery–Kivelson solution, we shall investigate the effects of the external magnetic field. The latter appears in the Hamiltonian with the term:

$$H_{\text{M}} = g_{\text{imp}} h s_z + g_{\text{cond}} h \sum_{i=1}^{k} \int \mathrm{d}x J_i^z(x) \tag{28.50}$$

We shall mainly consider the most interesting case of equal impurity and conduction electron g-factors;

$$g_{\text{imp}} = g_{\text{cond}} = g$$

(though the case $g_{imp} \neq g_{cond}$ may also be relevant for physical applications: see below). The bosonized version of (28.50) reads:

$$H_M = gh \int dx \left[\delta(x) s_z + \sqrt{\frac{k}{2\pi}} \partial_x \phi_s(x) \right]$$

(28.51)

where the field ϕ_s corresponds to the total spin of the conduction electrons:

$$\phi_s(x) = \frac{1}{\sqrt{k}} \sum_{i=1}^{k} \phi_i(x)$$

The quantity we are interested in is the impurity magnetic susceptibility

$$\chi_{imp} = \chi - \chi_0$$

(28.52)

which is the difference between the total magnetic susceptibility χ (i.e. the susceptibility of the system with respect to the perturbation (28.50)) and the susceptibility of the conduction sea in the absence of the Kondo coupling

$$\chi_0 = \frac{kg^2}{2\pi v_F}$$

(28.53)

A priori, there is no reason for the impurity susceptibility χ_{imp} to coincide with the previously defined local susceptibility χ_l given by Eq. (28.27). It turns out, however, that there is a relation between the two susceptibilities, which we shall establish in what follows.

One can see from (28.51) that the only Bose field related to the conduction electrons that couples to the external magnetic field is the total spin field ϕ_s. Will this fact help us to determine χ? In order to answer this question we have to understand the dynamics of the field ϕ_s. Of course, ϕ_s participates in the longitudinal exchange coupling in (28.13), which is of the form:

$$I_z \sqrt{\frac{k}{2\pi}} s_z \partial_x \phi_s(0)$$

(28.54)

(We consider, for the sake of simplicity, the channel symmetric model.) Apart from (28.54), ϕ_s is also incorporated in the transverse components of the spin current:

$$J^-(x) = \frac{1}{2\pi a_0} \sum_{i=1}^{k} e^{i\sqrt{8\pi}\phi_i(x)}$$

(28.55)

Let us make this more explicit by defining an orthogonal set $\{\phi'_\alpha(x), \ \alpha = 1, ..., k-1\}$ of $k-1$ combinations of the fields $\phi_i(x)$, such that each ϕ'_α is orthogonal to ϕ_s. In terms of these new fields the transverse component of the current (28.55) can be rewritten as:

$$J^-(x) = e^{i\sqrt{8\pi/k}\phi_s(x)}\chi_k\left[\{\phi'_\alpha(x)\}\right] \tag{28.56}$$

where the operator χ_k is a local function of $\{\phi'_\alpha(x)\}$. For the two-channel case, the set $\{\phi'_\alpha\}$ is, of course, uniquely defined: it simply consists of a single field $\phi'_1 = \phi_{sf} = (\phi_1 - \phi_2)/\sqrt{2}$, and $\chi_2 - (1/\pi a_0)\cos\sqrt{4\pi}\phi_{sf}$. For $k > 2$ there are many equivalent sets satisfying the orthogonality conditions. This is, however, not important – the only thing which matters is that the operator χ_k does not depend on the total spin field ϕ_s.**

Analogously to (28.16), we can now eliminate ϕ_s from the transverse exchange coupling term in the Hamiltonian by means of the canonical transformation:

$$H_K + H_M \to U^\dagger_{(k)}\left(H_K + H_M\right)U_{(k)} = H_0[\phi_s] + \sum_{\alpha=1}^{k-1}H_0[\phi'_\alpha]$$

$$+ \frac{I_\perp}{2}\sum_{\alpha=1}^{k-1}\left\{s_+e^{i\sqrt{8\pi/k}\phi_s(0)}\chi_k\left[\{\phi'_\alpha(0)\}\right] + \text{H.c.}\right\} \tag{28.57}$$

$$+ \lambda_k\sqrt{\frac{k}{2\pi}}s_z\partial_x\phi_s(0) + gh\sqrt{\frac{k}{2\pi}}\int dx\partial_x\phi_s(x)$$

where

$$U_{(k)} = e^{i\sqrt{8\pi/k}\phi_s(0)s_z} \tag{28.58}$$

and

$$\lambda_k = I_z - \frac{4\pi}{k}v_F \tag{28.59}$$

Notice that for the two-channel case, the transformation (28.58) coincides with (28.16), $\lambda_2 = \lambda_+$ and $\lambda_2 = 0$ being the Emery–Kivelson line. We remind the reader of the relation

$$U^\dagger_{(k)}\partial_x\phi_s(x)U_{(k)} = \partial_x\phi_s(x) - \sqrt{\frac{2\pi}{k}}\delta(x)s_z$$

which made the ghs_z term in (28.57) disappear. Thus, the external magnetic field is now coupled to the conduction electrons only (i.e. decoupled from the impurity spin s_z). We would actually prefer it to be the other way around (for the sake of comparing χ_{imp} with χ_1), so we employ another

---
** The reader acquainted with the paper by Zamolodchikov and Fateev (1985) will at once recognize in χ_k a level-k parafermion.

canonical transformation:

$$U_h = \exp\left\{ i\frac{gh}{v_F}\sqrt{\frac{k}{2\pi}} \int dx \phi_s(x) \right\}$$ (28.60)

which shifts the ϕ_s field as

$$U_h^\dagger \partial_x \phi_s(x) U_h = \partial_x \phi_s(x) - \frac{gh}{2v_F}\sqrt{\frac{k}{2\pi}}$$

and therefore leads to:

$$U_h^\dagger (H_K + H_M) U_h = H_0[\phi_s] + \sum_{\alpha=1}^{k-1} H_0[\phi_\alpha'] + \frac{I_\perp}{2}\sum_{\alpha=1}^{k-1}\{s_+\chi_k \left[\{\phi_\alpha'(0)\}\right] + \text{H.c.}\}$$

$$+ \lambda_k \sqrt{\frac{k}{2\pi}} s_z \partial_x \phi_s(0) - \frac{gk\lambda_k}{4\pi v_F} h s_z + \text{const}$$ (28.61)

The constant in (28.61) is equal to $-(kg^2h^2/4\pi)\int dx$, which accounts for the bulk susceptibility χ_0 of the free electrons. With (28.61), we have achieved our aim – we have proven that a uniform magnetic field acts on the system in the same way as a local field does, up to a coefficient. Put another way, the magnetic field gets renormalized:

$$h_l = -\frac{gk\lambda_k}{4\pi v_F} h$$

Thus, the relation between the impurity susceptibility and the local susceptibility is simply:

$$\chi_{imp}(T) = \left(\frac{gk\lambda_k}{4\pi v_F}\right)^2 \chi_l(T)$$ (28.62)

This relation holds for any number of channels and, particularly, for $k=2$. Alternatively, (28.62) can be proven by making use of the Anderson–Yuval expansion (see Fabrizio *et al.* (1995) for the discussion of an extension of this expansion to the multi-channel case). As usually, the difference with the bosonization is only in the cut-off procedure. Namely, the longitudinal exchange coupling is replaced by the electron scattering phase shift. Or, equivalently,

$$\lambda_k \to \lambda_k = 8\delta v_F - \frac{4\pi}{k}v_F$$

with $\delta = \tan^{-1}(I_z/8\pi v_F)$. The relation (28.62) then reads:

$$\chi_{imp}(T) = g^2\left(1 - \frac{2k\delta}{\pi}\right)^2 \chi_l(T)$$ (28.63)

One can extend (28.63) to the case of a channel anisotropic model and different g-factors. Not dwelling on the proof (which can be achieved by an elementary generalization of the above approach), we quote the result:

$$\chi_{\text{imp}}(T) = \left(g_{\text{imp}} - \frac{2}{\pi} g_{\text{cond}} \sum_{a-1}^{k} \delta_a \right)^2 \chi_1(T) \qquad (28.64)$$

δ_a being the phase shift of channel (a) electrons.

The inspection of formula (28.62) leads us to an important exact result: the impurity susceptibility vanishes, provided that the parameters of the model are chosen in such a way that $\lambda_k = 0$. In the non-Abelian bosonization approach by Affleck and Ludwig (1991) this is the property of the conformal fixed point (see also discussion at the end of the next Section). The fact that the impurity susceptibility vanishes also at the Emery–Kivelson line (for the two-channel model) has been noticed by Clarke, Giamarchi and Shraiman (1993). As we shall see in the following paragraph, it is possible to consistently generalize the Emery–Kivelson approach to the four-channel model; the impurity susceptibility is again zero in the corresponding Toulouse limit. It is therefore tempting to define the Toulouse limit for the general-k case as:

$$\lambda_k = 0$$

i.e., to locate the Toulouse limit at the line where the impurity susceptibility vanishes (in the parameter space $I_\perp - I_z$).[††] The model (28.57) with $\lambda_k = 0$ can be mapped onto a free fermion model for the $k = 2$ case only and it is nontrivial otherwise; this model has not been systematically studied for larger ks.

Coming back to the two-channel model, we observe that, given (28.62), we do not need to carry out any additional calculations for χ_{imp}, as we have already found χ_l: (28.28) and (28.47).

In the channel symmetric case, combining (28.62) with (28.28) gives:

$$\chi_{\text{imp}}(T) \simeq \frac{g^2 \lambda_+^2}{4\pi^3 \Gamma v_F^2} \ln\left(\frac{\Gamma}{T}\right) \qquad (28.65)$$

[††] Although this definition does include the Emery–Kivelson line, it paradoxically excludes the original Toulouse limit for the single-channel model. There are no flavour fields in this case, but only the total spin field ϕ_s, identical to ϕ of (28.6). Under the transformation $U_{(1)}$, the one-channel Hamiltonian (28.6) takes the form:

$$U_{(1)}^\dagger H U_{(1)} = H_0[\phi_s] + \frac{I_\perp}{2\pi a_0} s_x + \frac{\lambda_1}{\sqrt{2\pi}} s_z \partial_x \phi_s(0) \,,$$

instead of (28.9). Notice that $\lambda_1 \to 0$ can only be achieved on the verge of the unitarity limit for the scattering phase: $\delta \to \pi/2$, or $I_z \to \infty$. At $\lambda_1 = 0$, the impurity spin is completely decoupled from the conduction electrons, so the $\lambda_1 = 0$ limit is referred to as the 'decoupling limit'.

IV.3   Wilson ratio

Since the pioneering work by Wilson (1975), the quantity

$$R_W = \frac{\chi_{imp}}{\chi_0} \frac{\partial C_0 / \partial T}{\partial C_{imp} / \partial T} \tag{28.66}$$

(referred to as the Wilson ratio) has attracted considerable attention. (We recall that $\chi_0 = g^2 k / 2\pi v_F$ is the bulk susceptibility and $\partial C_0 / \partial T = 2\pi k / 3 v_F$ is the bulk specific heat coefficient.) The point is that (28.66) has been found by Wilson (who, of course, only discussed the one-channel model at the time) to be a universal number: $R_W = 2$. For the k-channel model, Affleck and Ludwig (1991) obtained the exact universal Wilson ratio

$$R_W = \frac{(2+k)(2+k/2)}{18}$$

assuming the channel and the spin symmetry. Indeed, gathering the above results (28.65) and (28.40), one finds

$$R_W = \frac{8}{3}$$

for the channel symmetric two-channel model (Sengupta and Georges (1994)). The channel anisotropy drastically effects the Wilson ratio; it becomes nonuniversal and is, in fact, very small (Fabrizio *et al.* (1995)):

$$R_W \simeq \frac{2\gamma \lambda_+^2}{\pi^2 \Gamma v_F} \ln\left(\frac{\Gamma}{\gamma}\right)$$

for $\gamma \ll \Gamma$. Physically, this result is quite clear: the residual entropy $\ln \sqrt{2}$ must be quenched in a temperature range $\sim \gamma^{-1}$, implying $\partial C_{imp} / \partial T \sim 1/\gamma$, while the susceptibility just rounds off logarithmic singularity, $\chi_{imp} \sim \ln(\Gamma/\gamma)$. Indeed, this is not surprising, for the channel asymmetry is a relevant perturbation which completely changes the low-energy behaviour of the system. As to the spin anisotropy of the exchange coupling it appears not to affect R_W. We shall see in the next Section that this is not correct for $k > 2$.

V   The Toulouse limit for the four-channel Kondo model

The Abelian bosonization approach, which proved so simple and effective in studying the $k = 1$ and $k = 2$ Kondo models, turns out to be not very helpful in the general $k > 2$ case. There is, though, one exception: $k = 4$ (Fabrizio and Gogolin (1994)). In this Section, we shall study how the Toulouse limit can be generalized to the four-channel model and establish interesting connections with the boundary sine-Gordon problem.

First of all we notice that what matters is the dynamics (which is governed by the commutation relations) of the spin currents, and not their concrete representation. In the previous Sections we have used the most natural representation of the spin currents in terms of the conduction electron fields:

$$J^a(x) = \sum_i \; : \psi^\dagger_{is}(x)\tau^a_{ss'}\psi_{is'}(x) : \qquad (28.67)$$

τ^a being spin-1/2 matrices and $a = x, y, z$. This is a natural representation for applications to magnetic impurities in metals (but it is not for some other applications, like those to quantum dots). As far as the thermodynamic functions and the impurity spin correlation functions are concerned, we can make use of any convenient representation, preserving the commutation relations of J^a. The latter are the standard SU(2) level-k Kac–Moody commutation relations (7.4).

Using the representation (13.22), we obtain the following expression for the Kondo Hamiltonian:

$$H = H_0[\phi_s] + H_0[\phi_f] + \frac{I_\perp}{\sqrt{2}\pi a_0}\left\{s_+ e^{i\sqrt{2\pi}\phi_s(0)}\cos\left[\sqrt{6\pi}\phi_f(0)\right] + \text{H.c.}\right\} +$$

$$\sqrt{\frac{2}{\pi}}I_z s_z \partial_x \phi_s(0) + gh\left[s_z + \sqrt{\frac{2}{\pi}}\int dx \partial_x \phi_s(x)\right] \qquad (28.68)$$

where we have included the magnetic field term.

Notice that the expression (13.22) for the spin currents coincides with (28.56), provided that we identify $\chi_4 = (\sqrt{2}/\pi a_0)\cos(\sqrt{6\pi}\phi_f)$. Hence the transformation:

$$H \to U^\dagger H U, \quad U = \exp\left\{i\sqrt{2\pi}S_z\phi_s(0)\right\}\exp\left\{\frac{igh}{v_F}\sqrt{\frac{2}{\pi}}\int dx\phi_s(x)\right\} \quad (28.69)$$

The transformed Hamiltonian reads:

$$H = H_0[\phi_s] + H_0[\phi_f] + \frac{\sqrt{2}I_\perp}{\pi a_0}s_x\cos\left[\sqrt{6\pi}\phi_f(0)\right]$$

$$+ \frac{\lambda}{\sqrt{\pi}}s_z\partial_x\phi_s(0) - \frac{\lambda h}{\sqrt{2}\pi v_F}s_z \qquad (28.70)$$

with

$$\lambda = \sqrt{2}\,(I_z - \pi v_F)$$

The line $\lambda = 0$ is indeed analogous to the Emery–Kivelson line for the two-channel Kondo model. Along this line, the canonical transformation (28.69) leads to the decoupling of the impurity degrees of freedom from the conduction electron ones (the impurity spin component s_x commutes with the Hamiltonian and hence loses its dynamics). However, the Hamiltonian for the phase field ϕ_f remains nontrivial. Let us write it in the form:

$$H[\phi_f] = H_0[\phi_f] \pm \frac{I_\perp}{\sqrt{2}\pi a_0} \cos\left[\sqrt{8\pi\Delta}\phi_f(0)\right] \qquad (28.71)$$

where the $\pm$ sign refers to the conserved spin component s_x equal $\pm 1/2$. In our case $\Delta = 3/4$.

We observe that the model (28.71) is nothing but the boundary sine-Gordon model. The following relation is therefore established: the four-channel Kondo model is equivalent, in the Toulouse limit, to the boundary sine-Gordon model with the scaling dimension of the boundary cosine operator $\Delta = 3/4$. This fact was used by Tsvelik (1995) to obtain an exact solution of the model (28.70). In the rest of this section we shall explore this equivalence.

We have studied the boundary sine-Gordon model in Chapter 27. Let us briefly summarize the results which are needed for the discussion of the Kondo model. At this stage, it is most convenient to use the fermionic version of (28.71) which describes the impurity backscattering in the Tomonaga–Luttinger liquid:

$$H = H_{LL} \pm \frac{I_\perp}{\sqrt{2}} \left[\Psi_R^\dagger(0)\Psi_L(0) + \text{H.c.}\right] \qquad (28.72)$$

where the field $\Psi_{R(L)}$ refers to right- (left-) moving fermions, and H_{LL} is the interacting Tomonaga–Luttinger liquid Hamiltonian.[‡‡] Since in the specific case we are considering the scaling dimension $\Delta = 3/4 < 1$, the impurity scattering operator is relevant and $I_\perp$ flows to infinity under scaling transformation. The resulting (stable) fixed point describes a perfectly reflecting barrier, which cuts the system into two semi-infinite lines a and b. Exactly at the fixed point, that is for everything regarding zero energy and temperature properties, the two regions a and b are disconnected. At any finite temperature there is a residual tunneling across the barrier, described by the (irrelevant) operator

$$O = t \left[\Psi_a^\dagger(0)\Psi_b(0) + \text{H.c.}\right] \qquad (28.73)$$

where $\Psi_{a(b)}$ is the electron field referred to region $a(b)$ and t is a (nonuniversal) tunneling amplitude. The scaling dimension of this operator is:

$$\Delta_t = \frac{1}{\Delta} = \frac{4}{3} \qquad (28.74)$$

[‡‡] As we have seen in Chapter 27, the model (28.72) is the same as (28.71) upon application of standard bosonization rules.

According to (28.37), the tunneling operator contributes

$$\delta F^{(t)}(T) - \delta F^{(t)}(0) \sim t^2 T^{5/3} \tag{28.75}$$

to the free energy, and

$$\frac{\partial C^{(t)}_{imp}}{\partial T} \sim t^2 T^{-1/3} \tag{28.76}$$

to the impurity specific heat coefficient. Thus, we have determined the low-energy behaviour of the free energy in the Toulouse limit: $\lambda = 0$. As follows from (28.70), the impurity susceptibility $\chi_{imp} = 0$ for $\lambda = 0$ (in agreement with the general analysis of Section IV.2). One therefore needs to study the operator

$$\frac{\lambda}{\sqrt{\pi}} s_z \partial_x \phi_s(0) \tag{28.77}$$

which describes the deviations from the Toulouse limit. Since ϕ_s is a free field (for $\lambda = 0$), the correlation function of (28.77) factorizes. So, the only thing we ought to know is the scaling dimension of the impurity spin operator, which we shall denote x:

$$\langle\langle s_z(\tau) s_z(0)\rangle\rangle \sim \left[\frac{\pi T}{\sin(\pi\tau T)}\right]^{2x} \tag{28.78}$$

In order to determine x, we notice that the operator s_z is nothing but the sum of raising and lowering operators for the states labelled by the s_x component of the impurity spin. Therefore, the operator s_z acts as a sign changing operator with respect to the backscattering potential in the Hamiltonian (28.72). In Chapter 27 (see Problem XIV), we have readily calculated the scaling dimension of such an operator:

$$x = \frac{1}{4\Delta} = \frac{1}{3} \tag{28.79}$$

Hence the impurity magnetic susceptibility:

$$\chi_{imp} \sim \lambda^2 T^{-1/3} \tag{28.80}$$

The scaling dimension of the operator (28.77) clearly is

$$\Delta_\lambda = x + 1 = \frac{4}{3} \tag{28.81}$$

(notice that $\Delta_\lambda = \Delta_t$) and its contribution to the free energy

$$\delta F^{(\lambda)}(T) - \delta F^{(\lambda)}(0) \sim \lambda^2 T^{5/3} \tag{28.82}$$

leads to

$$\frac{\partial C_{\text{imp}}^{(\lambda)}}{\partial T} \sim t^2 T^{-1/3} \tag{28.83}$$

Thus, the impurity specific heat coefficient receives $T^{-1/3}$-divergent contributions both from the operator (28.73), responsible for the renormalization of the system towards the $I_{\perp} = \infty$ fixed point along the Toulouse limit, and from the operator (28.77), describing deviations from the Toulouse limit. This is to be compared with the Emery–Kivelson solution for the two-channel model when it is the λ-operator which alone accounts for the log-divergent specific heat coefficient, while the temperature dependence of the free energy at the Emery–Kivelson line delivers sub-leading corrections only. On the other hand, one always needs to step away from the Toulouse limit in order to obtain the impurity magnetic susceptibility: (28.80).

We note that a nonuniversal Wilson ratio for the $k = 4$ model follows, the nonuniversality being the consequence of the spin exchange coupling anisotropy. Recently, Ye (1996) argued that the Wilson ratio is in fact nonuniversal for any $k > 2$ spin anisotropic Kondo model.

To conclude, in this Chapter we have presented explicit analytic solutions for the $k = 2$ and $k = 4$ multi-channel Kondo models. These solutions connect the weak- and the strong-coupling fixed points via operator transformations based on various free-field representations of the spin currents. It is not a coincidence that the conformal charge of the corresponding Kac–Moody algebras ($SU_k(2)$) is an (half-) integer ($c = 3/2$ and $c = 2$ respectively). Since the conformal charge is not an (half-) integer in the general-k case, the free-field representations do not offer much hope to obtain an explicit solution for this general case, as $c = 1/2$ is, of course, the lowest conformal charge which can be still realized by a free theory: a Majorana fermion. (A notable exception is the $k = 10$ case, when $c = 5/2$; we would like to encourage the reader to think about this case.) Yet it is possible to obtain a somewhat less explicit description of the arbitrary k model by phenomenologically identifying leading irrelevant operators allowed at the strong-coupling fixed point by means of symmetry-type arguments (rather than tracing them from the weak coupling via operator transformations). This programme was brilliantly achieved by Affleck and Ludwig (1991). The understanding of the latter approach would, however, require more knowledge of the conformal field theory (and, in particular, of the boundary conformal field theory) than we have chosen to introduce in this volume. We therefore abandon the discussion of the general multi-channel Kondo problem at this point, referring the interested reader to the Affleck and Ludwig paper in conjunction with the review by Cardy (1990) (see references to the previous Chapter). Instead we shall turn to some interesting physical applications of the $k = 2, 4$ solutions on the remaining pages of the book.

VI   Coulomb blockade

Recently, investigating the Coulomb blockade effect in semiconductor structures, Matveev (1994) discovered an interesting realization of multichannel Kondo models in the Toulouse limit. This realization involves tunneling of electrons between a reservoir (a two-dimensional normal metal) and a quantum dot via a narrow channel (point contact) (see Fig. 28.2).

The so-called Coulomb blockade is caused by suppession of the electron tunneling due to charging effects in the dot. The charging process is described by the following term in the Hamiltonian

$$H_c = \frac{(Q - eN)^2}{2C_d} \tag{28.84}$$

where N is a parameter proportional to the gate voltage V_g and C_d is the capacitance of the dot. (We note here that the Hamiltonian (28.84) is essentially phenomenological; a realistic description of the charging process involving dynamics and mutual capacitances of surrounding metals would require a rather complicated theory.)

Transferring a single electron from the reservoir to the dot costs the charging energy $\sim E_c = e^2/2C_d$. The charging energy is to be overcome by the energy of the electron in the external field $eNQ/C_d \sim QV_g$. If the point contact acts as a weak link (i.e. its conductance is very small, $G \ll e^2/h$) then the number of the electrons in the dot is quantized and the charge $\langle Q \rangle$ of the dot is a 'staircase' function of the gate voltage as shown in Fig. 28.3.

However, in the case of a good contact ($G \sim e^2/h$), the electron wave functions are shared by both the reservoir and the dot so that the charge depends linearly on the voltage, $\langle Q \rangle = eN$. The cross over between these two regimes (in the weak link limit – the smearing of the staircase steps due to the quantum tunneling) is the subject of the Coulomb blockade theory.

In this section we demonstrate how, under certain assumptions (a narrow channel and a large dot), the Coloumb blocked problem turns out to be equivalent to the two-channel Kondo problem in the Toulouse limit.

VI.1   One-dimensional electrons in point contacts

The electron scattering from the point contact of interest for the Coulomb blockade in semiconductor devices can be described in the framework

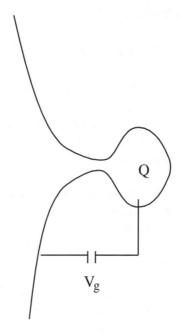

Fig. 28.2.   A quantum dot (on the right) connected to a reservoir (on the left) via a point contact.

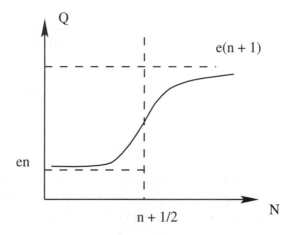

Fig. 28.3.   A qualitative dependence of the average charge of the quantum dot on the gate voltage.

of an effective Hamiltonian involving one-dimensional electron fields (see Matveev (1994) and references therein). Indeed, in the semiconductor devices the confining potential (which is responsible for the shape of the contact) is due to external metal gates usually situated well above the conducting plane. The distances involved are normally much bigger than the inverse Fermi momentum of the two-dimensional electron gas. Consequently the scattering of the electrons from the point contact can be considered quasiclassically. In particular, in the close vicinity of the contact, the confining potential may be approximated as a harmonic one

$$U(x, y) = U_0 - \frac{1}{2}m\omega_x^2 x^2 + \frac{1}{2}m\omega_y^2 y^2 \qquad (28.85)$$

m being the effective electron mass (we shall set $m = 1$ in what follows). The parameters $\omega_{x(y)}$ characterize the curvatures of the confining potential in the transverse (y) and in the longitudinal (x) directions.

The Schrödinger equation with the potential (28.85) can trivially be solved (see Landau and Lifshits (1982)). The variables separate so that the wave functions take the form $\Psi(x, y) = \phi_{nk}(x)\phi_n(y)$, k being the transverse momentum of the electron. The functions $\phi_n(y)$ are Hermit polynomials and the functions $\phi_{nk}(x)$ are the scattering wave functions related to the parabolic cylinder functions, the explicit form of which is of no interest here. The integer $n = 0, 1, 2, \ldots$ numbers the transverse channels. The reflection coefficient in each channel is given by

$$R_{nk} = \frac{1}{1 + e^{2\pi\epsilon_{nk}}} \qquad (28.86)$$

with $\epsilon_{nk} = [k^2/2 - \omega_y(n + 1/2) - U_0]/\omega_x$. As one can see from this expression, if the Fermi energy ϵ_F is tuned in such a way that $U_0 + \omega_y/2 < \epsilon_F < U_0 + 3\omega_y/2$ (as we shall always assume in what follows) then the channel $n = 0$ has an exponentially small reflection coefficient while the electrons in the other channels are almost perfectly reflected. Consequently we can disregard $n > 0$ channels and focus on the channel $n = 0$.

Let us for a moment completely neglect the reflection probability in the $n = 0$ channel. Then, linearizing as usual the electron spectrum around the Fermi points $k = \pm k_F + p$ and assigning a second quantization annihilation operator $a_{R\sigma p}$ to the wave function ϕ_{0,k_F+p} and $a_{L\sigma p}$ to $\phi_{0,-k_F+p}$ we obtain the effective Hamiltonian for the $n = 0$ channel electrons

$$H_0 = \sum_p v_F p \left(a_{R\sigma p}^\dagger a_{R\sigma p} - a_{L\sigma p}^\dagger a_{L\sigma p}\right) \qquad (28.87)$$

where v_F is the Fermi velocity and the energy is accounted for from the Fermi energy (σ is the spin-1/2 index, summation over which is implied). Next we introduce the Fourier transforms $R_\sigma(x)$ and $L_\sigma(x)$ of $a_{R\sigma p}$ and $a_{L\sigma p}$ respectively. The fields $R_\sigma(x)$ and $L_\sigma(x)$ are the usual chiral Fermi fields (note that, from now on, the coordinate x bears no direct relation to the 'physical' coordinate in (28.85)). In terms of these chiral fields the Hamiltonian (28.87) takes the standard form

$$H_0 = iv_F \int dx \left[R_\sigma^\dagger(x)\partial_x R_\sigma(x) - L_\sigma^\dagger(x)\partial_x L_\sigma(x) \right] \qquad (28.88)$$

The finite reflection probability (28.86) can be accounted for by adding a backscattering term to the Hamiltonian:

$$V = v_F r R_\sigma^\dagger(0) L_\sigma(0) + \text{H.c.} \qquad (28.89)$$

where $|r|^2$ is proportional to the reflection coefficient. We have multiplied (28.89) by v_F in order to make the quantity r adimensional. Without loss of generality we shall assume r to be real in what follows (if it is not real it can always be made such by a gauge transformation).

In fact, the Hamiltonian Eqs. (28.88, 28.89) can be understood as a phenomenological model for a single-channel point contact (so that r need not be related to the reflection coefficient of the harmonic model (28.85) and can include, e.g., a non-adiabatic contribution from a short-range impurity scatterer close to the point contact).

VI.2   Coulomb blockade and two-channel Kondo model

Thus, the problem is described by the effective Hamiltonian defined by (28.84), (28.88), and (28.89)

$$H = H_0 + V + H_c \qquad (28.90)$$

The total charge of the dot, entering H_c, takes the following form in terms of the chiral Fermi fields

$$Q = e \int_0^\infty dx \left[R_\sigma^\dagger(x) R_\sigma(x) + L_\sigma^\dagger(x) L_\sigma(x) \right] \qquad (28.91)$$

Notice that the model (28.90, 28.91) is only applicable to not too small dots. Indeed, this model neglects all the effects occurring on the energy scale of the order of the level spacing $\Delta \sim v_F/L$ in the dot (L being the linear size of the dot). This is acceptable if charging energy E_c lies well above this energy scale. Thus, for the above model to be relevant the size of the dot must be $L \gg e^2/v_F C_d$. This is the case for most semiconductor devices unless a special effort has been taken to make the dot as small as the electron Bohr radius for a given material. (We note that the model description of small dots is completely different – they rather act as Anderson impurities.)

We now proceed with the model (28.90, 28.91) by bosonizing the Fermi fields in a standard way. To the chiral Fermi fields $R_\sigma(x)$ and $L_\sigma(x)$

correspond the chiral Bose fields $\phi_{\sigma R}(x)$ and $\phi_{\sigma L}(x)$

$$R_\sigma(x) \simeq \frac{1}{\sqrt{2\pi a_0}} e^{i\sqrt{4\pi}\phi_{\sigma R}(x)}, \quad L_\sigma(x) \simeq \frac{1}{\sqrt{2\pi a_0}} e^{-i\sqrt{4\pi}\phi_{\sigma L}(x)}.$$

These fields combine to give the charge and the spin phase fields

$$\Phi_{c(s)}(x) = \frac{1}{\sqrt{2}} \left[\phi_{R\uparrow}(x) + \phi_{L\uparrow}(x) \pm \phi_{R\downarrow}(x) \pm \phi_{L\downarrow}(x) \right]$$

Next we closely follow the logic of Section II.1 (previous Chapter). Namely, we introduce the symmetric and the antisymmetric combinations of the Bose fields with respect to the origin

$$\Phi_{\pm c(s)}(x) = \frac{1}{\sqrt{2}} \left[\Phi_{c(s)}(x) \pm \Phi_{c(s)}(-x) \right]$$

and observe that, just like in Section II.1, only the symmetric combination enters the interacting part of the Hamiltonian. Consequently the problem can be formulated in terms of the chiral fields defined by

$$\phi_{c(s)}(x) = \begin{cases} \phi_{R+c(s)}(x), & \text{for} \quad x > 0 \\ \phi_{L+c(s)}(-x), & \text{for} \quad x < 0 \end{cases}$$

where $\phi_{R(L)+c(s)}(x)$ are the right- and the left-moving components of $\Phi_{\pm c(s)}(x)$.

In terms of the chiral Bose fields the relevant part of the Hamiltonian (28.90, 28.91) describing the Coulomb blockade problem takes the form

$$H[\phi_c, \phi_s] = H_0[\phi_c] + H_0[\phi_s] + \frac{2e^2}{\pi C_d} \left[\phi_c(0) - \frac{\sqrt{\pi}}{2} N \right]^2$$

$$+ \frac{2r}{\pi a_0} \cos \left[\sqrt{4\pi}\phi_c(0) \right] \cos \left[\sqrt{4\pi}\phi_s(0) \right] \tag{28.92}$$

where

$$H_0[\phi] = v_F \int dx \, (\partial_x \phi)^2$$

As one can see from (28.92), the charging energy is a relevant operator pinning the charge field at the average value

$$\langle \phi_c(0) \rangle = \frac{\sqrt{\pi}}{2} N$$

the fluctuations around which are almost harmonic. We can therefore replace

$$\cos \left[\sqrt{4\pi}\phi_c(0) \right] \rightarrow \lambda \cos(\pi N) \tag{28.93}$$

where λ is an adimensional nonuniversal constant (it is determined by the ratio of the charging energy to the electron bandwidth).

The replacement (28.93) allows one to formulate an effective Hamiltonian for the spin field

$$H_{\text{eff}}[\phi_s] = H_0[\phi_s] + \frac{2r\lambda v_F}{\pi a_0} \cos(\pi N) \cos\left[\sqrt{4\pi}\phi_s(0)\right] \qquad (28.94)$$

The Hamiltonian (28.94) is of course familiar to the reader. We have first encountered this Hamiltonian in Section IV (Chapter 27) – it describes an impurity in a spinless Tomonaga–Luttinger liquid at the exactly solvable point. It also corresponds to the quantum dissipation problem at the Guinea point (Section V of the same Chapter). Finally, this Hamiltonian appears in the Emery–Kivelson solution of the two-channel Kondo problem (Section IV). Before commenting on this last analogy, we would like to present the solution to (28.94). The Hamiltonian (28.94) has been refermionized and diagonalized in Section IV. Its ground state energy (or rather the correction to the ground state energy due to the impurity operator) is given by formula (28.29) at zero temperature:

$$\Delta E(N) = \int\limits_{v_F/a_0}^{0} \frac{d\omega}{2\pi} \tan^{-1}\left[\frac{\Gamma(N)}{\omega}\right] \qquad (28.95)$$

We have had to regularize the energy integral in (28.95) by a bandwidth cut-off. The parameter $\Gamma(N)$, which corresponds to the width of the Kondo resonance, is given by

$$\Gamma(N) = \frac{r^2\lambda^2 v_F}{\pi a_0} \cos^2(\pi N) \qquad (28.96)$$

The average electron charge on the dot can now be found by minimizing the energy with respect to the parameter N:

$$\langle Q \rangle = eN - \frac{r^2\lambda^2 ev_F}{2\pi a_0} \sin(2\pi N) \ln\left[\frac{\pi}{r^2\lambda^2 \cos^2(\pi N)}\right] \qquad (28.97)$$

The first term in this expression is due to the energy of the charge field while the second term is calculated by deriving (28.95) with respect to N. The second term describes the cross over from the linear charge–voltage dependence to the staircase function of Fig. 28.3. The steps occur at the half-integer values of N where the resonance width $\Gamma(N)$ tends to zero. Close to the steps, the charge exhibits a nonanalytic dependence on the voltage:

$$\frac{\partial\langle Q \rangle}{\partial \bar{N}} \sim e \ln\left(\frac{1}{\bar{N}}\right) \qquad (28.98)$$

where $\bar{N}$ is the deviation of N from a half-integer value.

The very appearance of the Kondo Hamiltonian in studying the Coulomb blockade problem can easily be understood. Indeed, the role of the pseudospin degree of freedom is played by the charging state of the dot with the electron tunneling processes corresponding to the Kondo exchange. It is perhaps more surprising that the effective Kondo model happens to be exactly in Toulouse limit. The number of channels (two) is dictated by the properties of the contact (which allows for a single transverse conducting channel) and the fact that there are two projections of the physical electron spin. The non-Fermi-liquid nature of the ground state of the two-channel Kondo problem is reflected by the nonanalyticity of the function $Q(N)$ close to the steps: (28.98).

In is quite clear that different (more complicated than in Fig. 28.2) experimental setups shall correspond to different versions of the multi-channel Kondo model. The description of such setups (which may involve more than one point contacts and quantum dots) is outside the scope of this book. We only mention here the paper by Furusaki and Matveev (1995), where the Coulomb blockade problem for a quantum dot with two symmetric point contacts has been mapped onto the four-channel Kondo Hamiltonian in the Toulouse limit (which we have discussed in Section V).

Notice also that in this Chapter we have only considered the 'canonical' multi-channel Kondo model (28.1), leaving behind numerous possibilities to complicate the model. Perhaps the most interesting among them is to consider a magnetic impurity in an interacting metallic host, e.g. a Tomonaga–Luttinger liquid. We refer the reader interested in this topic to the papers by Furusaki and Nagaosa (1992) and by Fröjdh and Johannesson (1996).

Problem I. Derive the β-function (28.3). For this purpose follow the considerations of the Appendix to Chapter 17. Namely, expand the S-matrix

$$S = T \exp\left\{-iI \int dt \mathbf{s}(t) \mathbf{J}(t)\right\}$$

using OPE of the form

$$J^a(t)J^b(0) = \frac{\epsilon^{abc} J^c(0)}{2\pi v_F t} - \frac{k\delta^{ab}}{8\pi^2 v_F^2 t^2} + \cdots$$

Problem II. Generalize the argument of Section II to the case of the impurity spin $S > 1/2$. In particular, show that for $k < 2S$ the coupling I' is ferromagnetic so that the $\lambda_K = \infty$ fixed is stable.

Problem III. Consider the Coulomb blockade effect in the same geometry as in Section VI but for spinless electrons. In particular, show that the average charge is, in this case, given by

$$\langle Q \rangle - eN \sim er \sin(2\pi N)$$

References

A. A. Abrikosov and A. B. Migdal, *Low Temp. Phys.* **3**, 519 (1970).

I. Affleck and A. W. W. Ludwig, *Nucl. Phys.* **B360**, 641 (1991).

N. Andrei, K. Furuya and J. H. Lowenstein, *Rev. Mod. Phys.* **55**, 331 (1983).

N. Andrei and A. Jerez, *Phys. Rev. Letters* **74**, 4507 (1995).

D. G. Clarke, T. Giamarchi, and B. I. Shraiman, *Phys. Rev.* **B48**, 7070 (1993).

V. J. Emery and S. Kivelson, *Phys. Rev.* **B47**, 10 812 (1992).

M. Fabrizio and A. O. Gogolin, *Phys. Rev.* **50**, 17 732 (1994).

M. Fabrizio, A. O. Gogolin and P. Nozières, *Phys. Rev.* **B51**, 16 088 (1995).

P. Fröjdh and H. Johannesson, *Phys. Rev.* **B53**, 3211 (1996).

A. Furusaki and K. A. Matveev, *Phys. Rev.* **B52**, 16 676 (1995).

A. Furusaki and N. Nagaosa, *Phys. Rev. Letters* **69**, 3378 (1992).

J. Gan, N. Andrei and P. Coleman, *Phys. Rev. Letters*, **70**, 686 (1993).

A. C. Hewson, *The Kondo Problem to Heavy Fermions*, Cambridge University Press (1993).

L. D. Landau and E. M. Lifshits, *Quantum Mechanics*, Pergamon Press, Oxford, (1982).

K. A. Matveev, *Phys. Rev.* **B51**, 1743 (1994).

P. Nozières, *J. Low Temp. Phys.* **17**, 31 (1974).

P. Nozières and A. Blandin, *J. Phys. (Paris)* **41**, 193 (1980).

P. Nozières and C. De Dominicis, *Phys. Rev.* **178**, 1097 (1969).

P. Schlottmann, *J. Phys. (Paris)* **6**, 1486 (1978).

A. M. Sengupta and A. Georges, *Phys. Rev.* **B49**, 10 020 (1994).

G. Toulouse, *C. R. Acad. Sci.* **268**, 1200 (1969).

A. M. Tsvelik, *Phys. Rev.* **B51**, 9449 (1995).

A. M. Tsvelik and P. W. Wiegmann, *Adv. Phys.* **32**, 453 (1983).

P. B. Wiegmann and A. M. Finkelstein, *Sov. Phys. JETP* **48**, 102 (1978).

K. G. Wilson, *Rev. Mod. Phys.* **47**, 773 (1975).

J. Ye, *Phys. Rev. Letters* **77**, 3224 (1996).

G. Yuval and P. W. Anderson, *Phys. Rev.* **B1**, 1522 (1970).

A. B. Zamolodchikov and V. A. Fateev, *Sov. Phys. JETP* **62**, 215 (1985).

General bibliography

Bosonization, collection of papers ed. by M. D. Stone, World Scientific (1993).

J. Cardy, *Scaling and Renormalization in Statistical Physics*, Cambridge University Press (1996).

Conformal Invariance and Applications to Statistical Mechanics, ed. by C. Itsykson, H. Saleur and J.-B. Zuber, World Scientific (1988).

P. Di Francesco, P. Mathieu and D. Senechal, *Conformal Field Theory*, Springer (1997).

E. Fradkin, *Field Theories of Condensed Matter Systems*, Addison-Wesley (1991).

J. Fuchs, *Affine Lie Algebras and Quantum Groups*, Cambridge University Press (1992).

I. S. Gradstein and I. M. Ryzhik, *Tables of Integrals, Series and Products*, Academic Press, Inc. (1980).

C. Itsykson and J.-M. Drouffe, *Statistical Field Theory*, Cambridge University Press (1989).

L.D. Landau and E.M. Lifshits, *Quantum Mechanics*, Pergamon Press, Oxford, (1982).

Les Houches 1988, *Fields, Strings and Critical Phenomena*, Session XLIX, ed. by E. Brezin and J. Zinn-Justin, North Holland (1990).

B. M. McCoy and T. T. Wu, *The Two-dimensional Ising Model*, Harvard University Press (1973).

V. N. Popov, *Functional Integrals and Collective Excitations*, Cambridge University Press (1990).

F. A. Smirnov, *Form Factors in Completely Integrable Models of Quantum Field Theory*, World Scientific (1992).

A. M. Tsvelik, *Quantum Field Theory in Condensed Matter Physics*, Cambridge University Press (1995).

J. Zinn-Justin, *Quantum Field Theory and Critical Phenomena*, second edition, Oxford University Press (1993).

Index